VOLUME ONE HUNDRED AND TWENTY NINE

ADVANCES IN
PROTEIN CHEMISTRY AND STRUCTURAL BIOLOGY

Immunotherapeutics

VOLUME ONE HUNDRED AND TWENTY NINE

ADVANCES IN
PROTEIN CHEMISTRY AND STRUCTURAL BIOLOGY

Immunotherapeutics

Edited by

ROSSEN DONEV
MicroPharm Ltd,
United Kingdom

ELSEVIER

ACADEMIC PRESS
An imprint of Elsevier

Academic Press is an imprint of Elsevier
50 Hampshire Street, 5th Floor, Cambridge, MA 02139, United States
525 B Street, Suite 1650, San Diego, CA 92101, United States
The Boulevard, Langford Lane, Kidlington, Oxford OX5 1GB, United Kingdom
125 London Wall, London, EC2Y 5AS, United Kingdom

First edition 2022

ISBN: 978-0-323-99227-5
ISSN: 1876-1623

For information on all Academic Press publications
visit our website at https://www.elsevier.com/books-and-journals

Publisher: Zoe Kruze
Acquisitions Editor: Ashlie Jackman
Developmental Editor: Naiza Ermin Mendoza
Production Project Manager: Abdulla Sait
Cover Designer: Matthew Limbert

Typeset by STRAIVE, India

Contents

Contributors

Rajaram Abhirami
Computer Aided Drug Design and Molecular Modelling Lab, Department of Bioinformatics, Science Block, Alagappa University, Karaikudi, India

Celeste Abreu
Department of Biochemistry, Faculty of Science, Charles University, Prague, Czech Republic

Rachel A. Adams
Department of Biomedical Sciences, Cardiff Metropolitan University, Cardiff, United Kingdom

Faiz Abdulaziz Alfaiz
Department of Biology, College of Science in Zulfi, Majmaah University, Majmaah, Saudi Arabia

Sogand Amiri
Student Research Committee; Department of Pharmaceutical Biotechnology, School of Pharmacy; Pharmaceutical Sciences Research Center, Shiraz University of Medical Sciences, Shiraz, Iran

Ashkan Bagheri
Department of Pharmaceutical Biotechnology, School of Pharmacy; Pharmaceutical Sciences Research Center, Shiraz University of Medical Sciences, Shiraz, Iran

Ambritha Balasundaram
School of Biosciences and Technology, Vellore Institute of Technology, Vellore, TN, India

Colin Berry
School of Bioscience, Cardiff University, Cardiff, United Kingdom

Fiona Bolton
Department of Research and Development, MicroPharm Ltd, Newcastle Emlyn, United Kingdom

Sasanka Chakrabarti
Department of Biochemistry and Central Research Cell, Maharishi Markandeshwar Institute of Medical Sciences and Research, Maharishi Markandeshwar University (Deemed to be), Mullana, India

Anjali Dhall
Department of Computational Biology, Indraprastha Institute of Information Technology, Okhla Phase 3, New Delhi, India

Seyed Amirreza Fatemi
Student Research Committee; Department of Pharmaceutical Biotechnology, School of Pharmacy; Pharmaceutical Sciences Research Center, Shiraz University of Medical Sciences, Shiraz, Iran

Shirin Fathi
Student Research Committee; Department of Pharmaceutical Biotechnology, School of Pharmacy; Pharmaceutical Sciences Research Center, Shiraz University of Medical Sciences, Shiraz, Iran

Upasana Ganguly
Department of Biochemistry and Central Research Cell, Maharishi Markandeshwar Institute of Medical Sciences and Research, Maharishi Markandeshwar University (Deemed to be), Mullana, India

C. George Priya Doss
School of Biosciences and Technology, Vellore Institute of Technology, Vellore, TN, India

Joanna Giles
MicroPharm Ltd, Newcastle Emlyn, United Kingdom

R. Gnanasambandan
School of BioSciences and Technology, Vellore Institute of Technology, Vellore, TN, India

Owen R. Griffiths
Micropharm Ltd, Carmarthenshire; Department of Biomedical Sciences, Cardiff Metropolitan University, Cardiff, United Kingdom

Shipra Jain
Department of Computational Biology, Indraprastha Institute of Information Technology, Okhla Phase 3, New Delhi, India

Philip E. James
Department of Biomedical Sciences, Cardiff Metropolitan University, Cardiff, United Kingdom

Barbora Kalousková
Department of Biochemistry, Faculty of Science, Charles University, Prague, Czech Republic

Bhavneet Kaur
Department of Biochemistry, PostGraduate Institute of Medical Education and Research, Chandigarh, India

Dilraj Kaur
Department of Computational Biology, Indraprastha Institute of Information Technology, Okhla Phase 3, New Delhi, India

Maninder Kaur
Department of Biochemistry, PostGraduate Institute of Medical Education and Research, Chandigarh, India

Monidipa Konar
Department of Biochemistry, PostGraduate Institute of Medical Education and Research, Chandigarh, India

M. Lakshmipriya
School of BioSciences and Technology, Vellore Institute of Technology, Vellore, TN, India

John Landon
Micropharm Ltd, Carmarthenshire, United Kingdom

R. Magesh
Department of Biotechnology, Sri Ramachandra Faculty of Biomedical Sciences and Technology, Sri Ramachandra Institute of Higher Education and Research (DU), Chennai, India

Aishwarya Mathew
School of BioSciences and Technology, Vellore Institute of Technology, Vellore, TN, India

Mender M. Mender
School of Bioscience, Cardiff University, Cardiff; Department of Research and Development, MicroPharm Ltd, Newcastle Emlyn, United Kingdom

Seyed Mohammad Iman Moezzi
Student Research Committee; Department of Pharmaceutical Biotechnology, School of Pharmacy; Pharmaceutical Sciences Research Center, Shiraz University of Medical Sciences, Shiraz, Iran

R. Keith Morris
Department of Biomedical Sciences, Cardiff Metropolitan University, Cardiff, United Kingdom

Leimarembi Devi Naorem
Department of Computational Biology, Indraprastha Institute of Information Technology, Okhla Phase 3, New Delhi, India

Manica Negahdaripour
Department of Pharmaceutical Biotechnology, School of Pharmacy; Pharmaceutical Sciences Research Center, Shiraz University of Medical Sciences, Shiraz, Iran

Shiva Nejadebrahim
Department of Biochemistry, Faculty of Science, Charles University, Prague, Czech Republic

Sumeet Patiyal
Department of Computational Biology, Indraprastha Institute of Information Technology, Okhla Phase 3, New Delhi, India

Gajendra P.S. Raghava
Department of Computational Biology, Indraprastha Institute of Information Technology, Okhla Phase 3, New Delhi, India

Shiva Rasekh
Student Research Committee; Department of Pharmaceutical Biotechnology, School of Pharmacy; Pharmaceutical Sciences Research Center, Shiraz University of Medical Sciences, Shiraz, Iran

April Roberts
Toxins Group, National Infection Service, Public Health England, Porton Down, United Kingdom

Adesh K. Saini
Department of Biotechnology, Maharishi Markandeshwar Engineering College, Maharishi Markandeshwar (Deemed to be University), Mullana, India

Reena V. Saini
Department of Biotechnology, Maharishi Markandeshwar Engineering College, Maharishi Markandeshwar (Deemed to be University), Mullana, India

P. Sathiyarajeswaran
Siddha Central Research Institute (CCRS), Chennai, India

Nadia Seifi
Student Research Committee; Department of Pharmaceutical Biotechnology, School of Pharmacy; Pharmaceutical Sciences Research Center, Shiraz University of Medical Sciences, Shiraz, Iran

Chandrabose Selvaraj
Computer Aided Drug Design and Molecular Modelling Lab, Department of Bioinformatics, Science Block, Alagappa University, Karaikudi, India

Neelam Sharma
Department of Computational Biology, Indraprastha Institute of Information Technology, Okhla Phase 3, New Delhi, India

Sadhna Sharma
Department of Biochemistry, PostGraduate Institute of Medical Education and Research, Chandigarh, India

Annie Sherlin
School of BioSciences and Technology, Vellore Institute of Technology, Vellore, TN, India

M.S. Shree Devi
Siddha Central Research Institute (CCRS), Chennai, India

Sanjeev Kumar Singh
Computer Aided Drug Design and Molecular Modelling Lab, Department of Bioinformatics, Science Block, Alagappa University, Karaikudi, India

Sukhpal Singh
Department of Biochemistry and Central Research Cell, Maharishi Markandeshwar Institute of Medical Sciences and Research, Maharishi Markandeshwar University (Deemed to be), Mullana, India

R. Siva
School of BioSciences and Technology, Vellore Institute of Technology, Vellore, TN, India

Ondřej Skořepa
Department of Biochemistry, Faculty of Science, Charles University, Prague, Czech Republic

D. Thirumal Kumar
Meenakshi Academy of Higher Education and Research (Deemed to be University), Chennai, India

S. Udhaya Kumar
School of Biosciences and Technology, Vellore Institute of Technology, Vellore, TN, India

Ondřej Vaněk
Department of Biochemistry, Faculty of Science, Charles University, Prague, Czech Republic

Rajendran Vijayakumar
Department of Biology, College of Science in Zulfi, Majmaah University, Majmaah, Saudi Arabia

Mark Young
School of Bioscience, Cardiff University, Cardiff, United Kingdom

CHAPTER ONE

In silico tools and databases for designing cancer immunotherapy

Anjali Dhall, Shipra Jain, Neelam Sharma, Leimarembi Devi Naorem, Dilraj Kaur, Sumeet Patiyal, and Gajendra P.S. Raghava*

Department of Computational Biology, Indraprastha Institute of Information Technology, Okhla Phase 3, New Delhi, India

*Corresponding author: e-mail address: raghava@iiitd.ac.in

Contents

Abstract

Immunotherapy is a rapidly growing therapy for cancer which have numerous benefits over conventional treatments like surgery, chemotherapy, and radiation. Overall survival of cancer patients has improved significantly due to the use of immunotherapy. It acts as

ISSN 1876-1623
https://doi.org/10.1016/bs.apcsb.2021.11.008

a novel pillar for treating different malignancies from their primary to the metastatic stage. Recent preferments in high-throughput sequencing and computational immunology leads to the development of targeted immunotherapy for precision oncology. In the last few decades, several computational methods and resources have been developed for designing immunotherapy against cancer. In this review, we have summarized cancer-associated genomic, transcriptomic, and mutation profile repositories. We have also enlisted in silico methods for the prediction of vaccine candidates, HLA binders, cytokines inducing peptides, and potential neoepitopes. Of note, we have incorporated the most important bioinformatics pipelines and resources for the designing of cancer immunotherapy. Moreover, to facilitate the scientific community, we have developed a web portal entitled ImmCancer (https://webs.iiitd.edu.in/raghava/immcancer/), comprises cancer immunotherapy tools and repositories.

1. Introduction

The formation of a tumor is evident by the uncontrolled growth of cells that characterize the disease cancer. Several genetic alterations and epigenetic modifications lead to apocryphal expression of genes that controls the growth and death of cells. The fundamental hurdle in cancer treatment is tumor heterogeneity and complexity. It rises due to the different combinations of genetic mutations in patients suffering from various cancers (Bailey et al., 2018; Baylin & Jones, 2016; Mansoori, Mohammadi, Davudian, Shirjang, & Baradaran, 2017). The heterogeneity in cancer is directly proportional to the progression of the disease. As per the GLOBOCAN the global cancer statistics 2020, 19.3 million new cases and 10 million deaths were reported, where female breast cancer is commonly diagnosed. Of note, the expected burden of cancer by 2040 is 28.4 million cases. Traditional cancer treatments include invasive approaches such as biopsy, followed by laboratory tests like the morphological and histological diagnosis of malignancy (Makki, 2015). Based on the clinical outcomes, treatments such as surgery, chemo treatment, and hormonal therapy have been designed for cancer patients. However, these treatments are not optimized and patient-specific, resulting in poor prognosis, i.e., the low survival rate in some aggressive cancer and also cause tumor relapse (Esfahani et al., 2020; Zacharakis et al., 2018). The panorama for treating cancer has revolutionized over the past few decades (Urruticoechea et al., 2010). Moreover, with the advancements in high-throughput sequencing, personalized therapies are designed, which improved the cancer patient's survival and quality of life. The traditional approaches for treating different cancers have been replaced by novel therapies such as immunotherapy, viral, and targeted therapy.

The era of cancer immunotherapy has changed dramatically after the discovery of cytotoxic T-lymphocyte-associated protein (CTLA-4) and programmed cell death protein 1/programmed cell death protein ligand 1 (PD-1/PD-L1) by Tasuku Honjo and James Allison (Huang & Chang, 2019; Kruger et al., 2019) in 2018. Cancer immunotherapy boosts our immune system to fight against cancer cells. There are immunotherapies based on immune checkpoint inhibitors (Panichnantakul, Bourgey, Montpetit, Bourque, & Riazalhosseini, 2016; Van Allen et al., 2015), monoclonal antibodies (Shih, Arkenau, & Infante, 2014), cytokines (Berraondo et al., 2019), immunomodulators (De Keersmaecker et al., 2014), cancer vaccines (van der Burg, Arens, Ossendorp, van Hall, & Melief, 2016), CAR T-cell therapy (Zheng et al., 2017), etc., as depicted in Fig. 1 (Varade, Magadan, & Gonzalez-Fernandez, 2021). For instance, renal cell carcinoma, melanoma have been effectively treated using immune checkpoint inhibitors (Joshi, Pal, & Drabick, 2016). Similarly, adoptive cell therapies, antigen receptor T-cells, and tumor vaccines are frequently used as cancer immunotherapy.

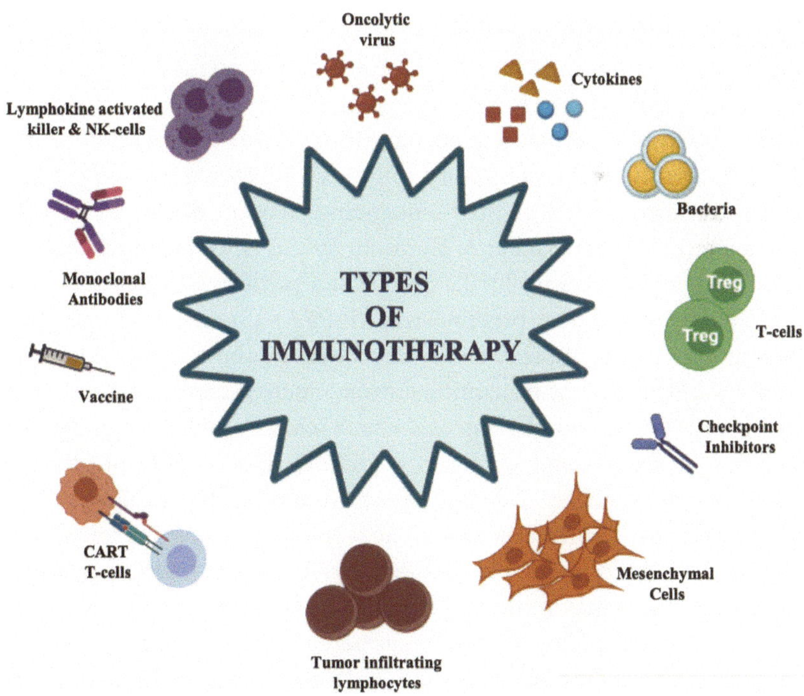

Fig. 1 Pictorial representation of different types of immunotherapies used for cancer treatment.

The basic principle behind these therapies is that T-cell triggers the immune response after recognizing neoantigens presented by human leucocytes antigens (HLA) on the surface of tumor cells (Jensen, 1999; Parham, 1999).

A plethora of in silico resources and tools have been established to design cancer therapy. The amalgamation of experimental and computational strategies improved the treatment of cancer. These computational tools used genomic, transcriptomic, and mutation profiles of the patient for better treatment. Of note, NGS pipelines and bioinformatics methodologies play a significant role in developing personalized therapies, HLA-peptide binding prediction, and neoantigen predictions. This chapter enlists several public repositories and in silico tools designed to analyze and predict the wide range of genomic data to design cancer immunotherapies.

2. Cancer associated genomic data

The genome profiles became popular in the last few decades as it acted as the key to many questions which were inexplicable before, such as how genes interact with each other or with the environment, why some people get a particular disease while others do not, why the reaction of a drug is different in various individuals. There are several methods and platforms for genomic profiling, like RNA sequencing (RNAseq) (Wang, Gerstein, & Snyder, 2009), microRNA sequencing (miRNAseq) (Bartel, 2009; Farazi, Spitzer, Morozov, & Tuschl, 2011), single-cell RNA sequencing (Paolillo, Londin, & Fortina, 2019), ChIP-sequencing (Rotem et al., 2015), DNA sequencing (DNAseq) (Sanger & Coulson, 1975), array-based DNA methylation sequencing (Laird, 2010), SNP-based platforms (McCarroll et al., 2008), and reverse-phase protein array (RPPA) (Stanislaus et al., 2008). These profiles can be utilized to develop novel methods for the prognosis, diagnosis, and treatment of various diseases, such as cancer. Cancer is the most enigmatic and complex disease, which leads to numerous progressive alterations in the genome (Sung et al., 2021). Due to advancements in sequencing technologies, a better understanding of the progression and putative therapies against cancer is rising. One of the most intriguing areas in the treatment of cancer is immunotherapy. It is an approach to reestablish and boost the patient's immune system to fight against cancer cells (Yang, 2015). Genome profiles can aid in prescreening the patients who will be responsive to the immunotherapies with maximum benefits and minimum side effects. Rapid growth in the technology leads to the inverse relationship between cost and accuracy of existed sequencing technologies such as WGS, WES, RNA-seq, and ChIP-seq. Additionally, new NGS technologies are also growing swiftly, such as NICHE-seq. These techniques provide spatial information and the single-cell RNA seq data (Medaglia et al., 2017). It helps researchers to understand the diseases in a better way to develop efficient immunotherapies. The resources of the genome profiles in today's era can be the goldmine in identifying prognostic and diagnostic signatures. In the past, the development of databases with genome profiles has grown exponentially (Table 1). These repositories can be further classified into three categories, which are described as follows.

2.1 Genome profile repositories

The most comprehensive, successful, and widely used program for cancer genomics is The Cancer Genome Atlas (TCGA) (Tomczak, Czerwinska, & Wiznerowicz, 2015) that has produced, analyzed, and shared clinical, microsatellite instability, miRNA, mRNA, and protein expression data on over 20,000 samples traversing 33 types of cancer. TCGA comprised more than 2.5 PB of multi-omics data, including genomic, transcriptomic, and proteomic data. The most crucial portal or site to access the multi-omics data related to cancer is Genomic Data Commons (GDC) data portal. It comprises data from around 68 projects, including TCGA. By August 2021, the GDC data portal stores, analyzes, and shares the RNA-seq, WES, ATAC-seq, WGS, scRNA-seq, etc., data from more than 84,500 samples. Another significant database is Gene Expression Omnibus (GEO) that freely shares high-throughput gene expression data and other forms of functional genomics data. GEO delivers information for >1.5 lakh studies involving around 45 lakh samples and provides more than 4000 datasets. Moreover, GEO provides the tools for the analysis and visualization of the data.

NCBI Sequence Read Archive (SRA) is the primary next-generation sequence data archive. Most of the genome profile repositories concentrated on the profiles of the diseased samples. On the contrary, GTEx provides gene expression, histology images, and quantitative trait locus (QTL) data. Another open-access platform for genomics data is the cBio Cancer Genomics Portal (cBioPortal), which comprises >1.3 lakh tumor samples spanning 316 cancer studies. It majorly provides the mutation, copy number alterations, RNA-seq, microarray-derived mRNA expression. Most of the repositories mentioned in Table 1 offers tissue-specific genome profiles, but

Table 1 List of cancer-associated repositories developed for maintaining different types of genomic data.

Name [Reference]	Description [Web link]
Genomic and transcriptomic repositories	
Oncomine 3.0 (Rhodes et al., 2007)	A collection of genes, pathways, and networks of gene expression profiles https://www.oncomine.org/resource/login.html
NCBI dbGaP (Mailman et al., 2007)	Individual-level phenotype, genotype, and NGS data https://www.ncbi.nlm.nih.gov/gap/
caBIG (caBIG Strategic Planning Workspace, 2007)	The Cancer Biomedical Informatics Grid https://biospecimens.cancer.gov/caBigTools.asp
NCBI SRA Database (Leinonen et al., 2011)	Primary high-throughput sequencing data http://www.ncbi.nlm.nih.gov/Traces/sra
CCLE (Barretina et al., 2012)	Compilation of genomic profiles of human cancer cell lines https://sites.broadinstitute.org/ccle/
cBioPortal (Cerami et al., 2012)	Open platform for exploring cancer genomics data https://www.cbioportal.org/
SurvExpress (Aguirre-Gamboa et al., 2013)	A database for cancer gene expression and survival analysis http://bioinformatica.mty.itesm.mx:8080/Biomatec/SurvivaX.jsp
GTEx (GTEx Consortium, 2013)	A platform for understanding tissue-specific gene expression https://gtexportal.org/home/
TCGA (Tomczak et al., 2015)	A comprehensive resource on cancer https://www.cancer.gov/about-nci/organization/ccg/research/structural-genomics/tcga
GEO (Clough & Barrett, 2016)	High-throughput gene expression and functional genomics data sets http://www.ncbi.nlm.nih.gov/geo/
GDC Data portal (Jensen, Ferretti, Grossman, & Staudt, 2017)	Multi-omics and clinical data available for different types of cancers https://portal.gdc.cancer.gov/

Table 1 List of cancer-associated repositories developed for maintaining different types of genomic data.—cont'd

Name [Reference]	Description [Web link]
TCIA (Feng, Li, Wagner, & Li, 2018)	A comprehensive data on immunogenomic analyses of cancers https://tcia.at
CancerEnD (Kumar, Lathwal, et al., 2020)	A resource on enhancer information on different types of cancers https://webs.iiitd.edu.in/raghava/cancerend/
NGDC (National Genomics Data Center Members and Partners, 2020)	An open access to a suite of cancer resources https://ngdc.cncb.ac.cn
Mutation profiles databases	
KMDB/MutationView (Minoshima et al., 2001)	A database of mutations in disease-associated genes https://mutationview.jp/MutationView/jsp/index.jsp
MutationView/KMcancerDB (Shimizu, Ohtsubo, & Minoshima, 2007)	A database of cancer-associated mutations http://mutview.dmb.med.keio.ac.jp/
ClinVar (Landrum et al., 2016)	Public archive developed to maintain clinically relevant variants https://www.ncbi.nlm.nih.gov/clinvar/
CanProVar 2.0 (Zhang, Wang, et al., 2017)	A resource on human cancer proteome variation http://canprovar2.zhang-lab.org/
HGMD (Stenson et al., 2017)	The human gene mutation database http://www.hgmd.org
BioMuta (Dingerdissen et al., 2018)	A knowledgebase of cancer-associated mutation and expression https://hive.biochemistry.gwu.edu/tools/biomuta/biomuta.php
CANCER3D 2.0 (Sedova et al., 2019)	Database to analyze cancer missense mutations http://www.cancer3d.org
COSMIC (Tate et al., 2019)	A repository of somatic mutations and associated information https://cancer.sanger.ac.uk/cosmic

Continued

Table 1 List of cancer-associated repositories developed for maintaining different types of genomic data.—cont'd

Name [Reference]	Description [Web link]
ICGC Data Portal (Zhang, Bajari, et al., 2019)	Cancer genome consortium for visualization, analysis, and download https://dcc.icgc.org
HotSpotAnnotations (Trevino, 2020)	Hotspot mutations and cancer annotations resource http://bioinformatica.mty.itesm.mx/HotSpotsAnnotations

Cancer Cell Line Encyclopedia (CCLE) delivers RNA splicing, genetic, DNA methylation, miRNA expression, and RPPA data for 1072 cell lines. The Cancer Immunome Atlas (TCIA) can be an essential resource to design novel immunotherapy against human cancer and offers gene expressions of immune-related genes, the cellular composition of immune cells, tumor-associated antigens (neoantigens), cancer-germline antigens, HLA types, and heterogeneity in tumors.

2.2 Mutation profile databases

These databases comprise the information regarding the mutation profile explicitly. Some repositories like TCGA, GDC Data Portal, cBioPortal contain the patient's mutation profile and genome profile information. The Catalogue of Somatic Mutations In Cancer (COSMIC) is the most comprehensive and extensively used database for traversing the impact of somatic mutation in cancer. The latest release (May 2021) of COSMIC comprises more than 38 million coding mutations across 1.5 million tumor samples, curated from over 28,000 articles. ClinVar is another important archive that captures the information and impact of genomic variations on human health (Landrum et al., 2016). This database encompasses the germline as well as somatic variants of any type, size, or loci. Another highly notable database is the Human Gene Mutation Database (HGMD) which contains more than 320,000 mutations (Stenson et al., 2017). The International Cancer Genome Consortium (ICGC) is one of the most popular databases, as it is a comprehensive catalog of mutational aberrations in significant types of tumor (Zhang et al., 2019). Another salient feature of this database is that the county-wise data is available for 16 different countries; along with that, more than 81 million somatic mutations are reported.

2.3 Cancer biomarker repositories

Over the past decades, a number of researchers, including our research groups have been focusing on the study of new cancer biomarkers and their functions. Several biomarker databases gathered the information about biomarkers. For example, OncoMX is a knowledgebase that allows the exploration of cancer biomarkers in the context of cancer and healthy data. The Global Online Biomarker Database gathered information from various sources covering therapeutic utilities. There are several other more specific databases such as Gastric Cancer (Biomarkers) Knowledgebase (Lee et al., 2007) and Liver Cancer bioMarker Reference Into Function (LiverCancerMarkerRIF) (Dai et al., 2014) database. These databases provide enormous information for both researchers and clinicians. The colorectal cancer biomarker database (CBD) comprises colorectal cancer biomarkers information collected from original articles in PubMed (Zhang et al., 2018). MarkerDB is a freely available database of molecular biomarkers that consolidate information for diagnostic, prognostic, exposure, and predictive biomarkers or biomarker panels (Wishart et al., 2021). Recently, our research group has devoted to develop various cancer-specific biomarker databases, namely CancerLivER and ProCanBio database of diagnostic, prognostic, exposure, and predictive biomarker data of liver and prostate cancer (Dikscha Sapra, Dhall, & Raghava, 2021; Kaur, Bhalla, Kaur, & Raghava, 2020). The list of cancer biomarker databases is provided on https://webs.iiitd.edu.in/raghava/immcancer/csbrepo.php.

3. Immunological database

The Immune Epitope Database and Analysis Resource (IEDB-AR) (Dhanda et al., 2019) is a widely used repository for developing various epitope-based novel vaccines and therapeutics for several diseases, including cancer. This repository includes a vast amount of experimentally validated epitopes data and prediction tools. Besides this, it incorporates T-cell, B-cell, cytokines, chemokines, MHC-induced epitopes, etc. The immunological data of IEDB can be used for designing immunotherapy candidates for the treatment of cancer and other diseases. IPD-IMGT/HLA Database (Robinson et al., 2020) was developed to maintain more than 25,000 experimentally validated HLA allele sequences and covers 45 HLA coding genes of the human genome. Similarly, one of the oldest and most efficient MHC-binding and nonbinding epitopes is MHCBN (Bhasin, Singh, &

Raghava, 2003). This database consists of 17,129 binders and 2648 nonbinders of MHC molecules. VDJdb (Shugay et al., 2018) is a repository developed to compile antigen-specific TCR sequences. It also annotates TCR repertoire data and visualizes antigen-specify TCR sequence motifs. Tumor-associated antigens (TAAs) are the most critical targets in diagnosing and treating various carcinomas. They are also employed for the development of immunotherapy against different types of cancers.

In the past, few tumor antigen databases developed for the treatment of several diseases including cancer. For example, the Human Potential Tumor-Associated Antigen database (HPtaa) is one of the most efficient repositories and compiles 3518 potential TAAs that can target various melanogenesis (Wang et al., 2006). Additionally, TANTIGEN 2.0 (Zhang, Chitkushev, Olsen, Keskin, & Brusic, 2021) is an updated version of the TANTIGEN knowledge base. It is a comprehensive data resource of tumor-associated T-cells antigens, and neoepitopes. It includes 403 unique tumor antigens, 4296 antigen variants, and over 1500 T-cell epitopes and HLA ligands. Besides this, another most essential factor in targeting cancer is immunoglobins or antibodies. The tumor-epitope-binding immunoglobins are very important for the immunotherapy responses. The major public resource for immunoglobulins is CIG-DB, which incorporates 2081 cancer-associated immunoglobulin genes and T-cell receptors (Nakamura, Komiyama, Furue, Gojobori, & Akiyama, 2010). The list of immunological resources is available in Table 2. The complete list of resources is available at https://webs.iiitd.edu.in/raghava/immcancer/adbrepo.php.

4. Analysis of genomic profiles

According to National Cancer Institute, "Genome profiling is a technique for deciphering genetic information about a single person or cell type as well as how their genes interact with one another and with the environment." High throughput genomic technologies such as microarrays, next-generation sequencing, and single-cell-based technologies emerged as a powerful tool that revolutionized genomic research (Hwang, Lee, & Bang, 2018). They have rapidly accelerated the generation of cancer-related genomic, molecular, and clinical data available in both public repositories and individual datasets. A plethora of large-scale datasets offers an excellent opportunity for understanding the cancer genome and biology. In addition, they enable to decipher the molecular alterations such as mutations, copy

Table 2 List of public repositories that contain immunological data.

Name [Reference]	Description [Web link]
MHCBN (Bhasin et al., 2003)	A database of MHC binding and nonbinding peptides https://webs.iiitd.edu.in/raghava/mhcbn/
JenPep 2.0 (McSparron, Blythe, Zygouri, Doytchinova, & Flower, 2003)	A comprehensive resource for immunobiology and vaccinology http://www.jenner.ac.uk/JenPep
Bcipep (Kumar, Bhasin, Natt, & Raghava, 2005)	Database of B-Cell epitopes https://webs.iiitd.edu.in/raghava/bcipep/
Epitome (Schlessinger, Ofran, Yachdav, & Rost, 2006)	Database of structurally inferred antigenic epitopes in proteins https://www.rostlab.org/services/epitome/
SuperHapten (Wang, Yang, et al., 2017)	Database of small immunogenic compounds https://bioinformatics.charite.de/superhapten/
Ctdatabase (Almeida et al., 2009)	Knowledge-base of high-throughput data on cancer-testis antigens http://www.cta.lncc.br
AntigenDB (Ansari, Flower, & Raghava, 2010)	A repository of experimentally validated antigens https://webs.iiitd.edu.in/raghava/antigendb/
Protegen (Yang, Sayers, Xiang, & He, 2011)	Protective antigen database and analysis system http://www.violinet.org/protegen/
DFRMLI (Zhang, Lin, Keskin, Reinherz, & Brusic, 2011)	Dana-Farber repository for machine learning in immunology http://bio.dfci.harvard.edu/DFRMLI/
AgAbDb (Kulkarni-Kale, Raskar-Renuse, Natekar-Kalantre, & Saxena, 2014)	A database of antigen-antibody interactions http://bioinfo.net.in/AgAbDb.htm
VDJdb (Shugay et al., 2018)	Database of T-cell receptor sequences with known antigen specificity https://vdjdb.cdr3.net

Continued

Table 2 List of public repositories that contain immunological data.—cont'd

Name [Reference]	Description [Web link]
IEDB-AR (Dhanda et al., 2019)	Immune epitope database—analysis resource in 2019 http://tools.iedb.org/
IPD-IMGT/HLA (Robinson et al., 2020)	The database of sequences of the human leukocyte antigens https://www.ebi.ac.uk/ipd/imgt/hla/
TANTIGEN 2.0 (Zhang, Chitkushev, et al., 2021)	A knowledge base of tumor T-cell antigens and epitopes http://projects.met-hilab.org/tadb/

number variations, translocations, and fusions in multiple genes (Mayakonda, Lin, Assenov, Plass, & Koeffler, 2018). Thus, it is essential to transform this useful data into user-friendly computational tools that can be used by the scientific community and researchers. These tools have made significant progress in closing the gap between complex genomic data and researchers.

In this section, we have listed several web-based tools to analyze genomic data of cancer; see Table 3 and https://webs.iiitd.edu.in/raghava/immcancer/genometool.php. We highlighted some of the most widely used tools, such as the UCSC Cancer Genomics Browser. It is an online tool facilitating an integrative, visualizing, and comprehensive analyses of cancer genomics and its associated clinical information. The browser currently contains genome-wide experiments of over 227,000 samples obtained from TCGA, CCLE, Connectivity Map (Lamb, 2007). It provides a wide range of options to users include: (i) download and upload clinical data, (ii) create dynamic Kaplan–Meier graphs, (iii) export data directly into Galaxy, and (iv) save relevant data views for themselves or others (Afgan et al., 2016). Another tool, ANNOVAR is an efficient tool used to annotate genetic variants, including single nucleotide variants (SNVs) and insertions/deletions detected from diverse genomes (Wang et al., 2010).

5. Prediction of cancer biomarkers

A biomarker is a measurable substance or characteristic that detects or confirms an organism's normal or diseased condition (Goossens, Nakagawa, Sun, & Hoshida, 2015). The study conducted by Grand View Research Inc.,

Table 3 List of computational tools developed for the analysis of genome data.

Name [Reference]	Description [Web link]
UCSC Genome Browser (Zhu et al., 2009)	A graphical viewer for the display of genome sequences and their annotations https://genome.ucsc.edu/
Annovar (Wang, Li, & Hakonarson, 2010)	Functional annotation of genetic variants from NGS data http://www.openbioinformatics.org/annovar/
GATK (McKenna et al., 2010)	Identification of variant in high-throughput sequencing data https://gatk.broadinstitute.org/hc/en-us
ExomeCNV (Sathirapongsasuti et al., 2011)	Detection of copy-number variation and loss of heterozygosity https://cran.r-project.org/src/contrib/Archive/ExomeCNV/
ICGCArray Express (Zhang, Baran, et al., 2011)	Cancer genomics data sets visualization and analysis https://dcc.icgc.org/
SomaticSniper (Larson et al., 2012)	Identification of somatic point mutations in whole genome sequencing data http://gmt.genome.wustl.edu/somatic-sniper/current/
VarScan 2 (Koboldt et al., 2012)	Identification of somatic mutation and copy number alteration in cancer http://varscan.sourceforge.net/
ADTEx (Amarasinghe et al., 2014)	Detection of somatic copy number variations and genotypes in tumor samples http://adtex.sourceforge.net
Oncotator (Ramos et al., 2015)	Annotation of cancer-specific variants https://github.com/broadinstitute/oncotator
VEP (McLaren et al., 2016)	A set of tools for analysis, annotation, and prioritization of genomic variants https://asia.ensembl.org/info/docs/tools/vep/
Maftools (Mayakonda et al., 2018)	A comprehensive analysis of somatic variants in cancer https://bioconductor.org/packages/maftools/
MethSurv (Modhukur et al., 2018)	A web-based tool for survival analysis based on DNA methylation https://biit.cs.ut.ee/methsurv/

Continued

Table 3 List of computational tools developed for the analysis of genome data.—cont'd

Name [Reference]	Description [Web link]
Cancer3D 2.0 (Sedova et al., 2019)	Interactive analysis of 3D patterns of cancer mutations in cancer subsets http://www.cancer3d.org/search
CoMutPlotter (Huang et al., 2019)	A web server for visualization of mutations in cancer cohorts http://tardis.cgu.edu.tw/comutplotter
VIC (He et al., 2019)	A tool for assessing clinical impacts of somatic variants https://github.com/HGLab/VIC/
SNPnexus (Oscanoa et al., 2020)	An annotation tool for functional annotation of human genome sequence variation http://www.snp-nexus.org
UCSC Xena (Goldman et al., 2020)	A tool for visualizing and interpreting genomic data of cancers https://xena.ucsc.edu/

revealed that by 2027 the global biomarkers market size will exceed over \$129.4 billion. Currently, biomarkers used in clinical medicine are mainly proteins, nucleic acids, antibodies, peptides, metabolites, and others (Classen, Staratschek-Jox, & Schultze, 2008). Recent advances have demonstrated that a collection of gene signatures, proteomic and metabolomic signatures, can also reveal important biomarker information. Furthermore, they can be found in the blood, plasma, serum, sputum, saliva, nipple discharge, urine, and stool (Goossens et al., 2015; Wang, Kaczor-Urbanowicz, & Wong, 2017). Already, several biomarkers have been incorporated into clinical practice, while others have only recently been recognized and is still being explored. Thus, biomarkers have been widely used in early cancer diagnosis, disease prevention, prognostics, drug development, etc. (Carlson & Roth, 2013; Chen, Huang, & Kerr, 2011; Imperiale, Ransohoff, & Itzkowitz, 2014; Sweet, Zhang, & Pinilla-Ibarz, 2013). Above all, recently, digital biomarkers derived from body sensors, image processing, health platforms are transforming cancer research and are increasingly being used in clinical decision-making.

A collection of several bioinformatics tools for biomarker identification have been developed due to the rise in genomics data. For instance, PROGmiR is a web-based online tool to study the prognostic importance of miRNAs in different cancers (Goswami & Nakshatri, 2012). CancerCSP, for early and late stage prediction of clear cell renal cell carcinoma using gene expression data derived from RNAseq experiments (Bhalla, Chaudhary, et al., 2017). A list of biomarkers tools is available, including CancerLSP (Kaur, Bhalla, & Raghava, 2019), and HCCPred (Kaur, Dhall, Kumar, & Raghava, 2019). We have highlighted the tools developed for cancer-specific biomarker prediction tools in Table 4 and https://webs.iiitd.edu.in/raghava/immcancer/csbtool.php.

6. Identification of vaccine candidates

The immune system of the human body is competent to recognize and destroy the damaged, infected, and cancerous cells. Peptide or epitope-based vaccine candidates have tremendous significance in treating various diseases including cancer. This field fascinates the scientific community and has great potential for vaccine development, diagnosis, prevention, and treatment. In the last few decades, several bioinformatics tools and resources have been developed to predict and compile B-cell, T-cell, MHC binding, and cytokine-inducing epitopes. In addition, a number of neoepitope prediction methods are also available for the detection of tumor-specific therapeutic targets. In this section, we have discussed the epitope-based prediction methods that can be used for designing potential immunotherapeutic and vaccine candidates.

6.1 B-cell epitope

B-cells play a prominent role in producing antibodies against any infectious disease. This property of B-cells has been widely used for treating cancer. Antibody-based cancer immunotherapy has transformed the treatment procedure for different cancers (Davda et al., 2019). Some of the monoclonal antibody therapies, namely, durvalumab, pembrolizumab, ipilimumab, avelumab, nivolumab, are FDA approved and used as immune checkpoint inhibitors. Also, Trastuzumab (Herceptin) is a monoclonal antibody that is extensively used for the treatment of breast and stomach cancer. Numerous bioinformatics methods such as ABCpred (Saha & Raghava, 2006b), BepiPred (Jespersen, Peters, Nielsen, & Marcatili, 2017), CBtope (Ansari & Raghava, 2010), and DiscoTope (Kringelum, Lundegaard, Lund, & Nielsen, 2012) facilitate the researchers to identify the linear, continuous, and discontinuous B-cell epitopes. In addition to these tools, PEASE (Sela-Culang et al., 2014), IgPred (Gupta, Ansari, et al., 2013), EpIC

Table 4 A list of computation tools developed for predicting prognostic, diagnostic, and predictive biomarkers for cancer.

Name [Reference]	Description [Web link]
SurvExpress (Aguirre-Gamboa et al., 2013)	An online tool for validation of survival associated biomarkers http://bioinformatica.mty.itesm.mx/SurvExpress
CancerCSP (Bhalla, Chaudhary, et al., 2017)	Gene expression-based prediction of early stage of clear cell renal cell carcinoma https://webs.iiitd.edu.in/raghava/cancercsp/
GSCAlite (Liu et al., 2018)	Web server for gene set cancer analysis http://bioinfo.life.hust.edu.cn/web/GSCALite/
CancerLSP (Kaur, Bhalla, & Raghava, 2019)	Prediction of early stage of liver cancer for transcriptomics and epigenomics data http://webs.iiitd.edu.in/raghava/cancerlsp/
CancerSPP (Bhalla, Kaur, Dhall, & Raghava, 2019)	Skin cancer progression prediction from genomic profiles http://webs.iiitd.edu.in/raghava/cancerspp/
CRCRpred (Lathwal, Arora, & Raghava, 2019)	Prediction of risk scores of colorectal cancer patients from protein concentration https://webs.iiitd.edu.in/raghava/crcrpred
OSblca (Zhang, Wang, et al., 2019)	A web server to investigate prognostic biomarkers of bladder cancer patients http://bioinfo.henu.edu.cn/BLCA/BLCAList.jsp
HCCPred (Kaur, Dhall, et al., 2019)	Classification of liver hepatocellular carcinoma and normal liver samples http://webs.iiitd.edu.in/raghava/hccpred/
BBcancer (Zuo et al., 2020)	Expression atlas of blood-based biomarkers in the early diagnosis of cancers http://bbcancer.renlab.org/
CancerTSP (Bhalla, Kaur, Kaur, Sharma, & Raghava, 2020)	Classification of early and late-stage samples of thyroid carcinoma http://webs.iiitd.edu.in/raghava/cancertsp/
OSluca (Yan et al., 2020)	A web server to evaluate prognostic biomarkers for lung cancer http://bioinfo.henu.edu.cn/LUCA/LUCAList.jsp
SKCMhrp (Dhall et al., 2020)	Prediction of high risk skin cutaneous melanoma patients from HLA-superallele https://webs.iiitd.edu.in/raghava/skcmhrp/
CMcrpred (Arora, Kaur, Lathwal, & Raghava, 2020)	Risk prediction of cutaneous melanoma patients based on the clinical features https://webs.iiitd.edu.in/raghava/cmcrpred/

(Marciniuk, Trost, & Napper, 2015) are some methods that can be used for peptide-based vaccine candidate prediction.

6.2 MHC binders

Prediction of MHC binders (class I and II) is the indirect method that provides more specificity than T-cell epitope and attracts more attention from researchers. NetMHCpan 4.0 (Jurtz et al., 2017), SMMPMBEC (Kim, Sidney, Pinilla, Sette, & Peters, 2009), PEPVAC (Reche & Reinherz, 2005) predicts the binding to MHC class I molecules. Class I Immunogenicity (Calis et al., 2013) method predicts the immunogenicity of a class I peptide MHC complex. The binding groove of MHC class I is closed and well-defined from both sides, whereas it is open for MHC class II. Thus, the prediction of the MHC class II binder is challenging as compared to MHC class I binders. In past methods like EpiDOCK (Atanasova, Patronov, Dimitrov, Flower, & Doytchinova, 2013), MHCMIR (Nielsen, Lundegaard, & Lund, 2007), SVRMHC (Wan et al., 2006) have been developed to predict promiscuous MHC-II binders (Table 5). Several in silico tools have also been developed to predict binding between HLAs and peptides. For accurate and reliable prediction, a number of prediction methods are available such as ProPred1 (Singh & Raghava, 2003), PSSMHCpan (Liu et al., 2017), ConvMHC (Han & Kim, 2017), Hlaffy (Mukherjee, Bhattacharyya, & Chandra, 2016), nHLAPred (Bhasin & Raghava, 2007), HLA-CNN (Vang & Xie, 2017) and PickPocket 1.1 (Zhang, Lund, & Nielsen, 2009) are massively used tools for predicting HLA class-I binding peptides. Methods such as MHCII3D (Laimer & Lackner, 2020), HLA-DR4Pred (Bhasin & Raghava, 2004b), ProPred (Singh & Raghava, 2001) and MHC2Pred (Lata, Bhasin, & Raghava, 2007) are available to predict and identify HLA-class II binding peptides accurately.

6.3 T-cell epitope

The cell-mediated immunity part is the major arm of the immune system to be activated for cancer immunotherapy. The primary step is to predict the T-cell epitopes that can be used for cancer therapy. T-cells (CD8+ and CD4+) play a vital role in cell-mediated immunity (Zhao & Cao, 2019). The cytotoxic T-cells (CD8+) can identify and kill malignant or tumor cells by recognizing the peptides which are presented on the surface of cell by the MHC class I molecules. In contrast, helper T-cells (CD4+) recognizes the

Table 5 List of computational methods developed for predicting potential targets for vaccine/immunotherapy.

Name [Reference]	**Description [Web link]**
T-cell epitopes prediction tools	
CTLPred (Bhasin & Raghava, 2004a)	Direct method for predicting CTL epitopes http://webs.iiitd.edu.in/raghava/ctlpred/
NetCTL 1.2 Server (Larsen et al., 2007)	Prediction of human cytotoxic T-Lymphocyte epitopes http://www.cbs.dtu.dk/services/NetCTL/
NetTepi (Trolle & Nielsen, 2014)	Prediction of T-cell epitopes http://www.cbs.dtu.dk/services/NetTepi/
Expitope 2.0 (Jaravine, Mosch, Raffegerst, Schendel, & Frishman, 2017)	Assessment of an epitope cross-reactivity in human tissues http://webclu.bio.wzw.tum.de/expitope2/
ProTECT (Rao et al., 2020)	Prediction of T-cell epitopes for cancer therapy https://github.com/BD2KGenomics/protect
MHC class I and class II binding prediction tools	
ProPred (Singh & Raghava, 2001)	Prediction of HLA-DR binders https://webs.iiitd.edu.in/raghava/propred/
MHCPred (Guan, Doytchinova, Zygouri, & Flower, 2003)	Quantitatively predict peptide binding to MHC http://www.ddg-pharmfac.net/mhcpred/MHCPred/
ProPred1 (Singh & Raghava, 2003)	Prediction of MHC-I binders http://webs.iiitd.edu.in/raghava/propred1/
HLA-DR4Pred (Bhasin & Raghava, 2004b)	Prediction of MHC Class II alleles (HLA-DRB1*0401) binding peptides http://webs.iiitd.edu.in/raghava/hladr4pred/

Table 5 List of computational methods developed for predicting potential targets for vaccine/immunotherapy.—cont'd

Name [Reference]	Description [Web link]
MHCMIR (Nielsen et al., 2007)	Prediction of the binding affinity of MHC-II peptides http://ailab.ist.psu.edu/mhcmir/predict.html
nHLAPred (Bhasin & Raghava, 2007)	Prediction of MHC Class I binders http://webs.iiitd.edu.in/raghava/nhlapred/
EpiDOCK (Atanasova et al., 2013)	Prediction of MHC-II binders http://www.ddg-pharmfac.net/epidock/EpiDockPage.html
PSSMHCpan 1.0 (Liu et al., 2017)	PSSM-based method for predicting binders for HLA class I alleles https://github.com/BGI2016/PSSMHCpan
NetMHCpan 4.0 (Jurtz et al., 2017)	Cancer neoantigens predictions, eluted peptide binding affinity data http://www.cbs.dtu.dk/services/NetMHCpan/
Cytokines prediction tools	
IFNepitope (Dhanda, Vir, & Raghava, 2013)	Prediction of interferon-gamma inducing MHC class-II binders https://webs.iiitd.edu.in/raghava/ifnepitope/
ProInflam (Gupta, Madhu, Sharma, & Sharma, 2016)	Prediction of proinflammatory epitopes http://metabiosys.iiserb.ac.in/proinflam/
IL10Pred (Nagpal et al., 2017)	A tool for identification of IL-10 inducing peptides https://webs.iiitd.edu.in/raghava/il10pred/
IL17eScan (Gupta, Mittal, Madhu, & Sharma, 2017)	A tool for identification of peptides that can induce IL-17 response http://metagenomics.iiserb.ac.in/IL17eScan/

Continued

Table 5 List of computational methods developed for predicting potential targets for vaccine/immunotherapy.—cont'd

Name [Reference]	Description [Web link]
AntiInflam (Gupta, Sharma, Shastri, Madhu, & Sharma, 2017)	Prediction of anti-inflammatory epitopes http://metagenomics.iiserb.ac.in/antiinflam/
PIP-EL (Manavalan, Shin, Kim, & Lee, 2018)	An ensemble learning method for proinflammatory peptide predictions http://www.thegleelab.org/PIP-EL/
IL6Pred (Dhall, Patiyal, Sharma, Usmani, & Raghava, 2021)	Computer-aided prediction and design of IL-6 inducing peptides https://webs.iiitd.edu.in/raghava/il6pred/
Neoepitope prediction tools	
NetMHCcons (Karosiene, Lundegaard, Lund, & Nielsen, 2012)	Consensus method for predicting MHC class I binders http://www.cbs.dtu.dk/services/NetMHCcons/
NetMHCstabpan (Rasmussen et al., 2016)	ANN based prediction of cancer neoantigens http://www.cbs.dtu.dk/services/NetMHCstabpan/
MixMHCpred (Bassani-Sternberg et al., 2017)	HLA-I motifs across HLA peptidomes improves neoantigen predictions https://github.com/GfellerLab/MixMHCpred
MHCSeqNet (Phloyphisut, Pornputtapong, Sriswasdi, & Chuangsuwanich, 2019)	Deep neural network model for universal MHC binding prediction https://github.com/cmbcu/MHCSeqNet
DeepHLApan (Wu et al., 2019)	Deep learning approach for neoantigen prediction https://github.com/jiujiezz/deephlapan
MHCflurry (O'Donnell, Rubinsteyn, & Laserson, 2020)	Prediction of peptides presented on MHC class I proteins https://github.com/openvax/mhcflurry
MHCnuggets (Shao et al., 2020)	Prediction of binding between neoantigen and MHC https://github.com/KarchinLab/mhcnuggets

peptides presented by MHC class II molecules (Mosch, Raffegerst, Weis, Schendel, & Frishman, 2019). Based on the ability of T-cells to act powerfully against cancer cells, some cellular immunotherapies have been designed. ProTECT is a method that can be used to identify the potential neoepitopes candidates that can guide vaccine-based or adoptive T-cell therapies (Rao et al., 2020). CTLPred (Bhasin & Raghava, 2004a), NetTepi (Trolle & Nielsen, 2014) are direct methods for predicting the cytotoxic T lymphocyte epitopes. Tools like Expitope 2.0 (Jaravine et al., 2017) and Expitope (Haase, Raffegerst, Schendel, & Frishman, 2015) can be used to assess an epitope cross-reactivity in human tissues. Machine learning-based prediction tools such as AntiCP 2.0 (Agrawal, Bhagat, Mahalwal, Sharma, & Raghava, 2020), and EnACP (Ge et al., 2020) are used for the prediction of anticancer peptides, which can be directly utilized in cancer immunotherapies. Table 5 depicts the information of the tools which are widely used for T-cell epitope prediction.

6.4 Cytokines inducing peptides

Besides some direct and indirect prediction methods for T-cell, MHC class I and class II binders, various approaches have been proposed to predict cytokines and interleukins. NK cells therapy can be used as a promising target to kill cancer cells by releasing specific cytokines. IL2Pred (Anjali Lathwal, Kaur, & Raghava, 2021), IL6Pred (Dhall, Patiyal, Sharma, Usmani, & Raghava, 2021), IL10Pred (Nagpal et al., 2017), IL4Pred (Dhanda, Gupta, Vir, & Raghava, 2013) and IL17eScan (Gupta, Mittal, et al., 2017) are some cytokines-specific methods that can be used to predict cytokines specific peptides. Other than that, PIP-EL (Manavalan et al., 2018), AntiInflam (Gupta, Sharma, et al., 2017), ProInflam (Gupta, Madhu, et al., 2016) are machine learning-based prediction tools that can assist in designing peptide-based cancer immunotherapy (Table 5).

6.5 Neoepitope for cancer

Tumor specific MHC class I peptide/protein prediction have proven to be important in cancer immunotherapy field. NetCTLpan method was developed in 2010 for identifying pan-specific MHC class I epitope prediction with their binding affinity (Stranzl, Larsen, Lundegaard, & Nielsen, 2010). A recent version of the aforementioned method is also available NetMHCpan-4.0 which can predict cancer neoantigens, and T-cell epitopes (Jurtz et al., 2017). Another method is NetMHC-4.0 based on artificial neural networks for the prediction of peptide-MHC class I binding affinity. This technique uses gapped sequence alignment to allow insertion and deletion in alignment (Andreatta & Nielsen, 2016). MHCflurry 2.0 is an integrated platform that combines NetMHCpan 4.0 and MixMHCpred 2.0 to predict MHC class I binding and antigen processing (O'Donnell et al., 2020). MHCnuggets is a high-throughput method to predict binding of rare alleles of MHC class I/II using neural network architecture (Shao et al., 2020). Another freely available tool using deep learning approach is DeepHLApan, for the identification of neoepitopes, HLA binding affinity and the potential immunogenicity of peptide-HLA complex (pHLA) (Wu et al., 2019). Here we focus on various other tools for neoepitope prediction that were proposed in recent years using different technology/ coding languages. The complete list of epitope prediction tools is available at https://webs.iiitd.edu.in/raghava/immcancer/ebptool.php.

7. Important pipelines

7.1 HLA typing

Human leukocyte antigen (HLA) is a highly polymorphic region in the human genome and located at chromosome 6. It is a complex set of genes that encode proteins and involved in antigen presentation (Wieczorek et al., 2017). In general, HLA class-I molecules present tumor antigens or peptides which were recognized by CD8 + T-cells; on the other hand, epitopes presented by HLA class-II molecules are recognized by CD4 + T-cells (Boegel, Lower, Bukur, Sahin, & Castle, 2014). Most important types of cancer immunotherapy, such as adoptive cell therapy and immune checkpoint inhibitors, require T-cells to recognize tumor-specific peptides presented by HLAs. Recent advancements in whole-genome and exome sequence studies show that the high frequency of somatic mutations in HLA class-I genes occurs in many cancer types (Lohr et al., 2012; Shukla et al., 2015). HLA genotype can be used as an immunotherapy biomarker. With a better understanding of HLA types, the clinician can design an appropriate personalized therapy or immunotherapy for cancer patients.

In the past, several algorithms and computational pipelines have been developed for the accurate and precise genotyping of HLA alleles using the human genome. In this book chapter we have incorporated all HLA-typing pipelines as mentioned in Table 6. For instance, seq2HLA (Boegel et al., 2012), HLA-VBSeq (Wang et al., 2019), and xHLA (Xie et al., 2017) tools take RNA-Seq reads for the class I and II HLA typing. In addition, HLA-typing tools such as ATHLATES (Liu et al., 2013),

Table 6 List of HLA typing and neoepitope prediction pipelines using NGS data.

Name [Reference]	**[Web link]** Description
HLA typing software and pipelines	
seq2HLA (Boegel et al., 2012)	Computational pipeline to determine HLA from RNA-seq reads https://github.com/TRON-Bioinformatics/seq2HLA
HLAminer (Warren et al., 2012)	Pipeline for class I and II HLA typing from shotgun sequence data https://github.com/bcgsc/HLAminer
ATHLATES (Liu et al., 2013)	Accurate typing of class I HLA through WES data https://www.broadinstitute.org/viral-genomics/athlates
OptiType (Szolek et al., 2014)	Class I HLA genotyping from NGS data https://github.com/FRED-2/OptiType
HLA-VBSeq (Nariai et al., 2015)	Full resolution (eight-digit) class I HLA typing using WGS data http://nagasakilab.csml.org/hla/
Polysolver (Shukla et al., 2015)	Pipeline for HLA typing and detection of mutations in HLA genes https://software.broadinstitute.org/cancer/cga/polysolver
xHLA (Xie et al., 2017)	Fast and accurate HLA typing from NGS data https://github.com/humanlongevity/HLA
Kourami (Lee & Kingsford, 2018)	Graph-guided assembly for novel HLA discovery using WGS data https://github.com/Kingsford-Group/kourami
HLA*LA (Dilthey et al., 2019)	HLA typing WGS, and WES data https://github.com/DiltheyLab/HLA-LA
HISAT-genotype (Kim, Paggi, Park, Bennett, & Salzberg, 2019)	HLA typing using WGS data https://daehwankimlab.github.io/hisat-genotype/
Neoepitope prediction pipelines	
Epi-Seq (Duan et al., 2014)	Genomic and bioinformatic profiling of mutational neoepitopes https://dna.engr.uconn.edu/?page_id=470
FRED 2 (Schubert et al., 2016)	Immunoinformatic framework for python http://fred-2.github.io

Continued

Table 6 List of HLA typing and neoepitope prediction pipelines using NGS data.—cont'd

Name [Reference]	[Web link] Description
JUMPg (Li et al., 2016)	Proteogenomic pipeline for annotating proteins in human brain and cancer cells https://github.com/gatechatl/JUMPg
CloudNeo (Bais, Namburi, Gatti, Zhang, & Chuang, 2017)	A cloud-based pipeline for patient-specific tumor neoantigens detection https://github.com/TheJacksonLaboratory/CloudNeo
MuPeXI (Bjerregaard, Nielsen, Hadrup, Szallasi, & Eklund, 2017)	Identifies tumor-specific peptides and neoepitopes http://www.cbs.dtu.dk/services/MuPeXI
TIminer (Tappeiner et al., 2017)	NGS data mining pipeline for cancer immunology and immunotherapy http://icbi.i-med.ac.at/software/timiner/timiner.shtml
INTEGRATE-neo (Zhang, Mardis, & Maher, 2017)	A pipeline for gene fusion and neoantigen discovery https://github.com/ChrisMaherLab/INTEGRATE-Neo
TSNAD (Zhou et al., 2017)	Pipeline for cancer somatic mutation and neoantigen detection https://github.com/jiujiezz/tsnad
Neopepsee (Kim et al., 2018)	Genome-level neoantigens prediction https://sourceforge.net/projects/neopepsee/
pTuneos (Zhou, Wei, et al., 2019)	Prioritizing tumor neoantigens from NGS data https://github.com/bm2-lab/pTuneos
ProGeo-neo (Li et al., 2020)	Customized proteogenomic workflow for neoantigen prediction https://github.com/kbvstmd/ProGeo-neo
pVACtools (Hundal et al., 2020)	A toolkit to identify and visualize cancer neoantigens http://www.pvactools.org
ASNEO (Zhang, Zhou, et al., 2020)	Computational pipeline to identify personalized tumor neoantigens https://github.com/bm2-lab/ASNEO
TruNeo (Tang et al., 2020)	Personalized pipeline for tumor neoantigen detection https://github.com/yucebio/TruNeo

PHLAT (Bai, Wang, & Fury, 2018), and HLA*LA (Dilthey et al., 2019) use whole-exome sequencing data. Some computational methods such as OptiType (Szolek et al., 2014), HLA*LA (Dilthey et al., 2019), and HISAT-genotype (Kim et al., 2019) use whole-genome sequencing data for accurate HLA typing. Polysolver (Shukla et al., 2015) is the most popular algorithm, which identifies HLA-alleles and corresponding somatic mutations from the WES data of cancer patients. HLA-IMPUTER (Shen et al., 2019) is a web platform developed on HLA genotype imputation with attribute bagging (HIBAG) algorithm and utilizes SNP genotype data for the HLA imputation. The complete list of HLA typing tools is available at https://webs.iiitd.edu.in/raghava/immcancer/hlabtool.php.

7.2 Neoantigen

Unlike other diseases, immunogenicity of each tumor is distinctive, making cancer a patient specific disease (Brennick, George, Corwin, Srivastava, & Ebrahimi-Nik, 2017). Thus, paving a way for novel strategies which targets epitopes of the mutated genes as personalized cancer immunotherapy (Vormehr, Tureci, & Sahin, 2019). In this direction, neoepitopes play a very important role and act as cancer specific mutated peptides. Our immune system detects neoepitopes as targets of T-cells, which further activates immune response against cancer (Katsnelson, 2016). Recent studies in literature have shown that neoepitopes are an alluring target in cancer immunotherapy due to its highly tumor specific nature (Wickstrom et al., 2019). Researchers are using these neoantigens or cancer specific peptides for diagnosis and treatment including vaccine designing for several cancers (Teku & Vihinen, 2018). Accurate and fast prediction of neoepitopes is vital in effective cancer immunotherapy (Zhou, Wei, et al., 2019). In this section, we are focusing on the frequently used in silico pipelines developed so far for neoepitopes prediction. Fig. 2 represents a schematic view of different computational approaches for the detection and prioritization of cancer associated neoepitopes for cancer immunotherapy.

One of the popular bioinformatics pipelines for neoepitope prediction is pVACtools, it is a computational framework integrated with the genomics pipeline for identification of neoantigens. This tool effectively identifies altered peptides with point mutations, in-frame and frameshift mutations and gene fusions. This tool can prioritize the predicted peptides and user can also visualize and interpret results using web interface (Hundal et al., 2020). pTuneos is

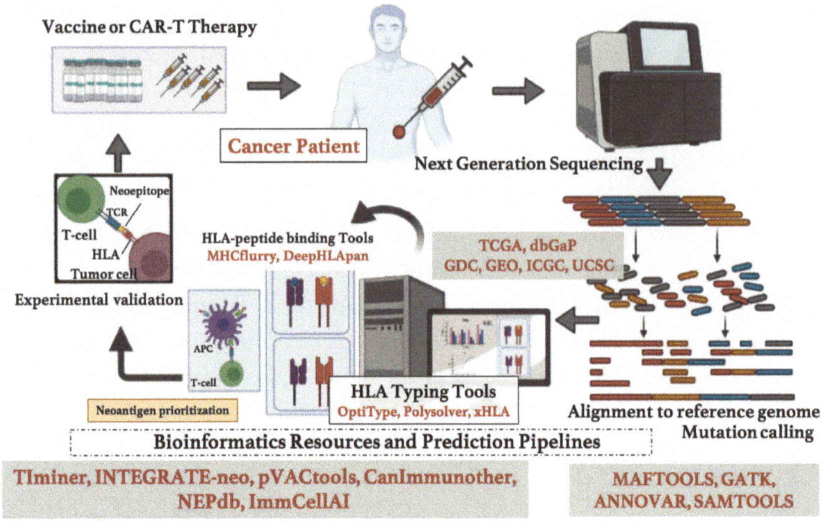

Fig. 2 Schematic representation of bioinformatics pipeline for the designing of personalized cancer immunotherapy using NGS data.

a comprehensive computational pipeline for prioritizing cancer neoantigens using NGS data. This tool is able to predict the immunogenicity of SNV based neopeptides, and T-cell recognition ability considering natural processing and presentation (Zhou, Wei, et al., 2019). In addition, TIminer is a computational pipeline for extracting tumor-immune cell interactions from NGS data. This tool allows user to do integrative immunogenomic analyses, such as HLA typing, neoepitopes prediction, and quantification of tumor immunogenicity (Tappeiner et al., 2017). MuPeXI, is a mutant peptide extractor and informer tool used to assess the potential of tumor-specific peptides as neoepitopes targets (Bjerregaard et al., 2017).

Neopepsee is a ML-based tool used for the prediction of neoepitopes using the NGS data. User can provide RNA-seq data and somatic mutations as input file and this program calculates mutated peptide sequences and gene expression levels (Kim et al., 2018). CloudNeo is a cloud-based pipeline used for identifying tumor neoantigens from patient specific NGS data. This tool uses Polysolver or HLAminer software for HLA typing and mutant detection and NetMHCpan for neoantigen prediction (Bais et al., 2017). An open-source immunoinformatic framework implemented in Python named FRED 2 for epitope prediction was developed in 2016 (Schubert et al., 2016).

With increasing parallel sequencing and proteomics profiling, researchers proposed proteogenomics pipelines to predict neoepitopes. JUMPg, an integrative proteogenomics pipeline which performs various steps such as customized database construction, tag-based database search, and data visualization to predict potential neoantigens (Li et al., 2016). Another group proposed ProGeo-neo, a proteogenomics pipeline for prediction of neoantigens (Zhou, Wei, et al., 2019). In order to identify personalized Alternative Splicing based NEOantigens using RNA-seq data, a research group proposed ASNEO, an integrated computational pipeline. They showed that this pipeline could detect MHC class I neoepitopes through mass spectrometry data validation (Zhang, Zhou, et al., 2020).

8. Cancer-specific immunotherapy

Cancer immunotherapy is one of the most emerging and rapidly growing fields from last few decades. Its primary aim is to target cancer cells by utilizing the immune system (Hegde & Chen, 2020). There are many possibilities to boost a patient's immune response against tumor, and one of them is adoptive cell therapy or T-cell receptor-mediated therapies (Mosch et al., 2019). This therapy involves combining the human T-cells with some artificial changes to cure cancer. T-cell based immunotherapies include chimeric antigen receptor (CAR) T-cell therapy, engineered T-cell receptor (TCR) therapy, tumor-infiltrating lymphocyte (TIL) therapy and Natural Killer (NK) Cell Therapy. These therapies take advantage of the natural potential of the T-cells to enhance the anticancer immune system (Zhao et al., 2021). CAR T-cell therapy is a type of immunotherapy in which T-cells are extracted from the patient and are modified in the laboratory. An artificially synthesized receptor named the chimeric antigen receptor is attached to the T-cell. These CAR T-cells are grown in the lab and then administered to the patient. It binds to the naturally occurring antigens on the surface of cancer cells and enhances the patient immune response to combat the cancer. TCR therapy also works on the same principle as CAR T-cell therapy. It increases the personalization of the treatment and contributes a higher potential for positive outcomes for patients. TIL therapy is widely used for treating solid tumors. The patient's lymphocytes that have earlier infiltrated the tumor cells as damaged or abnormal cells are TILs. This therapy harvests these trained TILs in large numbers and then reinfused into the patient, which aids the immune system to attack the tumor cells and destroys them. NK cell therapy is an emerging immunotherapy with a propitious therapeutic approach for treating tumors and malignancies (Minetto et al., 2019). Hence these cells can be proficiently used in immunotherapy to treat cancer.

Recently, several computational and experimental repositories have developed to design immunotherapeutic candidates for cancer patients (Table 7). CanImmunother is a comprehensive manually curated repository created to collect experimentally validated cancer immunotherapy association data of cancer patients (Zhang, Zeng, et al., 2021). The database incorporates 484 immunotherapies with 642 biomarkers, 108 targets, 121 control therapies, and 3267 experimentally validated relationships among 218 cancer subtypes across 34 body parts. Checkpoint targets are essential for designing cancer immunotherapy and play a significant function in tumor-mediated immune escape. Checkpoint therapeutic target database (CKTTD) is one of the leading repositories developed for the collection of checkpoint targets and their modulators in cancer immunotherapy. CKTTD database includes 105 protein target checkpoints, 30 miRNAs, 53 small-molecules and antibody modulators, and 18 long noncoding RNAs from an extensive literature review (Zhang, Yao, et al., 2020). Another critical resource for cancer immunotherapy is TCMIO (Liu et al., 2020). It majorly concentrated on immuno-oncology targets, traditional Chinese herbs/ingredients which play significant immunomodulatory roles, and antitumor effects. Various chemoinformatic and bioinformatics tools identify effective relationships between traditional Chinese medicine and cancer immunotherapy candidates (Liu et al., 2020). CancerImmunityQTL is a comprehensive resource of genetic variants such as tumor-infiltrating immune cells, which play a crucial role in the tumor progression, prognosis, and immunotherapy response in human cancer. It uses TCGA and CIBERSORT to identify 913 immunity quantitative trait loci (ImmunQTLs) among 33 cancer patients (Tian et al., 2021).

In addition to this, neoepitopes are the most important factors and play crucial role in cancer immunology and immunotherapy. In order to design cancer specific and personalized vaccine or immunotherapy several databases have developed. For example, TCLP is a cancer cell line resource that incorporates HLA type, predicted neoepitopes, gene expression and antigenic mutations data of cancer patients (Scholtalbers et al., 2015). TSNAdb is a tumor-specific neoantigen database, which includes 7748 tumor samples and 16 tumor types. Tumor-specific neoantigen can be utilized as biomarkers for immune checkpoint blockade therapy and potential cancer immunotherapy targets (Wu et al., 2018). Another essential database for

Table 7 List of databases and tools used for designing cancer immunotherapy.

Name [Reference]	Description [Web link]
TCLP (Scholtalbers et al., 2015)	A catalog of HLA type, neoepitopes, gene expression data http://celllines.tron-mainz.de
CancerTope (Gupta, Chaudhary, et al., 2016)	In silico tool for designing personalized immunotherapy https://webs.iiitd.edu.in/raghava/cancertope/
TSNAdb (Wu et al., 2018)	A database for tumor-specific neoantigens from immunogenomics data http://biopharm.zju.edu.cn/tsnadb/
NeoPeptide (Zhou, Qu, et al., 2019)	Immunoinformatic database of T-cell-defined neoantigens https://github.com/lyotvincent/NeoPeptide
dbPepNeo (Tan et al., 2020)	A manually curated database of tumor neoantigen peptides www.biostatistics.online/dbPepNeo/
CKTTD (Zhang, Yao, et al., 2020)	A comprehensive database of checkpoint targets in immunotherapy http://www.ckttdb.org/
OvirusTdb (Lathwal et al., 2020)	Oncolytic viruses database for the advancement of therapeutics https://webs.iiitd.edu.in/raghava/ovirustdb/
CanImmunother (Zhang, Zeng, et al., 2021)	A manually-curated database for precision cancer immunotherapy http://www.biomedical-web.com/cancerit/
CancerImmunityQTL (Tian et al., 2021)	A database for evaluating genetic variants on immune infiltration http://www.cancerimmunityqtl-hust.com/
NEPdb (Xia et al., 2021)	Compilation of T-cell neoepitopes for cancer immunotherapy http://nep.whu.edu.cn
ImmuCellAI (Miao et al., 2020)	A tool to estimate the abundance of immune cells from gene expression data http://bioinfo.life.hust.edu.cn/web/ImmuCellAI/
TCMIO (Liu et al., 2020)	A database of traditional Chinese medicine http://tcmio.xielab.net

Continued

Table 7 List of databases and tools used for designing cancer immunotherapy.—cont'd

Name [Reference]	Description [Web link]
NEPdb (Xia et al., 2021)	A database of T-cell neoepitopes for cancer immunotherapy http://nep.whu.edu.cn/
VLCvirus (Lathwal, Kumar, & Raghava, 2021)	In silico identification of subunit vaccine candidates https://webs.iiitd.edu.in/raghava/vlcvirus/
HLA Ligand Atlas (Marcu et al., 2021)	A resource of HLA ligands presented on benign tissues https://hla-ligand-atlas.org/

cancer immunotherapy is NEPdb (Xia et al., 2021), which includes experimentally validated and predicted pan-cancer neoantigens of cancer patients. dbPepNeo (Tan et al., 2020), NeoPeptide (Zhou, Qu, et al., 2019) are immunoinformatic databases consist of human tumor neoantigen peptides. Additionally, most recently, HLA Ligand Atlas, a benign reference repository of the HLA-presented peptides has been developed to improve T-cell-based cancer immunotherapy (such as CAR T-cells, affinity-enhanced T-cell transfer, cancer vaccines and immune checkpoint inhibition) (Marcu et al., 2021). caAtlas is a virtual repository for prioritizing HLA-binding neoepitopes/peptides to test immunogenicity and cancer immunotherapy development. Of note, oncolytic virus-based immunotherapy is one of the newest cancer-fighting therapies. T-vec, an oncolytic virus, was recently authorized by the FDA for the treatment of melanoma. OvirusTdb is a manually curated collection of oncolytic viruses and contains data on 9 different forms of DNA and 15 different types of RNA. It also incorporate 300 recombinant and 9 wild-type viral strains that have been tested against 124 cancer types and 427 cancer cell lines (Lathwal, Kumar, & Raghava, 2020). OvirusTdb, may help the scientific community for the designing of cancer immunotherapy.

9. Miscellaneous

Along with the above-mentioned immunotherapy-based tools and databases, some methods are also available that can be used to design immunotherapy. For example, CancerDR is a cancer drug resistance database that contains information on 148 anticancer medicines and their pharmacological profile on 952 cancer cell lines. This database is extremely valuable in identifying genetic changes in medication target genes and the residues that cause drug resistance (Kumar et al., 2013). Another important anticancer peptide/protein database is CancerPPD which comprises 3491 anticancer peptides and 121 proteins. CancerPPD also contains information on about 249 different cancer cell lines and 16 different assays for evaluating anticancer peptides (Tyagi et al., 2015). Another important resource, CancerPDF (Cancer Peptidome Database of BioFluids) is a database of endogenous cancer biomarker peptides found in human biofluids. From 56 high-throughput investigations for 27 cancer types, CancerPDF includes 14,367 entries with 9692 distinct peptide sequences corresponding to 2230 unique precursor proteins (Bhalla, Verma, et al., 2017). SAPdb is a manually curated resource of self-assembling peptides which have therapeutic applications, and used as drug delivery vehicle (Mathur, Kaur, Dhall, Sharma, & Raghava, 2021). In addition to this, some recent studies have shown that linear ncRNAs, such as long ncRNAs and microRNAs, and circRNAs play a significant role in regulating tumor immunity and immunotherapy (Xu, Li, Fan, & Wu, 2018). circRNAs are a new kind of regulatory noncoding RNA discovered and cataloged recently. While a rising number of circRNAs have biological activities, it is important to combine all of the data in order to assist research on circRNA functions and regulatory networks in human illnesses. Databases like CircNet, Circ2Disease, LncRNADisease2.0, ExoRbase (Vo et al., 2019; Yao et al., 2018) are beneficial and heavily used resources for RNAs. Genome-wide association studies (GWAS) deliver an effective approach for recognizing genetic variation relationships with the disease susceptibility, prognosis, survival, and predictive drug response (Buniello et al., 2019). In addition, genome alignment tools (Table 8), can be utilized for personal and comparative genomics in order to discover sequence conservation and variation. The complete list of miscellaneous resources is available at https://webs.iiitd.edu.in/raghava/immcancer/mistool.php.

10. Discussion and conclusion

Cancer immunotherapies, such as immune checkpoint inhibitors and adoptive cell therapy, kill cancer cells by manipulating the immune system. These immunotherapies have modified the medication landscape for numerous solid and hematological cancers (Kennedy & Salama, 2020). With the improvement in these therapies, patients' survival and life span enhanced dramatically. In the last 10 years, several FDA-approved

Table 8 List of miscellaneous resources for cancer research.

Name [Reference]	**Description [Web link]**
BLAT (Kent, 2002)	Rapidly align sequences to the genome http://genome.ucsc.edu
Samtools (Li et al., 2009)	Sequence Alignment/Map Format Specification https://github.com/samtools/hts-specs
Cgaln (Nakato & Gotoh, 2010)	Fast and space-efficient whole-genome alignment tool https://github.com/rnakato/Cgaln
Circ2Traits (Ghosal, Das, Sen, Basak, & Chakrabarti, 2013)	A database maintaining circular RNA of 105 diseases and cancer http://gyanxet-beta.com/circdb
CancerDR (Kumar et al., 2013)	Cancer drug resistance database https://webs.iiitd.edu.in/raghava/cancerdr/
SCUDO (Lauria, Moyseos, & Priami, 2015)	Tool for signature-based clustering of expression profiles https://www.cosbi.eu/research/prototypes
SAMmate (Xu et al., 2011)	A GUI tool for processing short read alignments in SAM/BAM format http://sammate.sourceforge.net
CancerPPD (Tyagi et al., 2015)	A repository of experimentally ACPs and anticancer proteins https://webs.iiitd.edu.in/raghava/cancerppd/
CancerPDF (Bhalla, Verma, et al., 2017)	A manually curated database of cancer-associated peptides https://webs.iiitd.edu.in/raghava/cancerpdf/
ExoRBase (Li et al., 2018)	Database of circRNA, lncRNA and mRNA in human blood exosomes http://www.exorbase.org/
Circ2Disesase (Yao et al., 2018)	A manually curated database of experimentally validated circRNAs http://bioinformatics.zju.edu.cn/Circ2Disease/index.html
GWAS (Buniello et al., 2019)	A catalog of human genome-wide association studies https://www.ebi.ac.uk/gwas/

immunotherapies are available for different cancer types. FDA-approved immunotherapies such as atezolizumab (Tecentriq), ipilimumab (Yervoy), interleukin-2 (IL-2) and talimogene laherparepvec (Imlygic) (Bommareddy, Patel, Hossain, & Kaufman, 2017) are available to treat melanoma. Similarly, two FDA-approved drugs like pembrolizumab (Keytruda) and nivolumab (Opdivo), effectively treat head and neck cancer, Hodgkin lymphoma, lung, and kidney cancer (Vaddepally, Kharel, Pandey, Garje, & Chandra, 2020). In addition to this, atezolizumab (Tecentriq) can be used for the treatment of bladder carcinoma. In 2021, pembrolizumab has been approved for the high-risk early-stage triple-negative breast cancer treatment. These FDA-approved drugs are used to target and inhibit PD-1, PD-L1 and CTLA-4 immune checkpoints in cancer patients. By blocking these proteins, the immune system response against cancer cells (Gonzalez, Hagerling, & Werb, 2018). Although immunotherapy is effective in treating several cancers, however, it causes some immunorelated adverse effects (Waldman, Fritz, & Lenardo, 2020). In some conditions, the immune system kills the healthy cells which is the major cause of immunotherapy side effects. Some side effects are mild and moderate, but in few cases, they can be severe and cause life-threatening complications (Kroschinsky et al., 2017). The commonly occurred side-effects are dermatitis, pruritus, fever, chills, fatigue, and diarrhea. On the other hand, few rarely occurred adverse effects includes hepatitis, endocrinopathies, thyroiditis, myocarditis, as shown in Fig. 3.

Clinicians and doctors must recognize these side-effects and manage the toxicities for the better treatment and cure of the cancer patients. Since immunotherapy is a vanguard for cancer treatment, several challenges are still there for effectively targeting the tumor cells with reduced toxicity. Number of toxicities induced by the drug response, such as hemotoxicity, allergenicity, immunotoxicity and cytotoxicity (Yang, Sun, Li, Liu, & Tang, 2018). The detection of toxicity of different drugs is labor extensive and time-consuming process, therefore in last few years several computational tools are developed for the prediction of toxicity of drugs, vaccine candidates and peptides. For instance, ToxinPred can predict the toxicity of the vaccine candidates (Gupta, Kapoor, et al., 2013), ToxiM (Sharma, Srivastava, Roy, & Sharma, 2017), and ToxTree (Patlewicz, Jeliazkova, Safford, Worth, & Aleksiev, 2008) detect the toxicity of chemical molecules. In order to predict the allergenicity of chemical compounds and peptides in silico tools such as AlgPred (Saha & Raghava, 2006a), AlgPred 2.0 (Sharma et al., 2020), AllerTOP (Dimitrov, Flower, & Doytchinova,

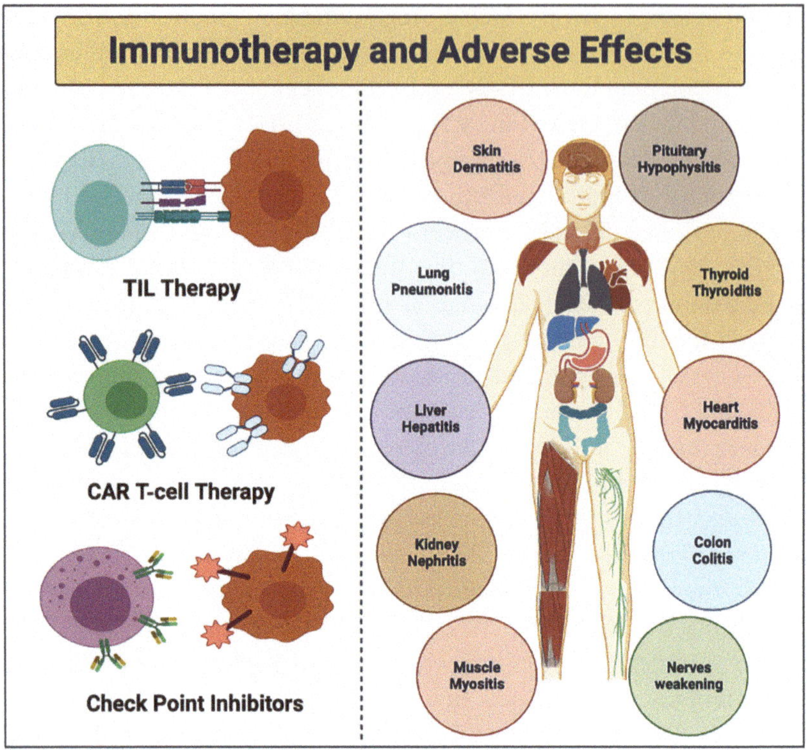

Fig. 3 Adverse effects of different types of immunotherapy.

2013), ChalPred (Sharma, Patiyal, Dhall, Devi, & Raghava, 2021) can be used by the clinicians. Moreover, HemoPI (Chaudhary et al., 2016), HemoPI-MOD (Kumar, Kumar, Agrawal, Patiyal, & Raghava, 2020) can predict the hemotoxicity of the peptides and chemically modified peptides which can be used as vaccine candidates. Recent studies revels that the upregulation of STAT3 transcription factor, is directly linked with the production of PD-L1 in number of tumors, which results into the tumor cell escape mechanism (Attili et al., 2018). Currently, STAT3In a computational tool was developed for the prediction and designing of STAT3 inhibitors (Dhall, Patiyal, Sharma, Devi, & Raghava, 2021).

The available bioinformatics tools can transform the treatment strategy and provide therapeutically efficient methods for predicting immunotherapeutic targets. For instance, in the case of neoantigen vaccine, HLA-typing, epitope prediction, cancer biomarker detection bioinformatics plays a significant role. These powerful tools can detect essential patterns in the genomic and mutation profiles of cancer patients. Moreover, the integration of experimental and computational approaches can lead to novel insights into cancer pathogenesis to improve prognosis, diagnosis, and therapeutic targets. This chapter discusses the latest computational resources, tools, and pipelines for designing patient-specific immunotherapy candidates.

Acknowledgments

A.D. and N.S. are thankful to the Department of Science and Technology (DST-INSPIRE) and S.P. is thankful to Department of Biotechnology (DBT) for providing senior research fellowships. N.L.D. is thankful to Department of Biotechnology (DBT) for providing Research Associate fellowship. S.P. and D.K. are thankful to Department of Computational Biology, IIITD New Delhi for fellowship, infrastructure, and facilities.

References

Afgan, E., Baker, D., van den Beek, M., Blankenberg, D., Bouvier, D., Cech, M., et al. (2016). The Galaxy platform for accessible, reproducible and collaborative biomedical analyses: 2016 update. *Nucleic Acids Research*, *44*(W1), W3–W10. https://doi.org/10.1093/nar/gkw343.

Agrawal, P., Bhagat, D., Mahalwal, M., Sharma, N., & Raghava, G. P. S. (2020). AntiCP 2.0: An updated model for predicting anticancer peptides. *Briefings in Bioinformatics*. https://doi.org/10.1093/bib/bbaa153.

Aguirre-Gamboa, R., Gomez-Rueda, H., Martinez-Ledesma, E., Martinez-Torteya, A., Chacolla-Huaringa, R., Rodriguez-Barrientos, A., et al. (2013). SurvExpress: An online biomarker validation tool and database for cancer gene expression data using survival analysis. *PLoS One*, *8*(9), e74250. https://doi.org/10.1371/journal.pone.0074250.

Almeida, L. G., Sakabe, N. J., deOliveira, A. R., Silva, M. C., Mundstein, A. S., Cohen, T., et al. (2009). CTdatabase: A knowledge-base of high-throughput and curated data on cancer-testis antigens. *Nucleic Acids Research*, *37*(Database issue), D816–D819. https://doi.org/10.1093/nar/gkn673.

Amarasinghe, K. C., Li, J., Hunter, S. M., Ryland, G. L., Cowin, P. A., Campbell, I. G., et al. (2014). Inferring copy number and genotype in tumour exome data. *BMC Genomics*, *15*, 732. https://doi.org/10.1186/1471-2164-15-732.

Andreatta, M., & Nielsen, M. (2016). Gapped sequence alignment using artificial neural networks: Application to the MHC class I system. *Bioinformatics*, *32*(4), 511–517. https://doi.org/10.1093/bioinformatics/btv639.

Anjali Lathwal, R. K., Kaur, D., & Raghava, G. P. S. (2021). In silico model for predicting IL-2 inducing peptides in human. *bioRxiv*. https://doi.org/10.1101/2021.06.20.449146.

Ansari, H. R., Flower, D. R., & Raghava, G. P. (2010). AntigenDB: An immunoinformatics database of pathogen antigens. *Nucleic Acids Research*, *38*(Database issue), D847–D853. https://doi.org/10.1093/nar/gkp830.

Ansari, H. R., & Raghava, G. P. (2010). Identification of conformational B-cell Epitopes in an antigen from its primary sequence. *Immunome Research*, *6*, 6. https://doi.org/10.1186/1745-7580-6-6.

Arora, C., Kaur, D., Lathwal, A., & Raghava, G. P. S. (2020). Risk prediction in cutaneous melanoma patients from their clinico-pathological features: Superiority of clinical data over gene expression data. *Heliyon*, *6*(8), e04811. https://doi.org/10.1016/j.heliyon.2020.e04811.

Atanasova, M., Patronov, A., Dimitrov, I., Flower, D. R., & Doytchinova, I. (2013). EpiDOCK: A molecular docking-based tool for MHC class II binding prediction. *Protein Engineering, Design & Selection*, *26*(10), 631–634. https://doi.org/10.1093/protein/gzt018.

Attili, I., Karachaliou, N., Bonanno, L., Berenguer, J., Bracht, J., Codony-Servat, J., et al. (2018). STAT3 as a potential immunotherapy biomarker in oncogene-addicted non-small cell lung cancer. *Therapeutic Advances in Medical Oncology*, *10*, 1758835918763744. https://doi.org/10.1177/1758835918763744.

Bai, Y., Wang, D., & Fury, W. (2018). PHLAT: Inference of high-resolution HLA types from RNA and whole exome sequencing. *Methods in Molecular Biology*, *1802*, 193–201. https://doi.org/10.1007/978-1-4939-8546-3_13.

Bailey, M. H., Tokheim, C., Porta-Pardo, E., Sengupta, S., Bertrand, D., Weerasinghe, A., et al. (2018). Comprehensive characterization of cancer driver genes and mutations. *Cell*, *174*(4), 1034–1035. https://doi.org/10.1016/j.cell.2018.07.034.

Bais, P., Namburi, S., Gatti, D. M., Zhang, X., & Chuang, J. H. (2017). CloudNeo: A cloud pipeline for identifying patient-specific tumor neoantigens. *Bioinformatics*, *33*(19), 3110–3112. https://doi.org/10.1093/bioinformatics/btx375.

Barretina, J., Caponigro, G., Stransky, N., Venkatesan, K., Margolin, A. A., Kim, S., et al. (2012). The Cancer Cell Line Encyclopedia enables predictive modelling of anticancer drug sensitivity. *Nature*, *483*(7391), 603–607. https://doi.org/10.1038/nature11003.

Bartel, D. P. (2009). MicroRNAs: Target recognition and regulatory functions. *Cell*, *136*(2), 215–233. https://doi.org/10.1016/j.cell.2009.01.002.

Bassani-Sternberg, M., Chong, C., Guillaume, P., Solleder, M., Pak, H., Gannon, P. O., et al. (2017). Deciphering HLA-I motifs across HLA peptidomes improves neo-antigen predictions and identifies allostery regulating HLA specificity. *PLoS Computational Biology*, *13*(8), e1005725. https://doi.org/10.1371/journal.pcbi.1005725.

Baylin, S. B., & Jones, P. A. (2016). Epigenetic determinants of cancer. *Cold Spring Harbor Perspectives in Biology*, *8*(9). https://doi.org/10.1101/cshperspect.a019505.

Berraondo, P., Sanmamed, M. F., Ochoa, M. C., Etxeberria, I., Aznar, M. A., Perez-Gracia, J. L., et al. (2019). Cytokines in clinical cancer immunotherapy. *British Journal of Cancer*, *120*(1), 6–15. https://doi.org/10.1038/s41416-018-0328-y.

Bhalla, S., Chaudhary, K., Kumar, R., Sehgal, M., Kaur, H., Sharma, S., et al. (2017). Gene expression-based biomarkers for discriminating early and late stage of clear cell renal cancer. *Scientific Reports*, 7, 44997. https://doi.org/10.1038/srep44997.

Bhalla, S., Kaur, H., Dhall, A., & Raghava, G. P. S. (2019). Prediction and analysis of skin cancer progression using genomics profiles of patients. *Scientific Reports*, *9*(1), 15790. https://doi.org/10.1038/s41598-019-52134-4.

Bhalla, S., Kaur, H., Kaur, R., Sharma, S., & Raghava, G. P. S. (2020). Expression based biomarkers and models to classify early and late-stage samples of Papillary Thyroid Carcinoma. *PLoS One*, *15*(4), e0231629. https://doi.org/10.1371/journal.pone.0231629.

Bhalla, S., Verma, R., Kaur, H., Kumar, R., Usmani, S. S., Sharma, S., et al. (2017). CancerPDF: A repository of cancer-associated peptidome found in human biofluids. *Scientific Reports*, 7(1), 1511. https://doi.org/10.1038/s41598-017-01633-3.

Bhasin, M., & Raghava, G. P. (2004a). Prediction of CTL epitopes using QM, SVM and ANN techniques. *Vaccine*, *22*(23–24), 3195–3204. https://doi.org/10.1016/j.vaccine.2004.02.005.

Bhasin, M., & Raghava, G. P. (2004b). SVM based method for predicting HLA-DRB1*0401 binding peptides in an antigen sequence. *Bioinformatics*, *20*(3), 421–423. https://doi.org/10.1093/bioinformatics/btg424.

Bhasin, M., & Raghava, G. P. (2007). A hybrid approach for predicting promiscuous MHC class I restricted T cell epitopes. *Journal of Biosciences*, *32*(1), 31–42. https://doi.org/10.1007/s12038-007-0004-5.

Bhasin, M., Singh, H., & Raghava, G. P. (2003). MHCBN: A comprehensive database of MHC binding and non-binding peptides. *Bioinformatics*, *19*(5), 665–666. https://doi.org/10.1093/bioinformatics/btg055.

Bjerregaard, A. M., Nielsen, M., Hadrup, S. R., Szallasi, Z., & Eklund, A. C. (2017). MuPeXI: Prediction of neo-epitopes from tumor sequencing data. *Cancer Immunology, Immunotherapy, 66*(9), 1123–1130. https://doi.org/10.1007/s00262-017-2001-3.

Boegel, S., Lower, M., Bukur, T., Sahin, U., & Castle, J. C. (2014). A catalog of HLA type, HLA expression, and neo-epitope candidates in human cancer cell lines. *Oncoimmunology, 3*(8), e954893. https://doi.org/10.4161/21624011.2014.954893.

Boegel, S., Lower, M., Schafer, M., Bukur, T., de Graaf, J., Boisguerin, V., et al. (2012). HLA typing from RNA-Seq sequence reads. *Genome Medicine, 4*(12), 102. https://doi.org/10.1186/gm403.

Bommareddy, P. K., Patel, A., Hossain, S., & Kaufman, H. L. (2017). Talimogene laherparepvec (T-VEC) and other oncolytic viruses for the treatment of melanoma. *American Journal of Clinical Dermatology, 18*(1), 1–15. https://doi.org/10.1007/s40257-016-0238-9.

Brennick, C. A., George, M. M., Corwin, W. L., Srivastava, P. K., & Ebrahimi-Nik, H. (2017). Neoepitopes as cancer immunotherapy targets: Key challenges and opportunities. *Immunotherapy, 9*(4), 361–371. https://doi.org/10.2217/imt-2016-0146.

Buniello, A., MacArthur, J. A. L., Cerezo, M., Harris, L. W., Hayhurst, J., Malangone, C., et al. (2019). The NHGRI-EBI GWAS Catalog of published genome-wide association studies, targeted arrays and summary statistics 2019. *Nucleic Acids Research, 47*(D1), D1005–D1012. https://doi.org/10.1093/nar/gky1120.

caBIG Strategic Planning Workspace. (2007). The Cancer Biomedical Informatics Grid (caBIG): Infrastructure and applications for a worldwide research community. *Studies in Health Technology and Informatics, 129*(Pt. 1), 330–334. https://www.ncbi.nlm.nih.gov/pubmed/17911733.

Calis, J. J., Maybeno, M., Greenbaum, J. A., Weiskopf, D., De Silva, A. D., Sette, A., et al. (2013). Properties of MHC class I presented peptides that enhance immunogenicity. *PLoS Computational Biology, 9*(10), e1003266. https://doi.org/10.1371/journal.pcbi.1003266.

Carlson, J. J., & Roth, J. A. (2013). The impact of the Oncotype Dx breast cancer assay in clinical practice: A systematic review and meta-analysis. *Breast Cancer Research and Treatment, 141*(1), 13–22. https://doi.org/10.1007/s10549-013-2666-z.

Cerami, E., Gao, J., Dogrusoz, U., Gross, B. E., Sumer, S. O., Aksoy, B. A., et al. (2012). The cBio cancer genomics portal: An open platform for exploring multidimensional cancer genomics data. *Cancer Discovery, 2*(5), 401–404. https://doi.org/10.1158/2159-8290.CD-12-0095.

Chaudhary, K., Kumar, R., Singh, S., Tuknait, A., Gautam, A., Mathur, D., et al. (2016). A web server and mobile app for computing hemolytic potency of peptides. *Scientific Reports, 6*, 22843. https://doi.org/10.1038/srep22843.

Chen, X. H., Huang, S., & Kerr, D. (2011). Biomarkers in clinical medicine. *IARC Scientific Publications*, (163), 303–322. https://www.ncbi.nlm.nih.gov/pubmed/22997869.

Classen, S., Staratschek-Jox, A., & Schultze, J. L. (2008). Use of genome-wide high-throughput technologies in biomarker development. *Biomarkers in Medicine, 2*(5), 509–524. https://doi.org/10.2217/17520363.2.5.509.

Clough, E., & Barrett, T. (2016). The Gene Expression Omnibus Database. *Methods in Molecular Biology, 1418*, 93–110. https://doi.org/10.1007/978-1-4939-3578-9_5.

Dai, H. J., Wu, J. C., Lin, W. S., Reyes, A. J., Dela Rosa, M. A., Syed-Abdul, S., et al. (2014). LiverCancerMarkerRIF: A liver cancer biomarker interactive curation system combining text mining and expert annotations. *Database: The Journal of Biological Databases and Curation, 2014*. https://doi.org/10.1093/database/bau085.

Davda, J., Declerck, P., Hu-Lieskovan, S., Hickling, T. P., Jacobs, I. A., Chou, J., et al. (2019). Immunogenicity of immunomodulatory, antibody-based, oncology therapeutics. *Journal for Immunotherapy of Cancer*, 7(1), 105. https://doi.org/10.1186/s40425-019-0586-0.

De Keersmaecker, B., Fostier, K., Corthals, J., Wilgenhof, S., Heirman, C., Aerts, J. L., et al. (2014). Immunomodulatory drugs improve the immune environment for dendritic cell-based immunotherapy in multiple myeloma patients after autologous stem cell transplantation. *Cancer Immunology, Immunotherapy*, *63*(10), 1023–1036. https://doi.org/10.1007/s00262-014-1571-6.

Dhall, A., Patiyal, S., Kaur, H., Bhalla, S., Arora, C., & Raghava, G. P. S. (2020). Computing skin cutaneous melanoma outcome from the HLA-alleles and clinical characteristics. *Frontiers in Genetics*, *11*, 221. https://doi.org/10.3389/fgene.2020.00221.

Dhall, A., Patiyal, S., Sharma, N., Devi, N. L., & Raghava, G. P. S. (2021). Computer-aided prediction of inhibitors against STAT3 for managing COVID-19 associated cytokine storm. *Computers in Biology and Medicine*, *137*, 104780. https://doi.org/10.1016/j.compbiomed.2021.104780.

Dhall, A., Patiyal, S., Sharma, N., Usmani, S. S., & Raghava, G. P. S. (2021). Computer-aided prediction and design of IL-6 inducing peptides: IL-6 plays a crucial role in COVID-19. *Briefings in Bioinformatics*, *22*(2), 936–945. https://doi.org/10.1093/bib/bbaa259.

Dhanda, S. K., Gupta, S., Vir, P., & Raghava, G. P. (2013). Prediction of IL4 inducing peptides. *Clinical & Developmental Immunology*, *2013*, 263952. https://doi.org/10.1155/2013/263952.

Dhanda, S. K., Mahajan, S., Paul, S., Yan, Z., Kim, H., Jespersen, M. C., et al. (2019). IEDB-AR: Immune epitope database-analysis resource in 2019. *Nucleic Acids Research*, *47*(W1), W502–W506. https://doi.org/10.1093/nar/gkz452.

Dhanda, S. K., Vir, P., & Raghava, G. P. (2013). Designing of interferon-gamma inducing MHC class-II binders. *Biology Direct*, *8*, 30. https://doi.org/10.1186/1745-6150-8-30.

Dikscha Sapra, H. K., Dhall, A., & Raghava, G. P. S. (2021). ProCanBio: A database of manually curated biomarkers for Prostate Cancer. *bioRxiv*.

Dilthey, A. T., Mentzer, A. J., Carapito, R., Cutland, C., Cereb, N., Madhi, S. A., et al. (2019). HLA*LA-HLA typing from linearly projected graph alignments. *Bioinformatics*, *35*(21), 4394–4396. https://doi.org/10.1093/bioinformatics/btz235.

Dimitrov, I., Flower, D. R., & Doytchinova, I. (2013). AllerTOP—A server for in silico prediction of allergens. *BMC Bioinformatics*, *14*(Suppl. 6), S4. https://doi.org/10.1186/1471-2105-14-S6-S4.

Dingerdissen, H. M., Torcivia-Rodriguez, J., Hu, Y., Chang, T. C., Mazumder, R., & Kahsay, R. (2018). BioMuta and BioXpress: Mutation and expression knowledgebases for cancer biomarker discovery. *Nucleic Acids Research*, *46*(D1), D1128–D1136. https://doi.org/10.1093/nar/gkx907.

Duan, F., Duitama, J., Al Seesi, S., Ayres, C. M., Corcelli, S. A., Pawashe, A. P., et al. (2014). Genomic and bioinformatic profiling of mutational neoepitopes reveals new rules to predict anticancer immunogenicity. *The Journal of Experimental Medicine*, *211*(11), 2231–2248. https://doi.org/10.1084/jem.20141308.

Esfahani, K., Roudaia, L., Buhlaiga, N., Del Rincon, S. V., Papneja, N., & Miller, W. H., Jr. (2020). A review of cancer immunotherapy: From the past, to the present, to the future. *Current Oncology*, *27*(Suppl. 2), S87–S97. https://doi.org/10.3747/co.27.5223.

Farazi, T. A., Spitzer, J. I., Morozov, P., & Tuschl, T. (2011). miRNAs in human cancer. *The Journal of Pathology*, *223*(2), 102–115. https://doi.org/10.1002/path.2806.

Feng, X., Li, L., Wagner, E. J., & Li, W. (2018). TC3A: The cancer 3' UTR atlas. *Nucleic Acids Research*, *46*(D1), D1027–D1030. https://doi.org/10.1093/nar/gkx892.

Ge, R., Feng, G., Jing, X., Zhang, R., Wang, P., & Wu, Q. (2020). EnACP: An ensemble learning model for identification of anticancer peptides. *Frontiers in Genetics*, *11*, 760. https://doi.org/10.3389/fgene.2020.00760.

Ghosal, S., Das, S., Sen, R., Basak, P., & Chakrabarti, J. (2013). Circ2Traits: A comprehensive database for circular RNA potentially associated with disease and traits. *Frontiers in Genetics*, *4*, 283. https://doi.org/10.3389/fgene.2013.00283.

Goldman, M. J., Craft, B., Hastie, M., Repecka, K., McDade, F., Kamath, A., et al. (2020). Visualizing and interpreting cancer genomics data via the Xena platform. *Nature Biotechnology*, *38*(6), 675–678. https://doi.org/10.1038/s41587-020-0546-8.

Gonzalez, H., Hagerling, C., & Werb, Z. (2018). Roles of the immune system in cancer: From tumor initiation to metastatic progression. *Genes & Development*, *32*(19–20), 1267–1284. https://doi.org/10.1101/gad.314617.118.

Goossens, N., Nakagawa, S., Sun, X., & Hoshida, Y. (2015). Cancer biomarker discovery and validation. *Translational Cancer Research*, *4*(3), 256–269. https://doi.org/10.3978/j.issn.2218-676X.2015.06.04.

Goswami, C. P., & Nakshatri, H. (2012). PROGmiR: A tool for identifying prognostic miRNA biomarkers in multiple cancers using publicly available data. *Journal of Clinical Bioinformatics*, *2*(1), 23. https://doi.org/10.1186/2043-9113-2-23.

GTEx Consortium. (2013). The Genotype-Tissue Expression (GTEx) project. *Nature Genetics*, *45*(6), 580–585. https://doi.org/10.1038/ng.2653.

Guan, P., Doytchinova, I. A., Zygouri, C., & Flower, D. R. (2003). MHCPred: A server for quantitative prediction of peptide-MHC binding. *Nucleic Acids Research*, *31*(13), 3621–3624. https://doi.org/10.1093/nar/gkg510.

Gupta, S., Ansari, H. R., Gautam, A., Open Source Drug Discovery Consortium, & Raghava, G. P. S. (2013). Identification of B-cell epitopes in an antigen for inducing specific class of antibodies. *Biology Direct*, *8*, 27. https://doi.org/10.1186/1745-6150-8-27.

Gupta, S., Chaudhary, K., Dhanda, S. K., Kumar, R., Kumar, S., Sehgal, M., et al. (2016). A platform for designing genome-based personalized immunotherapy or vaccine against cancer. *PLoS One*, *11*(11), e0166372. https://doi.org/10.1371/journal.pone.0166372.

Gupta, S., Kapoor, P., Chaudhary, K., Gautam, A., Kumar, R., Open Source Drug Discovery Consortium, et al. (2013). In silico approach for predicting toxicity of peptides and proteins. *PLoS One*, *8*(9), e73957. https://doi.org/10.1371/journal.pone.0073957.

Gupta, S., Madhu, M. K., Sharma, A. K., & Sharma, V. K. (2016). ProInflam: A webserver for the prediction of proinflammatory antigenicity of peptides and proteins. *Journal of Translational Medicine*, *14*(1), 178. https://doi.org/10.1186/s12967-016-0928-3.

Gupta, S., Mittal, P., Madhu, M. K., & Sharma, V. K. (2017). IL17eScan: A tool for the identification of peptides inducing IL-17 response. *Frontiers in Immunology*, *8*, 1430. https://doi.org/10.3389/fimmu.2017.01430.

Gupta, S., Sharma, A. K., Shastri, V., Madhu, M. K., & Sharma, V. K. (2017). Prediction of anti-inflammatory proteins/peptides: An insilico approach. *Journal of Translational Medicine*, *15*(1), 7. https://doi.org/10.1186/s12967-016-1103-6.

Haase, K., Raffegerst, S., Schendel, D. J., & Frishman, D. (2015). Expitope: A web server for epitope expression. *Bioinformatics*, *31*(11), 1854–1856. https://doi.org/10.1093/bioinformatics/btv068.

Han, Y., & Kim, D. (2017). Deep convolutional neural networks for pan-specific peptide-MHC class I binding prediction. *BMC Bioinformatics*, *18*(1), 585. https://doi.org/10.1186/s12859-017-1997-x.

He, M. M., Li, Q., Yan, M., Cao, H., Hu, Y., He, K. Y., et al. (2019). Variant Interpretation for Cancer (VIC): A computational tool for assessing clinical impacts of somatic variants. *Genome Medicine*, *11*(1), 53. https://doi.org/10.1186/s13073-019-0664-4.

Hegde, P. S., & Chen, D. S. (2020). Top 10 challenges in cancer immunotherapy. *Immunity*, *52*(1), 17–35. https://doi.org/10.1016/j.immuni.2019.12.011.

Huang, P. W., & Chang, J. W. (2019). Immune checkpoint inhibitors win the 2018 Nobel Prize. *Biomedical Journal*, *42*(5), 299–306. https://doi.org/10.1016/j.bj.2019.09.002.

Huang, P. J., Lin, H. H., Lee, C. C., Chiu, L. Y., Wu, S. M., Yeh, Y. M., et al. (2019). CoMutPlotter: A web tool for visual summary of mutations in cancer cohorts. *BMC Medical Genomics*, *12*(Suppl. 5), 99. https://doi.org/10.1186/s12920-019-0510-y.

Hundal, J., Kiwala, S., McMichael, J., Miller, C. A., Xia, H., Wollam, A. T., et al. (2020). pVACtools: A computational toolkit to identify and visualize cancer neoantigens. *Cancer Immunology Research, 8*(3), 409–420. https://doi.org/10.1158/2326-6066.CIR-19-0401.

Hwang, B., Lee, J. H., & Bang, D. (2018). Single-cell RNA sequencing technologies and bioinformatics pipelines. *Experimental & Molecular Medicine, 50*(8), 1–14. https://doi.org/10.1038/s12276-018-0071-8.

Imperiale, T. F., Ransohoff, D. F., & Itzkowitz, S. H. (2014). Multitarget stool DNA testing for colorectal-cancer screening. *The New England Journal of Medicine, 371*(2), 187–188. https://doi.org/10.1056/NEJMc1405215.

Jaravine, V., Mosch, A., Raffegerst, S., Schendel, D. J., & Frishman, D. (2017). Expitope 2.0: A tool to assess immunotherapeutic antigens for their potential cross-reactivity against naturally expressed proteins in human tissues. *BMC Cancer, 17*(1), 892. https://doi.org/10.1186/s12885-017-3854-8.

Jensen, P. E. (1999). Mechanisms of antigen presentation. *Clinical Chemistry and Laboratory Medicine, 37*(3), 179–186. https://doi.org/10.1515/CCLM.1999.034.

Jensen, M. A., Ferretti, V., Grossman, R. L., & Staudt, L. M. (2017). The NCI Genomic Data Commons as an engine for precision medicine. *Blood, 130*(4), 453–459. https://doi.org/10.1182/blood-2017-03-735654.

Jespersen, M. C., Peters, B., Nielsen, M., & Marcatili, P. (2017). BepiPred-2.0: Improving sequence-based B-cell epitope prediction using conformational epitopes. *Nucleic Acids Research, 45*(W1), W24–W29. https://doi.org/10.1093/nar/gkx346.

Joshi, M., Pal, S. K., & Drabick, J. J. (2016). Novel approaches in cancer immunotherapy—A light at the end of the tunnel. *Discovery Medicine, 21*(118), 479–487. https://www.ncbi.nlm.nih.gov/pubmed/27448784.

Jurtz, V., Paul, S., Andreatta, M., Marcatili, P., Peters, B., & Nielsen, M. (2017). NetMHCpan-4.0: Improved peptide-MHC class I interaction predictions integrating eluted ligand and peptide binding affinity data. *Journal of Immunology, 199*(9), 3360–3368. https://doi.org/10.4049/jimmunol.1700893.

Karosiene, E., Lundegaard, C., Lund, O., & Nielsen, M. (2012). NetMHCcons: A consensus method for the major histocompatibility complex class I predictions. *Immunogenetics, 64*(3), 177–186. https://doi.org/10.1007/s00251-011-0579-8.

Katsnelson, A. (2016). Mutations as munitions: Neoantigen vaccines get a closer look as cancer treatment. *Nature Medicine, 22*(2), 122–124. https://doi.org/10.1038/nm0216-122.

Kaur, H., Bhalla, S., Kaur, D., & Raghava, G. P. (2020). CancerLivER: A database of liver cancer gene expression resources and biomarkers. *Database: The Journal of Biological Databases and Curation, 2020*. https://doi.org/10.1093/database/baaa012.

Kaur, H., Bhalla, S., & Raghava, G. P. S. (2019). Classification of early and late stage liver hepatocellular carcinoma patients from their genomics and epigenomics profiles. *PLoS One, 14*(9), e0221476. https://doi.org/10.1371/journal.pone.0221476.

Kaur, H., Dhall, A., Kumar, R., & Raghava, G. P. S. (2019). Identification of platform-independent diagnostic biomarker panel for hepatocellular carcinoma using large-scale transcriptomics data. *Frontiers in Genetics, 10*, 1306. https://doi.org/10.3389/fgene.2019.01306.

Kennedy, L. B., & Salama, A. K. S. (2020). A review of cancer immunotherapy toxicity. *CA: A Cancer Journal for Clinicians, 70*(2), 86–104. https://doi.org/10.3322/caac.21596.

Kent, W. J. (2002). BLAT—The BLAST-like alignment tool. *Genome Research, 12*(4), 656–664. https://doi.org/10.1101/gr.229202.

Kim, S., Kim, H. S., Kim, E., Lee, M. G., Shin, E. C., Paik, S., et al. (2018). Neopepsee: Accurate genome-level prediction of neoantigens by harnessing sequence and amino acid immunogenicity information. *Annals of Oncology, 29*(4), 1030–1036. https://doi.org/10.1093/annonc/mdy022.

Kim, D., Paggi, J. M., Park, C., Bennett, C., & Salzberg, S. L. (2019). Graph-based genome alignment and genotyping with HISAT2 and HISAT-genotype. *Nature Biotechnology*, *37*(8), 907–915. https://doi.org/10.1038/s41587-019-0201-4.

Kim, Y., Sidney, J., Pinilla, C., Sette, A., & Peters, B. (2009). Derivation of an amino acid similarity matrix for peptide: MHC binding and its application as a Bayesian prior. *BMC Bioinformatics*, *10*, 394. https://doi.org/10.1186/1471-2105-10-394.

Koboldt, D. C., Zhang, Q., Larson, D. E., Shen, D., McLellan, M. D., Lin, L., et al. (2012). VarScan 2: Somatic mutation and copy number alteration discovery in cancer by exome sequencing. *Genome Research*, *22*(3), 568–576. https://doi.org/10.1101/gr.129684.111.

Kringelum, J. V., Lundegaard, C., Lund, O., & Nielsen, M. (2012). Reliable B cell epitope predictions: Impacts of method development and improved benchmarking. *PLoS Computational Biology*, *8*(12), e1002829. https://doi.org/10.1371/journal.pcbi.1002829.

Kroschinsky, F., Stolzel, F., von Bonin, S., Beutel, G., Kochanek, M., Kiehl, M., et al. (2017). New drugs, new toxicities: Severe side effects of modern targeted and immunotherapy of cancer and their management. *Critical Care*, *21*(1), 89. https://doi.org/10.1186/s13054-017-1678-1.

Kruger, S., Ilmer, M., Kobold, S., Cadilha, B. L., Endres, S., Ormanns, S., et al. (2019). Advances in cancer immunotherapy 2019—Latest trends. *Journal of Experimental & Clinical Cancer Research*, *38*(1), 268. https://doi.org/10.1186/s13046-019-1266-0.

Kulkarni-Kale, U., Raskar-Renuse, S., Natekar-Kalantre, G., & Saxena, S. A. (2014). Antigen-Antibody Interaction Database (AgAbDb): A compendium of antigen-antibody interactions. *Methods in Molecular Biology*, *1184*, 149–164. https://doi.org/10.1007/978-1-4939-1115-8_8.

Kumar, M., Bhasin, M., Natt, N. K., & Raghava, G. P. (2005). BhairPred: Prediction of beta-hairpins in a protein from multiple alignment information using ANN and SVM techniques. *Nucleic Acids Research*, *33*(Web Server issue), W154–W159. https://doi.org/10.1093/nar/gki588.

Kumar, R., Chaudhary, K., Gupta, S., Singh, H., Kumar, S., Gautam, A., et al. (2013). CancerDR: Cancer drug resistance database. *Scientific Reports*, *3*, 1445. https://doi.org/10.1038/srep01445.

Kumar, V., Kumar, R., Agrawal, P., Patiyal, S., & Raghava, G. P. S. (2020). A method for predicting hemolytic potency of chemically modified peptides from its structure. *Frontiers in Pharmacology*, *11*, 54. https://doi.org/10.3389/fphar.2020.00054.

Kumar, R., Lathwal, A., Kumar, V., Patiyal, S., Raghav, P. K., & Raghava, G. P. S. (2020). CancerEnD: A database of cancer associated enhancers. *Genomics*, *112*(5), 3696–3702. https://doi.org/10.1016/j.ygeno.2020.04.028.

Laimer, J., & Lackner, P. (2020). MHCII3D-robust structure based prediction of MHC II binding peptides. *International Journal of Molecular Sciences*, *22*(1). https://doi.org/10.3390/ijms22010012.

Laird, P. W. (2010). Principles and challenges of genomewide DNA methylation analysis. *Nature Reviews. Genetics*, *11*(3), 191–203. https://doi.org/10.1038/nrg2732.

Lamb, J. (2007). The Connectivity Map: A new tool for biomedical research. *Nature Reviews. Cancer*, 7(1), 54–60. https://doi.org/10.1038/nrc2044.

Landrum, M. J., Lee, J. M., Benson, M., Brown, G., Chao, C., Chitipiralla, S., et al. (2016). ClinVar: Public archive of interpretations of clinically relevant variants. *Nucleic Acids Research*, *44*(D1), D862–D868. https://doi.org/10.1093/nar/gkv1222.

Larsen, M. V., Lundegaard, C., Lamberth, K., Buus, S., Lund, O., & Nielsen, M. (2007). Large-scale validation of methods for cytotoxic T-lymphocyte epitope prediction. *BMC Bioinformatics*, *8*, 424. https://doi.org/10.1186/1471-2105-8-424.

Larson, D. E., Harris, C. C., Chen, K., Koboldt, D. C., Abbott, T. E., Dooling, D. J., et al. (2012). SomaticSniper: Identification of somatic point mutations in whole genome sequencing data. *Bioinformatics*, *28*(3), 311–317. https://doi.org/10.1093/bioinformatics/btr665.

Lata, S., Bhasin, M., & Raghava, G. P. (2007). Application of machine learning techniques in predicting MHC binders. *Methods in Molecular Biology*, *409*, 201–215. https://doi.org/10.1007/978-1-60327-118-9_14.

Lathwal, A., Arora, C., & Raghava, G. P. S. (2019). Prediction of risk scores for colorectal cancer patients from the concentration of proteins involved in mitochondrial apoptotic pathway. *PLoS One*, *14*(9), e0217527. https://doi.org/10.1371/journal.pone.0217527.

Lathwal, A., Kumar, R., & Raghava, G. P. S. (2020). OvirusTdb: A database of oncolytic viruses for the advancement of therapeutics in cancer. *Virology*, *548*, 109–116. https://doi.org/10.1016/j.virol.2020.05.016.

Lathwal, A., Kumar, R., & Raghava, G. P. S. (2021). In-silico identification of subunit vaccine candidates against lung cancer-associated oncogenic viruses. *Computers in Biology and Medicine*, *130*, 104215. https://doi.org/10.1016/j.compbiomed.2021.104215.

Lauria, M., Moyseos, P., & Priami, C. (2015). SCUDO: A tool for signature-based clustering of expression profiles. *Nucleic Acids Research*, *43*(W1), W188–W192. https://doi.org/10.1093/nar/gkv449.

Lee, H., & Kingsford, C. (2018). Kourami: Graph-guided assembly for novel human leukocyte antigen allele discovery. *Genome Biology*, *19*(1), 16. https://doi.org/10.1186/s13059-018-1388-2.

Lee, B. T., Song, C. M., Yeo, B. H., Chung, C. W., Chan, Y. L., Lim, T. T., et al. (2007). Gastric Cancer (Biomarkers) Knowledgebase (GCBKB): A curated and fully integrated knowledgebase of putative biomarkers related to gastric cancer. *Biomarker Insights*, *1*, 135–141. https://www.ncbi.nlm.nih.gov/pubmed/19690644.

Leinonen, R., Sugawara, H., Shumway, M., & International Nucleotide Sequence Database Collaboration. (2011). The sequence read archive. *Nucleic Acids Research*, *39*(Database issue), D19–D21. https://doi.org/10.1093/nar/gkq1019.

Li, H., Handsaker, B., Wysoker, A., Fennell, T., Ruan, J., Homer, N., et al. (2009). The Sequence Alignment/Map format and SAMtools. *Bioinformatics*, *25*(16), 2078–2079. https://doi.org/10.1093/bioinformatics/btp352.

Li, S., Li, Y., Chen, B., Zhao, J., Yu, S., Tang, Y., et al. (2018). exoRBase: A database of circRNA, lncRNA and mRNA in human blood exosomes. *Nucleic Acids Research*, *46*(D1), D106–D112. https://doi.org/10.1093/nar/gkx891.

Li, Y., Wang, X., Cho, J. H., Shaw, T. I., Wu, Z., Bai, B., et al. (2016). JUMPg: An integrative proteogenomics pipeline identifying unannotated proteins in human brain and cancer cells. *Journal of Proteome Research*, *15*(7), 2309–2320. https://doi.org/10.1021/acs.jproteome.6b00344.

Li, Y., Wang, G., Tan, X., Ouyang, J., Zhang, M., Song, X., et al. (2020). ProGeo-neo: A customized proteogenomic workflow for neoantigen prediction and selection. *BMC Medical Genomics*, *13*(Suppl. 5), 52. https://doi.org/10.1186/s12920-020-0683-4.

Liu, Z., Cai, C., Du, J., Liu, B., Cui, L., Fan, X., et al. (2020). TCMIO: A comprehensive database of Traditional Chinese Medicine on Immuno-Oncology. *Frontiers in Pharmacology*, *11*, 439. https://doi.org/10.3389/fphar.2020.00439.

Liu, C. J., Hu, F. F., Xia, M. X., Han, L., Zhang, Q., & Guo, A. Y. (2018). GSCALite: A web server for gene set cancer analysis. *Bioinformatics*, *34*(21), 3771–3772. https://doi.org/10.1093/bioinformatics/bty411.

Liu, G., Li, D., Li, Z., Qiu, S., Li, W., Chao, C. C., et al. (2017). PSSMHCpan: A novel PSSM-based software for predicting class I peptide-HLA binding affinity. *Gigascience*, *6*(5), 1–11. https://doi.org/10.1093/gigascience/gix017.

Liu, C., Yang, X., Duffy, B., Mohanakumar, T., Mitra, R. D., Zody, M. C., et al. (2013). ATHLATES: Accurate typing of human leukocyte antigen through exome sequencing. *Nucleic Acids Research*, *41*(14), e142. https://doi.org/10.1093/nar/gkt481.

Lohr, J. G., Stojanov, P., Lawrence, M. S., Auclair, D., Chapuy, B., Sougnez, C., et al. (2012). Discovery and prioritization of somatic mutations in diffuse large B-cell lymphoma (DLBCL) by whole-exome sequencing. *Proceedings of the National Academy of Sciences of the United States of America*, *109*(10), 3879–3884. https://doi.org/10.1073/pnas.1121343109.

Mailman, M. D., Feolo, M., Jin, Y., Kimura, M., Tryka, K., Bagoutdinov, R., et al. (2007). The NCBI dbGaP database of genotypes and phenotypes. *Nature Genetics*, *39*(10), 1181–1186. https://doi.org/10.1038/ng1007-1181.

Makki, J. (2015). Diversity of breast carcinoma: Histological subtypes and clinical relevance. *Clinical Medicine Insights: Pathology*, *8*, 23–31. https://doi.org/10.4137/CPath.S31563.

Manavalan, B., Shin, T. H., Kim, M. O., & Lee, G. (2018). PIP-EL: A new ensemble learning method for improved proinflammatory peptide predictions. *Frontiers in Immunology*, *9*, 1783. https://doi.org/10.3389/fimmu.2018.01783.

Mansoori, B., Mohammadi, A., Davudian, S., Shirjang, S., & Baradaran, B. (2017). The different mechanisms of cancer drug resistance: A brief review. *Advanced Pharmaceutical Bulletin*, 7(3), 339–348. https://doi.org/10.15171/apb.2017.041.

Marciniuk, K., Trost, B., & Napper, S. (2015). EpIC: A rational pipeline for epitope immunogenicity characterization. *Bioinformatics*, *31*(14), 2388–2390. https://doi.org/10.1093/bioinformatics/btv136.

Marcu, A., Bichmann, L., Kuchenbecker, L., Kowalewski, D. J., Freudenmann, L. K., Backert, L., et al. (2021). HLA Ligand Atlas: A benign reference of HLA-presented peptides to improve T-cell-based cancer immunotherapy. *Journal for Immunotherapy of Cancer*, *9*(4). https://doi.org/10.1136/jitc-2020-002071.

Mathur, D., Kaur, H., Dhall, A., Sharma, N., & Raghava, G. P. S. (2021). SAPdb: A database of short peptides and the corresponding nanostructures formed by self-assembly. *Computers in Biology and Medicine*, *133*, 104391. https://doi.org/10.1016/j.compbiomed.2021.104391.

Mayakonda, A., Lin, D. C., Assenov, Y., Plass, C., & Koeffler, H. P. (2018). Maftools: Efficient and comprehensive analysis of somatic variants in cancer. *Genome Research*, *28*(11), 1747–1756. https://doi.org/10.1101/gr.239244.118.

McCarroll, S. A., Kuruvilla, F. G., Korn, J. M., Cawley, S., Nemesh, J., Wysoker, A., et al. (2008). Integrated detection and population-genetic analysis of SNPs and copy number variation. *Nature Genetics*, *40*(10), 1166–1174. https://doi.org/10.1038/ng.238.

McKenna, A., Hanna, M., Banks, E., Sivachenko, A., Cibulskis, K., Kernytsky, A., et al. (2010). The Genome Analysis Toolkit: A MapReduce framework for analyzing next-generation DNA sequencing data. *Genome Research*, *20*(9), 1297–1303. https://doi.org/10.1101/gr.107524.110.

McLaren, W., Gil, L., Hunt, S. E., Riat, H. S., Ritchie, G. R., Thormann, A., et al. (2016). The Ensembl Variant Effect Predictor. *Genome Biology*, *17*(1), 122. https://doi.org/10.1186/s13059-016-0974-4.

McSparron, H., Blythe, M. J., Zygouri, C., Doytchinova, I. A., & Flower, D. R. (2003). JenPep: A novel computational information resource for immunobiology and vaccinology. *Journal of Chemical Information and Computer Sciences*, *43*(4), 1276–1287. https://doi.org/10.1021/ci030461e.

Medaglia, C., Giladi, A., Stoler-Barak, L., De Giovanni, M., Salame, T. M., Biram, A., et al. (2017). Spatial reconstruction of immune niches by combining photoactivatable reporters and scRNA-seq. *Science*, *358*(6370), 1622–1626. https://doi.org/10.1126/science.aao4277.

Miao, Y. R., Zhang, Q., Lei, Q., Luo, M., Xie, G. Y., Wang, H., et al. (2020). ImmuCellAI: A unique method for comprehensive T-cell subsets abundance prediction and its application in cancer immunotherapy. *Advanced Science (Weinheim, Baden-Württemberg, Germany)*, 7(7), 1902880. https://doi.org/10.1002/advs.201902880.

Minetto, P., Guolo, F., Pesce, S., Greppi, M., Obino, V., Ferretti, E., et al. (2019). Harnessing NK cells for cancer treatment. *Frontiers in Immunology*, *10*, 2836. https://doi.org/10.3389/fimmu.2019.02836.

Minoshima, S., Mitsuyama, S., Ohtsubo, M., Kawamura, T., Ito, S., Shibamoto, S., et al. (2001). The KMDB/MutationView: A mutation database for human disease genes. *Nucleic Acids Research*, *29*(1), 327–328. https://doi.org/10.1093/nar/29.1.327.

Modhukur, V., Iljasenko, T., Metsalu, T., Lokk, K., Laisk-Podar, T., & Vilo, J. (2018). MethSurv: A web tool to perform multivariable survival analysis using DNA methylation data. *Epigenomics*, *10*(3), 277–288. https://doi.org/10.2217/epi-2017-0118.

Mosch, A., Raffegerst, S., Weis, M., Schendel, D. J., & Frishman, D. (2019). Machine learning for cancer immunotherapies based on epitope recognition by T cell receptors. *Frontiers in Genetics*, *10*, 1141. https://doi.org/10.3389/fgene.2019.01141.

Mukherjee, S., Bhattacharyya, C., & Chandra, N. (2016). HLaffy: Estimating peptide affinities for Class-1 HLA molecules by learning position-specific pair potentials. *Bioinformatics*, *32*(15), 2297–2305. https://doi.org/10.1093/bioinformatics/btw156.

Nagpal, G., Usmani, S. S., Dhanda, S. K., Kaur, H., Singh, S., Sharma, M., et al. (2017). Computer-aided designing of immunosuppressive peptides based on IL-10 inducing potential. *Scientific Reports*, 7, 42851. https://doi.org/10.1038/srep42851.

Nakamura, Y., Komiyama, T., Furue, M., Gojobori, T., & Akiyama, Y. (2010). CIG-DB: The database for human or mouse immunoglobulin and T cell receptor genes available for cancer studies. *BMC Bioinformatics*, *11*, 398. https://doi.org/10.1186/1471-2105-11-398.

Nakato, R., & Gotoh, O. (2010). Cgaln: Fast and space-efficient whole-genome alignment. *BMC Bioinformatics*, *11*, 224. https://doi.org/10.1186/1471-2105-11-224.

Nariai, N., Kojima, K., Saito, S., Mimori, T., Sato, Y., Kawai, Y., et al. (2015). HLA-VBSeq: Accurate HLA typing at full resolution from whole-genome sequencing data. *BMC Genomics*, *16*(Suppl. 2), S7. https://doi.org/10.1186/1471-2164-16-S2-S7.

National Genomics Data Center Members and Partners. (2020). Database resources of the National Genomics Data Center in 2020. *Nucleic Acids Research*, *48*(D1), D24–D33. https://doi.org/10.1093/nar/gkz913.

Nielsen, M., Lundegaard, C., & Lund, O. (2007). Prediction of MHC class II binding affinity using SMM-align, a novel stabilization matrix alignment method. *BMC Bioinformatics*, *8*, 238. https://doi.org/10.1186/1471-2105-8-238.

O'Donnell, T. J., Rubinsteyn, A., & Laserson, U. (2020). MHCflurry 2.0: Improved pan-allele prediction of MHC class I-presented peptides by incorporating antigen processing. *Cell Systems*, *11*(1), 42–48.e7. https://doi.org/10.1016/j.cels.2020.06.010.

Oscanoa, J., Sivapalan, L., Gadaleta, E., Dayem Ullah, A. Z., Lemoine, N. R., & Chelala, C. (2020). SNPnexus: A web server for functional annotation of human genome sequence variation (2020 update). *Nucleic Acids Research*, *48*(W1), W185–W192. https://doi.org/10.1093/nar/gkaa420.

Panichnantakul, P., Bourgey, M., Montpetit, A., Bourque, G., & Riazalhosseini, Y. (2016). RNA-Seq as a tool to study the tumor microenvironment. *Methods in Molecular Biology*, *1458*, 311–337. https://doi.org/10.1007/978-1-4939-3801-8_22.

Paolillo, C., Londin, E., & Fortina, P. (2019). Single-cell genomics. *Clinical Chemistry*, *65*(8), 972–985. https://doi.org/10.1373/clinchem.2017.283895.

Parham, P. (1999). Virtual reality in the MHC. *Immunological Reviews*, *167*, 5–15. https://doi.org/10.1111/j.1600-065x.1999.tb01378.x.

Patlewicz, G., Jeliazkova, N., Safford, R. J., Worth, A. P., & Aleksiev, B. (2008). An evaluation of the implementation of the Cramer classification scheme in the Toxtree software. *SAR and QSAR in Environmental Research*, *19*(5–6), 495–524. https://doi.org/10.1080/10629360802083871.

Phloyphisut, P., Pornputtapong, N., Sriswasdi, S., & Chuangsuwanich, E. (2019). MHCSeqNet: A deep neural network model for universal MHC binding prediction. *BMC Bioinformatics*, *20*(1), 270. https://doi.org/10.1186/s12859-019-2892-4.

Ramos, A. H., Lichtenstein, L., Gupta, M., Lawrence, M. S., Pugh, T. J., Saksena, G., et al. (2015). Oncotator: Cancer variant annotation tool. *Human Mutation*, *36*(4), E2423–E2429. https://doi.org/10.1002/humu.22771.

Rao, A. A., Madejska, A. A., Pfeil, J., Paten, B., Salama, S. R., & Haussler, D. (2020). ProTECT—Prediction of T-cell Epitopes for Cancer Therapy. *Frontiers in Immunology*, *11*, 483296. https://doi.org/10.3389/fimmu.2020.483296.

Rasmussen, M., Fenoy, E., Harndahl, M., Kristensen, A. B., Nielsen, I. K., Nielsen, M., et al. (2016). Pan-specific prediction of peptide-MHC class I complex stability, a correlate of T cell immunogenicity. *Journal of Immunology*, *197*(4), 1517–1524. https://doi.org/10.4049/jimmunol.1600582.

Reche, P. A., & Reinherz, E. L. (2005). PEPVAC: A web server for multi-epitope vaccine development based on the prediction of supertypic MHC ligands. *Nucleic Acids Research*, *33*(Web Server issue), W138–W142. https://doi.org/10.1093/nar/gki357.

Rhodes, D. R., Kalyana-Sundaram, S., Mahavisno, V., Varambally, R., Yu, J., Briggs, B. B., et al. (2007). Oncomine 3.0: Genes, pathways, and networks in a collection of 18,000 cancer gene expression profiles. *Neoplasia*, *9*(2), 166–180. https://doi.org/10.1593/neo.07112.

Robinson, J., Barker, D. J., Georgiou, X., Cooper, M. A., Flicek, P., & Marsh, S. G. E. (2020). IPD-IMGT/HLA Database. *Nucleic Acids Research*, *48*(D1), D948–D955. https://doi.org/10.1093/nar/gkz950.

Rotem, A., Ram, O., Shoresh, N., Sperling, R. A., Goren, A., Weitz, D. A., et al. (2015). Single-cell ChIP-seq reveals cell subpopulations defined by chromatin state. *Nature Biotechnology*, *33*(11), 1165–1172. https://doi.org/10.1038/nbt.3383.

Saha, S., & Raghava, G. P. (2006a). AlgPred: Prediction of allergenic proteins and mapping of IgE epitopes. *Nucleic Acids Research*, *34*(Web Server issue), W202–W209. https://doi.org/10.1093/nar/gkl343.

Saha, S., & Raghava, G. P. (2006b). Prediction of continuous B-cell epitopes in an antigen using recurrent neural network. *Proteins*, *65*(1), 40–48. https://doi.org/10.1002/prot.21078.

Sanger, F., & Coulson, A. R. (1975). A rapid method for determining sequences in DNA by primed synthesis with DNA polymerase. *Journal of Molecular Biology*, *94*(3), 441–448. https://doi.org/10.1016/0022-2836(75)90213-2.

Sathirapongsasuti, J. F., Lee, H., Horst, B. A., Brunner, G., Cochran, A. J., Binder, S., et al. (2011). Exome sequencing-based copy-number variation and loss of heterozygosity detection: ExomeCNV. *Bioinformatics*, *27*(19), 2648–2654. https://doi.org/10.1093/bioinformatics/btr462.

Schlessinger, A., Ofran, Y., Yachdav, G., & Rost, B. (2006). Epitome: Database of structure-inferred antigenic epitopes. *Nucleic Acids Research*, *34*(Database issue), D777–D780. https://doi.org/10.1093/nar/gkj053.

Scholtalbers, J., Boegel, S., Bukur, T., Byl, M., Goerges, S., Sorn, P., et al. (2015). TCLP: An online cancer cell line catalogue integrating HLA type, predicted neo-epitopes, virus and gene expression. *Genome Medicine*, 7, 118. https://doi.org/10.1186/s13073-015-0240-5.

Schubert, B., Walzer, M., Brachvogel, H. P., Szolek, A., Mohr, C., & Kohlbacher, O. (2016). FRED 2: An immunoinformatics framework for Python. *Bioinformatics*, *32*(13), 2044–2046. https://doi.org/10.1093/bioinformatics/btw113.

Sedova, M., Iyer, M., Li, Z., Jaroszewski, L., Post, K. W., Hrabe, T., et al. (2019). Cancer3D 2.0: Interactive analysis of 3D patterns of cancer mutations in cancer subsets. *Nucleic Acids Research*, *47*(D1), D895–D899. https://doi.org/10.1093/nar/gky1098.

Sela-Culang, I., Benhnia, M. R., Matho, M. H., Kaever, T., Maybeno, M., Schlossman, A., et al. (2014). Using a combined computational-experimental approach to predict antibody-specific B cell epitopes. *Structure, 22*(4), 646–657. https://doi.org/10.1016/j.str.2014.02.003.

Shao, X. M., Bhattacharya, R., Huang, J., Sivakumar, I. K. A., Tokheim, C., Zheng, L., et al. (2020). High-throughput prediction of MHC class I and II neoantigens with MHCnuggets. *Cancer Immunology Research, 8*(3), 396–408. https://doi.org/10.1158/2326-6066.CIR-19-0464.

Sharma, N., Patiyal, S., Dhall, A., Devi, N. L., & Raghava, G. P. S. (2021). ChAlPred: A web server for prediction of allergenicity of chemical compounds. *Computers in Biology and Medicine, 136*, 104746. https://doi.org/10.1016/j.compbiomed.2021.104746.

Sharma, N., Patiyal, S., Dhall, A., Pande, A., Arora, C., & Raghava, G. P. S. (2020). AlgPred 2.0: An improved method for predicting allergenic proteins and mapping of IgE epitopes. *Briefings in Bioinformatics*. https://doi.org/10.1093/bib/bbaa294.

Sharma, A. K., Srivastava, G. N., Roy, A., & Sharma, V. K. (2017). ToxiM: A toxicity prediction tool for small molecules developed using machine learning and chemoinformatics approaches. *Frontiers in Pharmacology, 8*, 880. https://doi.org/10.3389/fphar.2017.00880.

Shen, J. J., Yang, C., Wang, Y. F., Wang, T. Y., Guo, M., Lau, Y. L., et al. (2019). HLA-IMPUTER: An easy to use web application for HLA imputation and association analysis using population-specific reference panels. *Bioinformatics, 35*(7), 1244–1246. https://doi.org/10.1093/bioinformatics/bty730.

Shih, K., Arkenau, H. T., & Infante, J. R. (2014). Clinical impact of checkpoint inhibitors as novel cancer therapies. *Drugs, 74*(17), 1993–2013. https://doi.org/10.1007/s40265-014-0305-6.

Shimizu, N., Ohtsubo, M., & Minoshima, S. (2007). MutationView/KMcancerDB: A database for cancer gene mutations. *Cancer Science, 98*(3), 259–267. https://doi.org/10.1111/j.1349-7006.2007.00405.x.

Shugay, M., Bagaev, D. V., Zvyagin, I. V., Vroomans, R. M., Crawford, J. C., Dolton, G., et al. (2018). VDJdb: A curated database of T-cell receptor sequences with known antigen specificity. *Nucleic Acids Research, 46*(D1), D419–D427. https://doi.org/10.1093/nar/gkx760.

Shukla, S. A., Rooney, M. S., Rajasagi, M., Tiao, G., Dixon, P. M., Lawrence, M. S., et al. (2015). Comprehensive analysis of cancer-associated somatic mutations in class I HLA genes. *Nature Biotechnology, 33*(11), 1152–1158. https://doi.org/10.1038/nbt.3344.

Singh, H., & Raghava, G. P. (2001). ProPred: Prediction of HLA-DR binding sites. *Bioinformatics, 17*(12), 1236–1237. https://doi.org/10.1093/bioinformatics/17.12.1236.

Singh, H., & Raghava, G. P. (2003). ProPred1: Prediction of promiscuous MHC Class-I binding sites. *Bioinformatics, 19*(8), 1009–1014. https://doi.org/10.1093/bioinformatics/btg108.

Stanislaus, R., Carey, M., Deus, H. F., Coombes, K., Hennessy, B. T., Mills, G. B., et al. (2008). RPPAML/RIMS: A metadata format and an information management system for reverse phase protein arrays. *BMC Bioinformatics, 9*, 555. https://doi.org/10.1186/1471-2105-9-555.

Stenson, P. D., Mort, M., Ball, E. V., Evans, K., Hayden, M., Heywood, S., et al. (2017). The Human Gene Mutation Database: Towards a comprehensive repository of inherited mutation data for medical research, genetic diagnosis and next-generation sequencing studies. *Human Genetics, 136*(6), 665–677. https://doi.org/10.1007/s00439-017-1779-6.

Stranzl, T., Larsen, M. V., Lundegaard, C., & Nielsen, M. (2010). NetCTLpan: Pan-specific MHC class I pathway epitope predictions. *Immunogenetics, 62*(6), 357–368. https://doi.org/10.1007/s00251-010-0441-4.

Sung, H., Ferlay, J., Siegel, R. L., Laversanne, M., Soerjomataram, I., Jemal, A., et al. (2021). Global cancer statistics 2020: GLOBOCAN estimates of incidence and mortality

worldwide for 36 cancers in 185 countries. *CA: A Cancer Journal for Clinicians, 71*(3), 209–249. https://doi.org/10.3322/caac.21660.
Sweet, K., Zhang, L., & Pinilla-Ibarz, J. (2013). Biomarkers for determining the prognosis in chronic myelogenous leukemia. *Journal of Hematology & Oncology, 6*, 54. https://doi.org/10.1186/1756-8722-6-54.
Szolek, A., Schubert, B., Mohr, C., Sturm, M., Feldhahn, M., & Kohlbacher, O. (2014). OptiType: Precision HLA typing from next-generation sequencing data. *Bioinformatics, 30*(23), 3310–3316. https://doi.org/10.1093/bioinformatics/btu548.
Tan, X., Li, D., Huang, P., Jian, X., Wan, H., Wang, G., et al. (2020). dbPepNeo: A manually curated database for human tumor neoantigen peptides. *Database: The Journal of Biological Databases and Curation, 2020*. https://doi.org/10.1093/database/baaa004.
Tang, Y., Wang, Y., Wang, J., Li, M., Peng, L., Wei, G., et al. (2020). TruNeo: An integrated pipeline improves personalized true tumor neoantigen identification. *BMC Bioinformatics, 21*(1), 532. https://doi.org/10.1186/s12859-020-03869-9.
Tappeiner, E., Finotello, F., Charoentong, P., Mayer, C., Rieder, D., & Trajanoski, Z. (2017). TIminer: NGS data mining pipeline for cancer immunology and immunotherapy. *Bioinformatics, 33*(19), 3140–3141. https://doi.org/10.1093/bioinformatics/btx377.
Tate, J. G., Bamford, S., Jubb, H. C., Sondka, Z., Beare, D. M., Bindal, N., et al. (2019). COSMIC: The Catalogue Of Somatic Mutations In Cancer. *Nucleic Acids Research, 47*(D1), D941–D947. https://doi.org/10.1093/nar/gky1015.
Teku, G. N., & Vihinen, M. (2018). Pan-cancer analysis of neoepitopes. *Scientific Reports, 8*(1), 12735. https://doi.org/10.1038/s41598-018-30724-y.
Tian, J., Cai, Y., Li, Y., Lu, Z., Huang, J., Deng, Y., et al. (2021). CancerImmunityQTL: A database to systematically evaluate the impact of genetic variants on immune infiltration in human cancer. *Nucleic Acids Research, 49*(D1), D1065–D1073. https://doi.org/10.1093/nar/gkaa805.
Tomczak, K., Czerwinska, P., & Wiznerowicz, M. (2015). The Cancer Genome Atlas (TCGA): An immeasurable source of knowledge. *Contemporary Oncology (Poznań, Poland), 19*(1A), A68–A77. https://doi.org/10.5114/wo.2014.47136.
Trevino, V. (2020). HotSpotAnnotations—A database for hotspot mutations and annotations in cancer. *Database: The Journal of Biological Databases and Curation, 2020*. https://doi.org/10.1093/database/baaa025.
Trolle, T., & Nielsen, M. (2014). NetTepi: An integrated method for the prediction of T cell epitopes. *Immunogenetics, 66*(7–8), 449–456. https://doi.org/10.1007/s00251-014-0779-0.
Tyagi, A., Tuknait, A., Anand, P., Gupta, S., Sharma, M., Mathur, D., et al. (2015). CancerPPD: A database of anticancer peptides and proteins. *Nucleic Acids Research, 43*(Database issue), D837–D843. https://doi.org/10.1093/nar/gku892.
Urruticoechea, A., Alemany, R., Balart, J., Villanueva, A., Vinals, F., & Capella, G. (2010). Recent advances in cancer therapy: An overview. *Current Pharmaceutical Design, 16*(1), 3–10. https://doi.org/10.2174/138161210789941847.
Vaddepally, R. K., Kharel, P., Pandey, R., Garje, R., & Chandra, A. B. (2020). Review of indications of FDA-approved immune checkpoint inhibitors per NCCN guidelines with the level of evidence. *Cancers (Basel), 12*(3). https://doi.org/10.3390/cancers12030738.
Van Allen, E. M., Miao, D., Schilling, B., Shukla, S. A., Blank, C., Zimmer, L., et al. (2015). Genomic correlates of response to CTLA-4 blockade in metastatic melanoma. *Science, 350*(6257), 207–211. https://doi.org/10.1126/science.aad0095.
van der Burg, S. H., Arens, R., Ossendorp, F., van Hall, T., & Melief, C. J. (2016). Vaccines for established cancer: Overcoming the challenges posed by immune evasion. *Nature Reviews. Cancer, 16*(4), 219–233. https://doi.org/10.1038/nrc.2016.16.

Vang, Y. S., & Xie, X. (2017). HLA class I binding prediction via convolutional neural networks. *Bioinformatics*, *33*(17), 2658–2665. https://doi.org/10.1093/bioinformatics/btx264.
Varade, J., Magadan, S., & Gonzalez-Fernandez, A. (2021). Human immunology and immunotherapy: Main achievements and challenges. *Cellular & Molecular Immunology*, *18*(4), 805–828. https://doi.org/10.1038/s41423-020-00530-6.
Vo, J. N., Cieslik, M., Zhang, Y., Shukla, S., Xiao, L., Zhang, Y., et al. (2019). The landscape of circular RNA in cancer. *Cell*, *176*(4). 869–881.e13 https://doi.org/10.1016/j.cell.2018.12.021.
Vormehr, M., Tureci, O., & Sahin, U. (2019). Harnessing tumor mutations for truly individualized cancer vaccines. *Annual Review of Medicine*, *70*, 395–407. https://doi.org/10.1146/annurev-med-042617-101816.
Waldman, A. D., Fritz, J. M., & Lenardo, M. J. (2020). A guide to cancer immunotherapy: From T cell basic science to clinical practice. *Nature Reviews. Immunology*, *20*(11), 651–668. https://doi.org/10.1038/s41577-020-0306-5.
Wan, J., Liu, W., Xu, Q., Ren, Y., Flower, D. R., & Li, T. (2006). SVRMHC prediction server for MHC-binding peptides. *BMC Bioinformatics*, *7*, 463. https://doi.org/10.1186/1471-2105-7-463.
Wang, Z., Gerstein, M., & Snyder, M. (2009). RNA-Seq: A revolutionary tool for transcriptomics. *Nature Reviews. Genetics*, *10*(1), 57–63. https://doi.org/10.1038/nrg2484.
Wang, X., Kaczor-Urbanowicz, K. E., & Wong, D. T. (2017). Salivary biomarkers in cancer detection. *Medical Oncology*, *34*(1), 7. https://doi.org/10.1007/s12032-016-0863-4.
Wang, K., Li, M., & Hakonarson, H. (2010). ANNOVAR: Functional annotation of genetic variants from high-throughput sequencing data. *Nucleic Acids Research*, *38*(16), e164. https://doi.org/10.1093/nar/gkq603.
Wang, Y. Y., Mimori, T., Khor, S. S., Gervais, O., Kawai, Y., Hitomi, Y., et al. (2019). HLA-VBSeq v2: Improved HLA calling accuracy with full-length Japanese class-I panel. *Human Genome Variation*, *6*, 29. https://doi.org/10.1038/s41439-019-0061-y.
Wang, D., Yang, L., Zhang, P., LaBaer, J., Hermjakob, H., Li, D., et al. (2017). AAgAtlas 1.0: A human autoantigen database. *Nucleic Acids Research*, *45*(D1), D769–D776. https://doi.org/10.1093/nar/gkw946.
Wang, X., Zhao, H., Xu, Q., Jin, W., Liu, C., Zhang, H., et al. (2006). HPtaa database-potential target genes for clinical diagnosis and immunotherapy of human carcinoma. *Nucleic Acids Research*, *34*(Database issue), D607–D612. https://doi.org/10.1093/nar/gkj082.
Warren, R. L., Choe, G., Freeman, D. J., Castellarin, M., Munro, S., Moore, R., et al. (2012). Derivation of HLA types from shotgun sequence datasets. *Genome Medicine*, *4*(12), 95. https://doi.org/10.1186/gm396.
Wickstrom, S. L., Lovgren, T., Volkmar, M., Reinhold, B., Duke-Cohan, J. S., Hartmann, L., et al. (2019). Cancer neoepitopes for immunotherapy: Discordance between tumor-infiltrating T cell reactivity and tumor MHC peptidome display. *Frontiers in Immunology*, *10*, 2766. https://doi.org/10.3389/fimmu.2019.02766.
Wieczorek, M., Abualrous, E. T., Sticht, J., Alvaro-Benito, M., Stolzenberg, S., Noe, F., et al. (2017). Major histocompatibility complex (MHC) class I and MHC class II proteins: Conformational plasticity in antigen presentation. *Frontiers in Immunology*, *8*, 292. https://doi.org/10.3389/fimmu.2017.00292.
Wishart, D. S., Bartok, B., Oler, E., Liang, K. Y. H., Budinski, Z., Berjanskii, M., et al. (2021). MarkerDB: An online database of molecular biomarkers. *Nucleic Acids Research*, *49*(D1), D1259–D1267. https://doi.org/10.1093/nar/gkaa1067.
Wu, J., Wang, W., Zhang, J., Zhou, B., Zhao, W., Su, Z., et al. (2019). DeepHLApan: A deep learning approach for neoantigen prediction considering both HLA-peptide binding and immunogenicity. *Frontiers in Immunology*, *10*, 2559. https://doi.org/10.3389/fimmu.2019.02559.

Wu, J., Zhao, W., Zhou, B., Su, Z., Gu, X., Zhou, Z., et al. (2018). TSNAdb: A database for tumor-specific neoantigens from immunogenomics data analysis. *Genomics, Proteomics & Bioinformatics, 16*(4), 276–282. https://doi.org/10.1016/j.gpb.2018.06.003.

Xia, J., Bai, P., Fan, W., Li, Q., Li, Y., Wang, D., et al. (2021). NEPdb: A database of T-cell experimentally-validated neoantigens and pan-cancer predicted neoepitopes for cancer immunotherapy. *Frontiers in Immunology, 12*, 644637. https://doi.org/10.3389/fimmu.2021.644637.

Xie, C., Yeo, Z. X., Wong, M., Piper, J., Long, T., Kirkness, E. F., et al. (2017). Fast and accurate HLA typing from short-read next-generation sequence data with xHLA. *Proceedings of the National Academy of Sciences of the United States of America, 114*(30), 8059–8064. https://doi.org/10.1073/pnas.1707945114.

Xu, G., Deng, N., Zhao, Z., Judeh, T., Flemington, E., & Zhu, D. (2011). SAMMate: A GUI tool for processing short read alignments in SAM/BAM format. *Source Code for Biology and Medicine, 6*(1), 2. https://doi.org/10.1186/1751-0473-6-2.

Xu, Z., Li, P., Fan, L., & Wu, M. (2018). The potential role of circRNA in tumor immunity regulation and immunotherapy. *Frontiers in Immunology, 9*, 9. https://doi.org/10.3389/fimmu.2018.00009.

Yan, Z., Wang, Q., Lu, Z., Sun, X., Song, P., Dang, Y., et al. (2020). OSluca: An interactive web server to evaluate prognostic biomarkers for lung cancer. *Frontiers in Genetics, 11*, 420. https://doi.org/10.3389/fgene.2020.00420.

Yang, Y. (2015). Cancer immunotherapy: Harnessing the immune system to battle cancer. *The Journal of Clinical Investigation, 125*(9), 3335–3337. https://doi.org/10.1172/JCI83871.

Yang, B., Sayers, S., Xiang, Z., & He, Y. (2011). Protegen: A web-based protective antigen database and analysis system. *Nucleic Acids Research, 39*(Database issue), D1073–D1078. https://doi.org/10.1093/nar/gkq944.

Yang, H., Sun, L., Li, W., Liu, G., & Tang, Y. (2018). In silico prediction of chemical toxicity for drug design using machine learning methods and structural alerts. *Frontiers in Chemistry, 6*, 30. https://doi.org/10.3389/fchem.2018.00030.

Yao, D., Zhang, L., Zheng, M., Sun, X., Lu, Y., & Liu, P. (2018). Circ2Disease: A manually curated database of experimentally validated circRNAs in human disease. *Scientific Reports, 8*(1), 11018. https://doi.org/10.1038/s41598-018-29360-3.

Zacharakis, N., Chinnasamy, H., Black, M., Xu, H., Lu, Y. C., Zheng, Z., et al. (2018). Immune recognition of somatic mutations leading to complete durable regression in metastatic breast cancer. *Nature Medicine, 24*(6), 724–730. https://doi.org/10.1038/s41591-018-0040-8.

Zhang, J., Bajari, R., Andric, D., Gerthoffert, F., Lepsa, A., Nahal-Bose, H., et al. (2019). The International Cancer Genome Consortium Data Portal. *Nature Biotechnology, 37*(4), 367–369. https://doi.org/10.1038/s41587-019-0055-9.

Zhang, J., Baran, J., Cros, A., Guberman, J. M., Haider, S., Hsu, J., et al. (2011). International Cancer Genome Consortium Data Portal—A one-stop shop for cancer genomics data. *Database (Oxford), 2011*, bar026. https://doi.org/10.1093/database/bar026.

Zhang, G., Chitkushev, L., Olsen, L. R., Keskin, D. B., & Brusic, V. (2021). TANTIGEN 2.0: A knowledge base of tumor T cell antigens and epitopes. *BMC Bioinformatics, 22*(Suppl. 8), 40. https://doi.org/10.1186/s12859-021-03962-7.

Zhang, G. L., Lin, H. H., Keskin, D. B., Reinherz, E. L., & Brusic, V. (2011). Dana-Farber repository for machine learning in immunology. *Journal of Immunological Methods, 374*(1-2), 18–25. https://doi.org/10.1016/j.jim.2011.07.007.

Zhang, H., Lund, O., & Nielsen, M. (2009). The PickPocket method for predicting binding specificities for receptors based on receptor pocket similarities: Application to MHC-peptide binding. *Bioinformatics, 25*(10), 1293–1299. https://doi.org/10.1093/bioinformatics/btp137.

Zhang, J., Mardis, E. R., & Maher, C. A. (2017). INTEGRATE-neo: A pipeline for personalized gene fusion neoantigen discovery. *Bioinformatics*, *33*(4), 555–557. https://doi.org/10.1093/bioinformatics/btw674.

Zhang, X., Sun, X. F., Cao, Y., Ye, B., Peng, Q., Liu, X., et al. (2018). CBD: A biomarker database for colorectal cancer. *Database: The Journal of Biological Databases and Curation*, *2018*. https://doi.org/10.1093/database/bay046.

Zhang, M., Wang, B., Xu, J., Wang, X., Xie, L., Zhang, B., et al. (2017). CanProVar 2.0: An updated database of human cancer proteome variation. *Journal of Proteome Research*, *16*(2), 421–432. https://doi.org/10.1021/acs.jproteome.6b00505.

Zhang, G., Wang, Q., Yang, M., Yuan, Q., Dang, Y., Sun, X., et al. (2019). OSblca: A web server for investigating prognostic biomarkers of bladder cancer patients. *Frontiers in Oncology*, *9*, 466. https://doi.org/10.3389/fonc.2019.00466.

Zhang, Y., Yao, Y., Chen, P., Liu, Y., Zhang, H., Liu, H., et al. (2020). Checkpoint therapeutic target database (CKTTD): The first comprehensive database for checkpoint targets and their modulators in cancer immunotherapy. *Journal for Immunotherapy of Cancer*, *8*(2). https://doi.org/10.1136/jitc-2020-001247.

Zhang, W., Zeng, B., Lin, H., Guan, W., Mo, J., Wu, S., et al. (2021). CanImmunother: A manually curated database for identification of cancer immunotherapies associating with biomarkers, targets, and clinical effects. *Oncoimmunology*, *10*(1), 1944553. https://doi.org/10.1080/2162402X.2021.1944553.

Zhang, Z., Zhou, C., Tang, L., Gong, Y., Wei, Z., Zhang, G., et al. (2020). ASNEO: Identification of personalized alternative splicing based neoantigens with RNA-seq. *Aging (Albany NY)*, *12*(14), 14633–14648. https://doi.org/10.18632/aging.103516.

Zhao, L., & Cao, Y. J. (2019). Engineered T cell therapy for cancer in the clinic. *Frontiers in Immunology*, *10*, 2250. https://doi.org/10.3389/fimmu.2019.02250.

Zhao, Q., Jiang, Y., Xiang, S., Kaboli, P. J., Shen, J., Zhao, Y., et al. (2021). Engineered TCR-T cell immunotherapy in anticancer precision medicine: Pros and cons. *Frontiers in Immunology*, *12*, 658753. https://doi.org/10.3389/fimmu.2021.658753.

Zheng, C., Zheng, L., Yoo, J. K., Guo, H., Zhang, Y., Guo, X., et al. (2017). Landscape of infiltrating T cells in liver cancer revealed by single-cell sequencing. *Cell*, *169*(7). 1342–1356.e.16 https://doi.org/10.1016/j.cell.2017.05.035.

Zhou, Z., Lyu, X., Wu, J., Yang, X., Wu, S., Zhou, J., et al. (2017). TSNAD: An integrated software for cancer somatic mutation and tumour-specific neoantigen detection. *Royal Society Open Science*, *4*(4), 170050. https://doi.org/10.1098/rsos.170050.

Zhou, W. J., Qu, Z., Song, C. Y., Sun, Y., Lai, A. L., Luo, M. Y., et al. (2019). NeoPeptide: An immunoinformatic database of T-cell-defined neoantigens. *Database: The Journal of Biological Databases and Curation*, *2019*. https://doi.org/10.1093/database/baz128.

Zhou, C., Wei, Z., Zhang, Z., Zhang, B., Zhu, C., Chen, K., et al. (2019). pTuneos: Prioritizing tumor neoantigens from next-generation sequencing data. *Genome Medicine*, *11*(1), 67. https://doi.org/10.1186/s13073-019-0679-x.

Zhu, J., Sanborn, J. Z., Benz, S., Szeto, C., Hsu, F., Kuhn, R. M., et al. (2009). The UCSC Cancer Genomics Browser. *Nature Methods*, *6*(4), 239–240. https://doi.org/10.1038/nmeth0409-239.

Zuo, Z., Hu, H., Xu, Q., Luo, X., Peng, D., Zhu, K., et al. (2020). BBCancer: An expression atlas of blood-based biomarkers in the early diagnosis of cancers. *Nucleic Acids Research*, *48*(D1), D789–D796. https://doi.org/10.1093/nar/gkz942.

CHAPTER TWO

Immunotherapeutic approaches for HPV-caused cervical cancer

Seyed Amirreza Fatemi[a,b,c,†], Nadia Seifi[a,b,c,†], Shiva Rasekh[a,b,c], Sogand Amiri[a,b,c], Seyed Mohammad Iman Moezzi[a,b,c], Ashkan Bagheri[b,c], Shirin Fathi[a,b,c], and Manica Negahdaripour[b,c,*]

[a]Student Research Committee, Shiraz University of Medical Sciences, Shiraz, Iran
[b]Department of Pharmaceutical Biotechnology, School of Pharmacy, Shiraz University of Medical Sciences, Shiraz, Iran
[c]Pharmaceutical Sciences Research Center, Shiraz University of Medical Sciences, Shiraz, Iran
*Corresponding author: e-mail address: negahdaripour@sums.ac.ir

Contents

Abstract

Cervical cancer, the fourth most frequent women cancer worldwide, is mostly (about 99%) associated with human papillomavirus (HPV). Despite availability of three effective prophylactic vaccines for more than one decade and some other preventive measures,

† Co-first authors.

Advances in Protein Chemistry and Structural Biology, Volume 129
ISSN 1876-1623
https://doi.org/10.1016/bs.apcsb.2021.11.002

it is still the fourth cause of cancer death among women globally. Thus, development of therapeutic vaccines seems essential, which has been vastly studied using different vaccine platforms. Even with very wide efforts during the past years, no therapeutic vaccine has been approved yet, which might be partly due to the complex events and interactions taken place in the tumor microenvironment.

On the other hand, immunotherapy has opened its way into the management plans of some cancers. The recent approval of pembrolizumab for the treatment of metastatic/recurrent cervical cancer brings new hopes to the management of this disease, while some other immunotherapeutic approaches are also under investigation either alone or in combination with vaccines.

Here, following a summary about HPV and its pathogenesis, cervical cancer therapeutic vaccines would be reviewed. Cell-based vaccines as well as immunomodulation and other modalities used along with vaccines would be also discussed.

1. Introduction

Cervical cancer (CxCa), the fourth most frequent cancer and cause of cancer death among women worldwide (Sung et al., 2021), is associated with human papillomavirus (HPV) in about 99% of cases (Bruni et al., 2016). Genital HPV is the most common sexually transmitted disease (STD) with an estimated prevalence of approximately 70% in both sexes on the lifetime. A higher burden of the disease is observed in low- and middle-income countries, where routine screenings such as Pap smear and HPV vaccination are not taken seriously (Schellenbacher, Roden, & Kirnbauer, 2017). However even in the US, despite some decline lately (Sung et al., 2021), it was the second cancer-related cause of death in women aged 20–39 years in 2018 (Siegal, Miller, & Jemal, 2014).

HPV prophylactic vaccines were introduced more than one decade ago. They have shown great success in CxCa reduction in societies with high vaccine coverage (Brisson et al., 2020; Patel et al., 2018). Though CxCa is considered a preventable disease owing to primary (vaccination) and secondary (screening) prophylactic modalities, it is still a global major health concern with a 5-year survival rate ranging about 20–66% in different countries (Ferrall, Lin, Roden, Hung, & Wu, 2021). Thus, efforts for developing therapeutic vaccines as well as prophylactic vaccines with improved features are underway.

On the other hand, immunotherapy has proven its role in treatment of some cancers. The FDA approval of pembrolizumab for metastatic/recurrent CxCa treatment (Ferrall et al., 2021) has opened the way for more therapeutic options for CxCa patients.

Here, following a summary about HPV and its pathogenesis, CxCa therapeutic vaccines are reviewed. Cell-based vaccines as well as immunomodulation and other modalities used along with vaccines in clinical trials or in vivo are also discussed.

2. HPV

2.1 Structure

HPV is a non-enveloped double-stranded DNA virus with an 8 kb genome. It has two types of proteins; two late (L) and six early (E) proteins. L1, the major protein, and L2, the minor one, are on the virus capsid. L1 protein (MW = 55 kDa) represents 80% of the viral capsid proteins (Panatto et al., 2015). E1 and E2 are responsible for viral genome replication. E4 stimulates virion release from keratinocytes (Huber, Wang, Roden, & Kirnbauer, 2021). E5 is a channel protein, regulating ion transfer, vesicle trafficking, viral genome entry, and virion production (Bruni et al., 2016). E6 and E7 oncogenic proteins dysregulate the cell cycle to guarantee viral replication. Moreover, the HPV genome has upstream regulatory regions (URR) as non-coding DNA, which contains regulatory elements for genome replication and transcription (Sung et al., 2021). Fig. 1 presents detailed function of HPV proteins and genes.

2.2 HPV types

HPV is a double-stranded DNA virus with more than 220 identified genotypes and five genera (α, β, γ, μ, and ν) based on L1 nucleotide sequence. HPV causes various diseases from benign skin warts to various carcinomas. HPVs are also classified as cutaneous and mucosal groups according to their infection site. Mucosal HPVs include low-risk (LR) and high-risk (HR) HPVs depending on their oncogenic potential (Huber et al., 2021).

HR-HPVs, which mostly belong to genus α, cause mainly CxCa, as well as head and neck cancer and the less common vulvar, vaginal, penile, and anal cancers. Chronic untreated infection by HR-HPV types may lead to cervical intraepithelial neoplasia (CIN) (Vici et al., 2016) including high-grade squamous intraepithelial lesions (HSIL), which can progress into invasive cancers (Schellenbacher et al., 2017) or CxCa. Among HR-HPVs, HPV16/18/31/33/35/39/45/51/52/56/58/59 are oncogenic, while probably-oncogenic types include HPV36/30/34/53/66/67/68/69/70/73/82/85/97

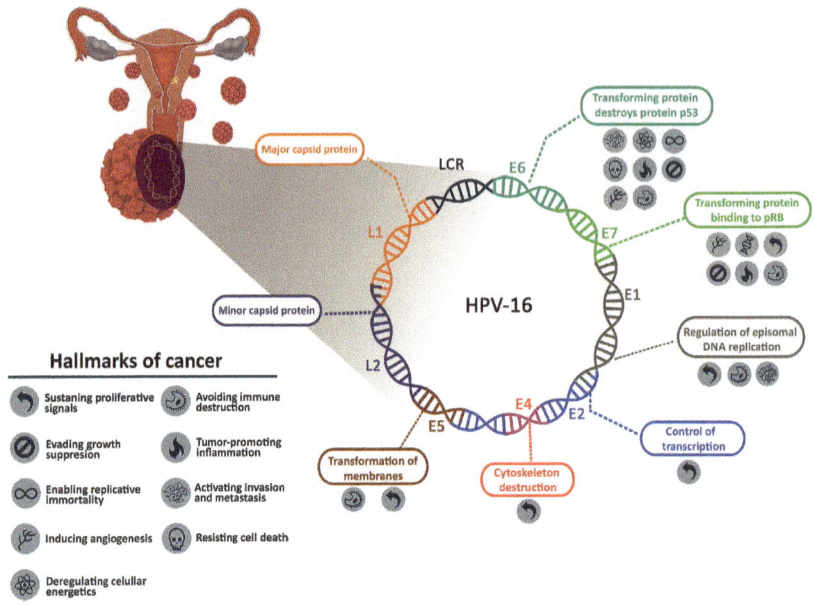

Fig. 1 Hallmarks of HPV cancer- Though high-risk HPV infection alone is not enough to cause cancer, it adjusts pathways that develop cancer hallmarks. These include sustaining proliferative signaling via E1, E2, E4, E5, E5, and E7; evading growth suppressors via E6 and E7; enabling replicative immortality via E6; inducing local angiogenesis via E6 and E7; deregulating cellular energetics via E6; avoiding immune destruction via E1, E5, E6, and E7; induction of interleukin-6 (IL-6) via E6 and E7 (which raises monocytes generating tumor-promoting inflammation); activating invasion and metastasis via E1 and E6; and resisting cell death via E6. LCR: Locus control region; pRb: Retinoblastoma protein.

(Bruni et al., 2016). About 50% and 20% of CxCa are associated with HPV16 and HPV18, respectively (De Sanjose et al., 2010).

LR-HPVs are common causes of benign genital warts and recurrent laryngeal papillomatosis, which are mostly self-limiting. Some LR-HPVs are HPV6/11/13/40/43/44/74. Most genital warts are caused by HPV6/11 (Siegal et al., 2014).

2.3 Viral life cycle and pathogenesis

With epithelial aberration, the basement membrane would expose, and HPV can reach basal keratinocytes. HPV L1 protein binds to heparan sulfate proteoglycan (HSPG), and the capsid conformation will change (Huber et al., 2021). This results in the exposure of conserved sites at L2 N-terminus, which are cleaved by extracellular furin. This cleavage again causes a conformational change that lets the virus bind to the cell receptors such as α6β4 integrin. Subsequently, the virus is internalized to the cell via endosomal vesicles. Here, the L2 protein is an important compartment during endosomal engagement with retromer translocating the virus to the nucleus. For efficient transcription of viral genome, L2 protein chaperones the viral genome into nuclear bodies (Vici et al., 2016).

Since HPV lacks polymerase enzymes, it employs host cell transcription and replication machinery to replicate its genome. Moreover, the viral genome can be inserted in random regions of the host genome, which is associated with HPV carcinogenicity (Ferreira, Ramalho, Marques, & Ribeiro, 2020). E1 and E2 are expressed first to regulate the cell cycle toward the S phase and producing required DNA polymerases. This leads to the production of 50–100 replicates of the HPV genome. After cell deviation, the viral genome segregates between daughter cells, and they start to make L proteins. Now, more than 1000 copies of HPV per cell would be produced, stored in squamous cells, and shed. Thus, the viral lifecycle would be repeated again (De Sanjose et al., 2010; Sung et al., 2021).

E proteins are produced only by infected basal epithelial cells, while L proteins are expressed in the epithelium before viral shedding. E5, E6, and E7, as HPV oncoproteins, drive keratinocytes into the S phase. E5 stimulates cell proliferation through the epidermal growth factor (EGF) pathway and modulates several pathways involved in cell adhesion, motility, and mitosis (Kumar, Biswas, & Jose, 2015). E6 degrades p53, the apoptosis regulator protein, by activation of the ubiquitin-proteasome mechanism. Moreover, E6 stimulates telomerase and MYC expression, which causes cell survival and longevity (Negahdaripour et al., 2018). E7 binds to the hyperphosphorylated form of retinoblastoma protein (pRb) (a tumor suppressor protein) and enhances its phosphorylation, leading the cell cycle to S phase, DNA synthesis, and proliferation (De Sanjose et al., 2010).

2.4 Immune response to HPV infection

HPV infections are generally asymptomatic and would be cleared within few months in 90% of cases, but sometimes the virus might escape from the immune systems mainly due to the presence of the immunogenic virions mostly at the outer sites of epithelium (Ferreira et al., 2020). This could lead to persistent infections and subsequent CxCa. To heal precancerous lesions, T-cells and macrophages infiltrate into the area and produce pro-inflammatory cytokines and neutralizing antibodies (nAbs). The essential cells for recognition of HPV infection and activation of cellular immunity are Langerhans cells (LCs), which are the main epithelial antigen-presenting cells (APCs) (Schiller & Lowy, 2009). The effector T-cells, which are activated against HPV-E proteins, can eliminate infected cells. L1-identifying T helper cells are responsible for nAb induction, which would inhibit re-infection and viral transmission. Though, such immune response might be weak and slow; besides, induced nAbs are only active against viral particles and not the infected cells. When the immune system fails to eliminate the infection, methylation on the E2 promoter would result in elevation of E6 and E7 expression, which leads to lesion progression (Zhou et al., 2020). Since the immune system cannot eradicate the tumor, it is proposed that an immunosuppressive microenvironment exists in the infected area (Negahdaripour et al., 2020).

The context of the immune system has a crucial role in determining HPV infection consequences; i.e., whether it develops to cancer or not. For example, HPV infection is more prevalent in immunocompromised patients, including those undergoing solid organ transplantation or human immunodeficiency virus-positive (HIV+) individuals (Bergman et al., 2019; Negahdaripour et al., 2017). The relationship between inflammation and cancer is well explained today. In fact, continuous inflammation would cause tumor progression via dysregulation of toll-like receptor (TLR) and transforming growth factor β (TGF-β) signaling pathways (Negahdaripour et al., 2017). HPV can manipulate the nuclear factor κB (NF-κB) pathway and regulate the cascades of inflammatory cytokines (Olczak & Roden, 2020). Moreover in CxCa, HPV induces the expression of TGF-β and interleukin-10 (IL-10), which inhibits the antitumor immune responses (Wang et al., 2020). Besides, attraction of immune suppressor cells such as regulatory T-cells (Tregs), tumor-associated macrophages, immature and tolerogenic dendritic cells (DCs), and myeloid-derived suppressor cells (MDSCs) are provoked (Negahdaripour et al., 2020).

HPV has several mechanisms to evade the immune system (Schiller, Castellsagué, & Garland, 2012). The non-lytic nature of HPV results in poor recognition of the infection by the immune system. Additionally, HPV-E7 inhibits TLR9 (an important ds-DNA sensor) expression via the NF-κB-p65 signaling pathway (Orbegoso, Murali, & Banerjee, 2018). Absence of pro-inflammatory cytokines prevents DC activation and migration to the infection site. Moreover, HPV does not enter the blood in its life cycle, so very few viruses would expose to the immune system (Rumfield, Roller, Pellom, Schlom, & Jochems, 2020). Since HPV cannot cross the basement membrane and the infection is produced in keratinocytes that stimulate only a local immunity, the infection cannot trigger the immune system properly (Kumar et al., 2015).

The genome instability caused by viral oncogenic proteins helps HPV to escape the immune system as well (Hancock, Hellner, & Dorrell, 2018). Moreover, HPV oncoproteins can regulate LCs differently. E6 expression would decrease E-cadherin, which is responsible for keratinocyte and LC adhesion. This will reduce the LC deposition in the infected area (Malonis, Lai, & Vergnolle, 2019). Both E6 and E7 disturb macrophage inflammatory protein-3 (MIP-3), causing reduction of immature LC migration and immunosurveillance (Yang et al., 2016). Thus, HPV can decrease the number of LCs by immunosuppressive mechanisms. Furthermore, E5 can suppress tumor necrosis factor (TNF) ligand (FasL)- and TNF-related apoptosis-inducing ligand (TRAIL)-mediated apoptosis leading to HPV-infected cell surveillance (Hung, Wu, Monie, & Roden, 2008).

3. Vaccines

Vaccines can produce strong immune responses by activating humoral and cellular immunities. Two types of vaccines are followed for HPV-caused CxCa: prophylactic and therapeutic. Prophylactic HPV vaccines are mostly designed using HPV capsid late proteins, L1 and L2 (Negahdaripour et al., 2018). Their main goal is nAb production stimulation (Schiller & Lowy, 2009); as nAbs bind to HPV and block viral entry to the basal cell layer, thereby inhibiting its proliferation site (Zhou et al., 2020).

Currently, three prophylactic L1-based vaccines are licensed by US FDA (Negahdaripour et al., 2020), which are all developed using HPV virus-like particles (VLPs) through recombinant DNA technology (Negahdaripour, Golkar, et al., 2017). 1) Cervarix® (GlaxoSmithKline), a bivalent (HPV16/18) vaccine; 2) Gardasil® (Merck), a quadrivalent vaccine also containing HPV6/11 VLPs; and 3) Gardasil9® (Merck), a nonavalent vaccine approved in 2014, which also includes HPV31/33/34/52/58 VLPs (Negahdaripour et al., 2020). Gardasil9® can protect against about 90% of oncogenic HPVs, but still routine screening is needed to prevent the remained 10%. Despite acceptable safety and high efficiency caused by repeated epitope presentation (Bergman et al., 2019), some issues limit usage of these vaccines, including type-restricted coverage (because of L1 low-conserved nature), costly production and transfer (due to the need for eucaryotic hosts and cold chain) (Negahdaripour, Eslami, et al., 2017).

These shortcomings have driven the vast research for finding other alternatives with wider coverage and easier production through using the highly conserved L2 protein, which could be produced in *Escherichia coli*. Detailed information regarding HPV prophylactic vaccines is found elsewhere (Huber et al., 2021; Olczak & Roden, 2020; Wang et al., 2020).

On the other hand, current prophylactic vaccines are unable to provide benefit for HPV-infected patients and cannot eliminate pre-existing infections (Schiller et al., 2012). While prophylactic vaccines induce response against capsid antigens, basal epithelial cells infected with HPV usually lack considerable amounts of capsid proteins (Orbegoso et al., 2018). Moreover, because of limited global coverage of HPV vaccines, new cases of HPV infection are found. These justify the need for HPV therapeutic vaccine development.

4. HPV therapeutic vaccines

Therapeutic vaccines aim to induce cell-mediated immunity through stimulation of pro-inflammatory cytokines to eradicate infected cells (Negahdaripour et al., 2020). E6 and E7 proteins are optimum targets for developing therapeutic vaccines due to their integral expression on tumor cells, essential role in malignant cell transformation (Bagheri, Nezafat, Eslami, Ghasemi, & Negahdaripour, 2021), and mere existence in infected cells in the body. Moreover, their antigen loss probability is low (Rumfield, Roller, et al., 2020). Different HPV therapeutic vaccines have been designed and studied using various platforms, though none has been approved yet (Hancock et al., 2018).

Therapeutic vaccine classification, as discussed in this review, is shown in Fig. 2.

4.1 Peptide-/protein-based vaccines

Subunit vaccines include antigenic peptides/proteins derived from virus antigens, which would be processed by DCs and presented on major histocompatibility complex (MHC) class I or II to start the immunity cascade (Malonis et al., 2019; Yang, Jeang, et al., 2016).

Peptide-based vaccines include synthetic long peptides (SLP) and short peptides (specific epitope peptides) or combinations of lipopeptides (Hancock et al., 2018; Panatto et al., 2015). Some advantages of these vaccines are high safety, stability, lack of active pathogen, low allergic response, and large-scale easy production (Hung et al., 2008; Pellom, Rumfield, &

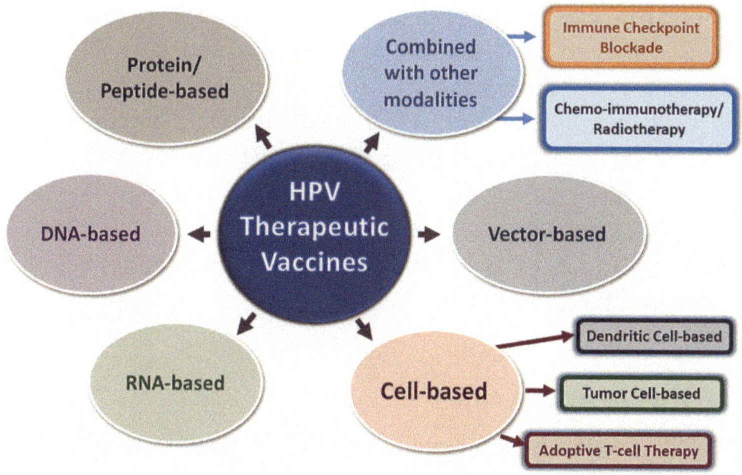

Fig. 2 HPV therapeutic vaccine classes. These therapeutic vaccines could be accompanied by immunotherapy or some other modalities to increase the outcomes.

Morillon, 2021). Moreover, possibility of using several immunodominant epitopes, even from various pathogens, widens their usage (Hansson, Nygren, & Stahl, 2000; Negahdaripour, Golkar, et al., 2017). However, their immunogenicity is weak, because short peptides might not evoke an effective immune response (Yang, Jeang, et al., 2016). Thus, adjuvants and immunostimulating molecules are regarded highly essential to enhance their immunogenicity in both adaptive and innate arms (Dorostkar et al., 2021; Pellom et al., 2021). Additionally, since peptide epitopes are MHC-specific, patients with different HLA genetic backgrounds identify different epitopes. Therefore, for the clinical use of peptide vaccines in society, at least, one epitope is required for each HLA class I allele (Kim, Chung, Lee, & Lim, 2014). To overcome these barriers, overlapping long-peptide vaccines could be used (Yang, Jeang, et al., 2016). For instance, whole HPV proteins (such as E6 or E7) do not need recognition of special HLA-specific epitopes, because they cover all possible HLA epitopes. Hence, protein-based vaccines could circumvent the limitation of peptide-based vaccines caused by MHC restriction (Skeate, Woodham, Einstein, Da Silva, & Kast, 2016; Yang, Jeang, et al., 2016). Furthermore, peptide vaccines could be prone to fast degradation, since they might not form stable tertiary conformations (Clark & Trimble, 2020), so structural consideration should be noted as well (Schiller & Lowy, 2009).

SGN-00101 (HSP-E7) is a protein-based therapeutic vaccine containing HPV16-E7 fused to *Mycobacterium bovis* recombinant heat shock protein 65 (HSP65). Regression of HSIL, increase in HPV-specific cytotoxic T-cells (CTLs), and decrease of HSIL incidence were among the clinical outcomes of this vaccine in a phase II clinical trial on CIN3, anal intra-epithelial neoplasia (AIN), and cervical HSIL (NCT00054041) (Rumfield, Roller, et al., 2020).

The clinical trials of therapeutic peptide-/protein-based vaccines are summarized in Table 1.

4.2 Vector-based vaccines

Live vector vaccines, which employ bacterial or viral vectors, have shown promising results in clinical trials (Franconi, Massa, Paolini, Vici, & Venuti, 2020). *Listeria monocytogenes*, *Lactobacillus lactis*, *Lactobacillus plantarum*, *Lactobacillus casei*, *E. coli*, *Salmonella typhimurium*, *Shigella dysenteriae*, and *M. bovis* are examples of investigated bacterial vectors for therapeutic HPV vaccines (Panatto et al., 2015). They can induce mucosal, humoral, and cellular immunity (Hancock et al., 2018). The humoral immune response involves mucosal IgA and IgG and serum IgG, while $CD8^+$ CTLs and Th17-, Th2-, and Th1- $CD4^+$ T-cells characterize the cell-mediated immunity (Kumar, 2019). Employing live bacteria as vaccines grants the chance of using the immune response of a specific physiological compartment, namely the mucosal system. Vaccines with mucosal delivery stimulate systemic and local immune responses, while formulations with other administration routes mainly stimulate systemic immunization (Silva, Zangirolami, Novo-Mansur, Giordano, & Martins, 2014).

ADXS11-001 (axalimogene filolisbac) is an irrecoverably live-attenuated recombinant *L. monocytogenes* bacteria bioengineered to express and secrete a fusion protein, which contains full-length HPV16-E7 (antigen) and a truncated, non-hemolytic fragment of listeriolysin O (LLO) (adjuvant) (Krupar et al., 2016), which is in phase III clinical trial for locally advanced CxCa.

Both RNA and DNA viruses could be employed as viral vectors (Ewer et al., 2016). For example, adenoviruses (Ad) (Gomez-Gutierrez et al., 2007), adeno-associated viruses (AAV) (Liu et al., 2000), and vaccinia virus (namely strain Wyeth (Malcolm Adams et al., 2001) and Modified-Vaccinia Ankara (MVA) (Graham et al., 2000)) have been investigated as DNA vectors in HPV therapeutic vaccines. While alphaviruses (such as Semliki Forest virus (SFV); Ip et al., 2015) or Venezuelan Equine Encephalitis virus (VEE)

Table 1 Clinical trials of peptide-/protein-based vaccines against CxCa ("U.S National Library of Medicine, ClinicalTrials.gov, https://clinicaltrials.gov/).

	Vaccine/Adj.	Phase	Disease	Admin. route	NCT#	Status/completion date (last update)
1	Four HPV16-E6 peptides/Candin® (yeast extract)	I	HSIL	Inj.	NCT01653249	Completed (2016)
2	Hespecta: HPV16-E6/Amplivant® (synthetic TLR2 ligand) (conjugated)	I	HPV16+ tumors or premalignant lesions	ID	NCT02821494	Completed (2021)
3	P16_37–63/Montanide™ ISA-51 VG (combined)	I/IIa	HPV-induced cancer	SC	NCT01462838	Completed (2015)
4	PepCan: HPV16-E6/Candin®	II	CIN	ID	NCT02481414	Active, not recruiting (2021)
5	TA-CIN	I	HPV16-associated CxCa	IM	NCT02405221	Recruiting Completion/Apr 2022
6	TVGV-1/GPI-0100	IIa	HSIL	–	NCT02576561	Unknown (2017)
7	ProCervix: CyaA-HPV16-E7 + CyaA-HPV18-E7/Aldara™ (imiquimod)	II	HPV genital infection	Inj/top adjuvant	NCT01957878	Completed (2016)
8	HspE7: E7/HSP65	II	CIN3	SC	NCT00054041	Completed (2013)
9	HspE7: E7/HSP65	II	CxCa with atypical squamous cells	SC	NCT00091130	Completed (2013)

R-DOTAP: R-enantiomer of 1,2-dioleoyl-3-trimethylammonium-propane chloride.

(Cassetti et al., 2004; Schmidt et al., 2019), and arenaviruses (Schmidt et al., 2019) are used as RNA vectors. To use these vectors, their viral structural genes are excluded and replaced with the genes encoding desired antigens. Therefore, they cannot revert back to the initial viruses and are considered as non-pathogenic agents (Lundstrom, 2018; Yang, Farmer, Wu, & Hung, 2016).

TA-HPV vaccine (NCT00002916), a live recombinant vaccinia virus encoding HPV16/18-E6/E7 proteins, was examined in several clinical trials. Despite feasibility, tolerability, safety, immunogenicity, and minor toxicity, limited clinical advantages were observed (Kaufmann et al., 2002; Peng, Ferrall, et al., 2021). Table 2 represents clinical trials of vector-based HPV vaccines.

Despite the high immunogenicity of live vector-based vaccines, some issues hinder their wide usage (Yang, Jeang, et al., 2016). Pre-existing immunity and induction of nAbs against the applied vectors could inhibit vector uptake into target cells and restrict the effectiveness of using multiple vaccines with same vector (Pellom et al., 2021; Yang, Jeang, et al., 2016). However, each succeeding vaccine booster dose can further enhance responses against the target gene (Hancock et al., 2018). As a solution to this problem, cyclooxygenase 2 inhibitors, namely celecoxib, could inhibit nAbs produced against vaccinia and prevent the issue of efficacy loss upon duplicate vaccine administrations (Chang, Ma, Pang, Wu, & Hung, 2009). Another disadvantage of live-vector vaccines is their potential safety risks in the immunosenescent or immunocompromised people (Schiller & Müller, 2015; Yang, Jeang, et al., 2016).

4.3 Nano-delivery systems

Nanoparticles such as nanotubes can also deliver diverse vaccine antigens (Foldvari, 2011), including DNA/RNA, synthetic peptides, recombinant proteins/peptides from oncogenic protein, and killed, inactivated, or the DNA of attenuated pathogens (Uddin, Kouzi, & Hussain, 2015).

Nano-delivery systems act as an immunostimulatory adjuvant, increase biological stability of antigens, and improve effectiveness in targeting APCs for stimulation of natural and specific immunity due to involvement in MHC-I and -II presentation (Parkin & Cohen, 2001; Uddin et al., 2015).

For instance, MPG-based nanoparticles (180–210 nm) were used as adjuvant for an HPV16-E7 DNA vaccine. MPG, a peptide-based gene delivery system, was applied with a model antigen in vivo and in vitro. Different

Table 2 Clinical trials of vector-based and nano-delivery therapeutic HPV vaccines (www.clinicaltrials.gov/).

	Vaccine	Vector/delivery platform	Phase	Disease	Administration route	NCT#	Status/completion date (last update)
1	Vvax001 (RNA-based)	Viral	I	CIN2/3	IM	NCT03141463	Completed (2018)
2	Ad26HPV16/18 & MVA-HPV16/18	Viral	I/II	HPV infections	IM	NCT03610581	Recruiting/Dec 2022
3	HB-201	Viral	I/II	HPV-related squamous cell carcinoma	IM/IV	NCT04180215	Recruiting/Jun 2022
4	RO5217790	Viral	II	CIN	SC	NCT01022346	Completed (2016)
5	TA-HPV	Viral	II	CxCa	–	NCT00002916	Completed (2012)
6	ADXS11-001	Bacterial	II	Persistent or recurrent CxCa	IV	NCT01266460	Active, not recruiting/Oct 2018
7	ADXS11–001	Bacterial	III	Locally advanced CxCa	IV	NCT02853604	Active, not recruiting/Oct 2024
8	DPX-E7: HPV16-$E7_{11-19}$ nanomer	Liposomal	Ib/II	HN, cervix, and anal cancers	SC	NCT02865135	Active, not recruiting/Dec 2023
9	PDS0101: R-DOTAP with HPV16-E6/E7 peptides	Liposomal	I	CIN1	SC	NCT02065973	Completed (2018)

MPG/E7DNA complexes, defined by nitrogen/phosphate (N/P) ratios, were prepared. In a murine tumor model, the nanoparticles at an N/P ratio of 10:1 provoked stronger interferon-γ (IFN-γ)-predominant response of Th1-cells than those induced by E7DNA alone (Saleh, Bolhassani, Shojaosadati, & Aghasadeghi, 2015).

Liposomes have been also studied widely to deliver proteins, peptides, and DNA/RNA (Chen & Huang, 2008) (Table 2). A Cuban trial examined a minuscule size proteoliposome system to deliver the HPV16-$E7_{86-93}$ peptide and indicated favorable clinical responses in CIN2/3 patients (Solares et al., 2011). DPX-E7, a short peptide-based therapeutic vaccine containing HPV16-$E7_{11-19}$ peptide and DepoVax® adjuvant was packed into liposomes. The vaccine was freeze-dried, and just before vaccination, it was resuspended in Montanide® ISA51 oil (Fakhr, Modic, & Cid-Arregui, 2021; Karkada, Quinton, Blackman, & Mansour, 2013). This vaccine is in phase Ib/II trial to test its efficacy and safety in treating incurable HPV 16-related cervical, anal, and oropharyngeal cancer in HLA-A*02 positive patients (NCT02865135) (Rumfield, Pellom, Morillon, Schlom, & Jochems, 2020; Rumfield, Roller, et al., 2020). RNA-lipoplex is the first RNA-based liposomal HPV16 therapeutic vaccine that could selectively direct antigens to DCs and induce $CD8^+$ T-cells responses strongly, without any extrinsic adjuvant. Upon administration in mice, RNA-lipoplex vaccine generated robust T-cell memory against E7, caused sustained recovery, and prevented tumor recurrence (Grunwitz et al., 2019).

Archeosomes, which are liposomes containing at least one total polar lipid extract (TPL) from archaea, could create a stronger immune response than traditional liposomes. Archeosomes are more stable in a wide range of temperatures and pH (Kaur, Garg, Rath, & Goyal, 2016). Using archeosomes in combination with L1/E1/E7 vaccines in mice enhanced tumor apoptosis 25–40% versus the plasmid DNA vaccine alone (Karimi, Soleimanjahi, Abdoli, & Banijamali, 2020).

4.4 DNA-based vaccines

DNA-based vaccines are a novel type of vaccines using plasmid structures that can encode target virus antigens after transfer into mammalian cells. E6 and E7 oncoproteins are important immune targets used for HPV therapeutic DNA-based vaccines (Table 3). The structure and content of the plasmid affect the activity of DNA vaccines (Ghaffarifar, 2018).

Table 3 Clinical trials of therapeutic HPV DNA/RNA-based vaccines.

	Vaccine	Class	Phase	Disease	Admin. route	NCT#	Status/Completion date (Last update)
1	ZFN-603 and ZFN-758	DNA	I	HPV-related malignant neoplasm	IT	NCT02800369	Unknown/Jul 2017
2	pNGVL4aCRTE6E7L2	DNA	I	HPV16+ CIN2/3	IM/EP	NCT04131413	Recruiting/Dec 2024
4	pnGVL4a-CRT/E7 (Detox)	DNA	I	HPV16+ CIN2/3	IM/EP/IL	NCT00988559	Completed (2018)
5	INO-3112	DNA	I/IIa	CxCa	IM/EP	NCT02172911	Completed (2021)
6	pNGVL4a-Sig/E7 (detox)/HSP70	DNA	I/II	HPV16+ CIN2/3	IM	NCT00121173	Completed (2018)
7	VB10.16	DNA	I/II	High grade HPV16; HSIL, CIN2/3	IM	NCT02529930	Completed (2019)
8	VGX 3100	DNA	II	HPV16/18+ CIN2/3	IM/EP	NCT01304524	Completed (2018)
9	VGX-3100	DNA	III	Cervical HSIL, CIN2/3	IM/EP	NCT03721978	Recruiting/Aug 2022
10	GX-188E	DNA	II	HPV16/18+ CIN2/3	IM/EP	NCT02596243	Unknown/Aug 2018
11	HARE 40 (anti-CD40)	RNA	I/II	Cervix, HN, penile neoplasms	ID	NCT03418480	Active, not recruiting/Dec 2023

EP: Electroporation; ID: Intradermal; IT: Intrathecal.

By introducing DNA-encoding antigens on APC surface, DNA-based vaccines can stimulate and activate both types of naive T-cells, i.e., $CD4^+$ and $CD8^+$ T-cells. Professional APCs, specifically DCs, have an important role in the activity of DNA-based vaccines and are responsible for antigen representation to naive T-cells via MHC molecules. Therefore, DNA-based vaccines are able to activate both antibody- and cell-mediated immune responses (Monie, Tsen, Hung, & Wu, 2009).

DNA-based vaccines have some advantages compared to other types, as they are temperature-stable, easily prepared with high purity, and have a long shelf-life. Moreover, unlike live-attenuated vaccines, they may not return to the virulent forms, and their administration could be repeated without losing efficacy. Their major disadvantage is their low immunogenicity (Lin, Roosinovich, Ma, Hung, & Wu, 2010).

To develop stronger therapeutic HPV DNA-based vaccines, some strategies could be employed including (1) increasing DCs' ability to absorb HPV antigens; DNA vaccine transportation techniques greatly affect vaccine efficacy. Administration via gene gun, electroporation, intradermal laser therapy or encapsulation by nanoparticle delivery systems can promote DNA vaccines immunogenicity, (2) enhancing the processing and representing potential of HPV antigens by promoting DC capability in antigen expression, especially through cross-presentation mechanism, and (3) improving interactions between DCs and T-cells via immune cytokines and also DC cross-presentation enhancement by TLR agonists (Cheng et al., 2018).

Recently, a novel non-viral delivery system, called PEI600-Tat, was found to boost DNA plasmid uptake. Tat is a prominent HIV type 1 domain, which can translocate DNA plasmid into the cells effectively as a cell-penetrating peptide (CPP). Studies suggested that this polymer-peptide hybrid transported DNA vaccines successfully and improved induction of both arms of the adaptive immunity (Bolhassani, Ghasemi, Servis, Taghikhani, & Rafati, 2009).

DNA vaccination with gene gun enhanced vaccine efficacy through E7-specific $CD8^+$ T-cell immune signaling and antitumor response improvement. Thus, intradermal DNA vaccine administration via gene guns could deliver the desired genes into APCs (Lin et al., 2009).

To enhance the vaccine potency, a combination of HPV oncoproteins and herpes virus glycoprotein D was studied in murine models. This strategy augmented induction of cytotoxic $CD8^+$ T-cell response due to enhancing co-activation mechanisms and blocking co-inhibitory signals (Diniz, Cariri, Aps, & Ferreira, 2013).

A combination of HPV-E6/E7 and vascular endothelial growth factor receptor 2 (VEGFR2) genes as a DNA plasmid generated significant immune stimulation and tumor proliferation inhibition and shrinkage. Co-expression of VEGFR2 and HPV-E6/E7 oncoproteins caused anti-angiogenesis and antitumor synergistic effects (Gao, Fan, Ma, & Xiao, 2016).

The first therapeutic vaccine that showed efficiency in HPV16/18-positive CIN2/3 was a DNA-based vaccine called VGX-3100. Its efficacy and safety were assessed in a phase III clinical trial by electroporation delivery (NCT03185013), which led to histopathological regression of CIN2/3 to CIN1, non-disease condition, and even viral clearance. Thus, VGX-3100 could probably get approval for CxCa soon (Trimble et al., 2015).

4.5 RNA-based vaccines

Recently, RNA-based vaccines have attracted much attention especially following the introduction of two mRNA-based vaccines for COVID-19 pandemic (Negahdaripour, 2021). DCs can uptake mRNA vaccines. Upon cell entry, mRNA vaccines encode the viral antigen proteins and generate cell-mediated and humoral immune responses through antigen presenting via MHC-I and -II, respectively (Pardi, Hogan, Porter, & Weissman, 2018). Furthermore, they are involved in activating the innate immune system by binding to TLR7 and TLR8.

HPV RNA-based therapeutic vaccines formed on self-replicon systems include negative or positive single-stranded RNA viruses and non-pathogenic viral vectors. Self-replication causes a large number of antigens to be encoded thereby generating highly efficient immune responses (Wang et al., 2020; Yang, Farmer, et al., 2016).

The advantages of RNA vaccines include easy production, high safety, efficient delivery, and high-level recombinant antigen expression. Furthermore, the possibility of incorporation into chromosome and cell transformation is not a concern for RNA replicons. However, they suffer from poor stability (Wang et al., 2020). Lipid nanoparticle (LNP) delivery systems could overcome the stability issue of mRNA vaccines appropriately (Zhang et al., 2021), as used in Comirnaty (Pfizer-Biotech) and Moderna COVID-19 vaccines (Negahdaripour et al., 2021). It also increases their immunogenicity thereby potentiating immune responses (Geall et al., 2012). Other alternatives for RNA-based vaccine delivery and increasing stability include liposomes (Grunwitz et al., 2019) and viral vectors (other RNA viruses namely alphaviruses and flaviviruses) (Schmidt et al., 2019) (see Tables 2 and 3).

4.6 Cell-based vaccines

4.6.1 Dendritic cell-based vaccines

The exclusive characteristics of DCs make them a perfect option for cell-based vaccines. DCs can stimulate naive T-cells against specific antigens several times more potent than other APCs and establish efficient immunological memory. DCs, as the primary APC, can cross-present. Based on this ability, exogenous antigens can be presented by MHC-I, which is essential for the differentiation of $CD8^+$ T-cells to CTLs (Cintolo, Datta, Mathew, & Czerniecki, 2012).

DC vaccines are developed using *ex-vivo* culturing. After separation and purification of harvested DC types, these progenitor cells are cultured with differentiation elements and growth factors such as granulocyte-macrophage colony-stimulating factor (GM-CSF) and IL-4 to become ready for antigen loading (Santos & Butterfield, 2018).

Molecules such as peptides/proteins, RNA, or even tumor cell lysates (autologous or allogeneic), could be loaded as antigens on the MHCs, directly or after processing. For appropriate functioning of DC vaccines, the antigen-loaded DCs must be matured with factors including TLR ligands such as LPS or poly:IC and cytokine cocktails (TNFα, IL-1β, IL-6, and PGE_2) (Sabado, Balan, & Bhardwaj, 2017).

DC vaccines can trigger the innate and adaptive immune systems with high immunogenicity (Wang et al., 2020). Vaccination with HPV-antigen-loaded DCs could extirpate tumors in 80% of mice (Adams et al., 2003). Restrictions of DC vaccines include lack of general agreement about the best DC preparation method, their short half-life due to T-cell-mediated apoptosis and lack of their lifespans (Vici et al., 2016; Wang et al., 2020).

Pleurotus ferulae polysaccharides (PFPS) is considered an optimal candidate adjuvant in DC vaccines, as it induced strong HPV-specific $CD8^+$ T-cell signals, enhanced the antitumor responses against TC-1 tumors, and inhibited tumor growth. Moreover, PFPS reduced immunosuppressive effects of myeloid-derived suppressor cells (MDSCs) (Li, Aipire, Zhao, Yuan, & Li, 2019). Another study suggested HPV16-E7 application along with the endosomal/lysosomal elements could promote DC vaccine effectiveness and significant *in-vitro* stimulation and *in-vivo* development of E7-specific T-cells (Kang et al., 2006). Co-culturing of DC and cytokine-induced killer cells also caused rapid proliferation, high cytotoxicity, and killing a wide range of tumors (Vici et al., 2016).

4.6.2 Tumor-cell-based vaccines

Whole tumor-cell vaccines contain the patient's tumor cells genetically modified *ex-vivo* through encoding cytokine genes such as IL-2, IL-12, or GM-CSF to promote the immune signals (Remic, Sersa, Ursic, Cemazar, & Kamensek, 2020). These genetic modifications are aimed to overcome the low immunogenic nature of these vaccines. Since these genes are producing immunomodulatory cytokines, they can stimulate naive T-cells to differentiate to effector and helper T-cells. Specifically, GM-CSF, as the most used gene in tumor cell-based vaccines, is a growth factor for DCs and can increase tumor immunogenicity and activate both humoral and cellular immunities (Chang et al., 2000).

The most important advantage of tumor-cell vaccines is ability to present all potential antigens to the immune system. Besides, genetic modifications can enhance the immune response against the employed antigens. Though, production of these vaccines is challenging and time-consuming. Moreover, tumor cell-based vaccines may potentially implant new cancers in patients (Yang, Jeang, et al., 2016). Because of these risks, the safety and efficacy of each vaccine should be evaluated individually. Thus, in the case of HPV that the exact tumor antigens are known, avoiding such risks by using safer vaccine platforms is usually preferred. Currently, such vaccines are not used for CIN primary stages due to safety issues (Vici et al., 2016). However, there are some studies about these vaccines that evaluate their efficacy in various conditions. For instance, pre-conditioning with tetanus toxoid following vaccination with a GM-CSF-overexpressing tumor cell-based vaccine enhanced immunogenicity and overall survival in CxCa mice models (Alson, Schuyler, Yan, Samimuthu, & Qiu, 2020). In another study, multiple vaccinations with a codon-modified GM-CSF tumor cell-based vaccine were investigated in mice. The three-dose regimen elicited the strongest immunity when compared to one- or five-dose protocols (Qiu et al., 2019). Table 4 represents other examples.

4.6.3 Adoptive T-cell therapy

Adoptive T-cell therapy (ACT) is an immunotherapy approach using injection of *ex-vivo* modified T-cells to fight against the tumors. ACT is generally based on two strategies: (1) separation of tumor-specific T-cells that naturally exist in the tumor mass, called tumor infiltering lymphocytes (TILs), and (2) genetic manipulation of T-cells isolated from the peripheral blood to express receptors that specifically cause tumor cell destruction, such as

Table 4 Preclinical studies for cell-based therapeutic HPV vaccines.

	Vaccine platform	Adjuvant	Employed antigens	Main findings	Refs.
1	DC vaccine	Not used	HPV16-E6/E7	-Efficient cellular responses -Potentiation of the immune system against TC-1 tumor cells by DC cells primed by codon-optimized HPV16 oncoproteins	Zhou et al. (2015)
2	DC vaccine	*Mycobacterium tuberculosis* HspX	HPV16-E6/E7	-Significant adjuvant effects of HspX in DC vaccine-based treatment -Enhanced tumor growth inhibition by strong induced cell-mediated immunity	Jung et al. (2014)
3	Tumor cell-based vaccine	IL-12	HPV16-E7/E6	-Abolishment of the pro-tumorigenic effects of senescent tumor cells through the IL-12 provided by tumor cell vaccine	Simova et al. (2016)
4	ACT	CD40/ Poly I:C	HPV16-E7	-CD40 activation could potentiate HPV16-E7-specific Tscms	Zhang et al. (2020)

Tscm: Stem-like memory CD8+ T-cells; CD40: Cluster of differentiation; Poly I:C: Polyinosinic: polycytidylic acid.

T-cell receptor (TCR) or chimeric antigen receptor (CAR), also called CAR T-cell therapy (Simova et al., 2016).

In the TIL-based ACT, harvested tumor-specific T-cells are expanded in the *in-vitro* culture by factors such as IL-2, then they are reinfused to the patient's body with more potency. Moreover, genetic engineering can be done on TILs (Perica, Varela, Oelke, & Schneck, 2015). Recent studies indicated treatment of malignant CxCa with the infusion of HPV-targeted TILs expanded by IL-2, generated objective tumor responses in three out of nine patients. Two of these three responses were complete and persistent for 15–22 months (Stern, 2021).

Based on a study on 11 patients with initial stage CxCa, the tumor-draining lymph nodes (TDLN) could provide a rich pool of polyclonal HPV16-specific T-cells, thus could be a desirable primary source for the ACT technique (van Poelgeest et al., 2016). Clinical examples are available in Table 5.

Even though ACT is considered an expensive method and not simple to perform, it can present an exquisite perspective for the treatment of malignant CxCa (Vici et al., 2016).

4.7 Therapeutic HPV vaccines in combination with therapeutic modalities

Despite the potential of therapeutic HPV vaccines against CxCa by delivering tumor antigens to APCs, multiple factors in the tumor microenvironment (TME) and probably on tumor cells cause immunosuppression and tumor evasion. These immunoediting routes can be categorized into four main mechanisms: (1) resistance to immune-stimulated apoptosis, (2) downregulation of MHC-I to suppress antigen presentation, (3) expression of immunosuppressant cytokines, and (4) attraction of immune-inhibiting cells (Piersma, 2011). More details about TME in CxCa and HPV immune escape mechanisms were published previously (Guo & Hua, 2020; Lechien et al., 2020; Liu, Li, Li, & Zhao, 2020). To defeat these impairing procedures, using other treatment modalities combined with HPV therapeutic vaccines could support boosting the antitumor responses and reducing immunosuppressive factors (Rumfield, Pellom, et al., 2020; Rumfield, Roller, et al., 2020).

4.7.1 Therapeutic HPV vaccines accompanied by immune checkpoint blockade

Immune checkpoint mechanisms avert inordinate and uncontrolled immune responses (Hancock et al., 2018). Programmed cell death-ligand 1 (PD-L1) is expressed by many tumor cells and significantly contributes to eradication of tumor reactive T-cells in the TME (Dong et al., 2002). PD-L1 binding to programmed cell death-1 (PD-1) leads to T-cell receptor signaling downregulation, then apoptosis and anergy of T-cells, and finally immunosuppression (Francisco, Sage, & Sharpe, 2010). Cytotoxic T-lymphocyte antigen 4 (CTLA-4) binding to CD80/86 ligands results in inhibiting T-cell multiplication and cytokine production as well as promoting T-cell anergy (Hancock et al., 2018). Ligation of CTLA-4 with B7 (CD80 (B7-1) and CD86 (B7-2); Knechtle, Marson, & Morris, 2019) leads

Table 5 Clinical trials of HPV therapeutic cell-based vaccines (www.clinicaltrials.gov).

	Vaccine/Adj.	Phase	Disease	Admin. route	NCT#	Status/completion date (last update)
1	HPV16-E7 peptide-pulsed autologous DCs	I	Recurrent CxCa	LN	NCT00155766	Unknown (2005)
2	DC vaccines targeting HPV-E6/E7 proteins	I	CIN1/2	Lymph node injection	NCT03870113	Not yet recruiting/Dec 2022
3	BVAC-C: Autologous B-cells and monocytes transfected with HPV-E6/E7 genes	I/IIa	HPV-16/18+ CxCa	IV	NCT02866006	Active, not recruiting/Mar 2022
4	DC-HPV16-E7 peptide vaccine	I	Recurrent or persistent CxCa	IV/SC	NCT00003977	Completed (2009)
5	HPV16-E6 or E7 pulsed PBMC	I	Advanced or recurrent HPV-related cancers.	IV	NCT00019110	Completed (2015)
6	Autologous T-cells genetically engineered to express an HPV16-E6-specific T-cell receptor	I	HSIL	IV	NCT03197025	Terminated (2021)
7	E7 TCR cell	I/II	HPV-related cancers	IV	NCT02858310	Recruiting/Jan 2026
8	T-cells genetically engineered with a TCR targeting HPV16-E6	I/II	HPV-related cancers	IV	NCT02280811	Completed (2017)
9	CxCa-specific CAR T-cells	I/II	HPV+ CxCa	IV	NCT03356795	Unknown (2019)
10	Autologous CTL generated and expanded by co-culturing with partially resected tumor combined with low-dose IL-2	II	HPV-related cancers	IV	NCT01585428	Completed (2018)

PBMC: Peripheral blood mononuclear cell.

to another important inhibitory pathway for T-cells. This ligation also prevents CD28 interaction with B7, and consequently disrupting co-stimulation (Postow, Callahan, & Wolchok, 2015). 4-1BB (TNFRSF9) is a counterpart to the co-inhibitory nature of CTLA-4, which is engaged in survival signaling in T-cells. Moreover, activation of tumor necrosis factor receptor superfamily member 9 (4-1BB) pathway increased the inflammatory cytokines IL-6 and IL-12 (Wang, Lin, McPherson, & Watts, 2009).

Introduction of immune checkpoint blockers (ICBs) has revamped treatment options in oncology. Some inhibitors of PD-1/PD-L1 and CTLA-4 are approved by FDA in cancers (Liu et al., 2019). Pembrolizumab was licensed in 2018 for metastatic/recurrent CxCa treatment (Ferrall et al., 2021).

ICBs block the pairing of checkpoint proteins to their partner proteins (Peng, Tan, et al., 2021); therefore, by preventing transmission of "off" signals, ICBs allow the T-cells to remove cancer cells. Other methods of cancer therapy, such as chemotherapy, radiotherapy, or oncogene-targeted therapies, can also modify the immune-suppressive TME and might potentiate ICB therapy as well (Postow et al., 2015).

According to studies, expression of checkpoints PDL1 and CTLA is associated with HPV-E6/E7 (Rumfield, Schlom, & Jochems, 2021). Multiple studies have investigated ICBs as immunomodulators to increase the benefits of therapeutic HPV vaccines (Peng, Ferrall, et al., 2021; Peng, Tan, et al., 2021; Rumfield, Pellom, et al., 2020; Rumfield, Roller, et al., 2020) (Tables 6 and 7). For example, an HPV16 peptide-based vaccine (ISA 101) was administered subcutaneously plus a PD-1 checkpoint inhibitor (nivolumab) in patients with recurrent HPV16-related cancers (NCT02426892). Better therapeutic outcomes than nivolumab alone with no synergistic adverse events were observed (Massarelli et al., 2019).

4.7.2 Therapeutic HPV vaccines along with chemo-immunotherapy and radiotherapy

Current strategies for the treatment of bulky CxCa are based upon conventional *in-situ* ablative modalities inclusive of radiotherapy (RT) and chemotherapeutic regimens (namely cisplatin) (Tseng et al., 2009). Some chemotherapeutics can improve the antitumor effect of therapeutic vaccines (Hancock et al., 2018). Though definite mechanisms of their synergistic effects are not entirely determined, it is assumed that chemotherapy increases T-cell infiltration and antigen density and sensitizes tumors to cytotoxic $CD8^+$ T-cells. All these events together change the TME in favor of vaccine functioning (Hancock et al., 2018).

Table 6 Preclinical studies of therapeutic vaccines combined with therapeutic modalities against HPV.

	Vaccine class/adjuvant or vector	Combined therapy	Employed antigens	Disease	Admin. route	Main finding	Refs./year
1	Peptide-based (Elim Biopharma)/NKT cell ligand αGalCer	CTLA-4 antagonist and 4-1BB agonist antibodies	HPV16-E6/E7	HPV+ SCCOP	IN	– Elevated CD8+ T-cell response and antitumor activity – Minimal toxicity	Dorta-Estremera et al. (2018)
2	Viral vector-based/Ad5-vector	PD-L1 blocker	HPV16-E6/E7	HPV-associated cancers	SC	– No change in CD8+ TILs – Decrease in PD-L1 expression on tumor cells – Decreased PD-1 + TILs	Rice et al. (2015)
3	RNA-liposomal-based (RNA-lipoplex)/pST1-A120 and pST1-MITD vector	PD-L1 blocker	HPV16-E7	HPV+ cancers	IV	– Notable survival advantage – Synergistic inhibition of tumor growth	Grunwitz et al. (2019)/2019

4	Protein/peptide-based/ CpG1826/PolyI:C & DEC205-specific single-chain variable fragment (scFv)	PD-L1 blocker	E7 long peptide	CxCa	SC	– Strong CTL antitumor responses by DC targeting – Synergistic improvement of antitumor effects	Liu, Zhou, Wang, Fu, and Zhu (2016)
5	Protein/peptide-based (Elim Biopharma)/ Alpha-galactosylceramide (αGalCer)	4-1BB agonist	HPV16-E6/$E7_{15–19}$	CxCa	IN	– Increased T-cell responses – Tumor regression	Bartkowiak et al. (2015)
6	Peptide-based (Celtek Bioscience)/QuilA®	Immune checkpoint molecules	HPV16-$E7_{49–57}$ + synthetic Th epitopes	CxCa	SC	Delayed therapy of anti–PD-1 did not enhance antitumor impacts of the vaccine and anti-OX40 regimen.	Shrimali et al. (2017)
7	Protein-based (Global BioPharma, Taiwan)	Anti-PD-L1 mAb	E6/E7	HPV-infected cancers	IP	– Enhanced antitumor activity – Increased survival	Lin et al. (2017)

Continued

Table 6 Preclinical studies of therapeutic vaccines combined with therapeutic modalities against HPV.—cont'd

	Vaccine class/adjuvant or vector	Combined therapy	Employed antigens	Disease	Admin. route	Main finding	Refs./year
8	DNA-based/HSV-1 glycoprotein D	Gemcitabine	HPV16-E7	CxCa	IM	Enhanced efficiency in established HPV-related tumor eradication	Ramos da Silva et al. (2021)
9	Peptide-based/CpG	Cisplatin	HPV16-$E7_{43-77}$ (synthetic long peptide)	CxCa	–	Unleashing synergistic mechanism of cisplatin with therapeutic vaccination for effective systemic antitumor T-cell responses.	van der Sluis et al. (2015)
10	Peptide-based: STxB-$E7_{43-57}$/Shiga toxin B-subunit variant & α-galactosylceramide	Radiation	$E7_{43-57}$	HPV-associated HNC	IP	– Complete tumor clearance in most cases – Induction of $CD8^+$ T-cells	Mondini et al. (2015)
11	DNA-based	(1) anti-PD-1, (2) NOX4 inhibitor (Setanaxib)]	HPV16-E7	HNSCC	DNA vaccine: IM/ Setanaxib: oral/Anti-PD-1: IP	Supporting immunotherapy by NOX4 inhibition through suppressing cancer-associated fibroblast-mediated exclusion of $CD8^+$ T-cells from tumors	Ford et al. (2020)

12	DNA-based/CCL21	Anti-PD-L1 antibody	HPV16-E7	CxCa	DNA vaccine: SC/ Anti-PD-L1 mAb: IP	Synergistic promotion of E7-specific cellular responses	Moeini et al. (2017)
13	Liposomal-based (PDS0101)	Immune modulators (Bintrafusp alfa + NHS-IL12)	HPV16-E6/E7 in ImmunoMAPK-RDOTAP	CxCa, HNSCC	PDS0101 and NHS-IL12: SC/ Bintrafusp alfa: IP	Potentiation of cell-mediated responses in the TME	Rumfield, Pellom, et al. (2020)
14	DNA-based (CRT/E7)/ calreticulin	Histone deacetylase inhibitor (HDACi): (1) AR-42 (vorinostat) (2) romidepsin	HPV16-E7	CxCa	DNA vaccine: gene gun/ AR-42 and Vorinostat: oral/ Romidepsin IP	– Enhanced potency of vaccine by AR-42 – Higher anti-tumor responses of AR-42 than other clinically available HDACi	Lee et al. (2013)/ 2013

HSV: Herpes simplex virus; CCL21: Chemokine ligand 21; SCCOP: Squamous cell carcinoma of oropharynx.

Table 7 Clinical trials of therapeutic HPV vaccines along with therapeutic modalities (www.clinicaltrials.gov).

	Vaccine class/name	Accompanied modality	Class of the combined modality	Phase	Disease	NCT#	Status/ completion date (last update)
1	Peptide-based/ ISA101	Nivolumab (anti-PD-1)	ICB	II	HPV16+ incurable solid tumors	NCT02426892	Active, not recruiting/Dec 2020
2	Peptide-based/ ISA101b	Cemiplimab (anti-PD-1)	ICB	II	HPV16+ OPC	NCT03669718	Recruiting/Nov 2022
3	ISA101b	Pembrolizumab (anti-PD-1)/ Radiation/ Cisplatin	ICB	II	Intermediate-risk HPV16-associated	NCT04369937	Recruiting/Jun 2022
4	ISA101b	Utomilumab (4-1BB/CD137 agonist)	ICB	II	HPV16+ incurable OPC	NCT03258008	Active not recruiting/Jun 2022
5	ISA101/ISA101b	Pegylated IFNα/ Carboplatin/ Paclitaxel/ Bevacizumab	Chemo-immunotherapy	I/II	HPV16+ advanced or recurrent CxCa	NCT02128126 (CervISA)	Completed (2019)
6	DNA-based/ MEDI0457 (INO-3112) (Adj: IL-12)	Durvalumab (anti-PD-L1)	ICB	II	R/M HPV-associated cancers	NCT03439085	Recruiting/Dec 2022

7	Peptide-based/ UCPVax	Atezolizumab (anti-PD-L1)	ICB	II	HPV+ cancers	NCT03946358 (VolATIL)	Recruiting/Sep 2022
8	Two vector-based vaccines: (1) MG1-E6E7, (2) Ad-E6E7	Atezolizumab (anti-PD-L1)	ICB	I	HPV16/18-associated cancers	NCT03618953	Active, not recruiting/Jun 2022
9	Cell-based/ SQZ-PBMC-HPV	Atezolizumab (anti-PD-L1)	ICB	I	HPV16+ recurrent, locally advanced or metastatic solid tumors	NCT04084951	Recruiting/Nov 2022
10	Viral-vector-based/TG4001	Avelumab (anti-PD-L1)	ICB	Ib/II	HPV16+ R/M cancers	NCT03260023	Recruiting/Dec 2024
11	MEDI0457 (INO-3112)	Durvalumab (anti-PD-L1)	ICB	Ib/IIa	HPV-associated R/M HNC	NCT03162224	Completed (2021)
12	GX188E	Pembrolizumab (anti-PD-1)	ICB	I/II	HPV16/18+ CxCa	NCT03444376	Recruiting/Dec 2023
13	Peptide-based in liposomal carrier/ PDS0101	Pembrolizumab (anti-PD-1)	ICB	II	HPV16+ R/M HNSCC	NCT04260126	Recruiting/Jul 2024
14	PDS0101	M7824 (Bintrafusp alfa) and NHS-IL12	Immunotherapy	I/II	Advanced HPV-associated malignancies	NCT04287868	Recruiting/Jan 2023

Continued

Table 7 Clinical trials of therapeutic HPV vaccines along with therapeutic modalities (www.clinicaltrials.gov).—cont'd

	Vaccine class/name	Accompanied modality	Class of the combined modality	Phase	Disease	NCT#	Status/ completion date (last update)
15	PDS0101	Cisplatin + Radiotherapy	Chemoradiation	IIA	HPV16+ IB3-IVA CxCa	NCT04580771/ (IMMUNOCERV)	Recruiting/Mar 2024
16	Bacterial-vector-based/ ADXS11–001	Durvalumab (anti-PD-L1)	ICB	I/II	CxCa/HPV+ HNC	NCT02291055	Active, not recruiting/May 2021
17	Peptide-based/ P16_37–63 (Adj: Montanide® ISA-51 VG)	Cisplatin	Chemo-immunotherapy	I	HPV+ cancers	NCT02526316 (VICORYX-2)	Completed (2017)
18	HPV-E7 & KLH (Keyhole Limpet Hemocyanin)/ Hiltonol (Poly-ICLC)	Fimaporfin	Photosensitizer	I	Healthy volunteers (safety, tolerability and immune response assessment)	NCT02947854	Completed (2019)
19	Cell-based (HPV16/18-E6/ E7-Specific T-Lymphocytes)	Nivolumab	Chemo-immunotherapy	I	HPV16/18-associated cancers	NCT02379520 (HESTIA)	Recruiting/Oct 2033

HNSCC: Head and Neck Squamous Cell Carcinoma; OPC: Oropharyngeal cancer; R/M: Recurrent and/or Metastatic.

In a multicenter phase I/II study, two therapeutic HPV16-E6/E7 long peptide vaccines (ISA101/ISA101b) were indicated with the standard-of-care regimens, such as carboplatin, paclitaxel, or bevacizumab (NCT02128126). After three doses of vaccine (2 weeks after cycle 2/3/4 of chemotherapy) IFN-γ + T-cell responses were raised showing relation with overall survival (Melief et al., 2020). Although the benefit of HPV immunization in treating CxCa patients receiving chemotherapy is still mainly unexplored, some studies reported the tolerability, safety, and induced immune responses of these treatments (Hasan et al., 2020).

An HPV-E7 subunit vaccine used with radiation in mice generated high cure rates and complete response of tumors and decreased recurrence rates contrasted with single therapy. This therapeutic synergy is probably caused through increasing tumor sensitivity to CTL-mediated killing and changes in radiated tumor's phenotype (Ye, Park, Park, Choi, & Sin, 2007). CRT/E7 DNA vaccine accompanied by low-dose RT generated notable antitumor responses preclinically in tumor-bearing mice (Garnett-Benson, Hodge, & Gameiro, 2015).

Combination of therapeutic vaccines with radiation or chemotherapy is being evaluated (Rumfield et al., 2021) (Tables 6 and 7).

4.7.3 Therapeutic HPV vaccines in combination with other treatment modalities

Some other modalities when combined with therapeutic vaccines improve clinical effectiveness. Acetylation and deacetylation of histones by histone acetyltransferases (HAT) and histone deacetylases (HDAC) regulates the gene expression epigenetically. HPV-E6 binding to p53 could affect these two biochemical reactions. Thus, HDAC inhibitors (HDACi) could reduce the development of HPV-associated tumors, especially when combined with other therapies (de Freitas et al., 2020). AR-42 (a pan-HDACi) administration with CRT/E7 DNA vaccine improved survival and generated E7-specific $CD8^+$ T-cell antitumor responses in tumor-bearing mice (Lee et al., 2013).

Agents that deplete Tregs in the TME by antibodies against CD25 marker were administered before vaccination with E7-HSP70 vaccine (Hancock et al., 2018), which prevented tumor cell growth and improved antitumor properties of therapeutic vaccines (Chabeda et al., 2018).

Recently, a triple combination of a liposomal-based HPV therapeutic vaccine (PDS0101) with two new immunomodulatory factors was studied in mice. These two agents included Bintrafusp alfa, an anti-PDL1 antibody fused with a TGFβ "trap" and NHS-IL12, an IL-12 provider, which augments T-cell immunity. More potent immune system promotion and augmented T-cell-mediated antitumor responses in the TME were observed (Rumfield, Pellom, et al., 2020).

Furthermore, ubiquitin sequence addition to E6/E7 DNA vaccines boosted antigen processing and presenting through MHC-I, thus improving the vaccine potency. This system can also be combined with immune checkpoint blockers to further improve the vaccine efficacy (Chandra et al., 2017). Linking HSP70 genes to the vaccine sequence could also promote APC activation, specifically for DCs, thereby enhancing the MHC-I pathway (Cheng et al., 2018). Higher efficacy of HPV16-E7 DNA vaccine fused to HSP70 was reported in mice (Farzanehpour et al., 2013). Combining the HPV-E7 peptide with a photosensitizer drug, fimaprofin, also improved drug targeting in a light-directed way (Selbo et al., 2019).

Examples of combinational strategies in preclinical/clinical studies are shown in Tables 6 and 7, respectively.

5. Conclusion

Despite availability of several efficient HPV prophylactic vaccines for more than one decade, CxCa has a poor prognosis and is still a major cause of women's death globally. Immunotherapeutic approaches, especially therapeutic vaccines, have garnered great attention in recent years, aiming to eliminate infected cells through cellular immune responses. HPV-E6/E7 proteins, as the main essential oncoproteins in commencement and continuation of malignant cell transformation, are the main targets for developing HPV therapeutic vaccines with diverse platforms. Regardless of the promising outcomes of few therapeutic vaccines in some studies, none have achieved approval so far. Therefore, novel candidate vaccines with potent adjuvant/delivery systems or new classes are under development to improve immunogenicity. On the other hand, with more understanding on the complex immunosuppressive events in the TME, immunomodulation along with therapeutic vaccines seem promising. Some impressive preliminary results and approval of several immunomodulatory agents for different cancers have further risen the hopes and encouraged clinical trials on monoclonal antibodies and immunomodulatory agents especially PD-1/PDL-1 and CTL-4 blockers or 4-1BB agonists. Considering patients' biomarkers and personalized approaches, as well as combination of several immunotherapeutic modalities could help to activate immunosurveillance. Moreover, the potential role of combined chemotherapy/radiation or cell-based immunotherapies should not be ignored.

Funding

None.

Conflicts of interest

No conflict of interests is declared.

References

Adams, M., Borysiewicz, L., Fiander, A., Man, S., Jasani, B., Navabi, H., et al. (2001). Clinical studies of human papilloma vaccines in pre-invasive and invasive cancer. *Vaccine*, *19*(17–19), 2549–2556.

Adams, M., Navabi, H., Jasani, B., Man, S., Fiander, A., Evans, A. S., et al. (2003). Dendritic cell (DC) based therapy for cervical cancer: Use of DC pulsed with tumour lysate and matured with a novel synthetic clinically non-toxic double stranded RNA analogue poly[I]:Poly [C12U] (Ampligen®). *Vaccine*, *21*(7), 787–790.

Alson, D., Schuyler, S. C., Yan, B. X., Samimuthu, K., & Qiu, J. T. (2020). Combination vaccination with tetanus toxoid and enhanced tumor-cell based vaccine against cervical cancer in a mouse model. *Frontiers in Immunology*, *11*, 927.

Bagheri, A., Nezafat, N., Eslami, M., Ghasemi, Y., & Negahdaripour, M. (2021). Designing a therapeutic and prophylactic candidate vaccine against human papillomavirus through vaccinomics approaches. *Infection, Genetics and Evolution*, *95*, 105084.

Bartkowiak, T., Singh, S., Yang, G., Galvan, G., Haria, D., Ai, M., et al. (2015). Unique potential of 4-1BB agonist antibody to promote durable regression of HPV+ tumors when combined with an E6/E7 peptide vaccine. *Proceedings of the National Academy of Sciences*, *112*(38), E5290–E5299.

Bergman, H., Buckley, B. S., Villanueva, G., Petkovic, J., Garritty, C., Lutje, V., et al. (2019). Comparison of different human papillomavirus (HPV) vaccine types and dose schedules for prevention of HPV-related disease in females and males. *Cochrane Database of Systematic Reviews*, *11*, CD013479.

Bolhassani, A., Ghasemi, N., Servis, C., Taghikhani, M., & Rafati, S. (2009). The efficiency of a novel delivery system (PEI600-tat) in development of potent DNA vaccine using HPV16 E7 as a model antigen. *Drug Delivery*, *16*(4), 196–204.

Brisson, M., Kim, J. J., Canfell, K., Drolet, M., Gingras, G., Burger, E. A., et al. (2020). Impact of HPV vaccination and cervical screening on cervical cancer elimination: A comparative modelling analysis in 78 low-income and lower-middle-income countries. *The Lancet*, *395*(10224), 575–590.

Bruni, L., Diaz, M., Barrionuevo-Rosas, L., Herrero, R., Bray, F., Bosch, F. X., et al. (2016). Global estimates of human papillomavirus vaccination coverage by region and income level: A pooled analysis. *The Lancet Global Health*, *4*(7), e453–e463.

Cassetti, M. C., McElhiney, S. P., Shahabi, V., Pullen, J. K., Le Poole, I. C., Eiben, G. L., et al. (2004). Antitumor efficacy of Venezuelan equine encephalitis virus replicon particles encoding mutated HPV16 E6 and E7 genes. *Vaccine*, *22*(3–4), 520–527.

Chabeda, A., Yanez, R. J. R., Lamprecht, R., Meyers, A. E., Rybicki, E. P., & Hitzeroth, I. (2018). Therapeutic vaccines for high-risk HPV-associated diseases. *Papillomavirus Research*, *5*, 46–58.

Chandra, J., Dutton, J. L., Li, B., Woo, W. P., Xu, Y., Tolley, L. K., et al. (2017). DNA vaccine encoding HPV16 oncogenes E6 and E7 induces potent cell-mediated and humoral immunity which protects in tumor challenge and drives E7-expressing skin graft rejection. *Journal of Immunotherapy*, *40*(2), 62–70.

Chang, E. Y., Chen, C. H., Ji, H., Wang, T. L., Hung, K., Lee, B. P., et al. (2000). Antigen-specific cancer immunotherapy using a GM-CSF secreting allogeneic tumor cell-based vaccine. *International Journal of Cancer*, *86*(5), 725–730.

Chang, C. L., Ma, B., Pang, X., Wu, T. C., & Hung, C. F. (2009). Treatment with cyclooxygenase-2 inhibitors enables repeated administration of vaccinia virus for control of ovarian cancer. *Molecular Therapy*, *17*(8), 1365–1372.

Chen, W., & Huang, L. (2008). Induction of cytotoxic T-lymphocytes and antitumor activity by a liposomal lipopeptide vaccine. *Molecular Pharmaceutics*, *5*(3), 464–471.

Cheng, M. A., Farmer, E., Huang, C., Lin, J., Hung, C. F., & Wu, T. C. (2018). Therapeutic DNA vaccines for human papillomavirus and associated diseases. *Human Gene Therapy*, *29*(9), 971–996.

Cintolo, J. A., Datta, J., Mathew, S. J., & Czerniecki, B. J. (2012). Dendritic cell-based vaccines: Barriers and opportunities. *Future Oncology*, *8*(10), 1273–1299.

Clark, K. T., & Trimble, C. L. (2020). Current status of therapeutic HPV vaccines. *Gynecologic Oncology*, *156*(2), 503–510.

de Freitas, N. L., Deberaldini, M. G., Gomes, D., Pavan, A. R., Sousa, Â., Dos Santos, J. L., et al. (2020). Histone deacetylase inhibitors as therapeutic interventions on cervical cancer induced by human papillomavirus. *Frontiers in Cell and Development Biology*, *8*, 592868.

De Sanjose, S., Quint, W. G., Alemany, L., Geraets, D. T., Klaustermeier, J. E., Lloveras, B., et al. (2010). Human papillomavirus genotype attribution in invasive cervical cancer: A retrospective cross-sectional worldwide study. *The Lancet Oncology*, *11*(11), 1048–1056.

Diniz, M. O., Cariri, F. A., Aps, L. R., & Ferreira, L. C. (2013). Enhanced therapeutic effects conferred by an experimental DNA vaccine targeting human papillomavirus-induced tumors. *Human Gene Therapy*, *24*(10), 861–870.

Dong, H., Strome, S. E., Salomao, D. R., Tamura, H., Hirano, F., Flies, D. B., et al. (2002). Tumor-associated B7-H1 promotes T-cell apoptosis: A potential mechanism of immune evasion. *Nature Medicine*, *8*(8), 793–800.

Dorostkar, F., Arashkia, A., Roohvand, F., Shoja, Z., Navari, M., Shirazi, M. M. A., et al. (2021). Co-administration of 2′3′-cGAMP STING activator and CpG-C adjuvants with a mutated form of HPV 16 E7 protein leads to tumor growth inhibition in the mouse model. *Infectious Agents and Cancer*, *16*(1), 1–10.

Dorta-Estremera, S., Chin, R. L., Sierra, G., Nicholas, C., Yanamandra, A. V., Nookala, S. M., et al. (2018). Mucosal HPV E6/E7 peptide vaccination in combination with immune checkpoint modulation induces regression of HPV+ oral cancers. *Cancer Research*, *78*(18), 5327–5339.

Ewer, K. J., Lambe, T., Rollier, C. S., Spencer, A. J., Hill, A. V., & Dorrell, L. (2016). Viral vectors as vaccine platforms: From immunogenicity to impact. *Current Opinion in Immunology*, *41*, 47–54.

Fakhr, E., Modic, Ž., & Cid-Arregui, A. (2021). Recent developments in immunotherapy of cancers caused by human papillomaviruses. *Immunology*, *163*(1), 33–45.

Farzanehpour, M., Soleimanjahi, H., Hassan, Z., Amanzadeh, A., Ghaemi, A., & Fazeli, M. (2013). HSP70 modified response against HPV based tumor. *Vectors*, *2*, 2.

Ferrall, L., Lin, K. Y., Roden, R. B., Hung, C. F., & Wu, T. C. (2021). Cervical cancer immunotherapy: Facts and hopes. *Clinical Cancer Research*, *27*(18), 4953–4973.

Ferreira, A. R., Ramalho, A. C., Marques, M., & Ribeiro, D. (2020). The interplay between antiviral signalling and carcinogenesis in human papillomavirus infections. *Cancers*, *12*(3), 646.

Foldvari, M. (2011). HPV infections: Can they be eradicated using nanotechnology? *Nanomedicine: Nanotechnology, Biology, and Medicine*, *8*(2), 131–135.

Ford, K., Hanley, C. J., Mellone, M., Szyndralewiez, C., Heitz, F., Wiesel, P., et al. (2020). NOX4 inhibition potentiates immunotherapy by overcoming cancer-associated fibroblast-mediated CD8 T-cell exclusion from tumors. *Cancer Research*, *80*(9), 1846–1860.

Francisco, L. M., Sage, P. T., & Sharpe, A. H. (2010). The PD-1 pathway in tolerance and autoimmunity. *Immunological Reviews*, *236*(1), 219–242.

Franconi, R., Massa, S., Paolini, F., Vici, P., & Venuti, A. (2020). Plant-derived natural compounds in genetic vaccination and therapy for HPV-associated cancers. *Cancers*, *12*(11), 3101.

Gao, J., Fan, L., Ma, W., & Xiao, H. (2016). Synergistic antitumor effect of a human papillomavirus DNA vaccine harboring E6E7 fusion gene and vascular endothelial growth factor receptor 2 gene. *Microbiology and Immunology*, *60*(9), 626–633.

Garnett-Benson, C., Hodge, J. W., & Gameiro, S. R. (2015). Combination regimens of radiation therapy and therapeutic cancer vaccines: Mechanisms and opportunities. *Seminars in Radiation Oncology*, *25*(1), 46–53.

Geall, A. J., Verma, A., Otten, G. R., Shaw, C. A., Hekele, A., Banerjee, K., et al. (2012). Nonviral delivery of self-amplifying RNA vaccines. *Proceedings Natl Acad Sci USA*, *109*(36), 14604–14609.

Ghaffarifar, F. (2018). Plasmid DNA vaccines: Where are we now? *Drugs Today*, *54*(5), 315–333.

Gomez-Gutierrez, J. G., Elpek, K. G., de Oca-Luna, R. M., Shirwan, H., Zhou, H. S., & McMasters, K. M. (2007). Vaccination with an adenoviral vector expressing calreticulin-human papillomavirus 16 E7 fusion protein eradicates E7 expressing established tumors in mice. *Cancer Immunology, Immunotherapy*, *56*(7), 997–1007.

Graham, V. V., Sutter, G., José, M. V., García-Carranca, A., Erfle, V., Mendoza, N. M., et al. (2000). Human tumor growth is inhibited by a vaccinia virus carrying the E2 gene of bovine papillomavirus. *Cancer*, *88*(7), 1650–1662.

Grunwitz, C., Salomon, N., Vascotto, F., Selmi, A., Bukur, T., Diken, M., et al. (2019). HPV16 RNA-LPX vaccine mediates complete regression of aggressively growing HPV-positive mouse tumors and establishes protective T cell memory. *Oncoimmunology*, *8*(9), e1629259.

Guo, L., & Hua, K. (2020). Cervical cancer: Emerging immune landscape and treatment. *Oncotargets and Therapy*, *13*, 8037.

Hancock, G., Hellner, K., & Dorrell, L. (2018). Therapeutic HPV vaccines. *Best Practice & Research. Clinical Obstetrics & Gynaecology*, *47*, 59–72.

Hansson, M., Nygren, P. A. K., & Stahl, S. (2000). Design and production of recombinant subunit vaccines. *Biotechnology and Applied Biochemistry*, *32*(2), 95–107.

Hasan, Y., Furtado, L., Tergas, A., Lee, N., Brooks, R., McCall, A., et al. (2020). A phase 1 trial assessing the safety and tolerability of a therapeutic DNA vaccination against HPV16 and HPV18 E6/E7 oncogenes after chemoradiation for cervical cancer. *International Journal of Radiation Oncology, Biology, Physics*, *107*(3), 487–498.

Huber, B., Wang, J. W., Roden, R., & Kirnbauer, R. (2021). RG1-VLP and other L2-based, broad-spectrum HPV vaccine candidates. *Journal of Clinical Medicine*, *10*(5), 1044.

Hung, C. F., Wu, T., Monie, A., & Roden, R. (2008). Antigen-specific immunotherapy of cervical and ovarian cancer. *Immunological Reviews*, *222*(1), 43–69.

Ip, P., Boerma, A., Walczak, M., Oosterhuis, K., Haanen, J., Schumacher, T., et al. (2015). Antigen design enhances the immunogenicity of Semliki Forest virus-based therapeutic human papillomavirus vaccines. *Gene Therapy*, *22*(7), 560–567.

Jung, I. D., Shin, S. J., Lee, M. G., Kang, T. H., Han, H. D., Lee, S. J., et al. (2014). Enhancement of tumor-specific T cell-mediated immunity in dendritic cell-based vaccines by mycobacterium tuberculosis heat shock protein X. *Journal of Immunology*, *193*(3), 1233–1245.

Kang, T. H., Lee, J. H., Bae, H. C., Noh, K. H., Kim, J. H., Song, C. K., et al. (2006). Enhancement of dendritic cell-based vaccine potency by targeting antigen to endosomal/lysosomal compartments. *Immunology Letters*, *106*(2), 126–134.

Karimi, H., Soleimanjahi, H., Abdoli, A., & Banijamali, R. S. (2020). Combination therapy using human papillomavirus L1/E6/E7 genes and archaeosome: A nanovaccine confer immuneadjuvanting effects to fight cervical cancer. *Scientific Reports*, *10*(1), 1–15.

Karkada, M., Quinton, T., Blackman, R., & Mansour, M. (2013). Tumor inhibition by DepoVax-based cancer vaccine is accompanied by reduced regulatory/suppressor cell proliferation and tumor infiltration. *International Scholarly Research Notices*, *2013*, 753427.

Kaufmann, A. M., Stern, P. L., Rankin, E. M., Sommer, H., Nuessler, V., Schneider, A., et al. (2002). Safety and immunogenicity of TA-HPV, a recombinant vaccinia virus expressing modified human papillomavirus (HPV)-16 and HPV-18 E6 and E7 genes, in women with progressive cervical cancer. *Clinical Cancer Research*, *8*(12), 3676–3685.

Kaur, G., Garg, T., Rath, G., & Goyal, A. K. (2016). Archaeosomes: An excellent carrier for drug and cell delivery. *Drug Delivery*, *23*(7), 2497–2512.

Kim, S., Chung, H. W., Lee, K. R., & Lim, J. B. (2014). Identification of novel epitopes from human papillomavirus type 18 E7 that can sensitize PBMCs of multiple HLA class I against human cervical cancer. *Journal of Translational Medicine*, *12*(1), 1–13.

Knechtle, S., Marson, L., & Morris, P. (2019). *Kidney transplantation: Principles and practice* (8th Edition). Elsevier Health Sciences.

Krupar, R., Imai, N., Miles, B., Genden, E., Misiukiewicz, K., Saenger, Y., et al. (2016). Abstract LB-095: HPV E7 antigen-expressing Listeria-based immunotherapy (ADXS11-001) prior to robotic surgery for HPV-positive oropharyngeal cancer enhances HPV-specific T cell immunity. *Cancer Research*, *76*(Suppl 14), am2016-lb-095.

Kumar, S. (2019). Live-attenuated bacterial vectors for delivery of mucosal vaccines, DNA vaccines, and cancer immunotherapy. *Pharmaceuticals from Microbes*, *26*, 39–73.

Kumar, S., Biswas, M., & Jose, T. (2015). HPV vaccine: Current status and future directions. *Medical Journal, Armed Forces India*, *71*(2), 171–177.

Lechien, J. R., Descamps, G., Seminerio, I., Furgiuele, S., Dequanter, D., Mouawad, F., et al. (2020). HPV involvement in the tumor microenvironment and immune treatment in head and neck squamous cell carcinomas. *Cancers*, *12*(5), 1060.

Lee, S. Y., Huang, Z., Kang, T. H., Soong, R. S., Knoff, J., Axenfeld, E., et al. (2013). Histone deacetylase inhibitor AR-42 enhances E7-specific CD8 + T cell-mediated antitumor immunity induced by therapeutic HPV DNA vaccination. *Journal of Molecular Medicine*, *91*(10), 1221–1231.

Li, J., Aipire, A., Zhao, H., Yuan, P., & Li, J. (2019). Pleurotus ferulae polysaccharides improve the antitumor efficacy of therapeutic human papillomavirus dendritic cell-based vaccine. *Human Vaccines & Immunotherapeutics*, *15*(3), 611–619.

Lin, P. L., Cheng, Y. M., Wu, D. W., Huang, Y. J., Lin, H. C., Chen, C. Y., et al. (2017). A combination of anti-PD-L1 mA b plus lm-LLO-E6 vaccine efficiently suppresses tumor growth and metastasis in HPV-infected cancers. *Cancer Medicine*, *6*(9), 2052–2062.

Lin, K., Roosinovich, E., Ma, B., Hung, C. F., & Wu, T. C. (2010). Therapeutic HPV DNA vaccines. *Immunologic Research*, *47*(1), 86–112.

Lin, C. T., Yen, C. F., Shaw, S. W., Yen, T. C., Chen, Y. J., Soong, Y. K., et al. (2009). Gene gun administration of therapeutic HPV DNA vaccination restores the efficacy of prolonged defrosted viral based vaccine. *Vaccine*, *27*(52), 7352–7358.

Liu, D. W., Tsao, Y. P., Kung, J. T., Ding, Y. A., Sytwu, H. K., Xiao, X., et al. (2000). Recombinant adeno-associated virus expressing human papillomavirus type 16 E7 peptide DNA fused with heat shock protein DNA as a potential vaccine for cervical cancer. *Journal of Virology*, *74*(6), 2888–2894.

Liu, Y., Li, L., Li, Y., & Zhao, X. (2020). Research progress on tumor-associated macrophages and inflammation in cervical cancer. *BioMed Research International*, *2020*, 6842963.

Liu, Y., Wu, L., Tong, R., Yang, F., Yin, L., Li, M., et al. (2019). PD-1/PD-L1 inhibitors in cervical cancer. *Frontiers in Pharmacology*, *10*, 65.

Liu, Z., Zhou, H., Wang, W., Fu, Y. X., & Zhu, M. (2016). A novel dendritic cell targeting HPV16 E7 synthetic vaccine in combination with PD-L1 blockade elicits therapeutic antitumor immunity in mice. *Oncoimmunology*, *5*(6), e1147641.

Lundstrom, K. (2018). Self-replicating RNA viruses for RNA therapeutics. *Molecules*, *23*(12), 3310.

Malonis, R. J., Lai, J. R., & Vergnolle, O. (2019). Peptide-based vaccines: Current progress and future challenges. *Chemical Reviews*, *120*(6), 3210–3229.

Massarelli, E., William, W., Johnson, F., Kies, M., Ferrarotto, R., Guo, M., et al. (2019). Combining immune checkpoint blockade and tumor-specific vaccine for patients with incurable human papillomavirus 16–related cancer: A phase 2 clinical trial. *JAMA Oncology*, *5*(1), 67–73.

Melief, C. J., Welters, M. J., Vergote, I., Kroep, J. R., Kenter, G. G., Ottevanger, P. B., et al. (2020). Strong vaccine responses during chemotherapy are associated with prolonged cancer survival. *Science Translational Medicine*, *12*(535).

Moeini, S., Saeidi, M., Fotouhi, F., Mondanizadeh, M., Shirian, S., Mohebi, A., et al. (2017). Synergistic effect of programmed cell death protein 1 blockade and secondary lymphoid tissue chemokine in the induction of anti-tumor immunity by a therapeutic cancer vaccine. *Archives of Virology*, *162*(2), 333–346.

Mondini, M., Nizard, M., Tran, T., Mauge, L., Loi, M., Clémenson, C., et al. (2015). Synergy of radiotherapy and a cancer vaccine for the treatment of HPV-associated head and neck cancer. *Molecular Cancer Therapeutics*, *14*(6), 1336–1345.

Monie, A., Tsen, S. W. D., Hung, C. F., & Wu, T. C. (2009). Therapeutic HPV DNA vaccines. *Expert Review of Vaccines*, *8*(9), 1221–1235.

Negahdaripour, M. (2021). COVID-19 vaccine global access is an urgency. *Iranian Journal of Medical Sciences*, *46*(2), 79–80.

Negahdaripour, M., Eslami, M., Nezafat, N., Hajighahramani, N., Ghoshoon, M. B., Shoolian, E., et al. (2017). A novel HPV prophylactic peptide vaccine, designed by immunoinformatics and structural vaccinology approaches. *Infection, Genetics and Evolution*, *54*, 402–416.

Negahdaripour, M., Golkar, N., Hajighahramani, N., Kianpour, S., Nezafat, N., & Ghasemi, Y. (2017). Harnessing self-assembled peptide nanoparticles in epitope vaccine design. *Biotechnology Advances*, *35*(5), 575–596.

Negahdaripour, M., Nezafat, N., Eslami, M., Ghoshoon, M. B., Shoolian, E., Najafipour, S., et al. (2018). Structural vaccinology considerations for in silico designing of a multi-epitope vaccine. *Infection, Genetics and Evolution*, *58*, 96–109.

Negahdaripour, M., Nezafat, N., Heidari, R., Erfani, N., Hajighahramani, N., Ghoshoon, M. B., et al. (2020). Production and preliminary in vivo evaluations of a novel in silico-designed L2-based potential HPV vaccine. *Current Pharmaceutical Biotechnology*, *21*(4), 316–324.

Negahdaripour, M., Shafiekhani, M., Moezzi, S. M. I., Amiri, S., Rasekh, S., Bagheri, A., et al. (2021). Administration of COVID-19 vaccines in immunocompromised patients. *International Immunopharmacology*, *108021*.

Olczak, P., & Roden, R. (2020). Progress in L2-based prophylactic vaccine development for protection against diverse human papillomavirus genotypes and associated diseases. *Vaccine*, *8*(4), 568.

Orbegoso, C., Murali, K., & Banerjee, S. (2018). The current status of immunotherapy for cervical cancer. *Reports of Practical Oncology and Radiotherapy*, *23*(6), 580–588.

Panatto, D., Amicizia, D., Bragazzi, N. L., Rizzitelli, E., Tramalloni, D., Valle, I., et al. (2015). Human papillomavirus vaccine: State of the art and future perspectives. *Advances in Protein Chemistry and Structural Biology*, *101*, 231–322.

Pardi, N., Hogan, M. J., Porter, F. W., & Weissman, D. (2018). mRNA vaccines-a new era in vaccinology. *Nature Reviews. Drug Discovery*, *17*(4), 261–279.

Parkin, J., & Cohen, B. (2001). An overview of the immune system. *The Lancet*, *357*(9270), 1777–1789.

Patel, C., Brotherton, J. M., Pillsbury, A., Jayasinghe, S., Donovan, B., Macartney, K., et al. (2018). The impact of 10 years of human papillomavirus (HPV) vaccination in Australia: What additional disease burden will a nonavalent vaccine prevent? *Eurosurveillance*, *23*(41), 1700737.

Pellom, S. T., Rumfield, C. S., & Morillon, Y. M. (2021). Characterization of recombinant gorilla adenovirus HPV therapeutic vaccine PRGN-2009. *JCI Insight*, *6*(7), e141912.

Peng, S., Ferrall, L., Gaillard, S., Wang, C., Chi, W. Y., Huang, C. H., et al. (2021). Development of DNA vaccine targeting E6 and E7 proteins of human papillomavirus 16 (HPV16) and HPV18 for immunotherapy in combination with recombinant vaccinia boost and PD-1 antibody. *MBio*, *12*(1), e03220–e03224.

Peng, S., Tan, M., Li, Y. D., Cheng, M. A., Farmer, E., Ferrall, L., et al. (2021). PD-1 blockade synergizes with intratumoral vaccination of a therapeutic HPV protein vaccine and elicits regression of tumor in a preclinical model. *Cancer Immunology, Immunotherapy*, *70*(4), 1049–1062.

Perica, K., Varela, J. C., Oelke, M., & Schneck, J. (2015). Adoptive T cell immunotherapy for cancer. *Rambam Maimonides Medical Journal*, *6*(1), e0004s.

Piersma, S. J. (2011). Immunosuppressive tumor microenvironment in cervical cancer patients. *Cancer Microenvironment*, *4*(3), 361–375.

Postow, M. A., Callahan, M. K., & Wolchok, J. D. (2015). Immune checkpoint blockade in cancer therapy. *Journal of Clinical Oncology*, *33*(17), 974.

Qiu, J. T., Alson, D., Lee, T. H., Tsai, C. C., Yu, T. W., Chen, Y. S., et al. (2019). Effect of multiple vaccinations with tumor cell-based vaccine with codon-modified GM-CSF on tumor growth in a mouse model. *Cancers*, *11*(3), 368.

Ramos da Silva, J., Ramos Moreno, A. C., Silva Sales, N., de Oliveira Silva, M., Aps, L. R., Porchia, B. F., et al. (2021). A therapeutic DNA vaccine and gemcitabine act synergistically to eradicate HPV-associated tumors in a preclinical model. *Oncoimmunology*, *10*(1), 1949896.

Remic, T., Sersa, G., Ursic, K., Cemazar, M., & Kamensek, U. (2020). Development of tumor cell-based vaccine with IL-12 gene Electrotransfer as adjuvant. *Vaccine*, *8*(1), 111.

Rice, A. E., Latchman, Y., Balint, J. P., Lee, J. H., Gabitzsch, E. S., & Jones, F. R. (2015). An HPV-E6/E7 immunotherapy plus PD-1 checkpoint inhibition results in tumor regression and reduction in PD-L1 expression. *Cancer Gene Therapy*, *22*(9), 454–462.

Rumfield, C. S., Pellom, S. T., Morillon, Y. M., Schlom, J., & Jochems, C. (2020). Immunomodulation to enhance the efficacy of an HPV therapeutic vaccine. *Journal for Immunotherapy of Cancer*, *8*(1), e000612.

Rumfield, C. S., Roller, N., Pellom, S. T., Schlom, J., & Jochems, C. (2020). Therapeutic vaccines for HPV-associated malignancies. *ImmunoTargets and Therapy*, *9*, 167.

Rumfield, C. S., Schlom, J., & Jochems, C. (2021). Combination therapies for HPV-associated malignancies. *Journal of Clinical & Cellular Immunology*, *12*(1), 608.

Sabado, R. L., Balan, S., & Bhardwaj, N. (2017). Dendritic cell-based immunotherapy. *Cell Research*, *27*(1), 74–95.

Saleh, T., Bolhassani, A., Shojaosadati, S. A., & Aghasadeghi, M. R. (2015). MPG-based nanoparticle: An efficient delivery system for enhancing the potency of DNA vaccine expressing HPV16E7. *Vaccine*, *33*(28), 3164–3170.

Santos, P. M., & Butterfield, L. H. (2018). Dendritic cell-based Cancer vaccines. *Journal of Immunology*, *200*(2), 443–449.

Schellenbacher, C., Roden, R. B., & Kirnbauer, R. (2017). Developments in L2-based human papillomavirus (HPV) vaccines. *Virus Research*, *231*, 166–175.

Schiller, J. T., Castellsagué, X., & Garland, S. M. (2012). A review of clinical trials of human papillomavirus prophylactic vaccines. *Vaccine, 30*, F123–F138.

Schiller, J. T., & Lowy, D. R. (2009). Immunogenicity testing in human papillomavirus virus-like-particle vaccine trials. *The Journal of Infectious Diseases, 200*(2), 166–171.

Schiller, J. T., & Müller, M. (2015). Next generation prophylactic human papillomavirus vaccines. *The Lancet Oncology, 16*(5), e217–e225.

Schmidt, S., Bonilla, W. V., Pauzuolis, M., Reiter, A., Kleissner, T., Oeler, D., et al. (2019). Live-attenuated LCMV-based vector for active immunotherapy of HPV16 + cancer. *Journal of Clinical Oncology, 37*(Suppl. 15), Abstract 1003.

Selbo, P., Janetzki, S., Welters, M., Håkerud, M., Nedberg, A., Edwards, V., et al. (2019). Phase I clinical study for validation of Fimaporfin-based photochemical internalisation: A novel technology for enhancing cellular immune responses important for therapeutic effect of peptide-and protein-based vaccines. *Annals of Oncology, 30*, xi40-xi41.

Shrimali, R. K., Ahmad, S., Verma, V., Zeng, P., Ananth, S., Gaur, P., et al. (2017). Concurrent PD-1 blockade negates the effects of OX40 agonist antibody in combination immunotherapy through inducing T-cell apoptosis. *Cancer Immunology Research, 5*(9), 755–766.

Siegal, R., Miller, K., & Jemal, A. (2014). Cancer statistics, 2012. *CA: A Cancer Journal for Clinicians, 64*(1), 9–29.

Silva, A. J., Zangirolami, T. C., Novo-Mansur, M. T. M., Giordano, R. D. C., & Martins, E. A. L. (2014). Live bacterial vaccine vectors: An overview. *Brazilian Journal of Microbiology, 45*, 1117–1129.

Simova, J., Sapega, O., Imrichova, T., Stepanek, I., Kyjacova, L., Mikyskova, R., et al. (2016). Tumor growth accelerated by chemotherapy-induced senescent cells is suppressed by treatment with IL-12 producing cellular vaccines. *Oncotarget*, 7(34), 54952–54964.

Skeate, J. G., Woodham, A. W., Einstein, M. H., Da Silva, D. M., & Kast, W. M. (2016). Current therapeutic vaccination and immunotherapy strategies for HPV-related diseases. *Human Vaccines & Immunotherapeutics, 12*(6), 1418–1429.

Solares, A. M., Baladron, I., Ramos, T., Valenzuela, C., Borbon, Z., Fanjull, S., et al. (2011). Safety and immunogenicity of a human papillomavirus peptide vaccine (CIGB-228) in women with high-grade cervical intraepithelial neoplasia: First-in-human, proof-of-concept trial. *ISRN Obstetrics and Gynecology, 2011*, 292951.

Stern, P. L. (2021). Harnessing immunity for therapy in human papillomavirus driven cancers. *Tumour Virus Research, 11*, 200212.

Sung, H., Ferlay, J., Siegel, R. L., Laversanne, M., Soerjomataram, I., & Jemal, A. (2021). Global cancer statistics 2020: GLOBOCAN estimates of incidence and mortality worldwide for 36 cancers in 185 countries. *Cancer Journal for Clinicians, 71*(3), 209–249.

Trimble, C. L., Morrow, M. P., Kraynyak, K. A., Shen, X., Dallas, M., Yan, J., et al. (2015). Safety, efficacy, and immunogenicity of VGX-3100, a therapeutic synthetic DNA vaccine targeting human papillomavirus 16 and 18 E6 and E7 proteins for cervical intraepithelial neoplasia 2/3: A randomised, double-blind, placebo-controlled phase 2b trial. *Lancet, 386*(10008), 2078–2088.

Tseng, C. W., Trimble, C., Zeng, Q., Monie, A., Alvarez, R. D., Huh, W. K., et al. (2009). Low-dose radiation enhances therapeutic HPV DNA vaccination in tumor-bearing hosts. *Cancer Immunology, Immunotherapy, 58*(5), 737–748.

Uddin, M. N., Kouzi, S. A., & Hussain, M. D. (2015). Strategies for developing oral vaccines for human papillomavirus (HPV) induced cancer using nanoparticle mediated delivery system. *Journal of Pharmacy & Pharmaceutical Sciences, 18*(2), 220–234.

van der Sluis, T. C., van Duikeren, S., Huppelschoten, S., Jordanova, E. S., Nejad, E., Sloots, A., et al. (2015). Vaccine-induced tumor necrosis factor–producing T cells synergize with cisplatin to promote tumor cell death. *Clinical Cancer Research, 21*(4), 781–794.

van Poelgeest, M. I. E., Visconti, V. V., Aghai, Z., van Ham, V. J., Heusinkveld, M., Zandvliet, M. L., et al. (2016). Potential use of lymph node-derived HPV-specific T cells for adoptive cell therapy of cervical cancer. *Cancer Immunology, Immunotherapy, 65*(12), 1451–1463.

Vici, P., Pizzuti, L., Mariani, L., Zampa, G., Santini, D., Di Lauro, L., et al. (2016). Targeting immune response with therapeutic vaccines in premalignant lesions and cervical cancer: Hope or reality from clinical studies. *Expert Review of Vaccines, 15*(10), 1327–1336.

Wang, C., Lin, G. H., McPherson, A. J., & Watts, T. H. (2009). Immune regulation by 4-1BB and 4-1BBL: Complexities and challenges. *Immunological Reviews, 229*(1), 192–215.

Wang, R., Pan, W., Jin, L., Huang, W., Li, Y., Wu, D., et al. (2020). Human papillomavirus vaccine against cervical cancer: Opportunity and challenge. *Cancer Letters, 471*, 88–102.

Yang, A., Farmer, E., Wu, T., & Hung, C. F. (2016). Perspectives for therapeutic HPV vaccine development. *Journal of Biomedical Science, 23*(1), 1–19.

Yang, A., Jeang, J., Cheng, K., Cheng, T., Yang, B., Wu, T. C., et al. (2016). Current state in the development of candidate therapeutic HPV vaccines. *Expert Review of Vaccines, 15*(8), 989–1007.

Ye, G. W., Park, J. B., Park, Y. J., Choi, Y. S., & Sin, J. I. (2007). Increased sensitivity of radiated murine cervical Cancer tumors to E7 subunit vaccine–driven CTL-mediated killing induces synergistic anti-tumor activity. *Molecular Therapy, 15*(8), 1564–1570.

Zhang, Y., Wang, N., Ding, M., Yang, Y., Wang, Z., Huang, L., et al. (2020). CD40 accelerates the antigen-specific stem-like memory CD8(+)T cells formation and human papilloma virus (HPV)-positive tumor eradication. *Frontiers in Immunology, 11*, 1012.

Zhang, Y., Xie, F., Yin, Y., Zhang, Q., Jin, H., Wu, Y., et al. (2021). Immunotherapy of tumor RNA-loaded lipid nanoparticles against hepatocellular carcinoma. *International Journal of Nanomedicine, 16*, 1553–1564.

Zhou, Z. X., Li, D., Guan, S. S., Zhao, C., Li, Z. L., & Zeng, Y. (2015). Immunotherapeutic effects of dendritic cells pulsed with a Coden-optimized HPV 16 E6 and E7 fusion gene in vivo and in vitro. *Asian Pacific Journal of Cancer Prevention, 16*(9), 3843–3847.

Zhou, X., Sun, L., Yao, X., Li, G., Wang, Y., & Lin, Y. (2020). Progress in vaccination of prophylactic human papillomavirus vaccine. *Frontiers in Immunology, 11*, 1434.

CHAPTER THREE

Natural killer cell-based strategies for immunotherapy of cancer

Ondřej Vaněk*, Barbora Kalousková, Celeste Abreu, Shiva Nejadebrahim, and Ondřej Skořepa
Department of Biochemistry, Faculty of Science, Charles University, Prague, Czech Republic
*Corresponding author: e-mail address: ondrej.vanek@natur.cuni.cz

Contents

Abstract

Natural killer (NK) cells are a family of lymphocytes with a natural ability to kill infected, harmed, or malignantly transformed cells. As these cells are part of the innate immunity, the cytotoxic mechanisms are activated upon recognizing specific patterns without prior antigen sensitization. This recognition is crucial for NK cell function in the maintenance of homeostasis and immunosurveillance. NK cells not only act directly toward malignant cells but also participate in the complex immune response by producing cytokines or cross-talk with other immune cells. Cancer may be seen as a break of all immune defenses when malignant cells escape the immunity and invade surrounding tissues creating a microenvironment supporting tumor progression. This process may be reverted by intervening immune response with immunotherapy, which may restore immune recognition. NK cells are important effector cells for immunotherapy. They may be used for adoptive cell transfer, genetically modified with chimeric antigen receptors, or triggered with appropriate antibodies and other antibody-fragment-based recombinant therapeutic proteins tailored specifically for NK cell engagement. NK cell receptors, responsible for target recognition and activation of cytotoxic response, could also be

Advances in Protein Chemistry and Structural Biology, Volume 129
ISSN 1876-1623
https://doi.org/10.1016/bs.apcsb.2022.02.001

targeted in immunotherapy, for example, by various bi-, tri-, or multi-specific fusion proteins designed to bridge the gap between tumor markers present on target cells and activation receptors expressed on NK cells. However, this kind of immunoactive therapeutics may be developed only with a deep functional and structural knowledge of NK cell receptor: ligand interactions. This review describes the recent developments in the fascinating protein-engineering field of NK cell immunotherapeutics.

Abbreviations

ADCC antibody-dependent cellular cytotoxicity
BiKE bispecific killer engager
CAR chimeric antigen receptor
mAbs monoclonal antibodies
NK cells natural killer cells
TriKE trispecific killer engager

1. Natural killer cells origin and functions

Last year marked the 45th anniversary of the discovery of a distinct group of large granular lymphocytes, presently known as "natural killer cells," or NK cells for short. Their discovery, and particularly the first description of their ability to eliminate tumor cells, was described in three publications between 1974 and 1975 (Greenberg & Playfair, 1974; Herberman, Nunn, & Lavrin, 1975; Kiessling, Klein, & Wigzell, 1975). Initially, it was found that the observed cytotoxicity to allogeneic tumors is not of T- or B-cell nature. This effect could be observed because NK cells do not need prior antigen sensitization for their function and, thanks to a very rapid response to malignantly transformed cells (Herberman, Nunn, Holden, & Lavrin, 1975; Kiessling et al., 1975), virus-infected cells (Hammer & Romagnani, 2017; Trinchieri, 1989) and parasites (Walk & Sauerwein, 2019), they are an essential part of our innate immune system.

NK cells are large granular lymphocytes derived from the common lymphoid progenitor (Kondo, Weissman, & Akashi, 1997). With approximately a 2-week blood circulation life span, they are estimated to constitute 5%–15% of circulating mononuclear cells (Blum & Pabst, 2007). NK cell development can be divided into several stages, resulting in tissue-resident or NK cells circulating in peripheral blood (Fig. 1). NK cell precursors (NKPs) differentiate from hematopoietic stem cells (HSCs) or early lymphoid progenitors (ELPs) in the bone marrow and thymus. Their further

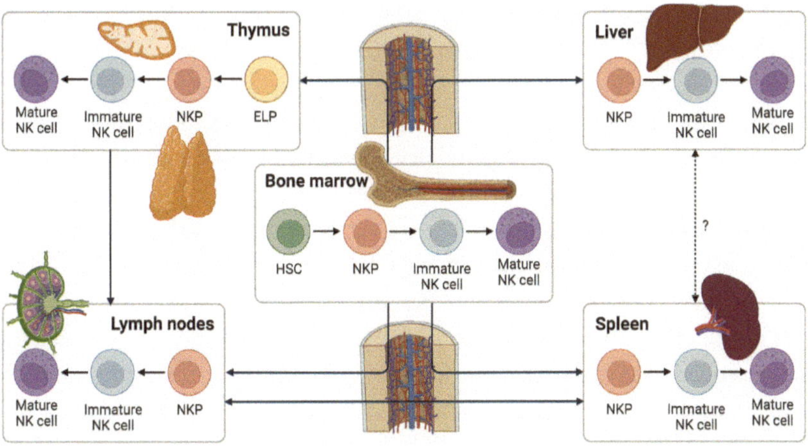

Fig. 1 NK cell development. Bone marrow, thymus, lymph nodes, spleen, and liver are the primary sites for NK cell development. Several maturation stages could be identified—NK cell precursors (NKP), immature NK cells, and mature NK cells. NKPs differentiate from hematopoietic stem cells (HSC) or early lymphoid progenitors (ELP).

development is not limited only to these sites but also to other tissues such as lymph nodes, spleen, and liver (Huntington, Vosshenrich, & Di Santo, 2007). NKPs develop into immature NK cells and later into mature, educated, and competent NK cells, which acquired effector functions and self-tolerance. Mature NK cells could remain in a resting state; activation of lytic competence is primed by the exposure to interferon-gamma (IFN-γ), interleukin 2 (IL-2), or interleukin 15 (IL-15) (Mak & Saunders, 2006).

NK cells lack T- and B-cell receptors, and their individual development stages are characterized by the expression of specific cell surface markers (Di Vito, Mikulak, & Mavilio, 2019; Montaldo et al., 2013). Mature NK cells are characterized by the expression of CD56, CD94 and CD161 (more precisely $CD34^{-}$ $CD117^{+/-}$ $CD94^{+}$ HLA-DR^{-} $CD10^{-}$ $CD122^{+}$ $CD94^{+}$ $NKp44^{low}$ $NKG2D^{+}$ $CD161^{+}$). Based on the expression of CD56, we can distinguish two main NK cell subsets: $CD56^{dim}$ ($CD56^{dim}$ $CD16^{+}$, 90% of circulating NK cells) cells with cytotoxic activity and $CD56^{bright}$ ($CD56^{bright}$ $CD16^{-}$) cells producing high amounts of cytokines and having a regulatory function (Cooper, Fehniger, & Caligiuri, 2001).

NK cells are close relatives of innate lymphoid cells (ILCs); sometimes, they are even listed as an independent group within ILCs (Artis & Spits, 2015). ILCs are lymphocytes without surface molecules specific to other immune cell types. The primary role of ILCs is regulation of the immune response by secretion of cytokines; they are non-cytotoxic. In contrast,

NK cells are primarily cytotoxic cells but additionally have a secretory function (CD56^{bright}). NK cells and ILCs share common ancestors during cell development (Vivier, van de Pavert, Cooper, & Belz, 2016).

NK cells play several essential roles in immunity. The most evident (hence their name) is the cytotoxicity toward stressed or transformed cells (Vivier et al., 2011). This cytotoxicity is independent of previous specific immunization (MHC-independent). On the other hand, NK cells must learn how to distinguish harmed cells from the healthy ones and ensure self-tolerance (Vivier, Tomasello, Baratin, Walzer, & Ugolini, 2008).

The mechanism by which activated NK cells directly eliminate target cells involves the secretion of cytotoxic granules containing perforin, granzymes, and Fas ligand (Bossi & Griffiths, 1999; de Saint Basile, Menasche, & Fischer, 2010; Voskoboinik, Whisstock, & Trapani, 2015). In addition, NK cells express the CD16 receptor (FcγRIIIa), through which they recognize IgG-opsonized targets (Leibson, 1997; Orange, 2013; Vivier et al., 2008), mediating antibody-dependent cellular cytotoxicity (ADCC). In addition to innate immune surveillance, NK cells can support and regulate immunity through interactions with dendritic cells (DCs), macrophages, and T lymphocytes. NK cells meet dendritic cells in secondary lymphoid organs and peripheral tissues, where they interact in two ways. Firstly, NK cells can influence homeostasis by eliminating immature DCs via a pathway dependent on the TNF-related apoptosis-inducing ligand (TRAIL) (Hayakawa et al., 2004). The second outcome of NK cell interaction with dendritic cells is the promotion of maturation by the action of IFN-γ and TNF (Vivier et al., 2008). These cytokines also induce differentiation of helper T cells into Th1 and Th2 cells. As a result, due to the production of IL-2, IL-12, IL-15, and IL-18 by phagocytosing dendritic cells and macrophages, NK cells proliferate, secrete more cytokines, and have a reduced threshold for triggering cytotoxicity, resulting in a higher cytolytic potential (Degli-Esposti & Smyth, 2005; Moretta, Bottino, et al., 2006; Moretta, Ferlazzo, et al., 2006; Walzer, Dalod, Robbins, Zitvogel, & Vivier, 2005). The overview of NK cell functions is shown in Fig. 2.

Recent studies revealed that NK cells also have an immunological memory (Paust & von Andrian, 2011; von Andrian, 2021). Despite lacking the mechanism of recombinase-dependent clonal antigen receptors, being traditionally considered responsible for immunological memory (the case of B and T cells), NK cells are capable of a fast and enhanced immune response upon rechallenge (Cerwenka & Lanier, 2016). There are two types of NK cell immunological memory. The first one is based on the interaction

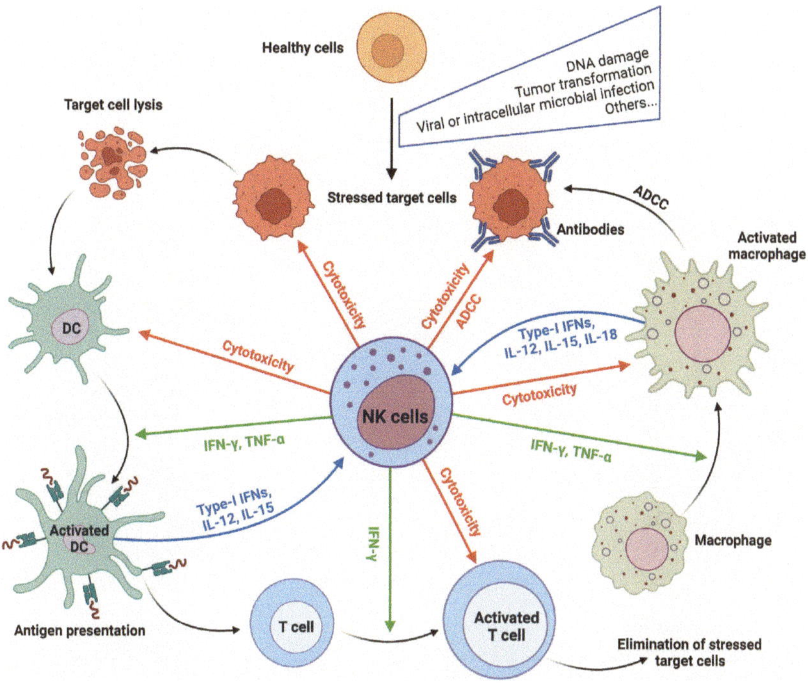

Fig. 2 Regulation of immune responses by NK cells. NK cells have effector functions (lysis of harmed cells and tolerance of healthy cells) and regulatory function (secretion of cytokines, cross-talk with other immune cells). Primed by several interleukins, NK cells can promote the maturation and activation of other lymphocytes (green arrows). However, at the same time, they can eliminate unmatured DCs, hyperreactive macrophages, and $CD4^+$ T lymphocytes (red arrows). Such functions are tightly regulated by the balance between signals received by activating and inhibitory receptors present on NK cells.

with antigens, resulting in short-term antigen-specific memory; the second one is antigen non-specific, as NK cells can recognize known inflammatory cytokine milieus.

2. NK cell recognition in health and disease

How do the NK cells recognize healthy cells from diseased ones? NK cells' surfaces are decorated by many different receptors, both activating and inhibiting the cytotoxicity and having other physiological functions (receptors for cytokines or molecules of cellular adhesion) (Vivier et al., 2011). Activating and inhibitory receptors engage a broad range of health and

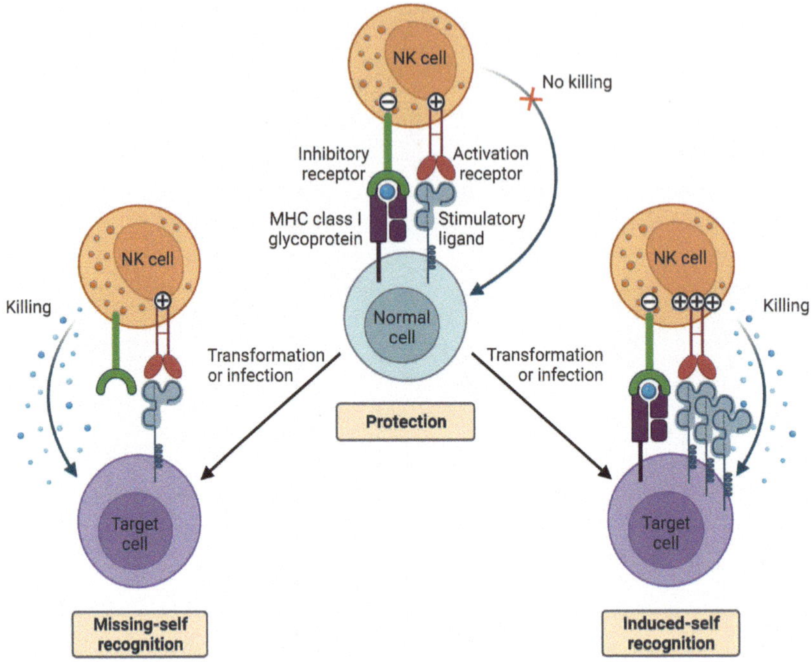

Fig. 3 Principles of "missing-self" and "induced-self" NK cell recognition. Normal healthy cells (middle) present a balanced mixture of activating and inhibitory signals on their surface, which results in self-tolerance. This equilibrium could be disrupted by the absence of the inhibitory signals ("missing-self," left) or the strong presence of the activating signals ("induced-self," right), both resulting in NK cell activation and target cell killing.

cellular disease markers, resulting in two modes of recognition, termed as "missing-self" and "induced-self" (Fig. 3) (Vivier & Ugolini, 2011). Major histocompatibility complex class I (MHC I) molecules are expressed on normal cells under physiological conditions and considered markers of good health, which determine the "self" phenotype. These molecules are ligands for inhibitory NK cell receptors (Ljunggren & Kärre, 1990). The "missing-self" recognition results in the killing of the cells where inhibitory markers are missing (Kärre, 2008; Lanier, 2008). On the other hand, "induced-self" recognition is based on the interaction with NK cell cytotoxicity activating ligands, typically expressed only under stress, infectious, or malignant conditions (Malmberg et al., 2017). NK cells learn to recognize self-MHC I molecules in the licensing process. Only licensed cells get inhibited by MHC I and become competent NK cells (Kim et al., 2005).

2.1 NK cell receptor families

Natural killer cell recognition and cytotoxicity against target cells are controlled by the expression of activating and inhibitory receptors encoded in two main gene clusters: the natural killer complex (NKC) and the leukocyte receptor complex (LRC) (Trowsdale et al., 2001). The cytotoxicity is regulated by balancing inhibitory and activation signaling triggered by the interactions between the NK cell receptors and target cell ligands (Diefenbach & Raulet, 2001; Moretta, Bottino, et al., 2006; Moretta, Ferlazzo, et al., 2006).

NK cell activating and inhibitory receptors use opposing signaling motifs to promote or inhibit activation (Pegram, Andrews, Smyth, Darcy, & Kershaw, 2011). Activating receptors signal through immunoreceptor tyrosine-based activating motifs (ITAMs) located in the cytoplasmic domain. ITAM-receptor activation triggers a signaling cascade stimulating degranulation by promoting cytoskeleton rearrangement and transcription of genes encoding cytokines and chemokines (Tomasello, Blery, Vely, & Vivier, 2000). On the contrary, inhibitory receptors signal through intracellular immunoreceptor tyrosine-based inhibitory motifs (ITIMs) located in the cytoplasmic tails. Inhibitory receptors recruit phosphatases to the NK cell:target cell interface and promote dephosphorylation of specific intracellular components linked with the activation of NK cell receptors. ITIM-receptor inhibition has been described to suppress degranulation (therefore inhibiting the release of perforin and granzymes) and cytokine production (Peterson & Long, 2008).

Activating and inhibitory receptors can be subdivided into three prominent receptor families: C-type lectin-like receptors, killer immunoglobulin-like receptors, and natural cytotoxicity receptors.

2.1.1 C-type lectin-like receptors

C-type lectin-like receptors (CTLRs) comprise a large superfamily of receptors encoded in the natural killer gene complex, recognizing a diverse range of ligands (Dambuza & Brown, 2015). The conserved motif responsible for binding their ligands was termed C-type lectin-like domain (CTLD) (Drickamer, 1999; Drickamer & Fadden, 2002). The name of C-type lectin-like receptors derives from C-type lectins. C-type lectins are a group of proteins in which carbohydrate-binding activity, mediated through a carbohydrate recognition domain (CRD), is Ca^{2+}-dependent. However, biochemical and functional analyses have shown that, despite having sequences resembling C-type lectin carbohydrate recognition domains, CTLRs cannot bind carbohydrates in a Ca^{2+}-dependent manner (Grobárová, Benson, Rozbeský, Novák, & Černý, 2013; Kavan et al., 2010; Kovalová et al., 2010; Renaudet et al., 2010; Rozbeský et al., 2014). Instead, CTLRs have been later shown to recognize non-carbohydrate ligands, such as lipids and proteins (Rozbeský et al., 2015; Zelensky & Gready, 2005).

C-type lectin-like receptors are primarily expressed in myeloid cells such as monocytes, macrophages, neutrophils, and dendritic cells (Dambuza & Brown, 2015). C-type lectin-like receptors comprise both inhibitory and activating receptors that can signal either directly, through integral signaling motifs, or indirectly, through the association with adaptor proteins (Chiffoleau, 2018). The functions of CTLRs in immunity range from host defense against fungi and bacteria to autoimmunity and tumor cytotoxicity. Growing evidence also indicates the involvement of C-type lectin-like receptors in the modulation of adaptive immune responses (Chiffoleau, 2018; Dambuza & Brown, 2015).

C-type lectin-like receptors typically are type II transmembrane glycoproteins with an N-terminal cytoplasmic domain, a transmembrane domain followed by a stalk domain, and an extracellular C-type lectin-like domain at the C-terminus. The CTLD comprises two α-helices and two antiparallel β-sheets stabilized by two or three conserved intramolecular disulfide bonds and several connecting loops involved in ligand recognition (Sovová et al., 2011). In addition, CTLRs usually dimerize in a homo- or hetero-manner via the cysteine residue present in their stalk region (Vaněk et al., 2008, 2019).

Members of the C-type lectin-like family include the NK cell receptors from the CLEC2 family (i.e., human CD69, KACL, AICL, and LLT1, or murine Clr ligands) (Kolenko et al., 2009; Li, Wang, Chen, Brown, & Mariuzza, 2013; Neuss et al., 2018; Skálová et al., 2012, 2015), the KLR family (i.e., human NKR-P1 or murine Ly49) (Bláha et al., 2017, 2021; Kolenko et al., 2011; Rahim & Makrigiannis, 2015; Rozbeský et al., 2011), and the standalone activating NKG2D receptor and various CD94/NKG2x heterodimeric receptors (Bartel, Bauer, & Steinle, 2013).

2.1.2 Killer immunoglobulin-like receptors

Killer immunoglobulin-like receptors (KIRs) are a family of transmembrane receptors encoded in the leukocyte receptor complex. The human KIR family is encoded in 14 polymorphic genes. Distinct receptors can transduce either activating or inhibitory signals. KIR classification is based on the number of extracellular immunoglobulin-like domains (D0, D1, and D2 domains) and cytoplasmic tail length (L or S for long or short tails, respectively) (Campbell & Purdy, 2011). The functional activity of KIRs can be predicted by analyzing the cytoplasmic tail length and the presence of a charged amino acid in the transmembrane region of the molecule. Inhibitory forms of KIRs are long-tailed, with a few exceptions, whereas KIRs with shorter cytoplasmic tails are activating receptors. Activating receptors possess a charged amino acid in the transmembrane domain necessary for positive signaling upon binding with an ITAM-possessing molecule (MacFarlane & Campbell, 2006).

Killer immunoglobulin-like receptor expression has been reported predominantly on NK cells and to a lesser extent on subpopulations of T lymphocytes. NK cells detect the absence or downregulation of MHC class I molecules by an array of specific complementary systems. Killer immunoglobulin-like receptors preferentially recognize human leukocyte antigen (HLA)-A, -B, -C, and in very particular cases, HLA-G allotypes (Boyton & Altmann, 2007). The highly variable and polymorphic system of KIRs is vital in the direct recognition of diverse MHC class I molecules, thus pointing to the critical roles of KIRs in the regulation of development, tolerance, and activation of NK cells (Campbell & Purdy, 2011; Vilches & Parham, 2002). KIRs have been implicated in both physiological and pathological immune mechanisms. The binding of KIRs to their MHC class I ligands on potential target cells results in the modulation of cytotoxicity and cytokine secretion by NK cells. Despite their role in mounting effective responses to infectious diseases and tumors, emerging evidence additionally implicates KIRs in human autoimmune pathogenesis (Bashirova, Martin, McVicar, & Carrington, 2006; Lanier, 2005).

2.1.3 Natural cytotoxicity receptors

Natural cytotoxicity receptors (NCRs) are a family of immunoglobulin-like receptors encoded in the leukocyte receptor complex. In the 1990s, three natural cytotoxicity receptors were first described on human NK cells: NKp46 (NCR1; CD335), NKp44 (NCR2; CD336), and NKp30 (NCR3; CD337). Natural cytotoxicity receptors were initially grouped due to their ability to activate natural killer cells and kill tumor cells *in vitro* (Pende et al., 1999). Despite functional similarities, the members of the NCR family share only a limited similarity in amino acid sequence or structure (Kruse, Matta, Ugolini, & Vivier, 2014).

NCRs are primarily expressed in NK cells, although recent evidence points to expression in T lymphocytes as well. Natural cytotoxicity receptors are type I transmembrane proteins comprised of an extracellular ligand-binding domain, a membrane-proximal stalk domain, a transmembrane domain, and a short cytosolic domain. In the case of receptor NKp30, the presence of both stalk domain and N-glycosylation is required for its oligomerization (Skořepa et al., 2020). NCRs oligomerization on the cell surface is not yet detailly described but is expected to enhance their signaling. NCRs lack a functional intracellular signaling domain and thus must associate with adaptor proteins via a charged residue in their transmembrane domain (Koch, Steinle, Watzl, & Mandelboim, 2013). Unlike other NK cell receptor families that preferentially bind self-MHC and MHC-related ligands, natural cytotoxicity receptors preferentially recognize pathogen-derived molecules and non-MHC self-markers (Hudspeth, Silva-Santos, & Mavilio, 2013).

NCRs functions in the immune system, although also described for tumor cytotoxicity, are strongly related to NK cell activation in antiviral and pathogen-derived immune responses (Koch et al., 2013). In addition to their cytotoxicity-promoting activities, NCRs can also mediate the production of proinflammatory cytokines by NK cells, thus promoting pathogen clearance and immune response modulation (Hudspeth et al., 2013).

3. NK cells as a tool of cancer immunotherapy

3.1 Cancer immunosurveillance

As the immune system tries to identify the malignancies in our body, it sometimes fails because of the low immunogenicity of some tumor cells and thus suppresses the immune response permitting tumor growth (Zitvogel, Tesniere, & Kroemer, 2006). The dual role of the immune system, the host-protective, and the tumor-promoting, is termed "cancer immunoediting," which is presented in Fig. 4. The mutual interaction between the immune system (not only cells, but humoral components are involved as well) and transformed cells may result in tumor elimination (immunosurveillance was successful), tumor escape (anti-tumor immune response failed), or may stay in the equilibrium (cancer cells are tolerated, not entirely eliminated, but further outgrowth is prevented) (Dunn, Old, & Schreiber, 2004). The equilibrium stage may be an outcome of immunoediting, when cancer cells are functionally dormant for the rest

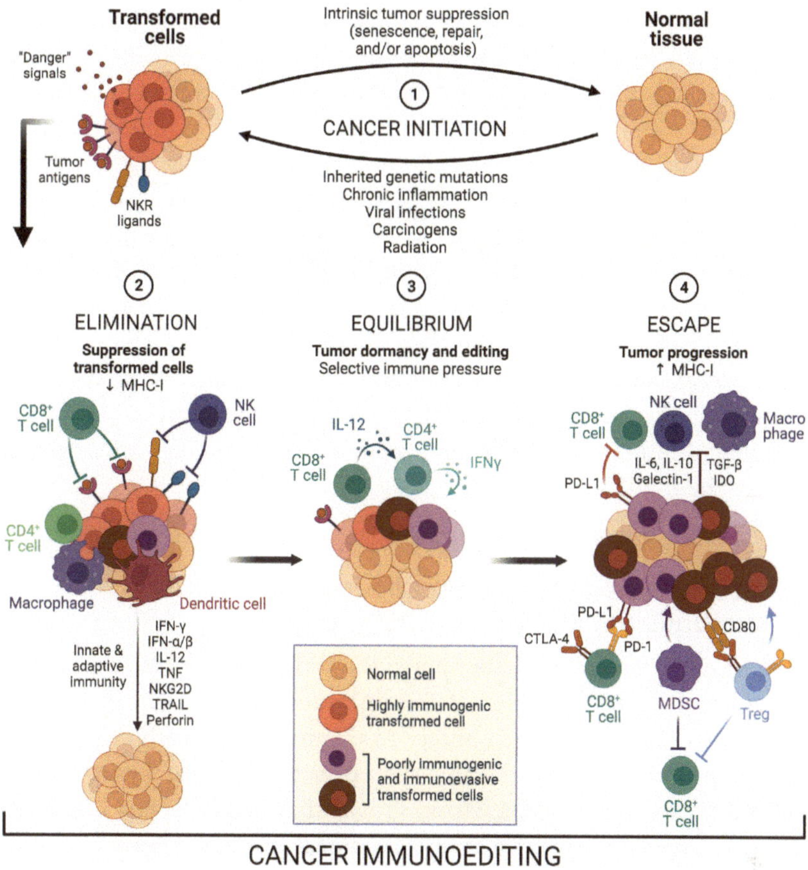

Fig. 4 Cancer immunoediting. Based on the immunogenicity of transformed cells, the immune system could successfully eliminate the tumor (highly immunogenic cells), or tumor cells escape the immunosurveillance mechanisms (poorly immunogenic and immunoevasive cells). In a particular case, the immune system may be in equilibrium with cancer cells when cancer is temporarily tolerated, and the editing process continues.

of their lives, without surrounding tissues' evasion (Vesely, Kershaw, Schreiber, & Smyth, 2011). The equilibrium phase explains a long period of latency for some tumors. On the other hand, tumor cells are under constant pressure, which may result in the selection of clones with lower immunogenicity and, thus, tumor escape (Schreiber, Old, & Smyth, 2011).

Tumor cells develop ingenious strategies for escaping immunity. There are three main tumor escape strategies—the loss of antigenicity, the loss of immunogenicity, and the creation of immune-suppressive tumor microenvironment (TME) (Beatty & Gladney, 2015). Loss of antigenicity is represented by the selection of tumor cells deficient in molecules of immune recognition (downregulation of MHC I (Khong, Wang, & Rosenberg, 2004), alterations in antigen processing machinery of MHC II (Sconocchia et al., 2014), and suppression of activating ligands (Stern-Ginossar et al., 2008)). Such cells are becoming undetectable by the immune system. The loss of immunogenicity is represented by the upregulation of immune inhibitory molecules (such as PD-L1 (Taube et al., 2012) or HLA-G antigens (Tripathi & Agrawal, 2006)), leading in consequence to the silencing of the immune response. TME is a specific niche created by the cancer cells when evading healthy tissue. It is constantly shaped by cancer cell activity, contains molecules that inhibit mechanisms of anti-tumor immunity (immunosuppressive environment), and promotes tumor growth (Hui & Chen, 2015).

3.2 Mechanisms of tumor escape from NK cell immunosurveillance

NK cells were described as lymphocytes with direct cytotoxic function toward tumor cells (Trinchieri, 1989). Although other NK cell functions were further discovered, their ability to eliminate malignant cells without prior antigen sensitization remains crucial in tumor immunosurveillance. Over the years, many studies have shown that NK cells are responsible for tumor rejection in experimental animal models (Vivier, Ugolini, Blaise, Chabannon, & Brossay, 2012). Triggering of this immunity is driven by the "missing-self" recognition mode, as the expression of MHC I is usually altered (mainly suppressed) on the surface of cancer cells (Kärre, Ljunggren, Piontek, & Kiessling, 1986). Inhibitory KIRs are responsible for the discrimination of modified cells (Shimasaki, Jain, & Campana, 2020). These receptors dampen undesirable cytotoxic responses toward healthy cells while they allow for attacking cancer cells lacking the corresponding KIR ligands, the markers of cell health. Apart from missing inhibition, NK cell cytotoxicity is also triggered directly by NK cell-activating receptors identifying specific cellular ligands expressed on tumor cells, such as B7-H6 and MICA (Brandt et al., 2009; El-Gazzar, Groh, & Spies, 2013; Li, Wang, & Mariuzza, 2011).

NK cells are tumor-infiltrating lymphocytes. Their presence in solid tumors is associated with a better prognosis (Habif, Crinier, André, Vivier, & Narni-Mancinelli, 2019). Although NK cells often mediate significant anti-tumor responses, many proposed therapies fail in the natural environment despite having promising results *in vitro*. This failure is because the suppressive microenvironment of solid tumors attenuates the efficacy of NK cells. In addition to hypoxia, which is very common in tumors and lowers NK cell activity, local immunosuppression results from the interaction with many soluble factors (e.g., TGF-β, macrophage migration inhibitory factor (Apte, Sinha, Mayhew, Wistow, & Niederkorn, 1998; Regis, Dondero, Caliendo, Bottino, & Castriconi, 2020), IL-4, indoleamine 2,3-dioxygenase (IDO) or prostaglandin E2 (PGE2)), suppressing NK cell function (Riggan, Shah, & O'Sullivan, 2021). These factors may be produced both by the tumor itself and by the lymphocytes that have been "attracted" by the tumor (including M2 macrophages, myeloid-derived suppressor cells, T-reg, and stromal cells) (Vitale, Cantoni, Pietra, Mingari, & Moretta, 2014). Ultimately, the final effect on NK cells is the down-regulation of activating receptors. NK cells that could normally eliminate activating ligand-expressing tumors (e.g., B7-H6 (Chen, Mo, Jia, & He, 2018)) are "disarmed" and unable to recognize the target (Pende et al., 2019). Another critical element of tumor escape is the expression of inhibitory checkpoint receptor PD-1 on NK and T cells and its corresponding ligand PD-L1 on tumor cells, whose engagement results, again, in inhibition (Pende et al., 2019; Pesce et al., 2017; Vacca et al., 2018).

Another way in which tumors escape immune surveillance is through shedding (proteolytic cleavage) and the internalization of activating receptors. Specific ligands for activating receptors, namely for NKG2D (ligands MICA and MICB (Kaiser et al., 2007)) and NKp30 (ligands B7-H6 (Schlecker et al., 2014) and BAG6 (Ponath et al., 2021)), are released by tumor cells through the action of specific proteases called ADAMs (a disintegrin and metalloprotease (Giebeler & Zigrino, 2016)) or are secreted on the surface of exosomes. The consequence of shedding is a reduction of the amount of activating ligands on the tumor surface and a decrease in NK cell responsiveness. In the context of therapeutic intervention, the use of antibodies that would prevent shedding by direct interaction with the activating ligand appears to be a promising approach (inhibition of proteases is not feasible due to their abundance). For example, therapy with a specific antibody against the α domain of MICA/B has shown promising results in this regard, resulting in increased levels of this activating ligand on the target cell surface and, thanks to the NKG2D receptor, to an overall increased NK cell responsiveness (Ben-Shmuel, Biber, & Barda-Saad, 2020; Ferrari de Andrade et al., 2020).

Tumor cells may also modify their MHC I repertoire (upregulation of inhibitory signals) to escape the recognition or express molecules preventing immunological synapse formation. As mentioned previously, activation of NK cells is driven by the overbalance of activating signals. However, tight cellular contact, involving the subsequent reorganization of many molecules, is needed to create a functional immune synapse at the interface of NK and target cell (Orange, 2008). This process is highly organized. It is initiated by the approach and cell recognition, promoted by the engagement of adhesive molecules and accumulation of receptors in supramolecular activation clusters (SMACs), resulting in cell polarization and exocytosis of lytic granules (Orange et al., 2003). An adhesive leukocyte function-associated molecule 1 (LFA-1, CD11a) recognizes the intercellular adhesion molecule 1 (ICAM-1, CD54) on target cells and initiates synapse formation as an early signal (Barber, Faure, & Long, 2004). Next, reorganization of the cytoskeleton is needed to promote cellular spreading, strengthening the cellular contact (Lagrue et al., 2013). A rearranged filament network also serves for the trafficking of vesicles containing perforin and granzymes toward the immune synapse, as they must be secreted in direct proximity of the target cell membrane.

To answer the question, which are the minimal requirements for NK cell activation, experiments with primary resting human NK cells and insect cells expressing activating ligands, as model cells substituting target cells, were performed, combining signals for polarization and degranulation (Bryceson, March, Ljunggren, & Long, 2006). Activating receptors NKG2D, DNAM-1, 2B4, CD2 can induce inside-out signaling needed for LFA-1 mediated adhesion, but only CD16 can induce degranulation on its own (Bryceson, Ljunggren, & Long, 2009). Only collective engagement of activating and co-activating receptors (as well as signaling molecules, cytoskeleton, and other players) results in a successful NK cell lytic strike.

4. NK cell-based immunotherapeutics

Cancer immunotherapy is a strategy to recruit endogenous anti-tumor mechanisms to restore the activity of immune cells against malignant cells (Farkona, Diamandis, & Blasutig, 2016). Several approaches have been developed so far, including classical boosting of adaptive mechanisms with anti-cancer vaccines (Yaddanapudi, Mitchell, & Eaton, 2013), adoptive cell transfer (ACT, administration of *ex vivo* re-educated immune cells)

(Hinrichs & Rosenberg, 2014), and modulation of immune response by delivering artificial activating signals or by blocking inhibitory signals (e.g., by mAbs) (Leach, Krummel, & Allison, 1996; Peipp et al., 2017; Shin, Park, & Choi, 2016). Some viruses have an oncolytic function and, thus, could be modified for use in immunotherapy (Tusell Wennier, Liu, & McFadden, 2012).

Although NK cells are tumor-killing effector lymphocytes, some tumor cells succeed in escaping. However, there are ways to restore NK cell activity and therapeutically modulate their natural function (Morvan & Lanier, 2016). The approaches of NK cell immunotherapy are schematically represented in Fig. 5. All strategies are based on patterns of NK cell recognition or implement principles of NK cell activation or inhibition. From the functional point of view, NK cell-based therapeutics focus on immune synapse stabilization, which promotes artificial recognition and develops into the functional lytic immune synapse (administration of polyspecific

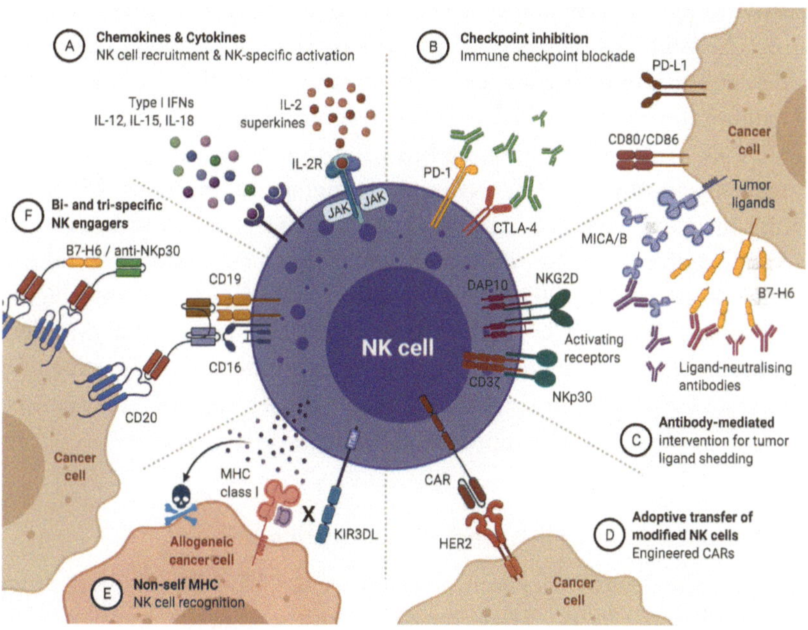

Fig. 5 Therapeutic approaches based on NK cell anti-tumor potential. (A) Activating NK cells and stimulating them to infiltrate the tumor site. (B) Restoring NK cell function by blocking inhibitory signals. (C) Scavenging soluble tumor ligands released by shedding to avoid the inhibition of NK cell function. (D) Transducing NK cells with engineered CARs. (E) Introducing mismatch between donor's KIRs and recipient MHC I. (F) Bridging the cellular interaction with bi- and tri-specific linkers.

linkers, mAbs, CAR-NK cells, NK cells modified with aptamers). However, the recognition is only effective when NK cells are active and infiltrate the tumor site, which may also be therapeutically modulated (administration of chemokines, interleukins, *ex vivo* activation of NK cells). Nowadays, NK cell-based immunotherapy is not just a concept, but there are already many therapeutics in different stages of clinical trials (Wu, Fu, Jiang, & Shao, 2020) and also private enterprises focusing mainly on harnessing NK cells for immunotherapy, such as Innate Pharma or Affimed (Garber, 2016; Khan, Arooj, & Wang, 2020).

Modern immunotherapeutics for NK cell activation are usually therapeutic proteins, beneficiating from the advantages of protein engineering. Such protein may contain several functional domains, creating not only poly-valent but also poly-competent molecules. This concept may be demonstrated by the TriKE concept (tri-specific killer engager) (Davis, Vallera, Miller, & Felices, 2017). This molecule combines NK-cell activating part (scFv targeting receptor CD16 and IL-15 responsible for NK cell priming directly *in situ*) with the tumor-targeting part (scFv specific for tumor antigens). The general drug design is usually based on the antibody recognition model, and antibody fragments are the most used functional domains, even for genetically modified NK cells expressing chimeric antigen receptors (CAR-NK). Ligands of activating NK cell receptors are another functional domain specifically targeting NK cells in therapeutic proteins. Moreover, affinity matured sequences of functional domains are used to increase the activation potential of the drug (e.g., engineered Fc fragment to increase ADCC (Klausz et al., 2020), affinity matured NK-cell activating ligands (Pekar et al., 2021)). The examples of drug design are described in detail within the following subchapters discussing the strategies for the generation of activated NK cells, therapeutic proteins harnessing NK cell cytotoxic functions, and genetically improved CAR-NK cells (Fig. 6).

4.1 Activated NK cell-based therapeutics

A classical therapeutic approach relies on using NK cells as "living drugs" (Shimasaki et al., 2020). These therapies use the adoptive transfer of NK cells, similar to the adoptive transfer of T cells, resulting in the administration of NK cells with enhanced anti-tumor properties. There are two types of such transfer, autologous (*in vitro* proliferation and activation of patient's own NK cells) or allogeneic (transfer of blood cells from allogeneic donors, haploidentical or MHC-matched) (Kundu, Gurney, & O'Dwyer, 2021).

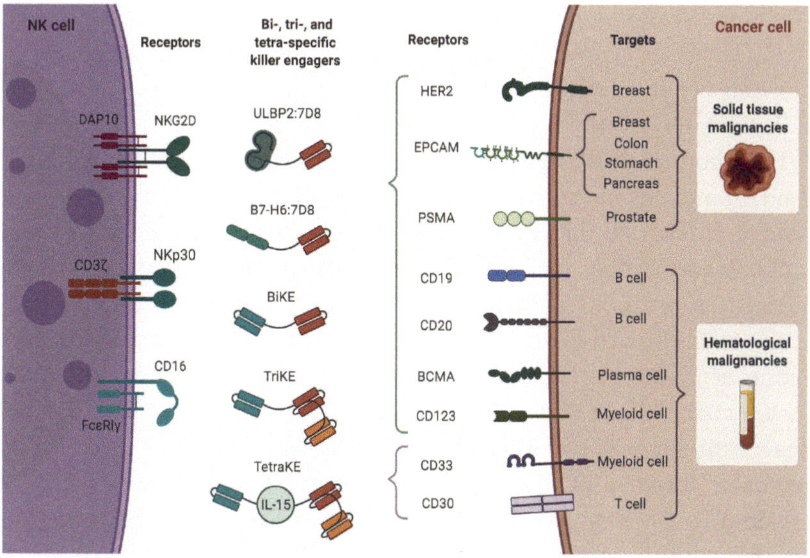

Fig. 6 Overview of the main NK cell activation receptors and tumor markers targeted by various experimental multi-specific recombinant therapeutic protein constructs. Tumor-associated antigens are targeted by one or two scFvs of the corresponding monoclonal antibodies (red and orange), while the activation receptors are ligated either by their natural ligands, such as ULBP2 for the NKG2D receptor or B7-H6 for the NKp30 receptor, or by the scFv of the corresponding antibody, such as anti-CD16 mAb.

In autologous transfer, the patient's blood cells are isolated, depleted from T cells, and the enriched NK cell population is expanded and activated *ex vivo* using IL-2 alone or combined with IL-12, IL-15, or IL-18. The K562 leukemic target cell line, well-susceptible to NK cell-mediated lysis, could be used as feeder cells to provide costimulatory signals. Depending on the origin of the cells (e.g., peripheral blood mononuclear cells—PBMCs, or cord-blood mononuclear cells) and the cytokines used to activate them, the resultant activated lymphocytes are called lymphokine-activated killers (LAKs) or cytokine-induced killers (CIKs). Featuring a mixed T- and NK cell-like phenotype, the mechanism of both LAK and CIK cells is distinctive from that of normal NK cells because they can lyse even cells that NK cells cannot, such as NK-resistant tumor cells. Activated cells are then administered back into the patient's peripheral blood (Liu, Meng, Zhang, Han, & Feng, 2021; Phan, Lee, Kim, & Cho, 2016; Rezaeifard et al., 2021). The undeniable advantage of the adoptive transfer of autologous NK cells is their natural source, which eliminates the need for immunosuppression in therapy due to the low risk of graft-versus-host disease

(Du, Guo, Wang, & Cui, 2021; Parkhurst, Riley, Dudley, & Rosenberg, 2011). *Ex vivo* manipulations are similar for allogeneic NK cells, but this approach benefits from the mismatch between donor KIRs and recipient HLA antigens, promoting immune reaction (Ruggeri, 2002).

In addition to NK cell lines (e.g., NK-92), peripheral and umbilical cord blood, or induced pluripotent stem cells (e.g., H9 hESC cell line) can also be a source of NK cells for expansion. These are differentiated by the action of cytokines and feeder cells into a uniform NK cell line that has high cytolytic activity (Du et al., 2021; Woll et al., 2009). A standard approach based on cytokines is the activation of NK cells *ex vivo* using native IL-2/15/18 (Leong et al., 2014; Ni, Miller, Stojanovic, Garbi, & Cerwenka, 2012; Romee et al., 2016) to increase and prolong the expression of CD25, which engages the high-affinity IL-2 receptor complex. The latter is sensitive to picomolar amounts of IL-2. Such pre-activated NK cells are suitable for low-dose IL-2 therapy, in which they preferentially proliferate and retain the capacity for increased stimulation for up to several weeks (Ni et al., 2012). Moreover, low-dose IL-2 is synergistic with IL-12 in stimulating INF-γ production (Leong et al., 2014). Recently, cyclic dinucleotides or CpG oligonucleotides serving as TLR ligands have been combined with a half-life extended IL-2 superkine H9 (also known as super-2) fused to human serum albumin (H9-MSA) to improve its serum half-life and reduce toxicity. In comparison to monotherapy, the combination therapy induced long-term remissions of both MHC I^+ and MHC I-deficient tumor transplant models by providing more prolonged and more intense NK cell activation, cytotoxicity, and expression of cytotoxic effector molecules (Wolf et al., 2021).

Analogous to T-cell therapy, the enhanced effects of chimeric immunocytokines are also being explored for NK cells. An example is ALT-803, a chimeric protein that finds its use both *in vivo* and *ex vivo*. It consists of two subunits of a human IL-15 analog that carries the N72D mutation, bound to a dimeric human IL-15RαSu on the human IgG1 Fc fragment. The aforesaid mutation on IL-15 increases its biological activity (Zhu et al., 2009). The whole complex also has a nearly 40-fold prolonged half-life (Han et al., 2011) and better distribution *in vivo* (Alter, Rhode, Jeng, & Wong, 2018; Rhode et al., 2016). It is worth mentioning that ALT-803 can also serve as a scaffold for further fusion with targeting antibodies, e.g., anti-CD20, and promote Fc fragment-mediated ADCC by activating NK cells through IL-15 at the same time (Rosario et al., 2016). A further strategy to stimulate NK cell activation independently on the high-affinity IL-2 receptor is using the so-called IL-2 superkine created by yeast-display affinity maturation to have a higher affinity for CD122, through which it signals toward robust proliferation not dependent on CD25 expression (Levin et al., 2012).

Another way to promote NK cell proliferation is, as already mentioned, the use of feeder cells. The K562 line is often used, which can be modified to express surface markers that further enhance the activation and expansion of the targeted NK cell line. For example, expressing a membrane-bound form of IL-15 and 41BB ligand (K562-mb15-41BBL) in K562 cells resulted in a 21.6-fold expansion of $CD56^{+}CD3^{-}$ NK cells isolated from peripheral blood (Fujisaki et al., 2009). Even better results were achieved using K562 feeder cells that expressed the OX40 ligand while being simultaneously briefly exposed to IL-21. Such pretreated cells achieved a 2000-fold expansion after 4 weeks of culture, compared to a 303-fold expansion achieved using the conventional K562 cells, while retaining similar surface receptor expression, ADCC cytotoxicity, cytokine secretion, and CD107 degranulation (Kweon et al., 2019). In another study, a feeder-based NK cell expansion approach utilizing OX40 ligand-expressing NK-92 cells secreting simultaneously neoleukin-2/15 (a hyper-stable mimetic with a high affinity to IL-2Rβγ) has been used. These modified NK-92-based feeder cells induced a 2180-fold expansion of NK cells that subsequently proved to be highly cytotoxic against Raji cells and several solid tumors *in vivo* (Guo et al., 2021). Other feeder cell variants include, for example, Jurkat cell lines (Lim et al., 2013) and Epstein-Barr virus-transformed lymphoblastoid cell lines (Granzin et al., 2016). Also of note is the newly established feeder $PC3^{PSCA}$ cell line expressing IL-2, 4-1BBL, and IL-15-mutDAP12 (mIL-15d) fusion protein. Such expanded NK cells exhibited upregulation of natural cytotoxicity receptors, DNAM-1 and NKG2C, and induced expression of the high-affinity IL-2 receptor, which were paralleled by attenuated KIR and increased NKG2A and ILT2 expression (Michen et al., 2020). The "NKF" feeder cells based on leukemia cell line OCI-AML3 transduced with mbIL-21 developed by Ojo *et al.* induce robust proliferation of NK cells that possess high cytotoxicity against various blood cancer and solid tumor cells. Membrane-bound IL-21 leads to activation of a STAT3/c-Myc pathway and increased NK cell metabolism with a shift toward aerobic glycolysis (Ojo et al., 2019). Generally, a similar approach to using feeder cells is priming with cell lysate obtained from, e.g., the CTV-1 leukemia cell line (Fehniger et al., 2018). Yet another method to overcome the low proliferative potential of NK cells is combining feeder cells with transduction of the expanded cells with telomerase reverse transcriptase (hTERT). These engineered NK cell cultures maintain a high percentage of cells in the S/G2 phase for an extended time after transduction. Data show that ectopic hTERT expression in NK cells enhances their activation and proliferation and extends their life span *in vitro* (Streltsova et al., 2021).

A considerable challenge for the adoptive transfer of NK cells is that only a fraction of NK cells reaches their target. A solution to overcome this could be found in nanotechnologies, such as magnetic nanoparticles targeting NK cells to the tumor site. For example, magnetic nanoparticles have been attached to the surface of NK cells using polymers (Sanz-Ortega, Rojas, Portilla, Perez-Yague, & Barber, 2019). It should then be possible to enhance tumor infiltration using an external magnetic field. Such mediated accumulation in the tumor has not yet been confirmed *in vivo*, but other advantages of this approach have been found, such as potentially reduced metastasis spreading (Sanz-Ortega, Portilla, Perez-Yague, & Barber, 2019). The use of magnetic biomimetic tumor cells is particularly interesting. These are magnetic nanoparticles coated with membranes bearing tumor markers, which can be used to activate NK cells (Wu et al., 2021). Another innovative approach is based on *ex vivo* labeling of NK cells using aptamers (single-strand oligonucleotide binders with specificity similar to antibodies) that attach to NK cells through a membrane anchor. The aptamer then provides recognition of a specific molecule on the target tumor cell, e.g., CD30 on CD30-expressing lymphomas (Yang et al., 2019).

4.2 Recombinant protein-based NK cell therapeutics

To assure self-tolerance, immune response activation is tightly regulated by robust inhibitory mechanisms, so-called immune checkpoints. However, malignant cells may upregulate the expression of checkpoint inhibitory receptor ligands and thus silence the immune response (Davis et al., 2017). NK cells share common checkpoint receptors with other lymphocytes (PD-1, CTLA-4, TIGIT, NKG2A/CD94, TIM-3, Lag-3, some KIR inhibitory receptors) (Chauhan, Koehl, & Kloess, 2020; Nayyar, Chu, & Cairo, 2019). A common approach to silence these receptors is the administration of mAbs, neutralizing these receptors or their ligands (or both) and preventing their interaction. The advantage is targeting NK cells and cytotoxic T-cells at the same time as well as blocking further inhibitory immune cell cross-talk.

Examples of mAbs targeting immune checkpoints are anti-PD-1 pembrolizumab (Kwok, Yau, Chiu, Tse, & Kwong, 2016), anti-PD-L1 avelumab (Gil-Bazo, 2017), anti-NKG2A monalizumab (André et al., 2018), or lirilumab blocking the interaction of KIR2DL1/L2/L3 with its ligands (Yalniz et al., 2018). This therapy may also benefit from the synergic effect targeting two checkpoints, as shown in the case of KLRG1 and PD-1 combination blockade (Tata et al., 2021). The use of mAb to block inhibitory signals is not limited only to checkpoint receptors. Antibodies may also be used to neutralize immune-suppressive modulators within the tumor microenvironment, as is the case of anti-TGF-β fresolimumab (de Gramont, Faivre, & Raymond, 2017).

Of course, NK cells may be alerted not only by blocking the inhibition but also directly by stimulation of activating pathways. The most effective activating pathway starts with the recruitment of the receptor CD16 (also known as FcγRIIIa or CD16a), which binds the Fc region of IgG antibodies and triggers ADCC (Shevtsov & Multhoff, 2016). Thus, NK cells can recognize cells decorated with antibodies produced by reactive B-lymphocytes or administrated therapeutically. Approved monoclonal antibodies targeting prototypical tumor antigens such as HER2 (trastuzumab), EGFR (cetuximab), CD20 (rituximab), and many others also contribute to NK cell stimulation, provoking ADCC (Kohrt, Rajasekaran, Chester, Yonezawa, & Zhao, 2015). Moreover, the Fc portion of antibodies may be modified, introducing mutations, or altering its N-glycosylation to increase the affinity (Abes & Teillaud, 2010; Saunders, 2019).

Another approach to target specific NK cell ligands on the surface of cancer cells is the fusion of the Fc-domain with the extracellular part of NK cell receptors. The receptor part assures specificity, and the Fc part activates NK cells (or other $CD16^+$ cytotoxic cells). NKp30-Fc fusion protein inhibited tumor growth in mice, resulting in complete tumor removal in 50% of cases (Arnon et al., 2008). A similar construct, NKp80-Fc, was successfully used to treat acute myeloid leukemia (Deng et al., 2015).

Fc fragment is a universal activating ligand for CD16. However, CD16 can also be targeted and activated by specific anti-CD16 antibodies or related scFv fragments (Felices, Lenvik, Davis, Miller, & Vallera, 2016). Moreover, the tools of recombinant expression enable the combination of two or more scFvs in one molecule, resulting in bi-, tri-, or poly-specific synthetic antibodies. Fragment of anti-CD16 antibody is, in fusion with antibody fragment targeting tumor-associated antigens, part of a bispecific killer engager (BiKE). These molecules are designed to strengthen the interaction between NK cells and target cells. Such tumor antigen is the CD33, a marker of myeloid-derived suppressor cells promoting the progress of myelodysplastic syndrome toward acute leukemia (Gleason et al., 2014) or HER2, breast carcinoma cell marker, targeted with tribody (three binding epitopes—two anti-HER2 scFv and one anti-CD16 scFv in a single molecule) (Oberg et al., 2018). These molecules are bispecific, genetically engineered on the antibody platform. The activity of BiKE was improved by the insertion of a third targeting moiety and the generation of a tri-specific killer engager (TriKE). Such moiety might be another scFv or IL-15, which results in cytokine boosting of NK cell activity at the right time and place (Cheng et al., 2020; Sarhan et al., 2018; Schmohl, Felices, Taras, Miller, & Vallera, 2016). Construction of a tetraspecific engager (TetraKE), a fusion protein consisting of anti-CD16 scFv, IL-15, anti-EpCAM scFv, and anti-CD133 scFv, is also possible and induces NK degranulation *in vitro* (Schmohl et al., 2016).

Synthetic molecules built from fragments of monoclonal antibodies designed to activate NK cell immune functions toward cancer are generally termed natural killer cell engagers (NKCEs). This group currently shows promising results in preclinical or even subsequent clinical trials (Demaria et al., 2019). Several biopharmaceutical companies developed unique high-throughput strategies (platforms) to prepare polyvalent molecules combining multiple functional scFvs (Demaria, Gauthier, Debroas, & Vivier, 2021; Myers & Miller, 2021). Some examples of such platforms are the already mentioned BiKEs and TriKEs (bi- and tri-specific killer engagers by GT Biopharma), TriNKETs (tri-specific NK cell engager therapies by Dragonfly Therapeutics), ROCK (redirected optimized cell-killing antibody platform by Affimed), ANKET (antibody-based NK cell engager technology by Innate Pharma), or StitchMabs (by Compass Therapeutics).

Therapeutic proteins based on the ROCK platform are built on different antibody formats, using two anti-CD16 scFv domains and at least one targeting domain to create bispecific binders for harnessing innate immunity (Ellwanger et al., 2019). There are generally four different ROCK families—Fc-less ROCK (dimeric proteins consisting only of variable domains in bivalent or trivalent format), Fc fusion ROCK (with optimized Fc fragment), IgG-like ROCK (human IgG molecules extended with anti-CD16 scFv), and antibody binding fragment (Fab) fusion ROCK (extended Fab molecules). An example of Fc-less ROCK is the molecule AFM13, tetravalent bispecific CD30/CD16a tandem diabody (TandAb), a promising therapeutic for the treatment of Hodgkin's lymphoma (Reusch et al., 2014). Ig-G-like ROCK molecule AFM24 (IgG_1-scFv fusion antibody; anti-CD16 IgG_1 and anti-EGFR scFv) demonstrated binding to EGFR-expressing tumor cells (regardless of the expression level and KRAS/BRAF mutational status) and was highly potent in induction of ADCC in NK cells (Wingert et al., 2021).

A prototypical ANKET molecule activating NK cells via receptor NKp46 (NKp46-ANKET) consists of Fab targeting NKp46, Fab directed against a tumor antigen, and Fc portion binding CD16. These molecules activate NK cells using two distinct synergic activation pathways without additional signs of cytotoxicity (Gauthier et al., 2019). Further, NKp46 expression is often conserved on infiltrating NK cells in most solid tumors.

The StitchMabs platform by Compass Therapeutics is creating novel antibody-based therapeutics by covalent linking of two or more mAbs (generated with a common light chain) or their fragments using recombinant expression of a whole protein or joining with complementary peptide tags. An example of NKCE built on this platform are molecules CTX-4419 (Draghi et al., 2019) and CTX-8573 (Watkins-Yoon et al., 2019) designed for the treatment of multiple myeloma by targeting BCMA (B-cell maturation antigen). Both activate NK cells by engaging CD16 with the Fc fragment, part of the molecule within the antibody frame, and anti-NKp30 scFv. NKp30 recruitment is sufficient for NK cell activation, which may be beneficent in CD16 activity loss due to receptor shedding or downregulation in the tumor microenvironment. CTX-8573 engager was constructed by appending two common-light chains compatible anti-NKp30 Fab fragments on the C-terminus of anti-BCMA IgG_1 antibody with afucosylated Fc portion.

Although mAbs fragments in NKCEs usually target CD16, activating receptors NKG2D and NKp30 are plausible alternatives. A bispecific antibody (biAb, prepared by fusion of two scFvs) CS1-NKG2D recruits NKG2D-positive cytotoxic cells (not only NK cells, but also $CD8^+$ T cells, $\gamma\delta$ T cells, and NKT cells) toward multiple myeloma cells (Chan et al., 2018). Bispecific Fab-like antibody anti-NKG2D $\times$ HER2 also exhibits anti-tumor activity. Moreover, this molecule does not compete with NKG2D ligand binding (Raynaud et al., 2020).

Combining monoclonal antibody fragments into polyspecific molecules is not the only way to reestablish and boost NK cell immune recognition. Fusion proteins containing soluble NK cell receptor ligands to trigger NK cells, or proteins utilizing NK cell receptor ectodomains to recognize tumor cells, are also an emerging group of promising therapeutics, as in the case of molecules based on a major NK cell activation receptor NKG2D and its ligands, jointly termed NKG2DL (Ding, Yang, & Wei, 2018). For example, NKG2D in fusion with Fc fragment, anti-CD16 scFv, or anti-CD3 scFv enhance the cytotoxic activity of $CD16^+$ cells or the number of effector T cells, respectively (Märklin et al., 2019). MICA, an activating NKG2D ligand, in fusion with CD24 targeting antibody mediates anti-tumor efficacy toward hepatocellular carcinoma, both *in vitro* and *in vivo* (Han et al., 2019). Another example of MICA fusion protein is 2A9-MICA, targeting BCMA expressed on malignant plasma cells in case of multiple myeloma (Wang et al., 2020). This malignancy could also be targeted by ULBP2:BB4, bispecific protein engaging NKG2D by its ligand UL-16 binding protein 2 (ULBP2) and CD138 antigen via scFv of the BB4 mAb (von Strandmann et al., 2006).

Engaging NKG2D or the NCR activation receptor NKp30 with ULBP2:7D8 or B7-H6:7D8 fusion proteins (where 7D8 is an scFv targeting CD20), respectively, also resulted in increased NK cell cytotoxicity, which was synergistically augmented when combined with anti-CD20 mAb rituximab or anti-CD38 mAb daratumumab (Kellner et al., 2016). Artificial coating of lymphoma cells with the NKp30-ligand B7-H6 results in increased NK cell cytotoxicity (Kellner et al., 2012). In the targeted therapy for $HER2^+$ breast cancer, fusion proteins B7-H6:HER2-scFv (activating NKp30 receptor), AICL:HER2-scFv (activating NKp80 receptor), and PVR:HER2-scFv (activating DNAM-1 receptor) all enhanced NK cell cytotoxicity (Peipp et al., 2015). This strategy was recently improved using an affinity-matured version of the B7-H6 ligand, firmly binding the NKp30 receptor, which replaced one arm of anti-EGFR mAb cetuximab (Pekar et al., 2021). In another recent study, ULBP2: HER2-scFv fusion protein triggered NK cell-mediated killing of $HER2^+$ breast cancer cells and synergistically sensitized these tumor cells for ADCC mediated by cetuximab simultaneously targeting EGFR as a second tumor marker (Kellner et al., 2021). Such a dual-targeting strategy that combines antibodies with fusion constructs targeting another tumor-associated antigen and engaging NKG2D as a second NK cell trigger molecule shows promising therapeutic potential. The resultant ADCC response may be enhanced even further by employing mAbs with glyco- and protein-engineered Fc fragment (Kellner et al., 2018; Mrázek et al., 2013; Rosskopf et al., 2020).

The approach to enhance NK cell-mediated cytotoxicity is complex and comprises a variety of fusion proteins that combine different functional segments (e.g., soluble NK cell receptors to target cancer cells selectively, soluble NK cell-activating ligands, scFv fragments of mAb, Fc portion of mAbs, cytokines). These proteins are usually bivalent, but polyvalent molecules created by the functionalization of biocompatible materials are also a strategy. Such molecules are referred to as nano-engagers (NE) (Ma, Xu, & Song, 2021). NE based on biocompatible poly(ethylene glycol)–*block*-poly(lactide-co-glycolide) (PEG-PLGA) functionalized with anti-CD16, anti-EGFR cetuximab, and anti-4-1BB antibodies to activate NK cells effectively inhibit EGFR-overexpressed cancer growth *in vivo*. Moreover, a chemotherapeutics epirubicin may be encapsulated within NE to release it in the proximity of $EGFR^+$ tumor cells (Au, Park, & Wang, 2020). Polymeric carriers might also be used to broaden the repertoire of NK cell enhancers as another versatile platform for assembling polyvalent particles using specific anchors. An example of such a therapeutic polymer is a biocompatible N-(2-hydroxypropyl)methacrylamide copolymer (pHPMA) carrier modified with coiled-coil peptide anchors, decorated with NK-cell activating ligand B7-H6 via complementary coiled-coil sequences (Kalousková et al., 2021). In general, pHPMA-based polymeric carriers show an excellent bioavailability in human plasma (Janisova et al., 2020) and are being tested and developed for therapeutic drug delivery in many settings, while the coiled-coil motif has been previously successfully utilized for the attachment of various mAb scFvs to these polymeric carriers (Pechar et al., 2011, 2014). The use of polymeric carriers represents an attractive, versatile alternative to the production of multi-specific protein constructs.

Large parts of mAbs may be replaced with smaller fragments; for example, nanobodies (Nbs, VHHs) derived from camelid antibodies, consisting only of the heavy-chain variable domain of IgG2 and IgG3, with the size of about 14 kDa (de Marco, 2020). Nanobody-based engagers targeting carcinoembryonic antigen (CEA) and CD16a, both with VHH fragments, inhibit tumor growth *in vivo* (Zhao et al., 2020). Nbs may also be used as immune checkpoint targeting molecules, equally to the full-size mAbs (Yu et al., 2021). Nbs broaden the repertoire of immunotherapeutic molecules when used independently or in fusion protein constructs (in combination with other Nb, scFv, mAbs, or activating ligands).

4.3 Gene-modified NK cell-based therapeutics

Adoptive cell therapy (ACT) using genetically engineered immune cells is at the current forefront in cancer immunotherapy. ACT relies on the modulation of endogenous immune mechanisms targeting highly specific tumor antigens. One way to achieve this is through the advanced engineering of immune cells—most commonly T lymphocytes but also NK cells—to express chimeric antigen receptors (CARs). CAR constructs typically consist of an extracellular antigen-recognition domain, which is usually a single-chain variable fragment, a flexible hinge linked to the transmembrane domain, and an intracellular signaling domain, commonly consisting of costimulatory domains and the CD3ζ chain of T lymphocytes (Pfefferle & Huntington, 2020). First-, second-, and third-generation CAR constructs differ solely in the number of stimulatory intracellular domains—from one to three, respectively. Third-generation CARs containing two costimulatory domains and a CD3ζ domain have demonstrated enhanced anti-tumor activity in comparison with constructs of previous generations, greatly due to their promotion of cytokine release, enhancement of immune cell survival, and overall increased tumor elimination (Zhong, Matsushita, Plotkin, Riviere, & Sadelain, 2010). T cells redirected for universal cytokine-mediated killing (TRUCKs), or fourth-generation CAR constructs, maintain the core structure from previous generations but with the additional engineering of a constitutive or inducible expression cassette for a transgenic product, such as a cytokine (e.g., IL-12), aimed at promoting a favorable microenvironment for the recruitment of other immune cells to the tumor tissue (Chmielewski & Abken, 2015).

A range of other next-generation constructs has been developed to address the known limitations of CAR-based immunotherapies. Excessive immunogenicity and toxicity, oftentimes presenting itself in the form of life-threatening cytokine release syndrome, have been addressed by developing ON-switch CARs that offer titratable, reversible, and temporary control of CAR activity via drug administration. ON-switch CAR constructs follow a split-receptor design where key components (extracellular antigen-binding domain and ITAM-containing CD3ζ domain) are physically separate and can only interact in the presence of a heterodimerizing small molecule, thus assuring the immune cell activation solely occurs in the event of ligand recognition and presence of an activating drug (Wu, Roybal, Puchner, Onuffer, & Lim, 2015). Tandem- and dual-CARs have addressed challenges in tumor escape, such as target antigen loss or downregulation, by simultaneously targeting different tumor markers. Multiple targeting of tumor cells can be achieved through various strategies; namely, simultaneous transduction of immune cells with two CAR constructs with distinct antigen-recognition domains; use of polycistronic vectors encoding two different CARs; or a single CAR construct with two differing scFv domains (Tong et al., 2020; van der Schans, van de Donk, & Mutis, 2020). The intrinsic ability of cancer cells to evade immune surveillance and promote an immunosuppressive microenvironment is a major challenge to cancer immunotherapy (Vinay et al., 2015). Inhibitory CARs provide a promising approach to these obstacles. The first designed inhibitory CARs combined the surface antigen recognition domain with the signaling domain of immune inhibitory receptors, such as CTLA-4 and PD-1, to limit cytokine secretion, cytotoxicity, and proliferation while increasing discrimination between on- and off-target cytotoxicity (Fedorov, Themeli, & Sadelain, 2013). Other novel strategies include the disruption of PD1 signaling through rational CAR design (Liu et al., 2016) and the genetic editing of CAR immune cells with CRISPR/Cas9 technology to disrupt T cell inhibitory receptors (Choi et al., 2019).

CAR-T immunotherapy has shown promising results in the treatment of lymphoid malignancies, especially with CARs targeting the CD19 surface antigen (Liu et al., 2020; Romanski et al., 2016). Thus far, the US Food and Drug Administration (FDA) has approved the commercial use of four therapies: tisagenlecleucel (Kymriah, Novartis), axicabtagene ciloleucel (Yescarta, Kite Pharma), brexucabtagene autoleucel (KTE-X19, Tecartus), and lisocabtagene maraleucel (Juno Therapeutics) (Daher, Melo Garcia, Li, & Rezvani, 2021). Despite the ever-increasing progress and proven success of CAR-T cell-based cellular therapies, many challenges are still faced regarding its safety—a direct result of inherent properties of T lymphocytes—thus prompting a shift toward the CAR engineering of NK cells. The unique biological features of NK cells—i.e., ability to target cells lacking self-antigens, directed cytotoxicity, cytokine secretion, and limited lifespan in circulation—allow for the combination of established CAR engineering with the safety of NK cells. CAR engineered NK cells can promote tumor elimination via three mechanisms: (1) CAR-dependent cytotoxicity; (2) CAR-independent cytotoxicity through innate NK cell activity (e.g., NCRs, NKG2D, activation KIRs); and (3) CD16-mediated antibody-dependent cellular cytotoxicity (ADCC), thus effectively targeting heterogeneous tumors through a combination of mechanisms

(Xie et al., 2020). Despite the promising characteristics of NK cell-based immunotherapies, challenges are encountered, namely regarding NK cell resistance to genetic engineering, limited *in vivo* persistence, and limited proliferative potential (Pfefferle & Huntington, 2020). These challenges have been extensively addressed; for example, by using NK cell lines such as NK-92 or iPSC-derived NK cells, the difficulties in NK cell expansion and genetic engineering were mitigated (Bachiller, Battram, Perez-Amill, & Martín-Antonio, 2020; Li, Hermanson, Moriarity, & Kaufman, 2018; Mehta & Rezvani, 2018). Additional approaches have been developed to overcome the difficulties encountered with NK cell engineering, namely, the use of electroporation (Ingegnere et al., 2019), transposons (Batchu et al., 2019; Li et al., 2018), and viral transduction of NK cells (Colamartino et al., 2019).

The field of CAR NK cell immunotherapy is in rapid progression due to the continuous development of new approaches to old problems. Innovative strategies include the development of an anti-CD19-CAR construct engineered with inducible caspase 9 (i.e., a suicide switch for prompt interruption of the therapy in the event of adverse effects) and interleukin-15 (for enhancement of NK cell expansion and persistence) (Liu et al., 2020); the blockade of CAR ubiquitination, followed by downmodulation and lysosomal degradation, through the mutation of all lysine residues in the CAR cytoplasmic domain leading to more effective long-term tumor elimination and improved *in vivo* persistence (Li et al., 2020); the increased efficacy of CAR-immune cell constructs in the treatment of solid tumors through the use of stringent hypoxia-sensing CARs (Kosti et al., 2021); and the optimization of differentiation, proliferation, and anti-tumor potency of CAR-immune cells through epigenetic programming via BET bromodomain protein inhibition (Kong et al., 2021).

CAR NK cell immunotherapy development has been expedited through the many advances in CAR T cell therapies, as evidenced by the 12 presently recruiting and five not-yet-recruiting ongoing CAR NK cell clinical trials. Other opportunities might open when novel concepts in T cell immunotherapy, such as the design of TCR-like chimeric receptors, will be employed for NK cells to create CAR NK-T cells. In this direction, it has been shown that replacing the ζ-chain with the ε-chain of the TCR complex improves CAR function (Salter et al., 2021) and that TCR-like chimeric receptors improve on the basic CAR design, for example, by fusing the scFv directly to the extracellular domain of CD3ε (termed TruC for T cell receptor fusion constructs) (Baeuerle et al., 2019) or by replacing the variable domains of the TCR with those in the scFv (termed STAR for synthetic T cell receptor and antigen receptor) (Liu et al., 2021). The latter construct showed superior antigen sensitivity compared to standard CAR designs due to its efficient use of accessory adhesion receptors suggesting that simultaneous ligation of adhesion receptors such as CD2 or LFA-1 would increase the sensitivity of the previous CAR designs as well (Burton et al., 2021). Thus, the full potential of NK cells for adoptive cell therapy can only be harnessed when the molecular and biological mechanisms of NK cell activation are considered during the design of the chimeric antigen receptor constructs.

5. Concluding remarks

NK cell immunotherapy of cancer holds great promise for future research and translation from the laboratory to clinical practice. The detailed understanding of the structural features and mechanisms of NK cell receptor: ligand interactions, and NK cell recognition in general, will enable rational development of biopharmaceuticals and cellular therapies fully exploring the cytotoxic potential of NK cells. While our knowledge of individual receptor:ligand complexes behavior in solution is quite substantial, elucidation of receptor dynamics on NK cell membrane, e.g., receptor oligomerization or clustering, and of signal transduction upon ligand binding may provide further possibilities for enhancing NK cell activity by careful design of multi-specific or polyvalent therapeutic proteins. Similarly, because NK cell integrates the incoming inhibitory and activation signals received through its many surface receptors, such multifunctional constructs have the potential to overcome the tumor immunosuppressive microenvironment inhibitory signaling by providing a combinatorial activation signal strong enough to trigger the desired anti-tumor cytotoxic response, as already evidenced for several successful examples discussed above. At the same time, NK cells themselves present a potent anti-tumor drug when properly stimulated and activated. The advances in T cell and NK cell gene manipulation, resulting in the creation of several successive generations of chimeric antigen receptors and TCR-like receptors, are the most recent and probably also the most promising for fast translation into the clinics and for the benefit of the patients. However, the potential of employing the cell-based and protein-based approaches into single combined, personalized immunotherapy remains so far largely unexplored.

Acknowledgments

This work was supported by the Czech Science Foundation (18-10687S), the Ministry of Education, Youth and Sports of the Czech Republic (LTC20078), and the Charles University (GAUK 1436220). Figures were created with BioRender.com.

References

Abes, R., & Teillaud, J. L. (2010). Impact of glycosylation on effector functions of therapeutic IgG. *Pharmaceuticals*, *3*(1), 146–157.

Alter, S., Rhode, P. R., Jeng, E. K., & Wong, H. C. (2018). Targeted IL-15-based protein fusion complexes as cancer immunotherapy approaches. *Journal of Immunological Sciences*, *2*(1), 15–18.

André, P., Denis, C., Soulas, C., Bourbon-Caillet, C., Lopez, J., Arnoux, T., et al. (2018). Anti-NKG2A mAb is a checkpoint inhibitor that promotes anti-tumor immunity by unleashing both T and NK cells. *Cell*, *175*(7), 1731–1743.

Apte, R. S., Sinha, D., Mayhew, E., Wistow, G. J., & Niederkorn, J. Y. (1998). Cutting edge: Role of macrophage migration inhibitory factor in inhibiting NK cell activity and preserving immune privilege. *Journal of Immunology*, *160*(12), 5693–5696.

Arnon, T. I., Markel, G., Bar-Ilan, A., Hanna, J., Fima, E., Benchetrit, F., et al. (2008). Harnessing soluble NK cell killer receptors for the generation of novel cancer immune therapy. *PLoS One*, *3*(5), e2150.

Artis, D., & Spits, H. (2015). The biology of innate lymphoid cells. *Nature*, *517*(7534), 293–301.

Au, K. M., Park, S. I., & Wang, A. Z. (2020). Trispecific natural killer cell nanoengagers for targeted chemoimmunotherapy. *Science Advances*, *6*(27), eaba8564.

Bachiller, M., Battram, A. M., Perez-Amill, L., & Martín-Antonio, B. (2020). Natural killer cells in immunotherapy: Are we nearly there? *Cancers*, *12*(11), 3139.

Baeuerle, P. A., Ding, J., Patel, E., Thorausch, N., Horton, H., Gierut, J., et al. (2019). Synthetic TRuC receptors engaging the complete T cell receptor for potent anti-tumor response. *Nature Communications*, *10*(1), 2087.

Barber, D. F., Faure, M., & Long, E. O. (2004). LFA-1 contributes an early signal for NK cell cytotoxicity. *Journal of Immunology*, *173*(6), 3653–3659.

Bartel, Y., Bauer, B., & Steinle, A. (2013). Modulation of NK cell function by genetically coupled C-type lectin-like receptor/ligand pairs encoded in the human natural killer gene complex. *Frontiers in Immunology*, *4*(11), 362.

Bashirova, A. A., Martin, M. P., McVicar, D. W., & Carrington, M. (2006). The killer immunoglobulin-like receptor gene cluster: Tuning the genome for defense. *Annual Review of Genomics & Human Genetics*, 7, 277–300.

Batchu, R. B., Gruzdyn, O. V., Tavva, P. S., Kolli, B. K., Dachepalli, R., Weaver, D. W., et al. (2019). Engraftment of mesothelin chimeric antigen receptor using a hybrid sleeping beauty/minicircle vector into NK-92MI cells for treatment of pancreatic cancer. *Surgery*, *166*(4), 503–508.

Beatty, G. L., & Gladney, W. L. (2015). Immune escape mechanisms as a guide for cancer immunotherapy. *Clinical Cancer Research*, *21*(4), 687–692.

Ben-Shmuel, A., Biber, G., & Barda-Saad, M. (2020). Unleashing natural killer cells in the tumor microenvironment-the next generation of immunotherapy? *Frontiers in Immunology*, *11*, 275.

Bláha, J., Kalousková, B., Skořepa, O., Pažický, S., Novák, P., & Vaněk, O. (2017). High-level expression and purification of soluble form of human natural killer cell receptor NKR-P1 in HEK293S GnTI^{-} cells. *Protein Expression and Purification*, *140*, 36–43.

Bláha, J., Skálová, T., Kalousková, B., Skořepa, O., Cmunt, D., Pažický, S., et al. (2021). Crystal structure of the human NKR-P1 bound to its lymphocyte ligand LLT1 reveals receptor clustering in the immune synapse. *bioRxiv*. https://doi.org/10.1101/2021.06.16.448687.

Blum, K. S., & Pabst, R. (2007). Lymphocyte numbers and subsets in the human blood. Do they mirror the situation in all organs? *Immunology Letters*, *108*(1), 45–51.

Bossi, G., & Griffiths, G. M. (1999). Degranulation plays an essential part in regulating cell surface expression of Fas ligand in T cells and natural killer cells. *Nature Medicine*, *5*(1), 90–96.

Boyton, R. J., & Altmann, D. M. (2007). Natural killer cells, killer immunoglobulin-like receptors and human leucocyte antigen class I in disease. *Clinical and Experimental Immunology*, *149*(1), 1–8.

Brandt, C. S., Baratin, M., Yi, E. C., Kennedy, J., Gao, Z., Fox, B., et al. (2009). The B7 family member B7-H6 is a tumor cell ligand for the activating natural killer cell receptor NKp30 in humans. *Journal of Experimental Medicine*, *206*(7), 1495–1503.

Bryceson, Y. T., Ljunggren, H.-G., & Long, E. O. (2009). Minimal requirement for induction of natural cytotoxicity and intersection of activation signals by inhibitory receptors. *Blood*, *114*(13), 2657–2666.

Bryceson, Y. T., March, M. E., Ljunggren, H.-G., & Long, E. O. (2006). Activation, co-activation, and co-stimulation of resting human NK cells. *Immunological Reviews*, *29*, 997–1003.

Burton, J., Siller-Farfán, J. A., Pettmann, J., Salzer, B., Kutuzov, M., van der Merwe, P. A., et al. (2021). Inefficient exploitation of accessory receptors reduces the sensitivity of chimeric antigen receptors. *bioRxiv*. https://doi.org/10.1101/2021.10.26.465853.

Campbell, K. S., & Purdy, A. K. (2011). Structure/function of human killer cell immunoglobulin-like receptors: Lessons from polymorphisms, evolution, crystal structures and mutations. *Immunology*, *132*(3), 315–325.

Cerwenka, A., & Lanier, L. L. (2016). Natural killer cell memory in infection, inflammation and cancer. *Nature Reviews Immunology*, *16*(2), 112–123.

Chan, W. K., Kang, S., Youssef, Y., Glankler, E. N., Barrett, E. R., Carter, A. M., et al. (2018). A CS1-NKG2D bispecific antibody collectively activates cytolytic immune cells against multiple myeloma. *Cancer Immunology Research*, *6*(7), 776–787.

Chauhan, S. K. S., Koehl, U., & Kloess, S. (2020). Harnessing NK cell checkpoint-modulating immunotherapies. *Cancers*, *12*(7), 1807.

Chen, Y., Mo, J., Jia, X., & He, Y. (2018). The B7 family member B7-H6: A new bane of tumor. *Pathology and Oncology Research*, *24*(4), 717–721.

Cheng, Y., Zheng, X., Wang, X., Chen, Y., Wei, H., Sun, R., et al. (2020). Trispecific killer engager 161519 enhances natural killer cell function and provides anti-tumor activity against CD19-positive cancers. *Cancer Biology and Medicine*, *17*(4), 1026–1038.

Chiffoleau, E. (2018). C-type lectin-like receptors as emerging orchestrators of sterile inflammation represent potential therapeutic targets. *Frontiers in Immunology*, *9*, 227.

Chmielewski, M., & Abken, H. (2015). TRUCKs: The fourth generation of CARs. *Expert Opinion on Biological Therapy*, *15*(8), 1145–1154.

Choi, B. D., Yu, X., Castano, A. P., Darr, H., Henderson, D. B., Bouffard, A. A., et al. (2019). CRISPR-Cas9 disruption of PD-1 enhances activity of universal EGFRvIII CAR T cells in a preclinical model of human glioblastoma. *Journal for Immunotherapy of Cancer*, 7(1), 304.

Colamartino, A. B. L., Lemieux, W., Bifsha, P., Nicoletti, S., Chakravarti, N., Sanz, J., et al. (2019). Efficient and robust NK-cell transduction with baboon envelope pseudotyped lentivector. *Frontiers in Immunology*, *10*, 2873.

Cooper, M. A., Fehniger, T. A., & Caligiuri, M. A. (2001). The biology of human natural killer-cell subsets. *Trends in Immunology*, *22*(11), 633–640.

Daher, M., Melo Garcia, L., Li, Y., & Rezvani, K. (2021). CAR-NK cells: The next wave of cellular therapy for cancer. *Clinical & Translational Immunology, 10*(4), e1274.

Dambuza, I. M., & Brown, G. D. (2015). C-type lectins in immunity: Recent developments. *Current Opinion in Immunology, 32*, 21–27.

Davis, Z. B., Vallera, D. A., Miller, J. S., & Felices, M. (2017). Natural killer cells unleashed: Checkpoint receptor blockade and BiKE/TriKE utilization in NK-mediated anti-tumor immunotherapy. *Seminars in Immunology, 31*, 64–75.

de Gramont, A., Faivre, S., & Raymond, E. (2017). Novel TGF- β inhibitors ready for prime time in onco-immunology. *OncoImmunology, 6*(1), e1257453.

de Marco, A. (2020). Recombinant expression of nanobodies and nanobody-derived immunoreagents. *Protein Expression and Purification, 172*, 105645.

de Saint Basile, G., Menasche, G., & Fischer, A. (2010). Molecular mechanisms of biogenesis and exocytosis of cytotoxic granules. *Nature Reviews Immunology, 10*(8), 568–579.

Degli-Esposti, M. A., & Smyth, M. J. (2005). Close encounters of different kinds: Dendritic cells and NK cells take centre stage. *Nature Reviews Immunology, 5*(2), 112–124.

Demaria, O., Cornen, S., Daëron, M., Morel, Y., Medzhitov, R., & Vivier, E. (2019). Harnessing innate immunity in cancer therapy. *Nature, 574*(7776), 45–56.

Demaria, O., Gauthier, L., Debroas, G., & Vivier, E. (2021). Natural killer cell engagers in cancer immunotherapy: Next generation of immuno-oncology treatments. *European Journal of Immunology, 51*(8), 1934–1942.

Deng, G., Zheng, X., Zhou, J., Wei, H., Tian, Z., & Sun, R. (2015). Generation and preclinical characterization of an NKp80-Fc fusion protein for redirected cytolysis of natural killer (NK) cells against leukemia. *Journal of Biological Chemistry, 290*(37), 22474–22484.

Di Vito, C., Mikulak, J., & Mavilio, D. (2019). On the way to become a natural killer cell. *Frontiers in Immunology, 10*(8), 1812.

Diefenbach, A., & Raulet, D. H. (2001). Strategies for target cell recognition by natural killer cells. *Immunological Reviews, 181*, 170–184.

Ding, H., Yang, X., & Wei, Y. (2018). Fusion proteins of NKG2D/NKG2DL in cancer immunotherapy. *International Journal of Molecular Sciences, 19*(1), 1–14.

Draghi, M., Schafer, J. L., Nelson, A., Frye, Z., Oliphant, A., Haserlat, S., et al. (2019). Abstract 4972: Preclinical development of a first-in-class NKp30xBCMA NK cell engager for the treatment of multiple myeloma. *Cancer Research, 79*(13 Suppl), 4972.

Drickamer, K. (1999). C-type lectin-like domains. *Current Opinion in Structural Biology, 9*(5), 585–590.

Drickamer, K., & Fadden, A. J. (2002). Genomic analysis of C-type lectins. *Biochemical Society Symposium, 69*, 59–72.

Du, N., Guo, F., Wang, Y., & Cui, J. (2021). NK cell therapy: A rising star in cancer treatment. *Cancers, 13*(16), 4129.

Dunn, G. P., Old, L. J., & Schreiber, R. D. (2004). The three es of cancer immunoediting. *Annual Review of Immunology, 22*(1), 329–360.

El-Gazzar, A., Groh, V., & Spies, T. (2013). Immunobiology and conflicting roles of the human NKG2D lymphocyte receptor and its ligands in cancer. *Journal of Immunology, 191*(4), 1509–1515.

Ellwanger, K., Reusch, U., Fucek, I., Wingert, S., Ross, T., Muller, T., et al. (2019). Redirected optimized cell killing (ROCK®): A highly versatile multispecific fit-for-purpose antibody platform for engaging innate immunity. *MAbs, 11*(5), 899–918.

Farkona, S., Diamandis, E. P., & Blasutig, I. M. (2016). Cancer immunotherapy: The beginning of the end of cancer? *BMC Medicine, 14*, 73.

Fedorov, V. D., Themeli, M., & Sadelain, M. (2013). PD-1-and CTLA-4-based inhibitory chimeric antigen receptors (iCARs) divert off-target immunotherapy responses. *Science Translational Medicine, 5*(215), 215ra172.

Fehniger, T. A., Miller, J. S., Stuart, R. K., Cooley, S., Salhotra, A., Curtsinger, J., et al. (2018). A phase 1 trial of CNDO-109-activated natural killer cells in patients with high-risk acute myeloid leukemia. *Biology of Blood and Marrow Transplantation, 24*(8), 1581–1589.

Felices, M., Lenvik, T. R., Davis, Z. B., Miller, J. S., & Vallera, D. A. (2016). Generation of BiKEs and TriKEs to improve NK cell-mediated targeting of tumor cells. *Methods in Molecular Biology, 1441*, 333–346.

Ferrari de Andrade, L., Kumar, S., Luoma, A. M., Ito, Y., Alves da Silva, P. H., Pan, D., et al. (2020). Inhibition of MICA and MICB shedding elicits NK-cell-mediated immunity against tumors resistant to cytotoxic T cells. *Cancer Immunology Research, 8*(6), 769–780.

Fujisaki, H., Kakuda, H., Shimasaki, N., Imai, C., Ma, J., Lockey, T., et al. (2009). Expansion of highly cytotoxic human natural killer cells for cancer cell therapy. *Cancer Research, 69*(9), 4010–4017.

Garber, K. (2016). Natural killer cells blaze into immuno-oncology. *Nature Biotechnology, 34*(3), 219–220.

Gauthier, L., Morel, A., Anceriz, N., Rossi, B., Blanchard-Alvarez, A., Grondin, G., et al. (2019). Multifunctional natural killer cell engagers targeting NKp46 trigger protective tumor immunity. *Cell, 177*(7), 1701–1713.

Giebeler, N., & Zigrino, P. (2016). A disintegrin and metalloprotease (ADAM): Historical overview of their functions. *Toxins, 8*(4), 122.

Gil-Bazo, I. (2017). Avelumab—A new programmed death-ligand 1 inhibitor against advanced non-small cell lung cancer. *Translational Lung Cancer Research, 6*(S1), S35–S38.

Gleason, M. K., Ross, J. A., Warlick, E. D., Lund, T. C., Verneris, M. R., Wiernik, A., et al. (2014). CD16xCD33 bispecific killer cell engager (BiKE) activates NK cells against primary MDS and MDSC $CD33^+$ targets. *Blood, 123*(19), 3016–3026.

Granzin, M., Stojanovic, A., Miller, M., Childs, R., Huppert, V., & Cerwenka, A. (2016). Highly efficient IL-21 and feeder cell-driven *ex vivo* expansion of human NK cells with therapeutic activity in a xenograft mouse model of melanoma. *OncoImmunology, 5*(9), e1219007.

Greenberg, A. H., & Playfair, J. H. (1974). Spontaneously arising cytotoxicity to the P-815-Y mastocytoma in NZB mice. *Clinical and Experimental Immunology, 16*(1), 99–109.

Grobárová, V., Benson, V., Rozbeský, D., Novák, P., & Černý, J. (2013). Re-evaluation of the involvement of NK cells and C-type lectin-like NK receptors in modulation of immune responses by multivalent GlcNAc-terminated oligosaccharides. *Immunology Letters, 156*(1–2), 110–117.

Guo, M., Sun, C., Qian, Y., Zhu, L., Ta, N., Wang, G., et al. (2021). Proliferation of highly cytotoxic human natural killer cells by OX40L armed NK-92 with secretory neoleukin-2/15 for cancer immunotherapy. *Frontiers in Oncology, 11*, 632540.

Habif, G., Crinier, A., André, P., Vivier, E., & Narni-Mancinelli, E. (2019). Targeting natural killer cells in solid tumors. *Cellular & Molecular Immunology, 16*(5), 415–422.

Hammer, Q., & Romagnani, C. (2017). About training and memory: NK-cell adaptation to viral infections. *Advances in Immunology, 133*, 171–207.

Han, Y., Sun, F., Zhang, X., Wang, T., Jiang, J., Cai, J., et al. (2019). CD24 targeting bi-specific antibody that simultaneously stimulates NKG2D enhances the efficacy of cancer immunotherapy. *Journal of Cancer Research and Clinical Oncology, 145*(5), 1179–1190.

Han, K. P., Zhu, X., Liu, B., Jeng, E., Kong, L., Yovandich, J. L., et al. (2011). IL-15:IL-15 receptor alpha superagonist complex: High-level co-expression in recombinant mammalian cells, purification and characterization. *Cytokine, 56*(3), 804–810.

Hayakawa, Y., Screpanti, V., Yagita, H., Grandien, A., Ljunggren, H. G., Smyth, M. J., et al. (2004). NK cell TRAIL eliminates immature dendritic cells in vivo and limits dendritic cell vaccination efficacy. *Journal of Immunology, 172*(1), 123–129.

Herberman, R. B., Nunn, M. E., Holden, H. T., & Lavrin, D. H. (1975). Natural cytotoxic reactivity of mouse lymphoid cells against syngeneic and allogeneic tumors. II. Characterization of effector cells. *International Journal of Cancer, 16*(2), 230–239.

Herberman, R. B., Nunn, M. E., & Lavrin, D. H. (1975). Natural cytotoxic reactivity of mouse lymphoid cells against syngeneic acid allogeneic tumors. I. Distribution of reactivity and specificity. *International Journal of Cancer, 16*(2), 216–229.

Hinrichs, C. S., & Rosenberg, S. A. (2014). Exploiting the curative potential of adoptive T-cell therapy for cancer. *Immunological Reviews, 257*(1), 56–71.

Hudspeth, K., Silva-Santos, B., & Mavilio, D. (2013). Natural cytotoxicity receptors: Broader expression patterns and functions in innate and adaptive immune cells. *Frontiers in Immunology, 4*(3), 69.

Hui, L., & Chen, Y. (2015). Tumor microenvironment: Sanctuary of the devil. *Cancer Letters, 368*(1), 7–13.

Huntington, N. D., Vosshenrich, C. A. J., & Di Santo, J. P. (2007). Developmental pathways that generate natural-killer-cell diversity in mice and humans. *Nature Reviews Immunology*, 7(9), 703–714.

Ingegnere, T., Mariotti, F. R., Pelosi, A., Quintarelli, C., De Angelis, B., Tumino, N., et al. (2019). Human CAR NK cells: A new non-viral method allowing high efficient transfection and strong tumor cell killing. *Frontiers in Immunology, 10*, 957.

Janisova, L., Gruzinov, A., Zaborova, O. V., Velychkivska, N., Vaněk, O., Chytil, P., et al. (2020). Molecular mechanisms of the interactions of N-(2-hydroxypropyl)methacrylamide copolymers designed for cancer therapy with blood plasma proteins. *Pharmaceutics, 12*(2), 106.

Kaiser, B. K., Yim, D., Chow, I. T., Gonzalez, S., Dai, Z., Mann, H. H., et al. (2007). Disulphide-isomerase-enabled shedding of tumour-associated NKG2D ligands. *Nature, 447*(7143), 482–486.

Kalousková, B., Skořepa, O., Cmunt, D., Abreu, C., Krejčová, K., Bláha, J., et al. (2021). Tumor marker B7-H6 bound to the coiled coil peptide-polymer conjugate enables targeted therapy by activating human natural killer cells. *Biomedicines, 9*(11), 1597.

Kärre, K. (2008). Natural killer cell recognition of missing self. *Nature Immunology, 9*(5), 477–480.

Kärre, K., Ljunggren, H. G., Piontek, G., & Kiessling, R. (1986). Selective rejection of H-2-deficient lymphoma variants suggests alternative immune defence strategy. *Nature, 319*(6055), 675–678.

Kavan, D., Kubíčková, M., Bílý, J., Vaněk, O., Hofbauerová, K., Mrázek, H., et al. (2010). Cooperation between subunits is essential for high-affinity binding of N-acetyl-D-hexosamines to dimeric soluble and dimeric cellular forms of human CD69. *Biochemistry, 49*(19), 4060–4067.

Kellner, C., Derer, S., Klausz, K., Rosskopf, S., Wirt, T., Rosner, T., et al. (2018). Fc Glyco- and Fc protein-engineering: Design of antibody variants with improved ADCC and CDC activity. *Methods in Molecular Biology, 1827*, 381–397.

Kellner, C., Günther, A., Humpe, A., Repp, R., Klausz, K., Derer, S., et al. (2016). Enhancing natural killer cell-mediated lysis of lymphoma cells by combining therapeutic antibodies with CD20-specific immunoligands engaging NKG2D or NKp30. *OncoImmunology, 5*(1), e1058459.

Kellner, C., Lutz, S., Oberg, H. H., Wesch, D., Otte, A., Diemer, K. J., et al. (2021). Tumor cell lysis and synergistically enhanced antibody-dependent cell-mediated cytotoxicity by NKG2D engagement with a bispecific immunoligand targeting the HER2 antigen. *Biological Chemistry*. https://doi.org/10.1515/hsz-2021-0229.

Kellner, C., Maurer, T., Hallack, D., Repp, R., van de Winkel, J. G. J., Parren, P. W. H. I., et al. (2012). Mimicking an induced self phenotype by coating lymphomas with the NKp30 ligand B7-H6 promotes NK cell cytotoxicity. *Journal of Immunology, 189*(10), 5037–5046.

Khan, M., Arooj, S., & Wang, H. (2020). NK cell-based immune checkpoint inhibition. *Frontiers in Immunology, 11*(2), 167.

Khong, H. T., Wang, Q. J., & Rosenberg, S. A. (2004). Identification of multiple antigens recognized by tumor-infiltrating lymphocytes from a single patient: Tumor escape by antigen loss and loss of MHC expression. *Journal of Immunotherapy, 27*(3), 184–190.

Kiessling, R., Klein, E., & Wigzell, H. (1975). "Natural" killer cells in the mouse. I. Cytotoxic cells with specificity for mouse Moloney leukemia cells. Specificity and distribution according to genotype. *European Journal of Immunology, 5*(2), 112–117.

Kim, S., Poursine-Laurent, J., Truscott, S. M., Lybarger, L., Song, Y. J., Yang, L., et al. (2005). Licensing of natural killer cells by host major histocompatibility complex class I molecules. *Nature, 436*(7051), 709–713.

Klausz, K., Cieker, M., Kellner, C., Rösner, T., Otte, A., Krohn, S., et al. (2020). Fc-engineering significantly improves the recruitment of immune effector cells by anti-ICAM-1 antibody MSH-TP15 for myeloma therapy. *Haematologica, 106*(7), 1857–1866.

Koch, J., Steinle, A., Watzl, C., & Mandelboim, O. (2013). Activating natural cytotoxicity receptors of natural killer cells in cancer and infection. *Trends in Immunology, 34*(4), 182–191.

Kohrt, H., Rajasekaran, N., Chester, C., Yonezawa, A., & Zhao, X. (2015). Enhancement of antibody-dependent cell mediated cytotoxicity: A new era in cancer treatment. *ImmunoTargets and Therapy, 4*, 91–100.

Kolenko, P., Rozbeský, D., Vaněk, O., Kopecký, V., Jr., Hofbauerová, K., Novák, P., et al. (2011). Molecular architecture of mouse activating NKR-P1 receptors. *Journal of Structural Biology, 175*(3), 434–441.

Kolenko, P., Skálová, T., Vaněk, O., Štěpánková, A., Dušková, J., Hašek, J., et al. (2009). The high-resolution structure of the extracellular domain of human CD69 using a novel polymer. *Acta Crystallographica. Section F, Structural Biology and Crystallization Communications, 65*(Pt. 12), 1258–1260.

Kondo, M., Weissman, I. L., & Akashi, K. (1997). Identification of clonogenic common lymphoid progenitors in mouse bone marrow. *Cell, 91*(5), 661–672.

Kong, W., Dimitri, A., Wang, W., Jung, I. Y., Ott, C. J., Fasolino, M., et al. (2021). BET bromodomain protein inhibition reverses chimeric antigen receptor extinction and reinvigorates exhausted T cells in chronic lymphocytic leukemia. *Journal of Clinical Investigation, 131*(16), e145459.

Kosti, P., Opzoomer, J. W., Larios-Martinez, K. I., Henley-Smith, R., Scudamore, C. L., Okesola, M., et al. (2021). Hypoxia-sensing CAR T cells provide safety and efficacy in treating solid tumors. *Cell Reports Medicine, 2*(4), 100227.

Kovalová, A., Ledvina, M., Šaman, D., Zyka, D., Kubíčková, M., Žídek, L., et al. (2010). Synthetic N-acetyl-D-glucosamine based fully branched tetrasaccharide, a mimetic of the endogenous ligand for CD69, activates $CD69^+$ killer lymphocytes upon dimerization via a hydrophilic flexible linker. *Journal of Medicinal Chemistry, 53*(10), 4050–4065.

Kruse, P. H., Matta, J., Ugolini, S., & Vivier, E. (2014). Natural cytotoxicity receptors and their ligands. *Immunology & Cell Biology, 92*(3), 221–229.

Kundu, S., Gurney, M., & O'Dwyer, M. (2021). Generating natural killer cells for adoptive transfer: Expanding horizons. *Cytotherapy, 23*(7), 559–566.

Kweon, S., Phan, M. T., Chun, S., Yu, H., Kim, J., Kim, S., et al. (2019). Expansion of human NK cells using K562 cells expressing OX40 ligand and short exposure to IL-21. *Frontiers in Immunology, 10*, 879.

Kwok, G., Yau, T. C. C., Chiu, J. W., Tse, E., & Kwong, Y.-L. (2016). Pembrolizumab (Keytruda). *Human Vaccines & Immunotherapeutics, 12*(11), 2777–2789.

Lagrue, K., Carisey, A., Oszmiana, A., Kennedy, P. R., Williamson, D. J., Cartwright, A., et al. (2013). The central role of the cytoskeleton in mechanisms and functions of the NK cell immune synapse. *Immunological Reviews, 256*(1), 203–221.

Lanier, L. L. (2005). NK cell recognition. *Annual Review of Immunology, 23*, 225–274.
Lanier, L. L. (2008). Up on the tightrope: Natural killer cell activation and inhibition. *Nature Immunology, 9*(5), 495–502.
Leach, D. R., Krummel, M. F., & Allison, J. P. (1996). Enhancement of antitumor immunity by CTLA-4 blockade. *Science, 271*(5256), 1734–1736.
Leibson, P. J. (1997). Signal transduction during natural killer cell activation: Inside the mind of a killer. *Immunity, 6*(6), 655–661.
Leong, J. W., Chase, J. M., Romee, R., Schneider, S. E., Sullivan, R. P., Cooper, M. A., et al. (2014). Preactivation with IL-12, IL-15, and IL-18 induces CD25 and a functional high-affinity IL-2 receptor on human cytokine-induced memory-like natural killer cells. *Biology of Blood and Marrow Transplantation, 20*(4), 463–473.
Levin, A. M., Bates, D. L., Ring, A. M., Krieg, C., Lin, J. T., Su, L., et al. (2012). Exploiting a natural conformational switch to engineer an interleukin-2 'superkine'. *Nature, 484*(7395), 529–533.
Li, Y., Hermanson, D. L., Moriarity, B. S., & Kaufman, D. S. (2018). Human iPSC-derived natural killer cells engineered with chimeric antigen receptors enhance anti-tumor activity. *Cell Stem Cell, 23*(2), 181–192.
Li, W., Qiu, S., Chen, J., Jiang, S., Chen, W., Jiang, J., et al. (2020). Chimeric antigen receptor designed to prevent ubiquitination and downregulation showed durable antitumor efficacy. *Immunity, 53*(2), 456–470.
Li, Y., Wang, Q., Chen, S., Brown, P. H., & Mariuzza, R. A. (2013). Structure of NKp65 bound to its keratinocyte ligand reveals basis for genetically linked recognition in natural killer gene complex. *Proceedings of the National Academy of Sciences of the United States of America, 110*(28), 11505–11510.
Li, Y., Wang, Q., & Mariuzza, R. A. (2011). Structure of the human activating natural cytotoxicity receptor NKp30 bound to its tumor cell ligand B7-H6. *Journal of Experimental Medicine, 208*(4), 703–714.
Lim, S. A., Kim, T. J., Lee, J. E., Sonn, C. H., Kim, K., Kim, J., et al. (2013). *Ex vivo* expansion of highly cytotoxic human NK cells by cocultivation with irradiated tumor cells for adoptive immunotherapy. *Cancer Research, 73*(8), 2598–2607.
Liu, Y., Liu, G., Wang, J., Zheng, Z. Y., Jia, L., Rui, W., et al. (2021). Chimeric STAR receptors using TCR machinery mediate robust responses against solid tumors. *Science Translational Medicine, 13*(586), eabb5191.
Liu, E., Marin, D., Banerjee, P., Macapinlac, H. A., Thompson, P., Basar, R., et al. (2020). Use of CAR-transduced natural killer cells in CD19-positive lymphoid tumors. *New England Journal of Medicine, 382*(6), 545–553.
Liu, M., Meng, Y., Zhang, L., Han, Z., & Feng, X. (2021). High-efficient generation of natural killer cells from peripheral blood with preferable cell vitality and enhanced cytotoxicity by combination of IL-2, IL-15 and IL-18. *Biochemical and Biophysical Research Communications, 534*, 149–156.
Liu, X., Ranganathan, R., Jiang, S., Fang, C., Sun, J., Kim, S., et al. (2016). A chimeric switch-receptor targeting PD1 augments the efficacy of second-generation CAR T cells in advanced solid tumors. *Cancer Research, 76*(6), 1578–1590.
Ljunggren, H.-G., & Kärre, K. (1990). In search of the 'missing self': MHC molecules and NK cell recognition. *Immunology Today, 11*(7), 237–244.
Ma, S., Xu, Y., & Song, W. (2021). Functional bionanomaterials for cell surface engineering in cancer immunotherapy. *APL Bioengineering, 5*(2), 021506.
MacFarlane, A. W., & Campbell, K. S. (2006). Signal transduction in natural killer cells. *Current Topics in Microbiology and Immunology, 298*, 23–57.
Mak, T. W., & Saunders, M. E. (2006). Bridging innate and adaptive immunity: NK, γδ T, and NKT cells. In T. W. Mak, & M. E. Saunders (Eds.), *The immune response* (pp. 517–552). Academic Press.

Malmberg, K.-J., Carlsten, M., Björklund, A., Sohlberg, E., Bryceson, Y. T., & Ljunggren, H.-G. (2017). Natural killer cell-mediated immunosurveillance of human cancer. *Seminars in Immunology*, *31*, 20–29.

Märklin, M., Hagelstein, I., Koerner, S. P., Rothfelder, K., Pfluegler, M. S., Schumacher, A., et al. (2019). Bispecific NKG2D-CD3 and NKG2D-CD16 fusion proteins for induction of NK and T cell reactivity against acute myeloid leukemia. *Journal for Immunotherapy of Cancer*, 7(1), 143.

Mehta, R. S., & Rezvani, K. (2018). Chimeric antigen receptor expressing natural killer cells for the immunotherapy of cancer. *Frontiers in Immunology*, *9*, 283.

Michen, S., Frosch, J., Fussel, M., Schackert, G., Momburg, F., & Temme, A. (2020). Artificial feeder cells expressing ligands for killer cell immunoglobulin-like receptors and CD94/NKG2A for expansion of functional primary natural killer cells with tolerance to self. *Cytotherapy*, *22*(7), 354–368.

Montaldo, E., Zotto, G. D., Chiesa, M. D., Mingari, M. C., Moretta, A., Maria, A. D., et al. (2013). Human NK cell receptors/markers: A tool to analyze NK cell development, subsets and function. *Cytometry Part A*, *83A*(8), 702–713.

Moretta, L., Bottino, C., Pende, D., Castriconi, R., Mingari, M. C., & Moretta, A. (2006). Surface NK receptors and their ligands on tumor cells. *Seminars in Immunology*, *18*(3), 151–158.

Moretta, L., Ferlazzo, G., Bottino, C., Vitale, M., Pende, D., Mingari, M. C., et al. (2006). Effector and regulatory events during natural killer-dendritic cell interactions. *Immunological Reviews*, *214*, 219–228.

Morvan, M. G., & Lanier, L. L. (2016). NK cells and cancer: You can teach innate cells new tricks. *Nature Reviews Cancer*, *16*(1), 7–19.

Mrázek, H., Weignerová, L., Bojarová, P., Novák, P., Vaněk, O., & Bezouška, K. (2013). Carbohydrate synthesis and biosynthesis technologies for cracking of the glycan code: Recent advances. *Biotechnology Advances*, *31*(1), 17–37.

Myers, J. A., & Miller, J. S. (2021). Exploring the NK cell platform for cancer immunotherapy. *Nature Reviews Clinical Oncology*, *18*(2), 85–100.

Nayyar, G., Chu, Y., & Cairo, M. S. (2019). Overcoming resistance to natural killer cell based immunotherapies for solid tumors. *Frontiers in Oncology*, *9*(2), 51.

Neuss, S., Bartel, Y., Born, C., Weil, S., Koch, J., Behrends, C., et al. (2018). Cellular mechanisms controlling surfacing of AICL glycoproteins, cognate ligands of the activating NK receptor NKp80. *Journal of Immunology*, *201*(4), 1275–1286.

Ni, J., Miller, M., Stojanovic, A., Garbi, N., & Cerwenka, A. (2012). Sustained effector function of IL-12/15/18-preactivated NK cells against established tumors. *Journal of Experimental Medicine*, *209*(13), 2351–2365.

Oberg, H. H., Kellner, C., Gonnermann, D., Sebens, S., Bauerschlag, D., Gramatzki, M., et al. (2018). Tribody [$(HER2)_2$xCD16] is more effective than trastuzumab in enhancing γδ T cell and natural killer cell cytotoxicity against HER2-expressing cancer cells. *Frontiers in Immunology*, *9*, 814.

Ojo, E. O., Sharma, A. A., Liu, R., Moreton, S., Checkley-Luttge, M. A., Gupta, K., et al. (2019). Membrane bound IL-21 based NK cell feeder cells drive robust expansion and metabolic activation of NK cells. *Scientific Reports*, *9*(1), 14916.

Orange, J. S. (2008). Formation and function of the lytic NK-cell immunological synapse. *Nature Reviews Immunology*, *8*(9), 713–725.

Orange, J. S. (2013). Natural killer cell deficiency. *Journal of Allergy and Clinical Immunology*, *132*(3), 515–525.

Orange, J. S., Harris, K. E., Andzelm, M. M., Valter, M. M., Geha, R. S., & Strominger, J. L. (2003). The mature activating natural killer cell immunologic synapse is formed in distinct stages. *Proceedings of the National Academy of Sciences of the United States of America*, *100*(24), 14151–14156.

Parkhurst, M. R., Riley, J. P., Dudley, M. E., & Rosenberg, S. A. (2011). Adoptive transfer of autologous natural killer cells leads to high levels of circulating natural killer cells but does not mediate tumor regression. *Clinical Cancer Research*, *17*(19), 6287–6297.

Paust, S., & von Andrian, U. H. (2011). Natural killer cell memory. *Nature Immunology*, *12*(6), 500–508.

Pechar, M., Pola, R., Laga, R., Braunová, A., Filippov, S. K., Bogomolova, A., et al. (2014). Coiled coil peptides and polymer-peptide conjugates: Synthesis, self-assembly, characterization and potential in drug delivery systems. *Biomacromolecules*, *15*(7), 2590–2599.

Pechar, M., Pola, R., Laga, R., Ulbrich, K., Bednárová, L., Maloň, P., et al. (2011). Coiled coil peptides as universal linkers for the attachment of recombinant proteins to polymer therapeutics. *Biomacromolecules*, *12*(10), 3645–3655.

Pegram, H. J., Andrews, D. M., Smyth, M. J., Darcy, P. K., & Kershaw, M. H. (2011). Activating and inhibitory receptors of natural killer cells. *Immunology and Cell Biology*, *89*(2), 216–224.

Peipp, M., Derer, S., Lohse, S., Staudinger, M., Klausz, K., Valerius, T., et al. (2015). HER2-specific immunoligands engaging NKp30 or NKp80 trigger NK-cell-mediated lysis of tumor cells and enhance antibody-dependent cell-mediated cytotoxicity. *Oncotarget*, *6*(31), 32075–32088.

Peipp, M., Wesch, D., Oberg, H. H., Lutz, S., Muskulus, A., van de Winkel, J. G. J., et al. (2017). CD20-specific immunoligands engaging NKG2D enhance γδ T cell-mediated lysis of lymphoma cells. *Scandinavian Journal of Immunology*, *86*(4), 196–206.

Pekar, L., Klausz, K., Busch, M., Valldorf, B., Kolmar, H., Wesch, D., et al. (2021). Affinity maturation of B7-H6 translates into enhanced NK cell–mediated tumor cell lysis and improved proinflammatory cytokine release of bispecific immunoligands via NKp30 engagement. *Journal of Immunology*, *206*(1), 225–236.

Pende, D., Falco, M., Vitale, M., Cantoni, C., Vitale, C., Munari, E., et al. (2019). Killer Ig-like receptors (KIRs): Their role in NK cell modulation and developments leading to their clinical exploitation. *Frontiers in Immunology*, *10*, 1179.

Pende, D., Parolini, S., Pessino, A., Sivori, S., Augugliaro, R., Morelli, L., et al. (1999). Identification and molecular characterization of NKp30, a novel triggering receptor involved in natural cytotoxicity mediated by human natural killer cells. *Journal of Experimental Medicine*, *190*(10), 1505–1516.

Pesce, S., Greppi, M., Tabellini, G., Rampinelli, F., Parolini, S., Olive, D., et al. (2017). Identification of a subset of human natural killer cells expressing high levels of programmed death 1: A phenotypic and functional characterization. *Journal of Allergy and Clinical Immunology*, *139*(1), 335–346.

Peterson, M. E., & Long, E. O. (2008). Inhibitory receptor signaling via tyrosine phosphorylation of the adaptor Crk. *Immunity*, *29*(4), 578–588.

Pfefferle, A., & Huntington, N. D. (2020). You have got a fast CAR: Chimeric antigen receptor NK cells in cancer therapy. *Cancers*, *12*(3), 706.

Phan, M.-T. T., Lee, S.-H., Kim, S.-K., & Cho, D. (2016). Expansion of NK cells using genetically engineered K562 feeder cells. In *Methods in Molecular Biology, 1441, 167–174.*

Ponath, V., Hoffmann, N., Bergmann, L., Mader, C., Alashkar Alhamwe, B., Preusser, C., et al. (2021). Secreted ligands of the NK cell receptor NKp30: B7-H6 is in contrast to BAG6 only marginally released via extracellular vesicles. *International Journal of Molecular Sciences*, *22*(4), 2189.

Rahim, M. M., & Makrigiannis, A. P. (2015). Ly49 receptors: Evolution, genetic diversity, and impact on immunity. *Immunological Reviews*, *267*(1), 137–147.

Raynaud, A., Desrumeaux, K., Vidard, L., Termine, E., Baty, D., Chames, P., et al. (2020). Anti-NKG2D single domain-based antibodies for the modulation of anti-tumor immune response. *OncoImmunology*, *10*(1), 1854529.

Regis, S., Dondero, A., Caliendo, F., Bottino, C., & Castriconi, R. (2020). NK cell function regulation by TGF-beta-induced epigenetic mechanisms. *Frontiers in Immunology*, *11*, 311.

Renaudet, O., Křenek, K., Bossu, I., Dumy, P., Kádek, A., Adámek, D., et al. (2010). Synthesis of multivalent glycoconjugates containing the immunoactive LELTE peptide: Effect of glycosylation on cellular activation and natural killing by human peripheral blood mononuclear cells. *Journal of the American Chemical Society*, *132*(19), 6800–6808.

Reusch, U., Burkhardt, C., Fucek, I., Le Gall, F., Le Gall, M., Hoffmann, K., et al. (2014). A novel tetravalent bispecific TandAb (CD30/CD16A) efficiently recruits NK cells for the lysis of $CD30^+$ tumor cells. *MAbs*, *6*(3), 727–738.

Rezaeifard, S., Heike, Y., Masuyama, J.-I., Rezvani, A., Vojdani, R., & Erfani, N. (2021). Autologous natural killer cell-enrichment for immune cell therapy: Preclinical setting phase, shiraz experience. *Iranian Journal of Allergy, Asthma and Immunology*, *20*(2), 233–243.

Rhode, P. R., Egan, J. O., Xu, W., Hong, H., Webb, G. M., Chen, X., et al. (2016). Comparison of the superagonist complex, ALT-803, to IL15 as cancer Immunotherapeutics in animal models. *Cancer Immunology Research*, *4*(1), 49–60.

Riggan, L., Shah, S., & O'Sullivan, T. E. (2021). Arrested development: Suppression of NK cell function in the tumor microenvironment. *Clinical & Translational Immunology*, *10*(1), 1–17.

Romanski, A., Uherek, C., Bug, G., Seifried, E., Klingemann, H., Wels, W. S., et al. (2016). CD19-CAR engineered NK-92 cells are sufficient to overcome NK cell resistance in B-cell malignancies. *Journal of Cellular and Molecular Medicine*, *20*(7), 1287–1294.

Romee, R., Rosario, M., Berrien-Elliott, M. M., Wagner, J. A., Jewell, B. A., Schappe, T., et al. (2016). Cytokine-induced memory-like natural killer cells exhibit enhanced responses against myeloid leukemia. *Science Translational Medicine*, *8*(357), 357ra123.

Rosario, M., Liu, B., Kong, L., Collins, L. I., Schneider, S. E., Chen, X., et al. (2016). The IL-15-based ALT-803 complex enhances FcgammaRIIIa-triggered NK cell responses and *in vivo* clearance of B cell lymphomas. *Clinical Cancer Research*, *22*(3), 596–608.

Rosskopf, S., Eichholz, K. M., Winterberg, D., Diemer, K. J., Lutz, S., Munnich, I. A., et al. (2020). Enhancing CDC and ADCC of CD19 antibodies by combining fc protein-engineering with fc Glyco-engineering. *Antibodies*, *9*(4), 63.

Rozbeský, D., Ivanova, L., Hernychová, L., Grobárová, V., Novák, P., & Černý, J. (2015). Nkrp1 family, from lectins to protein interacting molecules. *Molecules*, *20*(2), 3463–3478.

Rozbeský, D., Kavan, D., Chmelík, J., Novák, P., Vaněk, O., & Bezouška, K. (2011). High-level expression of soluble form of mouse natural killer cell receptor NKR-P1C (B6) in *Escherichia coli*. *Protein Expression and Purification*, 77(2), 178–184.

Rozbeský, D., Krejzová, J., Křenek, K., Prchal, J., Hrabal, R., Kožíšek, M., et al. (2014). Re-evaluation of binding properties of recombinant lymphocyte receptors NKR-P1A and CD69 to chemically synthesized glycans and peptides. *International Journal of Molecular Sciences*, *15*(1), 1271–1283.

Ruggeri, L. (2002). Effectiveness of donor natural killer cell alloreactivity in mismatched hematopoietic transplants. *Science*, *295*(5562), 2097–2100.

Salter, A. I., Rajan, A., Kennedy, J. J., Ivey, R. G., Shelby, S. A., Leung, I., et al. (2021). Comparative analysis of TCR and CAR signaling informs CAR designs with superior antigen sensitivity and in vivo function. *Science Signaling*, *14*(697), eabe2606.

Sanz-Ortega, L., Portilla, Y., Perez-Yague, S., & Barber, D. F. (2019). Magnetic targeting of adoptively transferred tumour-specific nanoparticle-loaded CD8(+) T cells does not improve their tumour infiltration in a mouse model of cancer but promotes the retention of these cells in tumour-draining lymph nodes. *Journal of Nanobiotechnology*, *17*(1), 87.

Sanz-Ortega, L., Rojas, J. M., Portilla, Y., Perez-Yague, S., & Barber, D. F. (2019). Magnetic nanoparticles attached to the NK cell surface for tumor targeting in adoptive transfer therapies does not affect cellular effector functions. *Frontiers in Immunology*, *10*, 2073.

Sarhan, D., Brandt, L., Felices, M., Guldevall, K., Lenvik, T., Hinderlie, P., et al. (2018). 161533 TriKE stimulates NK-cell function to overcome myeloid-derived suppressor cells in MDS. *Blood Advances*, *2*(12), 1459–1469.

Saunders, K. O. (2019). Conceptual approaches to modulating antibody effector functions and circulation half-life. *Frontiers in Immunology*, *10*, 1296.

Schlecker, E., Fiegler, N., Arnold, A., Altevogt, P., Rose-John, S., Moldenhauer, G., et al. (2014). Metalloprotease-mediated tumor cell shedding of B7-H6, the ligand of the natural killer cell-activating receptor NKp30. *Cancer Research*, *74*(13), 3429–3440.

Schmohl, J. U., Felices, M., Taras, E., Miller, J. S., & Vallera, D. A. (2016). Enhanced ADCC and NK cell activation of an anticarcinoma bispecific antibody by genetic insertion of a modified IL-15 cross-linker. *Molecular Therapy*, *24*(7), 1312–1322.

Schmohl, J. U., Felices, M., Todhunter, D., Taras, E., Miller, J. S., & Vallera, D. A. (2016). Tetraspecific scFv construct provides NK cell mediated ADCC and self-sustaining stimuli via insertion of IL-15 as a cross-linker. *Oncotarget*, 7(45), 73830–73844.

Schreiber, R. D., Old, L. J., & Smyth, M. J. (2011). Cancer immunoediting: Integrating Immunity's roles in cancer suppression and promotion. *Science*, *331*(6024), 1565–1570.

Sconocchia, G., Eppenberger-Castori, S., Zlobec, I., Karamitopoulou, E., Arriga, R., Coppola, A., et al. (2014). HLA class II antigen expression in colorectal carcinoma tumors as a favorable prognostic marker. *Neoplasia*, *16*(1), 31–42.

Shevtsov, M., & Multhoff, G. (2016). Immunological and translational aspects of NK cell-based antitumor immunotherapies. *Frontiers in Immunology*, 7(11), 462.

Shimasaki, N., Jain, A., & Campana, D. (2020). NK cells for cancer immunotherapy. *Nature Reviews Drug Discovery*, *19*(3), 200–218.

Shin, J. H., Park, H. B., & Choi, K. (2016). Enhanced anti-tumor reactivity of cytotoxic T lymphocytes expressing PD-1 decoy. *Immune Network*, *16*(2), 134.

Skálová, T., Bláha, J., Harlos, K., Dušková, J., Koval', T., Stránský, J., et al. (2015). Four crystal structures of human LLT1, a ligand of human NKR-P1, in varied glycosylation and oligomerization states. *Acta Crystallographica Section D Biological Crystallography*, *71*(Pt. 3), 578–591.

Skálová, T., Kotýnková, K., Dušková, J., Hašek, J., Koval', T., Kolenko, P., et al. (2012). Mouse Clr-g, a ligand for NK cell activation receptor NKR-P1F: Crystal structure and biophysical properties. *Journal of Immunology*, *189*(10), 4881–4889.

Skořepa, O., Pazicky, S., Kalousková, B., Bláha, J., Abreu, C., Ječmen, T., et al. (2020). Natural killer cell activation receptor NKp30 oligomerization depends on its *N*-glycosylation. *Cancers*, *12*(7), 1998.

Sovová, Ž., Kopecký, V., Jr., Pazderka, T., Hofbauerová, K., Rozbeský, D., Vaněk, O., et al. (2011). Structural analysis of natural killer cell receptor protein 1 (NKR-P1) extracellular domains suggests a conserved long loop region involved in ligand specificity. *Journal of Molecular Modeling*, *17*(6), 1353–1370.

Stern-Ginossar, N., Gur, C., Biton, M., Horwitz, E., Elboim, M., Stanietsky, N., et al. (2008). Human microRNAs regulate stress-induced immune responses mediated by the receptor NKG2D. *Nature Immunology*, *9*(9), 1065–1073.

Streltsova, M. A., Ustiuzhanina, M. O., Barsov, E. V., Kust, S. A., Velichinskii, R. A., & Kovalenko, E. I. (2021). Telomerase reverse transcriptase increases proliferation and lifespan of human NK cells without immortalization. *Biomedicines*, *9*(6), 662.

Tata, A., Dodard, G., Fugère, C., Leget, C., Ors, M., Rossi, B., et al. (2021). Combination blockade of KLRG1 and PD-1 promotes immune control of local and disseminated cancers. *OncoImmunology*, *10*(1), 1933808.

Taube, J. M., Anders, R. A., Young, G. D., Xu, H., Sharma, R., McMiller, T. L., et al. (2012). Colocalization of inflammatory response with B7-H1 expression in human melanocytic lesions supports an adaptive resistance mechanism of immune escape. *Science Translational Medicine*, *4*(127), 1–22.

Tomasello, E., Blery, M., Vely, F., & Vivier, E. (2000). Signaling pathways engaged by NK cell receptors: Double concerto for activating receptors, inhibitory receptors and NK cells. *Seminars in Immunology*, *12*(2), 139–147.

Tong, C., Zhang, Y., Liu, Y., Ji, X., Zhang, W., Guo, Y., et al. (2020). Optimized tandem CD19/CD20 CAR-engineered T cells in refractory/relapsed B-cell lymphoma. *Blood*, *136*(14), 1632–1644.

Trinchieri, G. (1989). Biology of natural killer cells. *Advances in Immunology*, *47*, 187–376.

Tripathi, P., & Agrawal, S. (2006). Non-classical HLA-G antigen and its role in the cancer progression. *Cancer Investigation*, *24*(2), 178–186.

Trowsdale, J., Barten, R., Haude, A., Stewart, C. A., Beck, S., & Wilson, M. J. (2001). The genomic context of natural killer receptor extended gene families. *Immunological Reviews*, *181*, 20–38.

Tusell Wennier, S., Liu, J., & McFadden, G. (2012). Bugs and drugs: Oncolytic virotherapy in combination with chemotherapy. *Current Pharmaceutical Biotechnology*, *13*(9), 1817–1833.

Vacca, P., Munari, E., Tumino, N., Moretta, F., Pietra, G., Vitale, M., et al. (2018). Human natural killer cells and other innate lymphoid cells in cancer: Friends or foes? *Immunology Letters*, *201*, 14–19.

van der Schans, J. J., van de Donk, N., & Mutis, T. (2020). Dual targeting to overcome current challenges in multiple myeloma CAR T-cell treatment. *Frontiers in Oncology*, *10*, 1362.

Vaněk, O., Celadova, P., Skořepa, O., Bláha, J., Kalousková, B., Dvorská, A., et al. (2019). Production of recombinant soluble dimeric C-type lectin-like receptors of rat natural killer cells. *Scientific Reports*, *9*(1), 17836.

Vaněk, O., Nálezková, M., Kavan, D., Borovičková, I., Pompach, P., Novák, P., et al. (2008). Soluble recombinant CD69 receptors optimized to have an exceptional physical and chemical stability display prolonged circulation and remain intact in the blood of mice. *FEBS Journal*, *275*(22), 5589–5606.

Vesely, M. D., Kershaw, M. H., Schreiber, R. D., & Smyth, M. J. (2011). Natural innate and adaptive immunity to cancer. *Annual Review of Immunology*, *29*(1), 235–271.

Vilches, C., & Parham, P. (2002). KIR: Diverse, rapidly evolving receptors of innate and adaptive immunity. *Annual Review of Immunology*, *20*, 217–251.

Vinay, D. S., Ryan, E. P., Pawelec, G., Talib, W. H., Stagg, J., Elkord, E., et al. (2015). Immune evasion in cancer: Mechanistic basis and therapeutic strategies. *Seminars in Cancer Biology*, *35*, S185–S198.

Vitale, M., Cantoni, C., Pietra, G., Mingari, M. C., & Moretta, L. (2014). Effect of tumor cells and tumor microenvironment on NK-cell function. *European Journal of Immunology*, *44*(6), 1582–1592.

Vivier, E., Raulet, D. H., Moretta, A., Caligiuri, M. A., Zitvogel, L., Lanier, L. L., et al. (2011). Innate or adaptive immunity? The example of natural killer cells. *Science*, *331*(6013), 44–49.

Vivier, E., Tomasello, E., Baratin, M., Walzer, T., & Ugolini, S. (2008). Functions of natural killer cells. *Nature Immunology*, *9*(5), 503–510.

Vivier, E., & Ugolini, S. (2011). Natural killer cells: From basic research to treatments. *Frontiers in Immunology*, *2*(6), 18.

Vivier, E., Ugolini, S., Blaise, D., Chabannon, C., & Brossay, L. (2012). Targeting natural killer cells and natural killer T cells in cancer. *Nature Reviews Immunology*, *12*(4), 239–252.

Vivier, E., van de Pavert, S. A., Cooper, M. D., & Belz, G. T. (2016). The evolution of innate lymphoid cells. *Nature Immunology*, *17*(7), 790–794.

von Andrian, U. H. (2021). NK cell memory: Discovery of a mystery. *Nature Immunology*, *22*(6), 669–671.

von Strandmann, E. P., Hansen, H. P., Reiners, K. S., Schnell, R., Borchmann, P., Merkert, S., et al. (2006). A novel bispecific protein (ULBP2-BB4) targeting the NKG2D receptor on natural killer (NK) cells and CD138 activates NK cells and has potent antitumor activity against human multiple myeloma in vitro and in vivo. *Blood*, *107*(5), 1955–1962.

Voskoboinik, I., Whisstock, J. C., & Trapani, J. A. (2015). Perforin and granzymes: Function, dysfunction and human pathology. *Nature Reviews Immunology*, *15*(6), 388–400.

Walk, J., & Sauerwein, R. W. (2019). Activatory receptor NKp30 predicts NK cell activation during controlled human malaria infection. *Frontiers in Immunology*, *10*, 2864.

Walzer, T., Dalod, M., Robbins, S. H., Zitvogel, L., & Vivier, E. (2005). Natural-killer cells and dendritic cells: "l'union fait la force". *Blood*, *106*(7), 2252–2258.

Wang, Y., Li, H., Xu, W., Pan, M., Qiao, C., Cai, J., et al. (2020). BCMA-targeting bispecific antibody that simultaneously stimulates NKG2D-enhanced efficacy against multiple myeloma. *Journal of Immunotherapy*, *43*(6), 175–188.

Watkins-Yoon, J., Guzman, W., Oliphant, A., Haserlat, S., Leung, A., Chottin, C., et al. (2019). CTX-8573, an innate-cell engager targeting BCMA, is a highly potent multispecific antibody for the treatment of multiple myeloma. *Blood*, *134*(Suppl. 1), 3182.

Wingert, S., Reusch, U., Knackmuss, S., Kluge, M., Damrat, M., Pahl, J., et al. (2021). Preclinical evaluation of AFM24, a novel CD16A-specific innate immune cell engager targeting EGFR positive tumors. *MAbs*, *13*(1), 1950264.

Wolf, N., Blaj, C., Picton, L., Snyder, G., Zhang, L., Nicolai, C. J., et al. (2021). Powerful synergistic effects of a STING agonist and an IL-2 superkine in cancer immunotherapy against MHC I-deficient and MHC I+ tumors. *bioRxiv*. https://doi.org/10.1101/2021.11.02.465022.

Woll, P. S., Grzywacz, B., Tian, X., Marcus, R. K., Knorr, D. A., Verneris, M. R., et al. (2009). Human embryonic stem cells differentiate into a homogeneous population of natural killer cells with potent in vivo antitumor activity. *Blood*, *113*(24), 6094–6101.

Wu, S. Y., Fu, T., Jiang, Y. Z., & Shao, Z. M. (2020). Natural killer cells in cancer biology and therapy. *Molecular Cancer*, *19*(1), 120.

Wu, C. Y., Roybal, K. T., Puchner, E. M., Onuffer, J., & Lim, W. A. (2015). Remote control of therapeutic T cells through a small molecule-gated chimeric receptor. *Science*, *350*(6258), aab4077.

Wu, D., Shou, X., Zhang, Y., Li, Z., Wu, G., Wu, D., et al. (2021). Cell membrane-encapsulated magnetic nanoparticles for enhancing natural killer cell-mediated cancer immunotherapy. *Nanomedicine*, *32*, 102333.

Xie, G., Dong, H., Liang, Y., Ham, J. D., Rizwan, R., & Chen, J. (2020). CAR-NK cells: A promising cellular immunotherapy for cancer. *eBioMedicine*, *59*, 102975.

Yaddanapudi, K., Mitchell, R. A., & Eaton, J. W. (2013). Cancer vaccines: Looking to the future. *OncoImmunology*, *2*(3), 37–41.

Yalniz, F. F., Daver, N., Rezvani, K., Kornblau, S., Ohanian, M., Borthakur, G., et al. (2018). A pilot trial of lirilumab with or without azacitidine for patients with myelodysplastic syndrome. *Clinical Lymphoma Myeloma and Leukemia*, *18*(10), 658–663.

Yang, S., Wen, J., Li, H., Xu, L., Liu, Y., Zhao, N., et al. (2019). Aptamer-engineered natural killer cells for cell-specific adaptive immunotherapy. *Small*, *15*(22), e1900903.

Yu, S., Xiong, G., Zhao, S., Tang, Y., Tang, H., Wang, K., et al. (2021). Nanobodies targeting immune checkpoint molecules for tumor immunotherapy and immunoimaging. *International Journal of Molecular Medicine*, *47*(2), 444–454.

Zelensky, A. N., & Gready, J. E. (2005). The C-type lectin-like domain superfamily. *FEBS Journal*, *272*(24), 6179–6217.

Zhao, Y., Li, Y., Wu, X., Li, L., Liu, J., Wang, Y., et al. (2020). Identification of anti-CD16a single domain antibodies and their application in bispecific antibodies. *Cancer Biology & Therapy*, *21*(1), 72–80.

Zhong, X. S., Matsushita, M., Plotkin, J., Riviere, I., & Sadelain, M. (2010). Chimeric antigen receptors combining 4-1BB and CD28 signaling domains augment PI_3kinase/AKT/Bcl-X_L activation and $CD8^+$ T cell-mediated tumor eradication. *Molecular Therapy*, *18*(2), 413–420.

Zhu, X., Marcus, W. D., Xu, W., Lee, H. I., Han, K., Egan, J. O., et al. (2009). Novel human interleukin-15 agonists. *Journal of Immunology*, *183*(6), 3598–3607.

Zitvogel, L., Tesniere, A., & Kroemer, G. (2006). Cancer despite immunosurveillance: Immunoselection and immunosubversion. *Nature Reviews Immunology*, *6*(10), 715–727.

CHAPTER FOUR

Noncoding RNAs as novel immunotherapeutic tools against cancer

Maninder Kaur, Bhavneet Kaur, Monidipa Konar, and Sadhna Sharma*
Department of Biochemistry, PostGraduate Institute of Medical Education and Research, Chandigarh, India
*Corresponding author: e-mail address: sadhnabiochem@gmail.com

Contents

Abstract

Immunotherapy is implemented as an important treatment strategy in various malignancies. In cancer, immunotherapy is employed for successful killing of tumor cells with high specificity and greater efficacy, with minimum side effects. Despite various available strategies, cellular immunotherapy including innate (NK cells, macrophages, dendritic cells) and adaptive (B cells and T cells) immune cells plays a critical role in tumor microenvironment. Since past few years, many drugs targeting immune checkpoint

Advances in Protein Chemistry and Structural Biology, Volume 129
ISSN 1876-1623
https://doi.org/10.1016/bs.apcsb.2021.11.011

proteins including CTLA-4 and PD-1/PD-L1 have been investigated as immunotherapy approach against cancer but complete effectiveness still remains a question, as diverse mechanisms involved in tumorigenesis may result in the development of cancer cell resistance. Number of evidences have highlighted the significant role of non-coding RNAs (ncRNAs) in regulating multiple stages of cancer initiation, progression & immunity. ncRNAs comprises 98% human transcriptome and are basically considered as dark genome. Among ncRNAs, miRNAs and lncRNAs have been extensively studied in regulating diverse processes of cancer tumorigenesis. Upregulation of oncogenic and downregulation of tumor suppressive miRNAs/lncRNAs has been reported to facilitate the cancer progression and invasiveness. This chapter summarizes how an interplay between ncRNAs and immune cells in cancer pathogenesis can be therapeutically targeted to improve current treatment regimen. Strategies should be employed to improve the efficacy and reduce off-target effects of ncRNA based immunotherapy. *Henceforth, combination of ncRNAs and available immunotherapy can be argued to enhance the efficacy of existing immunotherapeutic approaches against cancer to improve patient's survival.*

1. Introduction

Immune system is divided into innate and adaptive arms. Innate response is the first line of defense and employs macrophages, natural killer (NK) cells, monocytes and dendritic cells (DCs) to participate in the process of non-host entity recognition and further, presenting them to the adaptive arm (B and T cells) to generate immunity and memory cells for eliciting long lived immune responses. Immunotherapy is the process of harnessing the potential of an individual's self-immune system, both innate and adaptive responses to treat diseases. It engineers the immune cells by regulating the immune microenvironment to clear foreign entities.

2. Immunotherapy in human diseases

Immunotherapy is a promising treatment option in immune mediated diseases because of immune response. It is administered either indirectly in the form of immune components like monoclonal and polyclonal antibodies, cytokines, etc., or directly by targeting immune cells. Involvement of microbes or their products in form of vaccine was the first example of implication of immunotherapy in human diseases (bacterial infections). Vaccines have helped in eradication of small pox worldwide and currently, polio is also on the verge of extinction. Recently, SARS-CoV-2 pandemic highlighted the need of therapeutic vaccines in saving humankind. The role of immunotherapy has also been explored for viral infections (HIV, hepatitis, etc.) (Ramamurthy et al., 2021), tuberculosis (Mi et al., 2021; Mir & Sharma, 2014, 2020; Mir, Verma, & Sharma, 2014), etc., and auto-immune diseases (type 1 diabetes, multiple sclerosis, systemic lupus erythematosus, rheumatoid arthritis, etc.) (Wraith, 2017) where immune cells are guided to prevent an attack on self-cells by activation of dendritic cells, Treg and Breg cells. However, different aetiologies of multiple autoimmune diseases and variability in patient response have rendered the use of vaccines ineffective. Gradually, soluble antibodies and cytokines have replaced the use of vaccines for treating autoimmune disorders and cancers. Among these, the most extensive use of immunotherapy has been reported in cancer (Naran, Nundalall, Chetty, & Barth, 2018).

3. Immunotherapy in cancer

In cancer, immunotherapy is employed for the successful killing of tumor cells with high specificity and greater efficacy with minimum side effects. There are almost 10 different types of immunotherapy (Fig. 1) currently being administered in cancer patients (Varade et al., 2021). The first

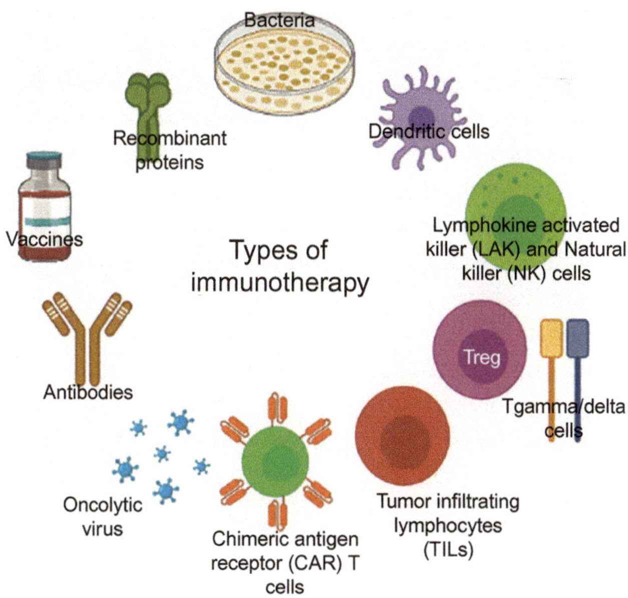

Fig. 1 Types of immunotherapy (Varade, Magadan, & Gonzalez-Fernandez, 2021). *Figure was created using BioRender.com.*

was using bacteria and oncolytic viruses (Davola & Mossman, 2019) as a therapeutic intervention against cancers, however, the non-specificity, poor capacity of tumor infiltration limits their use in immunotherapy. Then came the use of therapeutic vaccines, where tumor lysates, tumor protein or peptide fragments, and viral vectors (Hu, Ott, & Wu, 2018) were employed to elicit host immune memory which is free from any adverse reactions but variable patient response still remains a cause of concern. The third addition was use of monoclonal antibodies to specifically attack tumor cell by reducing the systemic toxicity (Akkın, Varan, & Bilensoy, 2021). But these are not effective against all cancer types due to variation in number of tumor cells in each patient, expression of MHC molecules, rate of mutations, tumor and immunosuppressive microenvironment, expression of checkpoint inhibitors, etc. The next intervention was the use of recombinant proteins like CTLA-4 chimeric protein, etanercept which captures soluble tumor necrosis factor to block the anti-inflammatory signal and fusion protein, abatacept that competitively inhibits the T-cell activation (Hassett et al., 2018).

Cellular immunotherapy mediated by NK cells, lymphokine activated killer cells, dendritic cells (Sabado, Balan, & Bhardwaj, 2017), T cells (Treg & Tgamma/delta), tumor infiltrating lymphocytes (TILs) and chimeric antigen receptor (CAR) T cells are the latest addition to cancer immunotherapy. However, hepatotoxicity and uncontrolled proliferation of Treg cells restricted the use of NK and lymphokine activated killer cells in clinical practice (Yron, Wood Jr., Spiess, & Rosenberg, 1980). Tgamma/delta cells were employed only for refractory hematological cancers and the use of Tregs is still in primitive stage. TILs have a success rate of 50–75%; however, optimization of the TIL count and its specificity against the type of tumor and tumor microenvironment is a big challenge (Xu et al., 2009). Although, initially CARs were designed for hematological cancers like lymphomas and leukemia, scientific advancements have now made it possible to use them for solid tumors (Casucci & Bondanza, 2011). Among all cell types, CAR T cells and DCs are most promising immunotherapeutic intervention strategies. Besides this, the use of noncoding genome to target immune cells in tumor microenvironment has emerged as a potential mode of immunotherapy.

4. Noncoding RNAs in cancer

The protein-coding genes of the human genome are well explored, but that accounts for only 2% of the human genome including the untranslated regions. Noncoding RNAs (ncRNAs) comprises 98% of the human transcriptome and are termed as the dark genome (Fig. 2).

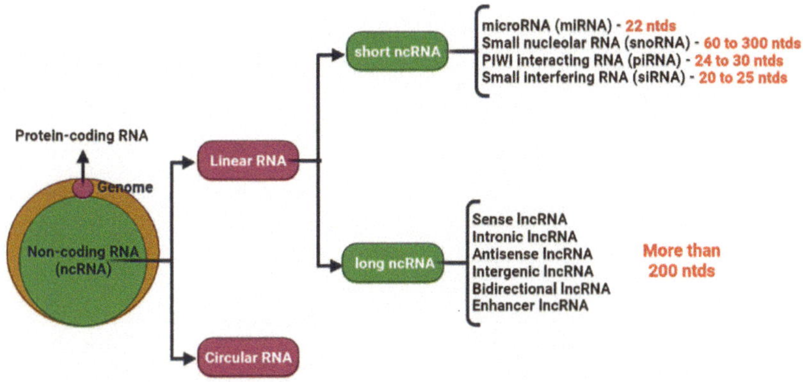

Fig. 2 Representation of the coding and non-coding segment of human genome and further classification of the non-coding genome, at first based on structure and then based on length (Zhang, Liu, et al., 2020). *Figure was created using BioRender.com.*

ncRNAs participate in regulating gene expression both at the transcriptional as well as translational levels. They have a crucial role in tumorigenesis owing to their potential in regulating tumoral gene expression. Among the ncRNAs, microRNAs and long noncoding RNAs have been most extensively studied and are still being explored because of their ability to modulate almost 60% of the protein encoding genes (Liu, Wang, et al., 2020). Fig. 3 comprehensively represents the involvement of ncRNAs in a myriad of cancers.

Tumor microenvironment (TME) consists of various cell populations of specific (adaptive) and non specific (innate) immune responses which comprise the tumor immune microenvironment (TIME). TIME primarily contain myeloid cells (neutrophils, macrophages and myeloid derived suppressor cells), lymphocytes (B cells, $CD4^+$ T helper cells, $CD8^+$ cytotoxic T lymphocytes and regulatory T cells), dendritic cells and natural killer cells. TIME affects the process of tumor immune evasion, response to chemo & immunotherapy and patient's survival rate. Studies reported that noncoding RNAs are significantly involved in regulating the multiple stages of cancer immunity. Role of miRNAs and lncRNAs has been explored extensively in regulating diverse processes like immune cell differentiation, maturation, recruitment to tumor site, antigen presentation activation, tumor immune surveillance and immune evasion (Liu, Wang, et al., 2020; Zhang, Liu, & Liao, 2020).

4.1 Reprogramming of macrophage differentiation & function by ncRNAs

Macrophages are a subset of leukocytes that are involved in providing non-specific defense (innate immunity) against various pathogens. Tissue

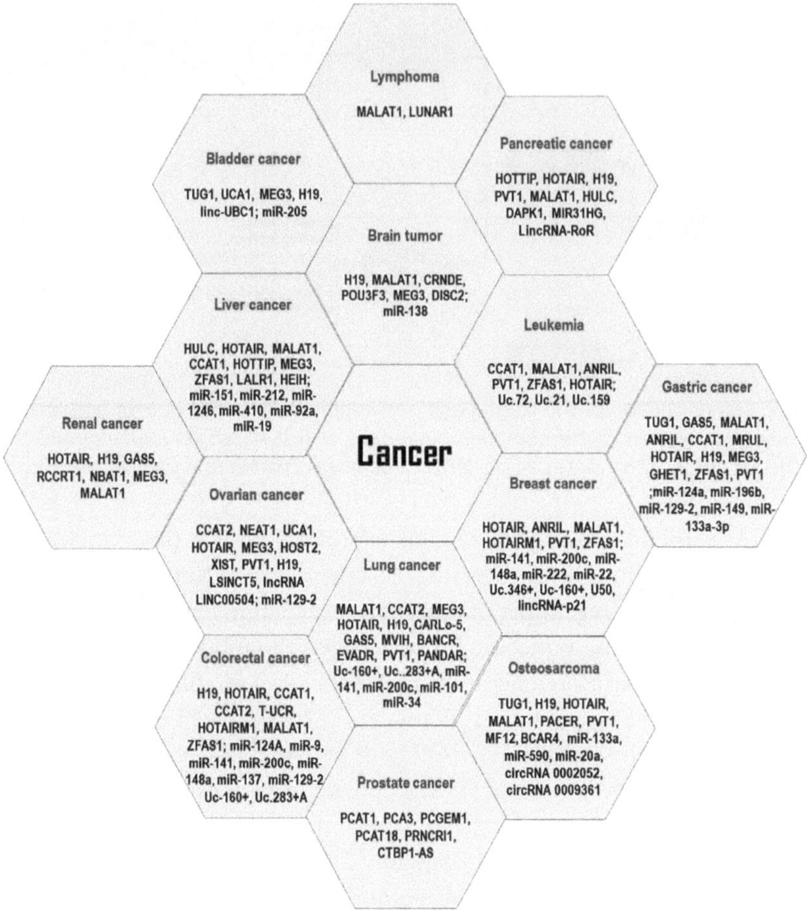

Fig. 3 A comprehensive representation of different noncoding RNAs involved in various types of cancer (Bhan, Soleimani, & Mandal, 2017).

resident macrophages play distinct roles in development, homeostasis, inflammation, phagocytic clearance of invading pathogens and antigen presentation to initiate specific defense mechanisms (adaptive immunity) (Ginhoux & Jung, 2014). In TME, various cytokines & chemokines secreted by malignant cells & immune cells, serve as a signal for recruitment of monocytes at the tumor site and their differentiation into tumor associated macrophages (TAMs) (Zhou et al., 2020). Multiple studies have demonstrated the direct or indirect involvement of macrophages in regulating various key features of malignant tumors including angiogenesis, invasiveness, metastasis, regulation of the tumor microenvironment and therapeutic resistance (Duan & Luo, 2021).

MicroRNAs are critical epigenetic regulators which control differentiation, maturation, polarization and functions of macrophages in the TME (Chatterjee et al., 2020). Overexpression of miR-146a promotes hematopoietic stem cell (HSC) differentiation while its downregulation inhibits the differentiation into macrophages during early zebra fish development (Ghani et al., 2011). lncRNAs are also crucial to regulate macrophage maturation, polarization and activation processes (Zhang, Liu, et al., 2020).

Macrophage polarization: Based on the stimulus received from tumor microenvironment, macrophages exhibit M1 and M2 like phenotype to perform various biochemical and cellular functions. M1 macrophages are proinflammatory in nature and are involved in anti-microbial and anti-tumoral responses. M2 polarization is the main phenotype of TAMs and activated by IL-10, glucocorticoids and IL-4 (Curtale, 2018). Downregulation of miR-340-5p promote macrophage recruitment to TME and M2 polarization of TAMs by directly targeting POSTN and LTBP-1 in glioblastoma multiforme (Liu, Li, et al., 2019). Transfection of pancreatic cell line with miR-155 and miR-125b-2 alters the exosome content leading to differential communication and reprogramming of macrophage toward M1 phenotype (Su, Aldawsari, & Amiji, 2016). Exosome derived miR-29a-3p promotes M2 subtype macrophage polarization by directly targeting SOCS1 in oral squamous cell carcinoma (Cai, Qiao, Gao, Lin, & He, 2019). Overexpression of let-7c also regulates macrophage polarization toward M2 phenotype. Not all miRNAs promote M2 polarization of TAMs, there are some miRNAs like miR-142-3p which inhibits M2 polarization and thus, the tumor growth in hepatocellular carcinoma (Chatterjee et al., 2020).

It has been demonstrated that lncRNA GNAS-AS1 is involved in the growth, proliferation and invasiveness of EGFR positive breast cancer cells via promoting M2 polarization through inhibition of miR-433-3p in tumor associated macrophages (Liu, Zhou, Dong, Guo, & Zhang, 2020). In addition to this, GNAS-AS1 was also found to switch macrophage polarization toward M2 phenotype in TME by directly targeting miR-4319 in non small cell lung cancer (Li, Feng, Guo, Hu, & Xie, 2020). LncRNA CCAT1 promote M1 macrophage phenotype to restrain cell growth & invasion of prostate cancer cells by inhibiting miR-148a which targets PKCζ, a key tumor suppressor gene (Liu, Ding, et al., 2019). T cell specific factor-4 regulates the expression of LncRNA-XIST which further boosts the M2 polarization of TAMs in lung cancer (Sun & Xu, 2019). LncRNA cox-2 reduces the tumor progression & metastasis in hepatocellular carcinoma (HCC) by promoting M1 phenotype in TAMs.

Antigen presentation: The other well known mechanism of tumor immune escape (TIE) is loss of tumor immunogenicity that can arise by inhibition of the antigen processing and presentation. Tumor antigen presentation is essential for T cells and NK cells to recognize and destroy malignant cells. Defects in antigen processing, presentation and recognition during tumor progression are regulated by miRNAs and lncRNAs. Increased expression of miR-27a (oncogenic miRNA) directly targets calreticulin and thus, decreases the expression of MHC-I complex on cell surface in colorectal cancer (Colangelo et al., 2016). Expression of miR-125a and miR-148-3p decreases transporter associated with antigen processing 2 (TAP2) and MHC-I which is required for antigen presentation in esophageal adenocarcinoma (Mari et al., 2018). The expression of miR-26-5p and -21-3p decreases the levels of transporter associated with antigen processing 1 (TAP1) with significant reduction in the expression of HLA class I cell surface antigens, leading to impairment in T-cell recognition (Lazaridou et al., 2020).

Exosomal microRNAs released by cancer cells: These miRNAs affect the cell's fate in TME by regulating TAM phenotypic switch, thus, promotes tumor progression and metastasis. Uptake of miR-1246 enriched exosomes by macrophages triggered their reprogramming to tumor supportive M2 phenotype in colon cancer cells. Further, miR-103a inhibits PTEN expression in macrophages which, in turn mediates immune suppressive effects via activating AKT/STAT3 pathways. Colorectal cancer cells derived extracellular vesicles enriched with miR-145 induce M2 polarization in macrophages via downregulation of histone deacetylase 11 (HDAC11) thus, promoting tumor progression (Chatterjee et al., 2020).

Extracellular microRNAs secreted by TAMs: Tumor associated macrophages reshape the TME by mediating tumor suppressive and tumor promoting events. TAMs derived exosomes enriched with miR-21-5p and -29a-3p targets STAT3 signaling in $CD4^+$ T cells and establish an immunosuppressive microenvironment which promotes ovarian cancer progression (Zhou et al., 2018). Uptake of miR-365 containing exosomes by PDAC cells confers resistance to gemcitabine and miR-223 induces chemoresistance in epithelial ovarian cancer cells (Binenbaum et al., 2018; Zhu, Shen, et al., 2019).

These studies signify the role of lncRNAs in controlling polarization, recruitment & functions of macrophages. miRNA and lncRNA mediate phenotypic transition of TAMs by promoting M2 to M1 polarization. Further, targeting of miRNAs involved in the inhibition of tumor antigen presentation can improve anticancer immune responses.

4.2 Modulation of natural killer cell mediated cytotoxicity by ncRNAs

Natural killer (NK) cells being cytotoxic in nature are critical in providing innate defense against various infections and to develop anticancer immune responses. The effector function of natural killer cell is determined by the balance between signals originated from activating NK receptor (aNKR) and inhibitory NK receptor (iNKR) (Del Zotto et al., 2017). It has been reported that tumor microenvironment may lead to phenotypic changes and impaired function in NK cells by downregulating the expression of aNKRs or by enhancing the expression of iNKRs (Bi & Tian, 2017; Di Vito et al., 2019). Besides T cell, NK cells also express PD-1 immune check point on the cell surface (Pesce et al., 2019) and raised levels of PD-1^+ NK cells were found in the patients with various cancers (Andre et al., 2018). Thus, recent literature suggests that noncoding RNAs have significant role in NK cell maturation & effector functions under homeostatic conditions as well as malignancies.

MicroRNAs can act as oncogenes or tumor suppressors in various cancers and they are known to alter antitumor activity of NK cells in TME. miR-583 inhibits the process of NK cell differentiation by targeting IL-2 receptor gamma expression (Yun et al., 2014). MicroRNAs are involved in regulating NK cell mediated cytotoxicity and cytokine production. It has been reported that miR-27a-5p negatively regulates the cytotoxic activity of NK cells by targeting both perforin & granzyme B (Kim et al., 2011). In addition, miR-378 also targets granzyme B, while miR-30e and -150 target perforin (Kim et al., 2014; Wang et al., 2012). In contrast, miR-362-5p increases the expression of perforin & granzyme B in human NK cells and promotes its cytotoxic functions (Ni et al., 2020). miR-155 induces natural killer cell activation and IFN-γ secretion by altering the expression of various genes and activating multiple signaling pathways (Sullivan et al., 2013). MicroRNA 218-5p suppresses the cytotoxic activity of NK cells against lung adenocarcinoma through serine hydroxymethyl transferase 1 (Yang et al., 2019). TGF-β induces miR-183 expression to target DAP12 mRNA thereby, decreasing the surface expression of activating NKp44 receptor (Donatelli et al., 2014). Furthermore, miR-182 enhances NK cell cytotoxicity against HCC by modulating expression of NKG2D and NKG2A receptors (Abdelrahman et al., 2016). In addition to microRNAs, lncRNAs also play significant role in modulating the NK cell cytotoxicity. LncRNA GAS5 expression was found to be reduced in NK cells in liver of cancer patients. Overexpression of lncRNA GAS5 inhibits the expression of miR-544 and enhanced IFN-γ secretion, RUNX3 expression and thus, the cytotoxic activity of NK cell (Fang et al., 2019). In contrast, upregulation of lncRNA SNHG12 promotes tumor cell proliferation and inhibit sensitivity to cisplatin in NK/T-cell lymphoma (Zhu, Li, et al., 2019). Upregulated expression of lncRNA CD56 was found to be associated with CD56 (surface marker of NK cells) thus, control development of NK cells (Zhang et al., 2016). NK cell receptor activation induces lncRNA IFNG-AS1 expression, which further promotes IFN-γ release by NK cells (Stein et al., 2019).

Studies cited above emphasize significant role of ncRNAs in regulation of NK cell effector functions and their reprogramming in the TME. Thus, ncRNAs have emerged as suitable targets to improve NK cell mediated antitumor immune responses.

4.3 Regulation of neutrophil plasticity & activity by ncRNAs

Neutrophils are the most abundant leukocytes which provide primary defense against cancer. Studies have revealed the involvement of tumor associated neutrophils (TANs) in the pathogenesis of various tumors. TANs exert antitumor effects by mediating direct cytotoxicity toward malignant cells and reducing tumor cell invasiveness. TANs can promote tumor progression by inducing the expression of angiogenic factors, increasing cell metastasis and by modulating immune cell function. TANs exhibit opposite functions during cancer inception and development attributed to their plasticity in tumor microenvironment (Zhang, Liu, et al., 2020). In TME, upon stimulation by various cytokines and chemokines neutrophils exhibit functional heterogeneity in terms of N1 and N2 polarization which is equivalent to monocyte polarization. N2 like TANs exhibit pro-tumorigenic activities and N1 exhibits cytotoxic activity against tumor cells in a similar manner as M1 macrophages do. Moreover, large scale meta-analysis of transcriptomics showed that TANs were associated with the poor prognosis of various solid tumors and emerged as critical regulator of tumor microenvironment (Mukaida, Sasaki, & Baba, 2020).

Considering the regulatory role of microRNAs, studies are being conducted to explore the link between miRNAs and neutrophils to discover the effector function played by neutrophils in the TME. Recently, miR-3938 and miR-4780 were differentially expressed between N2 type neutrophils (TGF-β1 stimulated) and neutrophils (non-stimulated) of colorectal carcinoma. MicroRNA-gene target networks established using bioinformatics tools revealed that miR-3938 and -4780 regulate ZNF197 and TUSC1.

Thus, uncovers the underlying mechanism of regulation of invasion & metastasis by N2 neutrophils in colorectal carcinoma (Wang et al., 2020). lncRNAs has also been reported to orchestrate neutrophil functions in the TIME. The knockdown of lncRNA MALAT1 reduces the inflammatory injury after lung transplant ischemia reperfusion via decreased activation and infiltration of neutrophils by inhibiting IL-8 (Wei, Li, Han, Chen, & Zhang, 2019). LncRNA XIST upregulates IL-12A expression by directly inhibiting miR-21 and enhances primary graft dysfunction by promoting neutrophil extracellular trap (NET) formation. Inhibition of XIST and NETs can be used as a strategy for the treatment of primary graft dysfunction (Li, Wei, Han, Chen, & Zhang, 2020). Lower levels of lncRNA TCL6 (T-cell leukemia/lymphoma 6) were correlated with neutrophil infiltration and poor prognosis in breast cancer patients (Zhang, Li, et al., 2020). Increased expression of lncRNA HOTTIP stimulated the production & secretion of IL-6 via binding to c-jun, thus resulted in increased PD-L1 expression in TANs. Therefore, lncRNA HOTTIP by inducing PD-L1 expression suppresses T cell activity & facilitates tumor immune evasion (Shang et al., 2019). These studies indicate that ncRNAs influences the tumor growth and progression by modulating neutrophils in TIME.

Considering the crucial role played by ncRNAs in the modulation of immune cell function in the TME, they can be targeted to improve tumor immunity and may serve as promising candidates to develop ncRNA based immunotherapeutics against cancer.

4.4 ncRNA mediated B cell regulation in TIME

B cells are the primary immune mediators in infections, autoimmunity and tumorigenesis (Danger, Braza, Giral, Soulillou, & Brouard, 2014). B cells are also part of TIME and influences the immune escape, immunotherapy response and, ultimately the patient survival. Tumor-infiltrating B cells (TIBs) are involved in tumor progression. Defects at both genetic and epigenetic levels of miRNA and in their processing machinery are hallmark of various diseases particularly the cancer (Zhang, Liu, et al., 2020).

miR-155 plays a substantial role in B-cell lymphoproliferative disorders, e.g., chronic lymphocytic leukemia (CLL) and Waldenstrom macroglobulinemia (WM). LNAs mediated inhibition of miR-155 reduces cell proliferation of WM and CLL (Zhang et al., 2012). Also, miR-155 expression was found to be high in lymphoma while its downregulation promoted cell apoptosis via arresting G0/G1 phase of cell cycle, showing that miR-155 targeting can be explored as treatment strategy for the diffuse large B-cell lymphoma treatment (DLBCL) (Zhu et al., 2016). Interestingly, miR-34a, miR-181a, miR-28 and miR-146a have also been discovered in B cell lymphomas as tumor suppressor genes (Zhang, Liu, et al., 2020). Studies concerning the involvement of B cell lncRNAs in the tumor microenvironment are limited. LncRNA BALR-2 and BALR-6 are shown to promote the progression in B-acute lymphoblastic leukemia (Fernando et al., 2015; Rodriguez-Malavc et al., 2015). LincRNA-p21 and LUNAR1 have been shown to predict clinical outcomes and impair tumorigenesis in DLBCL patients on chemotherapy (Peng & Feng, 2016; Peng, Wu, & Feng, 2017). It was observed that a positive feedback loop is formed by lncRNA-SNHG14/miR-5590-3p/ZEB1 axis leading to lymphoma progression through modulation of PD-1/PD-L1 proteins (Zhao, Liu, Zhang, Liu, & Qi, 2019). Targeting miR-195 by LncRNA-MALAT1 has been found to facilitate immune evasion of lymphoma (Wang, Lian, Song, Huang, & Gong, 2019). LncRNA-FIRRE activated by MYC enables progression of lymphoma by activation of Wnt/β-catenin pathway via regulation of nuclear translocation of β-catenin (Shi et al., 2019). In contrast, lncRNA PANDA induced by p53 via inactivation of MAPK/ERK signaling exerts tumor suppression in human lymphoma (Wang et al., 2017). Similarly, upregulation of lncRNA-OR3A4 modulates proliferation and apoptosis of cells by targeting the Wnt/β-catenin pathway (Meng, Zhao, & Wang, 2020; Zhao, Jiao, et al., 2019). lncRNAs are also associated with the prognosis of B cell related lymphoma, e.g., HOTAIR, PEG10, LncRNA HULC and LUNAR1 are shown to predict poor clinical outcomes, while HOTAIR, LUNAR1, and PEG10 lncRNAs demonstrated diagnostic value in DLBCL (Zhang, Liu, et al., 2020). LncRNA MALAT-1 was found to regulate chemotherapeutic sensitivity via modulating autophagy proteins in DLBCL (Li, Chai, Guo, Chu, & Zhang, 2017).

Overall, studies indicate that miRNAs have a significant role in B-cell lymphoproliferative disorders and lymphomas; however, its role as immunotherapeutic target needs to be explored. Further, limited studies are available on the role of lncRNA and B cells in the tumor microenvironment. Recent studies have shown their involvement in immune evasion and progression of tumor that can be explored for its prognostic value and immunotherapeutic potential in cancer.

4.5 Regulation of T cell effector function by ncRNAs

Chen and Mellman (2013) proposed the concept of cancer immunity cycle and the immune response against cancer, which illustrates the vital role played by T cells in tumorigenesis (Chen & Mellman, 2013). Regulatory T (Tregs) cells are immunosuppressive agents which play a significant role in TIE. Recently, ncRNAs are thought to be involved in TIE and considered to be potential targets against cancer (Liu, Wang, et al., 2020).

4.5.1 $CD4^{+}$ and $CD8^{+}$ T cells

Emerging evidence has suggested the role of ncRNAs against antitumor immune response mediated by T cells. This antitumor response is governed by various mechanisms, i.e., via controlling the function of Tregs, influencing the regulators of T cells, affecting immune checkpoints and regulating production of IFNγ. miR-17-92 cluster is known to control Th1 responses, particularly the loss of miR-17 and miR-19b in $CD4^{+}$ T cells, which results in tumor evasion and regulate Tregs by targeting Bim (Barron et al., 2010; Jiang et al., 2011). miR-491 can inhibit IFN-γ production by T cells via targeting $CD8^{+}$ T cells and subsequent inhibition of antitumor immunity (Yu et al., 2016). ncRNAs regulate T cell mediated immune responses and serve as critical players of TIME. LncRNA-NKILA influences tumor immune evasion by regulating T cell's sensitivity to activation induced cell death via inhibition of NF-κB activity (Huang et al., 2018). In contrast, lncRNA GM16343 decreases tumor growth via affecting the antitumor immune function of $CD8^{+}$ T cells, reducing tumor volume in mice and prolonged survival rate (Mao et al., 2019). Upregulated expression of lnc-Tim3 and activation of anti-apoptotic genes (MDM2 and Bcl-2) exaggerates $CD8^{+}$ T cell exhaustion and diminishes antitumor immunity in HCC patients (Ji et al., 2018). Furthermore, lncRNAs are reported to regulate $CD8^{+}$ T cell apoptosis and expansion via modulating PD-L1 expression, e.g., miR-195 sponged by LncRNA MALAT1 induces tumor cell and $CD8^{+}$ T cell interaction causing their apoptosis through PD-1/PDL1 thus, contributing to tumor immune evasion (Mao et al., 2019; Zhao, Liu, et al., 2019).

4.5.2 Treg cells

Human Treg cells are generally of two types: natural Tregs (nTregs) and induced Tregs (iTregs). nTregs facilitates suppression via cell contact dependent mechanisms and maintains peripheral immune tolerance while iTregs works by contact independent mechanisms and mediate expression of several immunosuppressive molecules, e.g., IL-10, TGF-β and coinhibitory factors like CTLA-4, PD-1 and PD-L1 (Zhang, Liu, et al., 2020). Tregs eventually lead to suppression of effector cells, including CTLs, $CD4^{+}$ T cells, NK cells and DCs perturbing the host's anti-tumor response

(Vignali, Collison, & Workman, 2008). miR-155 has been reported to contribute toward proliferative activity of Tregs via Foxp3 dependent regulation (Lu et al., 2009). TGF-β-miR-34a-CCL22 signaling has been demonstrated to promote Tregs recruitment and enhance venous metastasis, facilitating immune escape of HBV-positive HCC (Yang et al., 2012). LncRNA Flicr expressed in mature Tregs, negatively modulates Foxp3 via influencing Tregs differentiation, function, and stability and is linked with the colorectal cancer prognosis (Saito et al., 2016; Zemmour, Pratama, Loughhead, Mathis, & Benoist, 2017). LncRNA EGFR promotes differentiation of Tregs while inhibits CTLs leading to progression of hepatocellular carcinoma in an EGFR-dependent manner (Jiang et al., 2011). Exosomes comprising lncRNA RP11-323N12.5 secreted by tumor cells target TILs and stimulates differentiation of Treg thereby, promoting tumor growth (Wang et al., 2021). Acting as immunosuppressive agents, Tregs play vital role in TIE and evidences show that ncRNAs helps in modulating the function of T cells. It helps tumor cell to accomplish TIE, that can be effectively targeted to prevent TIE and hence, improving tumor immunotherapy. Extraordinary progress has been made in immune checkpoint blockade, which is another area of cancer immunotherapy. Generally, these checkpoint proteins are confined on surface of T cell and these proteins by functioning as brakes, stop attack on tumor cells. Despite regulating various processes in T cells, ncRNAs are also known to regulate immune checkpoint protein expression in various tumors.

5. ncRNAs in the regulation of immune checkpoint proteins

Immune checkpoint proteins and associated pathways serve as a major mechanism for tumor cells to evade immune recognition. Dysregulation of immune checkpoint proteins can lead to TIE which needs further exploration for targeting tumor progression.

5.1 Cytotoxic T lymphocyte antigen 4 (CTLA-4)

T cells are generally activated by either T cell receptor and MHC molecule interaction or by CD28 and B7-1/2 costimulatory signals (Liu, Wang, et al., 2020). Overexpression of CTLA-4 on the tumor infiltrating T cells will augment immune suppression by modulating T cells and promoting TIE. Expression of CTLA-4 was found to be inhibited by miR-155 and hence, shown to promote the antitumor immune response of the host (Huffaker et al., 2017).

5.2 Programmed cell death protein 1 (PD-1) and its ligand (PD-L1)

PD-1 is expressed primarily on activated T cells, B cells and dendritic cell's surface. PD-1 and PD-L1 interaction via downstream signaling leads to inhibition of the Ras/ERK signaling pathway and inhibits T cell activation. Consequently, it promotes T cell proliferation, apoptosis and cytokine secretion. Number of miRNAs and lncRNAs are documented to regulate the expression of PD-1. miR-138 targets PD-1 expression in mouse GL261 glioma cells and promotes T cell functioning and activity and inhibits tumor development (Wei et al., 2016). PD-L1 (B7-H1) is a transmembrane protein expressed on the surface of T cells, B cells and various tumor cells. Binding of PD-L1 with its receptor PD-1, inhibits cell proliferation and suppresses cytokines secretion by CTLs, thus, promoting TIE (Liu, Wang, et al., 2020). miR-34a, miR-138-5p, miR-25-93-106b cluster, miR-142-5p, miR-217, miR-152, miR-200, miR-221, miR-216a and miR-424 have been documented to inhibit PD-L1 via targeting its mRNA in various tumors, thereby, inhibiting metastasis and increasing antitumor immunity (Liu, Wang, et al., 2020; Zhang, Liu, et al., 2020).

lncRNAs and circular RNAs also regulate PD-1/PD-L1 pathway, e.g., LncRNA SNHG14 activates PD-L1 expression via ZEB1 and promote immune escape by miR-5590-3p in DLBCL (Zhao, Liu, et al., 2019). In lung carcinoma cells, expression of PD-L1 is suppressed at the level of transcription by lncRNA NKX2–1-AS that eventually inhibits immunosuppression and prevents immune evasion (Kathuria et al., 2018). Moreover, expression of hsa_circ_0020397 promotes PD-L1 expression which regulates the colorectal cancer cells invasion, viability and apoptosis (Zhang, Xu, & Wang, 2017).

Apart from these proteins, T cell immunoglobulin and mucin domain containing protein 3 (TIM-3) and B and T lymphocyte attenuator (BTLA) are less explored immune checkpoint proteins but reported to play role in immune escape. For instance, $CD8^+$ T cell exhaustion is stimulated by lncRNA Tim3 in hepatocellular carcinoma and was found to promote exhausted $CD8^+$ T cell survival (Ji et al., 2018). miR-28 inhibits T cells expression of PD-1, CTLA-4 and BTLA and TIM-3, causing reduced T cell exhaustion and increase secretion of TNF-α and IL-2 in melanoma (Li et al., 2016). Moreover, miR-155 has also been reported to inhibit the expression of BTLA which stimulates T cell activation (Liu et al., 2016).

Studies mentioned above indicate that ncRNAs regulate the expression of various immune checkpoint proteins and have an important impact on immune checkpoint blockade therapy. Therapeutic targeting of ncRNAs in combination to anti PD-L1/PD-1 or anti CTLA-4 immunotherapy will enhance their efficacy.

6. Noncoding RNAs as potential targets for cancer immunotherapy

MicroRNAs and lncRNAs acts as important regulators of the immune responses in tumor microenvironment and act by reprogramming of various intrinsic properties of TME, for instance, angiogenesis, EMT, immune escape, drug resistance, etc. (Liu, Wang, et al., 2020). Upregulation of oncogenic and downregulation of tumor suppressor miRNAs/lncRNAs facilitates enhanced cancer cell viability and invasiveness. Considering the influence of ncRNAs on the anticancer immune responses, they can be utilized as potential targets to develop ncRNA based therapeutics to enhance the efficacy of existing immunotherapy approaches. Two basic approaches are extensively used for targeting the ncRNA expression, i.e., oligonucleotides or virus vector constructs to inhibit oncogenic ncRNA or to replace the loss of tumor suppressor ncRNA (Bader, Brown, Stoudemire, & Lammers, 2011). ncRNAs can be therapeutically targeted using various strategies as described below:

6.1 ncRNA mimics: Restoration of tumor suppressor ncRNAs

ncRNA mimics are synthetic double stranded RNAs and acts as mature endogenous microRNAs after their transfection into the cells. This method has been studied in various cancers and successful progress has been recorded. MRX34, a mimic of miR-34a, was the first to enter the clinical trial. miR-34 is a tumor suppressor and its decreased expression was found to be associated with lung, liver, breast & colon cancer. Restoration of miR-34a expression via exogenous miR-34 mimics lead to the inhibition of tumor growth and progression. miR-34a also inhibits PD-L1 expression in acute myeloid leukemia (Wang et al., 2015). Moreover, restoration of miR-424 & -138 expression also interrupts the PD-L1/CD80 and CTLA4/PD1/Foxp3 signaling in tumor infiltrating $CD4^+$ T cells (Wei et al., 2016; Xu et al., 2016). LncRNA mimics are also used to increase the expression of pivotal tumor suppressing lncRNAs.

6.2 Antisense oligomers: Blockage of oncomiRs

Several methods have been investigated to block the binding of oncogenic ncRNAs to the target mRNAs, e.g., antisense oligonucleotides (ASO) & Locked Nucleic Acids (LNAs). LNAs are nucleic acid analogue in which ribose ring is locked by a methylene bridge that connects 2′-oxygen and 4′-carbon. Other chemical modifications (2′-O-methoxyethyl, phosphorothioate linkage, cholesterol-conjugation) have also been incorporated in the sequence of these oligonucleotides to enhance stability, specificity & binding affinity (Krutzfeldt et al., 2005; Weiler, Hunziker, & Hall, 2006). Cobomarsen (MRG106), a synthetic LNA based antimiR-155 has reached clinical trials and has been extensively explored for the inhibition of tumor growth and dissemination *in vivo* by directly targeting miR-155 (Van Roosbroeck et al., 2017). Blockade of miR-23a in engineered cytotoxic $CD8^+$ T cells restored the adoptive T cell transfer therapy and decreased tumor progression by normalizing the expression of transcription factor BLIMP-1 in a murine model of melanoma (Lin et al., 2014). LINK-A LNAs in combination with immune checkpoint blockade therapy significantly reduced the tumor growth, increased the survival rate and displayed synergistic efficacy between both therapies in a mice model (Hu et al., 2019).

6.3 miRNA sponges & miR masks

miRNA sponges contain multiple binding sites (2–7 consecutive nucleotides) for the target oncomiRs and acts as competitive inhibitors of microRNAs (Ebert & Sharp, 2010). miRNA sponge constructs developed against miR-155 and miR-19, inhibited cell growth and invasion by enhancing the expression of SOCS1 and p53 in human myeloma and murine leukemia cells (Mignacca et al., 2016). miRNA sponge increases the target protein HOXD-10 by effectively reducing miR-10b and inhibited the tumor growth, migration & invasiveness of breast cancer (Liang et al., 2016). Overall, miRNA sponges by antagonizing miRNA activity can resume tumor suppressor pathways in tumor cells. miR mask was developed to block the function of oncomiRs. These are single stranded 2′-O-methyl modified antisense oligomers which block the interaction of oncogenic miRNAs to the 3′ UTR of target mRNA and reactivates the genes that were repressed earlier (Li, Feng, Coukos, & Zhang, 2009). Breast cancer cell proliferation and invasiveness was suppressed by miR-196a2* expression but found to be increased by employing TP63 (direct target of miR-196a2*) target protector oligonucleotides

(Kim et al., 2011). However, despite the promising *in vitro* results, miR masks are still not extensively used as ncRNA therapeutics.

6.4 CRISPR/Cas9 systems: ncRNA biogenesis inhibition

CRISPR/Cas9 system becoming increasingly popular nowadays to develop ncRNA based cancer therapy by introducing mutations in the Drosha/Dicer processing site of miRNA precursors that lead to the inhibition of biogenesis and decreased mature miRNA levels in the cells. Inhibition of miR-17, -21, -141 and -3188 using CRISPR/Cas9 systems resulted in significant inhibition of proliferation, invasion, migration, epithelial-mesenchymal transition as well as increased induction of apoptosis in cancer cells (Huo et al., 2017; Yang et al., 2018). This strategy has also been proposed to mediate inhibition or activation of oncogenic lncRNA expression (Zhang, Li, et al., 2020). Increased specificity and stability of CRISPR/Cas9 system makes it a promising strategy to block particular oncogenic miRNAs.

Moreover, application of artificial ribonucleases, RNA zippers, small molecule inhibitors of ncRNAs, aptamers and RNA destabilizing elements also give promising results to target oncogenic ncRNAs (Gumireddy et al., 2008; Meng et al., 2017; Patutina et al., 2018; Li, Wei, et al., 2020).

7. Challenges of ncRNA based immunotherapeutics

Though preclinical studies involving ncRNA based therapeutics have been conducted over the years; only a few ncRNAs have reached the clinical development stage so far. Translation of ncRNA based therapeutics into clinical practice is majorly affected by immunogenicity, specificity and delivery routes. The off-targeting effects are attributed to either similarity in sequence or excessive uptake of ncRNAs in the cell. To intercept the deleterious effects of miRNA mimics & antimiRs in undesired cells, ncRNA therapeutics can be expressed using vectors comprising of promoter sequence whose expression is being upregulated in the cells of interest. Specificity can further be enhanced by directly targeting precursor microRNAs using peptides, nucleic acid oligonucleotides that bind to pre-miRNA and inhibit miRNA maturation. Circular RNAs act as miRNA sponges and are found to be more stable than the linear RNAs. Thus, synthetic circular RNAs can be explored to block the expression of oncogenic miRNAs in the near future (Garzon, Marcucci, & Croce, 2010; Winkle, El-Daly, Fabbri, & Calin, 2021).

Recognition of ncRNA oligonucleotides by Toll like receptors (TLRs) resulted in adverse immune effects, thus, leading to the termination of clinical trials, e.g., miR-34 mimic (MRX34) trial. MRX34 was found to have significant antitumor therapeutic effect in preclinical research. However, patients developed severe immune allied toxicities during the trial (Garzon et al., 2010; Winkle et al., 2021). Second phase I clinical trial of MRX34 showed that premedication with dexamethasone was necessary to manage adverse immune events. RNA based therapeutics are predominantly recognized by TLR signaling stimulation which is dependent on GU-rich RNA sequences and lead to the induction of proinflammatory immune response. Usage of chemically modified ncRNA based therapeutics was found to reduce the immunogenecity; however, even after these modifications in ncRNA oligos, the challenge is not overcome yet and needs to be explored further. Extensive screening of immune associated adverse reaction with TLR interaction will help in selecting ncRNA therapeutics having poor immunogenicity before the initiation of clinical trials. Furthermore, decreasing the length of antisense oligonucleotides (LNA with 7–8 nucleotides long cannot activate TLRs), use of limited doses of ncRNA for prolonged period of time and using ncRNA therapeutics in combination with other existing therapeutics will obviously circumvent the adverse immune responses (Garzon et al., 2010; Winkle et al., 2021).

Efficacy & potency of ncRNA therapeutics was also affected by instability of naked, chemically unmodified oligonucleotides and poor intracellular delivery due to lack of suitable delivery vehicles. To overcome this challenge, various delivery vehicles are being employed like viral vectors, lipid nanoparticles, polymer based delivery system, ligand oligonucleotide conjugates, cationic lipids, cell-penetrating peptides and metal based nanoparticles (Garzon et al., 2010; Winkle et al., 2021). Two alternative approaches that can serve as promising vehicle to be utilized clinically for the delivery of ncRNA therapeutics are exosomes and bacteriophage. Exosomes are cell derived membrane vesicles involved in cell to cell communication by delivering different cargos (protein, lipids, DNA, RNA & ncRNA). Thus, making exosomes more favorable to be considered as natural transporters of ncRNAs and may become an ideal delivery vehicle owing to their negligible immunogenecity, cytotoxicity and ability to evade endocytic pathway and phagocytosis. It has been demonstrated that bacteriophage can be used as safe and efficient delivery vehicle, e.g., pre-miR-146a covalently linked to modified MS2 bacteriophage considerably increased miR-146a expression and decreased target gene expression under both *ex vivo* & *in vivo* conditions (Garzon et al., 2010; Winkle et al., 2021).

8. Future perspectives

Non coding RNAs are critical for the remodeling of tumor immune environment by modulating diverse biological processes like macrophage polarization/activation, NK cell cytotoxicity, neutrophil plasticity and T cell effector function, etc., thereby regulating the underlying mechanisms of tumor immune evasion. Literature suggests that immune regulatory function of ncRNAs make them more efficacious therapeutic target for antitumor immunotherapies. Despite the continuous preclinical research for validating the therapeutic effects of ncRNAs, a huge gap still exists between preclinical findings & clinical application. Further research is required to explore suitable delivery vehicles and careful monitoring of toxicity at preclinical stage for the development of improved ncRNA based immunotherapeutics against cancer.

References

Abdelrahman, M. M., Fawzy, I. O., Bassiouni, A. A., Gomaa, A. I., Esmat, G., Waked, I., et al. (2016). Enhancing NK cell cytotoxicity by miR-182 in hepatocellular carcinoma. *Human Immunology*, 77(8), 667–673. https://doi.org/10.1016/j.humimm.2016.04.020.

Akkın, S., Varan, G., & Bilensoy, E. (2021). A review on cancer immunotherapy and applications of nanotechnology to chemoimmunotherapy of different cancers. *Molecules*, *26*, 3382–3405. https://doi.org/10.3390/molecules26113382.

Andre, P., Denis, C., Soulas, C., Bourbon-Caillet, C., Lopez, J., Arnoux, T., et al. (2018). Anti-NKG2A mAb is a checkpoint inhibitor that promotes anti-tumor immunity by unleashing both T and NK cells. *Cell*, *175*(7), 1731–1743. https://doi.org/10.1016/j.cell.2018.10.014.

Bader, A. G., Brown, D., Stoudemire, J., & Lammers, P. (2011). Developing therapeutic microRNAs for cancer. *Gene Therapy*, *18*(12), 1121–1126. https://doi.org/10.1038/gt.2011.79.

Barron, L., Dooms, H., Hoyer, K. K., Kuswanto, W., Hofmann, J., O'Gorman, W. E., et al. (2010). Cutting edge: Mechanisms of IL-2-dependent maintenance of functional regulatory T cells. *Journal of Immunology*, *185*(11), 6426–6430. https://doi.org/10.4049/jimmunol.0903940.

Bhan, A., Soleimani, M., & Mandal, S. S. (2017). Long noncoding RNA and cancer: A new paradigm. *Cancer Research*, 77(15), 3965–3981. https://doi.org/10.1158/0008-5472.CAN-16-2634.

Bi, J., & Tian, Z. (2017). NK cell exhaustion. *Frontiers in Immunology*, *8*, 760. https://doi.org/10.3389/fimmu.2017.00760.

Binenbaum, Y., Fridman, E., Yaari, Z., Milman, N., Schroeder, A., Ben David, G., et al. (2018). Transfer of miRNA in macrophage-derived exosomes induces drug resistance in pancreatic adenocarcinoma. *Cancer Research*, *78*(18), 5287–5299. https://doi.org/10.1158/0008-5472.CAN-18-0124.

Cai, J., Qiao, B., Gao, N., Lin, N., & He, W. (2019). Oral squamous cell carcinoma-derived exosomes promote M2 subtype macrophage polarization mediated by exosome-enclosed miR-29a-3p. *American Journal of Cell Physiology*, *316*(5), C731–C740. https://doi.org/10.1152/ajpcell.00366.2018.

Casucci, M., & Bondanza, A. (2011). Suicide gene therapy to increase the safety of chimeric antigen receptor-redirected T lymphocytes. *Journal of Cancer, 2*, 378–382. https://doi.org/10.7150/jca.2.378.

Chatterjee, B., Saha, P., Bose, S., Shukla, D., Chatterjee, N., Kumar, S., et al. (2020). MicroRNAs: As critical regulators of tumor-associated macrophages. *International Journal of Molecular Sciences, 21*(19). https://doi.org/10.3390/ijms21197117.

Chen, D. S., & Mellman, I. (2013). Oncology meets immunology: The cancer-immunity cycle. *Immunity, 39*(1), 1–10. https://doi.org/10.1016/j.immuni.2013.07.012.

Colangelo, T., Polcaro, G., Ziccardi, P., Pucci, B., Muccillo, L., Galgani, M., et al. (2016). Proteomic screening identifies calreticulin as a miR-27a direct target repressing MHC class I cell surface exposure in colorectal cancer. *Cell Death & Disease, 7*, e2120. https://doi.org/10.1038/cddis.2016.28.

Curtale, G. (2018). MiRNAs at the crossroads between innate immunity and cancer: Focus on macrophages. *Cell, 7*(2), 1–25. https://doi.org/10.3390/cells7020012.

Danger, R., Braza, F., Giral, M., Soulillou, J. P., & Brouard, S. (2014). MicroRNAs, major players in B cells homeostasis and function. *Frontiers in Immunology, 5*, 98–106. https://doi.org/10.3389/fimmu.2014.00098.

Davola, M. E., & Mossman, K. L. (2019). Oncolytic viruses: How "lytic" must they be for therapeutic efficacy? *Oncoimmunology, 8*(6), e1581528. https://doi.org/10.1080/2162402X.2019.1596006.

Del Zotto, G., Marcenaro, E., Vacca, P., Sivori, S., Pende, D., Della Chiesa, M., et al. (2017). Markers and function of human NK cells in normal and pathological conditions. *Cytometry. Part B, Clinical Cytometry, 92*(2), 100–114. https://doi.org/10.1002/cyto.b.21508.

Di Vito, C., Mikulak, J., Zaghi, E., Pesce, S., Marcenaro, E., & Mavilio, D. (2019). NK cells to cure cancer. *Seminars in Immunology, 41*, 101272–101282. https://doi.org/10.1016/j.smim.2019.03.004.

Donatelli, S. S., Zhou, J. M., Gilvary, D. L., Eksioglu, E. A., Chen, X., Cress, W. D., et al. (2014). TGF-beta-inducible microRNA-183 silences tumor-associated natural killer cells. *Proceedings of the National Academy of Sciences of the United States of America, 111*(11), 4203–4208. https://doi.org/10.1073/pnas.1319269111.

Duan, Z., & Luo, Y. (2021). Targeting macrophages in cancer immunotherapy. *Signal Transduction and Targeted Therapy, 6*(1), 127. https://doi.org/10.1038/s41392-021-00506-6.

Ebert, M. S., & Sharp, P. A. (2010). MicroRNA sponges: Progress and possibilities. *RNA, 16*(11), 2043–2050. https://doi.org/10.1261/rna.2414110.

Fang, P., Xiang, L., Chen, W., Li, S., Huang, S., Li, J., et al. (2019). LncRNA GAS5 enhanced the killing effect of NK cell on liver cancer through regulating miR-544/RUNX3. *Innate Immunity, 25*(2), 99–109. https://doi.org/10.1177/1753425919827632.

Fernando, T. R., Rodriguez-Malave, N. I., Waters, E. V., Yan, W., Casero, D., Basso, G., et al. (2015). LncRNA expression discriminates karyotype and predicts survival in B-lymphoblastic leukemia. *Molecular Cancer Research, 13*(5), 839–851. https://doi.org/10.1158/1541-7786.MCR-15-0006-T.

Garzon, R., Marcucci, G., & Croce, C. M. (2010). Targeting microRNAs in cancer: Rationale, strategies and challenges. *Nature Reviews. Drug Discovery, 9*(10), 775–789. https://doi.org/10.1038/nrd3179.

Ghani, S., Riemke, P., Schonheit, J., Lenze, D., Stumm, J., Hoogenkamp, M., et al. (2011). Macrophage development from HSCs requires PU.1-coordinated microRNA expression. *Blood, 118*(8), 2275–2284. https://doi.org/10.1182/blood-2011-02-335141.

Ginhoux, F., & Jung, S. (2014). Monocytes and macrophages: Developmental pathways and tissue homeostasis. *Nature Reviews. Immunology, 14*(6), 392–404. https://doi.org/10.1038/nri3671.

Gumireddy, K., Young, D. D., Xiong, X., Hogenesch, J. B., Huang, Q., & Deiters, A. (2008). Small-molecule inhibitors of microrna miR-21 function. *Angewandte Chemie (International Ed. in English)*, *47*(39), 7482–7484. https://doi.org/10.1002/anie.200801555.

Hassett, B., Singh, E., Mahgoub, E., O'Brien, J., Vicik, S. M., & Fitzpatrick, B. (2018). Manufacturing history of etanercept (Enbrel((R))): Consistency of product quality through major process revisions. *MAbs*, *10*(1), 159–165. https://doi.org/10.1080/19420862.2017.1388483.

Hu, Z., Ott, P. A., & Wu, C. J. (2018). Towards personalized, tumour-specific, therapeutic vaccines for cancer. *Nature Reviews. Immunology*, *18*(3), 168–182. https://doi.org/10.1038/nri.2017.131.

Hu, Q., Ye, Y., Chan, L. C., Li, Y., Liang, K., Lin, A., et al. (2019). Oncogenic lncRNA downregulates cancer cell antigen presentation and intrinsic tumor suppression. *Nature Immunology*, *20*(7), 835–851. https://doi.org/10.1038/s41590-019-0400-7.

Huang, D., Chen, J., Yang, L., Ouyang, Q., Li, J., Lao, L., et al. (2018). NKILA lncRNA promotes tumor immune evasion by sensitizing T cells to activation-induced cell death. *Nature Immunology*, *19*(10), 1112–1125. https://doi.org/10.1038/s41590-018-0207-y.

Huffaker, T. B., Lee, S. H., Tang, W. W., Wallace, J. A., Alexander, M., Runtsch, M. C., et al. (2017). Antitumor immunity is defective in T cell-specific microRNA-155-deficient mice and is rescued by immune checkpoint blockade. *The Journal of Biological Chemistry*, *292*(45), 18530–18541. https://doi.org/10.1074/jbc.M117.808121.

Huo, W., Zhao, G., Yin, J., Ouyang, X., Wang, Y., Yang, C., et al. (2017). Lentiviral CRISPR/Cas9 vector mediated miR-21 gene editing inhibits the epithelial to mesenchymal transition in ovarian cancer cells. *Journal of Cancer*, *8*(1), 57–64. https://doi.org/10.7150/jca.16723.

Ji, J., Yin, Y., Ju, H., Xu, X., Liu, W., Fu, Q., et al. (2018). Long non-coding RNA Lnc-Tim3 exacerbates CD8 T cell exhaustion via binding to Tim-3 and inducing nuclear translocation of Bat3 in HCC. *Cell Death & Disease*, *9*(5), 478. https://doi.org/10.1038/s41419-018-0528-7.

Jiang, S., Li, C., Olive, V., Lykken, E., Feng, F., Sevilla, J., et al. (2011). Molecular dissection of the miR-17-92 cluster's critical dual roles in promoting Th1 responses and preventing inducible Treg differentiation. *Blood*, *118*(20), 5487–5497. https://doi.org/10.1182/blood-2011-05-355644.

Kathuria, H., Millien, G., McNally, L., Gower, A. C., Tagne, J. B., Cao, Y., et al. (2018). NKX2-1-AS1 negatively regulates CD274/PD-L1, cell-cell interaction genes, and limits human lung carcinoma cell migration. *Scientific Reports*, *8*(1), 14418. https://doi.org/10.1038/s41598-018-32793-5.

Kim, N., Kim, M., Yun, S., Doh, J., Greenberg, P. D., Kim, T. D., et al. (2014). MicroRNA-150 regulates the cytotoxicity of natural killers by targeting perforin-1. *The Journal of Allergy and Clinical Immunology*, *134*(1), 195–203. https://doi.org/10.1016/j.jaci.2014.02.018.

Kim, T. D., Lee, S. U., Yun, S., Sun, H. N., Lee, S. H., Kim, J. W., et al. (2011). Human microRNA-27a* targets Prf1 and GzmB expression to regulate NK-cell cytotoxicity. *Blood*, *118*(20), 5476–5486. https://doi.org/10.1182/blood-2011-04-347526.

Krutzfeldt, J., Rajewsky, N., Braich, R., Rajeev, K. G., Tuschl, T., Manoharan, M., et al. (2005). Silencing of microRNAs in vivo with 'antagomirs'. *Nature*, *438*(7068), 685–689. https://doi.org/10.1038/nature04303.

Lazaridou, M. F., Massa, C., Handke, D., Mueller, A., Friedrich, M., Subbarayan, K., et al. (2020). Identification of microRNAs targeting the transporter associated with antigen processing TAP1 in melanoma. *Journal of Clinical Medicine*, *9*(9), 1–26. https://doi.org/10.3390/jcm9092690.

Li, L. J., Chai, Y., Guo, X. J., Chu, S. L., & Zhang, L. S. (2017). The effects of the long non-coding RNA MALAT-1 regulated autophagy-related signaling pathway on chemotherapy resistance in diffuse large B-cell lymphoma. *Biomedicine & Pharmacotherapy, 89*, 939–948. https://doi.org/10.1016/j.biopha.2017.02.011.

Li, C., Feng, Y., Coukos, G., & Zhang, L. (2009). Therapeutic microRNA strategies in human cancer. *The AAPS Journal, 11*(4), 747–757. https://doi.org/10.1208/s12248-009-9145-9.

Li, Z., Feng, C., Guo, J., Hu, X., & Xie, D. (2020). GNAS-AS1/miR-4319/NECAB3 axis promotes migration and invasion of non-small cell lung cancer cells by altering macrophage polarization. *Functional & Integrative Genomics, 20*(1), 17–28. https://doi.org/10.1007/s10142-019-00696-x.

Li, Q., Johnston, N., Zheng, X., Wang, H., Zhang, X., Gao, D., et al. (2016). miR-28 modulates exhaustive differentiation of T cells through silencing programmed cell death-1 and regulating cytokine secretion. *Oncotarget*, 7(33), 53735–53750. https://doi.org/10.18632/oncotarget.10731.

Li, J., Wei, L., Han, Z., Chen, Z., & Zhang, Q. (2020). Long non-coding RNA X-inactive specific transcript silencing ameliorates primary graft dysfunction following lung transplantation through microRNA-21-dependent mechanism. *eBioMedicine, 52*, 102600. https://doi.org/10.1016/j.ebiom.2019.102600.

Liang, A. L., Zhang, T. T., Zhou, N., Wu, C. Y., Lin, M. H., & Liu, Y. J. (2016). MiRNA-10b sponge: An anti-breast cancer study in vitro. *Oncology Reports, 35*(4), 1950–1958. https://doi.org/10.3892/or.2016.4596.

Lin, R., Chen, L., Chen, G., Hu, C., Jiang, S., Sevilla, J., et al. (2014). Targeting miR-23a in CD8+ cytotoxic T lymphocytes prevents tumor-dependent immunosuppression. *The Journal of Clinical Investigation, 124*(12), 5352–5367. https://doi.org/10.1172/JCI76561.

Liu, J., Ding, D., Jiang, Z., Du, T., Liu, J., & Kong, Z. (2019). Long non-coding RNA CCAT1/miR-148a/PKCzeta prevents cell migration of prostate cancer by altering macrophage polarization. *Prostate, 79*(1), 105–112. https://doi.org/10.1002/pros.23716.

Liu, Y., Li, X., Zhang, Y., Wang, H., Rong, X., Peng, J., et al. (2019). An miR-340-5p-macrophage feedback loop modulates the progression and tumor microenvironment of glioblastoma multiforme. *Oncogene, 38*(49), 7399–7415. https://doi.org/10.1038/s41388-019-0952-x.

Liu, Y., Nie, W., Jin, Y., Zhuo, A., Zang, Y., & Xiu, Q. (2016). B and T lymphocyte attenuator is a target of miR-155 during naive CD4 + T cell activation. *Iranian Journal of Immunology, 13*(2), 89–99.

Liu, L., Wang, Q., Qiu, Z., Kang, Y., Liu, J., Ning, S., et al. (2020). Noncoding RNAs: The shot callers in tumor immune escape. *Signal Transduction and Targeted Therapy, 5*(1), 102. https://doi.org/10.1038/s41392-020-0194-y.

Liu, S. Q., Zhou, Z. Y., Dong, X., Guo, L., & Zhang, K. J. (2020). LncRNA GNAS-AS1 facilitates ER+ breast cancer cells progression by promoting M2 macrophage polarization via regulating miR-433-3p/GATA3 axis. *Bioscience Reports, 40*(7), 1–14. https://doi.org/10.1042/BSR20200626.

Lu, L. F., Thai, T. H., Calado, D. P., Chaudhry, A., Kubo, M., Tanaka, K., et al. (2009). Foxp3-dependent microRNA155 confers competitive fitness to regulatory T cells by targeting SOCS1 protein. *Immunity, 30*(1), 80–91. https://doi.org/10.1016/j.immuni.2008.11.010.

Mao, D., Hu, C., Zhang, J., Feng, C., Zhang, Z., Wang, J., et al. (2019). Long noncoding RNA GM16343 promotes IL-36beta to regulate tumor microenvironment by CD8(+)T cells. *Technology in Cancer Research & Treatment, 18*, 1533033819883633. https://doi.org/10.1177/1533033819883633.

Mari, L., Hoefnagel, S. J. M., Zito, D., van de Meent, M., van Endert, P., Calpe, S., et al. (2018). microRNA 125a regulates MHC-I expression on esophageal adenocarcinoma cells, associated with suppression of antitumor immune response and poor outcomes of patients. *Gastroenterology, 155*(3), 784–798. https://doi.org/10.1053/j.gastro.2018.06.030.

Meng, L., Liu, C., Lu, J., Zhao, Q., Deng, S., Wang, G., et al. (2017). Small RNA zippers lock miRNA molecules and block miRNA function in mammalian cells. *Nature Communications*, *8*, 13964. https://doi.org/10.1038/ncomms13964.

Meng, H., Zhao, B., & Wang, Y. (2020). FOXM1-induced upregulation of lncRNA OR3A4 promotes the progression of diffuse large B-cell lymphoma via Wnt/beta-catenin signaling pathway. *Experimental and Molecular Pathology*, *115*, 104451. https://doi.org/10.1016/j.yexmp.2020.104451.

Mi, J., et al. (2021). The research progress in immunotherapy of tuberculosis. *Frontiers in Cellular and Infection Microbiology*, *11*. https://doi.org/10.3389/fcimb.2021.763591.

Mignacca, L., Saint-Germain, E., Benoit, A., Bourdeau, V., Moro, A., & Ferbeyre, G. (2016). Sponges against miR-19 and miR-155 reactivate the p53-Socs1 axis in hematopoietic cancers. *Cytokine*, *82*, 80–86. https://doi.org/10.1016/j.cyto.2016.01.015.

Mir, S. A., & Sharma, S. (2014). Immunotherapeutic potential of N-formylated peptides of ESAT-6 and glutamine synthetase in experimental tuberculosis. *International Immunopharmacology*, *18*(2), 298–303. https://doi.org/10.1016/j.intimp.2013.09.010.

Mir, S. A., & Sharma, S. (2020). Adjunctive immunotherapeutic efficacy of N-formylated internal peptide of mycobacterial glutamine synthetase in mouse model of tuberculosis. *Protein & Peptide Letters*, *27*(3), 236–242. https://doi.org/10.2174/0929866526666191028151615.

Mir, S. A., Verma, I., & Sharma, S. (2014). Immunotherapeutic potential of recombinant ESAT-6 protein in mouse model of experimental tuberculosis. *Immunology Letters*, *158*(1–2), 88–94. https://doi.org/10.1016/j.imlet.2013.12.007.

Mukaida, N., Sasaki, S. I., & Baba, T. (2020). Two-faced roles of tumor-associated neutrophils in cancer development and progression. *International Journal of Molecular Sciences*, *21*(10), 1–22. https://doi.org/10.3390/ijms21103457.

Naran, K., Nundalall, T., Chetty, S., & Barth, S. (2018). Principles of immunotherapy: Implications for treatment strategies in cancer and infectious diseases. *Frontiers in Microbiology*, *9*, 3158. https://doi.org/10.3389/fmicb.2018.03158.

Ni, C., Fang, Q. Q., Chen, W. Z., Jiang, J. X., Jiang, Z., Ye, J., et al. (2020). Breast cancer-derived exosomes transmit lncRNA SNHG16 to induce CD73+gammadelta1 Treg cells. *Signal Transduction and Targeted Therapy*, *5*(1), 41. https://doi.org/10.1038/s41392-020-0129-7.

Patutina, O. A., Bazhenov, M. A., Miroshnichenko, S. K., Mironova, N. L., Pyshnyi, D. V., Vlassov, V. V., et al. (2018). Peptide-oligonucleotide conjugates exhibiting pyrimidine-X cleavage specificity efficiently silence miRNA target acting synergistically with RNase H. *Scientific Reports*, *8*(1), 14990. https://doi.org/10.1038/s41598-018-33331-z.

Peng, W., & Feng, J. (2016). Long noncoding RNA LUNAR1 associates with cell proliferation and predicts a poor prognosis in diffuse large B-cell lymphoma. *Biomedicine & Pharmacotherapy*, 77, 65–71. https://doi.org/10.1016/j.biopha.2015.12.001.

Peng, W., Wu, J., & Feng, J. (2017). LincRNA-p21 predicts favorable clinical outcome and impairs tumorigenesis in diffuse large B cell lymphoma patients treated with R-CHOP chemotherapy. *Clinical and Experimental Medicine*, *17*(1), 1–8. https://doi.org/10.1007/s10238-015-0396-8.

Pesce, S., Greppi, M., Grossi, F., Del Zotto, G., Moretta, L., Sivori, S., et al. (2019). PD/1-PD-ls checkpoint: Insight on the potential role of NK cells. *Frontiers in Immunology*, *10*, 1242. https://doi.org/10.3389/fimmu.2019.01242.

Ramamurthy, D., Nundalall, T., Cingo, S., Mungra, N., Karaan, M., Naran, K., et al. (2021). Recent advances in immunotherapies against infectious diseases. *Immunotherapy Advances*, *1*(1), 1–27. https://doi.org/10.1093/immadv/ltaa007.

Rodriguez-Malave, N. I., Fernando, T. R., Patel, P. C., Contreras, J. R., Palanichamy, J. K., Tran, T. M., et al. (2015). BALR-6 regulates cell growth and cell survival in B-lymphoblastic leukemia. *Molecular Cancer*, *14*, 214. https://doi.org/10.1186/s12943-015-0485-z.

Sabado, R. L., Balan, S., & Bhardwaj, N. (2017). Dendritic cell-based immunotherapy. *Cell Research*, *27*(1), 74–95. https://doi.org/10.1038/cr.2016.157.

Saito, T., Nishikawa, H., Wada, H., Nagano, Y., Sugiyama, D., Atarashi, K., et al. (2016). Two FOXP3(+)CD4(+) T cell subpopulations distinctly control the prognosis of colorectal cancers. *Nature Medicine*, *22*(6), 679–684. https://doi.org/10.1038/nm.4086.

Shang, A., Wang, W., Gu, C., Chen, C., Zeng, B., Yang, Y., et al. (2019). Long non-coding RNA HOTTIP enhances IL-6 expression to potentiate immune escape of ovarian cancer cells by upregulating the expression of PD-L1 in neutrophils. *Journal of Experimental & Clinical Cancer Research*, *38*(1), 411. https://doi.org/10.1186/s13046-019-1394-6.

Shi, X., Cui, Z., Liu, X., Wu, S., Wu, Y., Fang, F., et al. (2019). LncRNA FIRRE is activated by MYC and promotes the development of diffuse large B-cell lymphoma via Wnt/beta-catenin signaling pathway. *Biochemical and Biophysical Research Communications*, *510*(4), 594–600. https://doi.org/10.1016/j.bbrc.2019.01.105.

Stein, N., Berhani, O., Schmiedel, D., Duev-Cohen, A., Seidel, E., Kol, I., et al. (2019). IFNG-AS1 enhances interferon gamma production in human natural killer cells. *iScience*, *11*, 466–473. https://doi.org/10.1016/j.isci.2018.12.034.

Su, M. J., Aldawsari, H., & Amiji, M. (2016). Pancreatic cancer cell exosome-mediated macrophage reprogramming and the role of MicroRNAs 155 and 125b2 transfection using nanoparticle delivery systems. *Scientific Reports*, *6*, 30110. https://doi.org/10.1038/srep30110.

Sullivan, R. P., Fogel, L. A., Leong, J. W., Schneider, S. E., Wong, R., Romee, R., et al. (2013). MicroRNA-155 tunes both the threshold and extent of NK cell activation via targeting of multiple signaling pathways. *Journal of Immunology*, *191*(12), 5904–5913. https://doi.org/10.4049/jimmunol.1301950.

Sun, Y., & Xu, J. (2019). TCF-4 regulated lncRNA-XIST promotes M2 polarization of macrophages and is associated with lung cancer. *Oncotargets and Therapy*, *12*, 8055–8062. https://doi.org/10.2147/OTT.S210952.

Van Roosbroeck, K., Fanini, F., Setoyama, T., Ivan, C., Rodriguez-Aguayo, C., Fuentes-Mattei, E., et al. (2017). Combining anti-Mir-155 with chemotherapy for the treatment of lung cancers. *Clinical Cancer Research*, *23*(11), 2891–2904. https://doi.org/10.1158/1078-0432.CCR-16-1025.

Varade, J., Magadan, S., & Gonzalez-Fernandez, A. (2021). Human immunology and immunotherapy: Main achievements and challenges. *Cellular & Molecular Immunology*, *18*(4), 805–828. https://doi.org/10.1038/s41423-020-00530-6.

Vignali, D. A., Collison, L. W., & Workman, C. J. (2008). How regulatory T cells work. *Nature Reviews. Immunology*, *8*(7), 523–532. https://doi.org/10.1038/nri2343.

Wang, P., Gu, Y., Zhang, Q., Han, Y., Hou, J., Lin, L., et al. (2012). Identification of resting and type I IFN-activated human NK cell miRNomes reveals microRNA-378 and microRNA-30e as negative regulators of NK cell cytotoxicity. *Journal of Immunology*, *189*(1), 211–221. https://doi.org/10.4049/jimmunol.1200609.

Wang, J., Huang, F., Shi, Y., Zhang, Q., Xu, S., Yao, Y., et al. (2021). RP11-323N12.5 promotes the malignancy and immunosuppression of human gastric cancer by increasing YAP1 transcription. *Gastric Cancer*, *24*(1), 85–102. https://doi.org/10.1007/s10120-020-01099-9.

Wang, X., Li, J., Dong, K., Lin, F., Long, M., Ouyang, Y., et al. (2015). Tumor suppressor miR-34a targets PD-L1 and functions as a potential immunotherapeutic target in acute myeloid leukemia. *Cellular Signalling*, *27*(3), 443–452. https://doi.org/10.1016/j.cellsig.2014.12.003.

Wang, Q. M., Lian, G. Y., Song, Y., Huang, Y. F., & Gong, Y. (2019). LncRNA MALAT1 promotes tumorigenesis and immune escape of diffuse large B cell lymphoma by sponging miR-195. *Life Sciences*, *231*, 116335. https://doi.org/10.1016/j.lfs.2019.03.040.

Wang, L., Yang, J., Huang, J., Wen, Z. Q., Xu, N., Liu, X., et al. (2020). miRNA expression profile in the N2 phenotype neutrophils of colorectal cancer and screen of putative key miRNAs. *Cancer Management and Research*, *12*, 5491–5503. https://doi.org/10.2147/CMAR.S251427.

Wang, Y., Zhang, M., Xu, H., Wang, Y., Li, Z., Chang, Y., et al. (2017). Discovery and validation of the tumor-suppressive function of long noncoding RNA PANDA in human diffuse large B-cell lymphoma through the inactivation of MAPK/ERK signaling pathway. *Oncotarget*, *8*(42), 72182–72196. https://doi.org/10.18632/oncotarget.20053.

Wei, L., Li, J., Han, Z., Chen, Z., & Zhang, Q. (2019). Silencing of lncRNA MALAT1 prevents inflammatory injury after lung transplant ischemia-reperfusion by downregulation of IL-8 via p300. *Molecular Therapy- -Nucleic Acids*, *18*, 285–297. https://doi.org/10.1016/j.omtn.2019.05.009.

Wei, J., Nduom, E. K., Kong, L. Y., Hashimoto, Y., Xu, S., Gabrusiewicz, K., et al. (2016). MiR-138 exerts anti-glioma efficacy by targeting immune checkpoints. *Neuro-Oncology*, *18*(5), 639–648. https://doi.org/10.1093/neuonc/nov292.

Weiler, J., Hunziker, J., & Hall, J. (2006). Anti-miRNA oligonucleotides (AMOs): Ammunition to target miRNAs implicated in human disease? *Gene Therapy*, *13*(6), 496–502. https://doi.org/10.1038/sj.gt.3302654.

Winkle, M., El-Daly, S. M., Fabbri, M., & Calin, G. A. (2021). Noncoding RNA therapeutics—Challenges and potential solutions. *Nature Reviews. Drug Discovery*, *20*(8), 629–651. https://doi.org/10.1038/s41573-021-00219-z.

Wraith, D. C. (2017). The future of immunotherapy: A 20-year perspective. *Frontiers in Immunology*, *8*, 1668. https://doi.org/10.3389/fimmu.2017.01668.

Xu, S., Tao, Z., Hai, B., Liang, H., Shi, Y., Wang, T., et al. (2016). miR-424(322) reverses chemoresistance via T-cell immune response activation by blocking the PD-L1 immune checkpoint. *Nature Communications*, 7, 11406. https://doi.org/10.1038/ncomms11406.

Xu, L., Xu, W., Jiang, Z., Zhang, F., Chu, Y., & Xiong, S. (2009). Depletion of CD4(+)CD25(high) regulatory T cells from tumor infiltrating lymphocytes predominantly induces Th1 type immune response in vivo which inhibits tumor growth in adoptive immunotherapy. *Cancer Biology & Therapy*, *8*(1), 66–72. https://doi.org/10.4161/cbt.8.1.7131.

Yang, P., Li, Q. J., Feng, Y., Zhang, Y., Markowitz, G. J., Ning, S., et al. (2012). TGF-beta-miR-34a-CCL22 signaling-induced Treg cell recruitment promotes venous metastases of HBV-positive hepatocellular carcinoma. *Cancer Cell*, *22*(3), 291–303. https://doi.org/10.1016/j.ccr.2012.07.023.

Yang, Q., Li, J., Hu, Y., Tang, X., Yu, L., Dong, L., et al. (2019). MiR-218-5p suppresses the killing effect of natural killer cell to lung adenocarcinoma by targeting SHMT1. *Yonsei Medical Journal*, *60*(6), 500–508. https://doi.org/10.3349/ymj.2019.60.6.500.

Yang, J., Meng, X., Pan, J., Jiang, N., Zhou, C., Wu, Z., et al. (2018). CRISPR/Cas9-mediated noncoding RNA editing in human cancers. *RNA Biology*, *15*(1), 35–43. https://doi.org/10.1080/15476286.2017.1391443.

Yron, I., Wood, T. A., Jr., Spiess, P. J., & Rosenberg, S. A. (1980). In vitro growth of murine T cells. V. the isolation and growth of lymphoid cells infiltrating syngeneic solid tumors. *Journal of Immunology*, *125*(1), 238–245. http://www.ncbi.nlm.nih.gov/pubmed/6966652.

Yu, T., Zuo, Q. F., Gong, L., Wang, L. N., Zou, Q. M., & Xiao, B. (2016). MicroRNA-491 regulates the proliferation and apoptosis of CD8(+) T cells. *Scientific Reports*, *6*, 30923. https://doi.org/10.1038/srep30923.

Yun, S., Lee, S. U., Kim, J. M., Lee, H. J., Song, H. Y., Kim, Y. K., et al. (2014). Integrated mRNA-microRNA profiling of human NK cell differentiation identifies MiR-583 as a negative regulator of IL2Rgamma expression. *PLoS One*, *9*(10), e108913. https://doi.org/10.1371/journal.pone.0108913.

Zemmour, D., Pratama, A., Loughhead, S. M., Mathis, D., & Benoist, C. (2017). Flicr, a long noncoding RNA, modulates Foxp3 expression and autoimmunity. *Proceedings of*

the National Academy of Sciences of the United States of America, 114(17), 3472–3480. https://doi.org/10.1073/pnas.1700946114.
Zhang, Y., Li, Z., Chen, M., Chen, H., Zhong, Q., Liang, L., et al. (2020). lncRNA TCL6 correlates with immune cell infiltration and indicates worse survival in breast cancer. *Breast Cancer, 27*(4), 573–585. https://doi.org/10.1007/s12282-020-01048-5.
Zhang, Y., Liu, Q., & Liao, Q. (2020). Long noncoding RNA: A dazzling dancer in tumor immune microenvironment. *Journal of Experimental & Clinical Cancer Research, 39*(1), 231. https://doi.org/10.1186/s13046-020-01727-3.
Zhang, Y., Roccaro, A. M., Rombaoa, C., Flores, L., Obad, S., Fernandes, S. M., et al. (2012). LNA-mediated anti-miR-155 silencing in low-grade B-cell lymphomas. *Blood, 120*(8), 1678–1686. https://doi.org/10.1182/blood-2012-02-410647.
Zhang, R., Ni, F., Fu, B., Wu, Y., Sun, R., Tian, Z., et al. (2016). A long noncoding RNA positively regulates CD56 in human natural killer cells. *Oncotarget*, 7(45), 72546–72558. https://doi.org/10.18632/oncotarget.12466.
Zhang, X. L., Xu, L. L., & Wang, F. (2017). Hsa_circ_0020397 regulates colorectal cancer cell viability, apoptosis and invasion by promoting the expression of the miR-138 targets TERT and PD-L1. *Cell Biology International, 41*(9), 1056–1064. https://doi.org/10.1002/cbin.10826.
Zhao, C. C., Jiao, Y., Zhang, Y. Y., Ning, J., Zhang, Y. R., Xu, J., et al. (2019). Lnc SMAD5-AS1 AS ceRNA inhibit proliferation of diffuse large B cell lymphoma via Wnt/beta-catenin pathway by sponging miR-135b-5p to elevate expression of APC. *Cell Death & Disease, 10*(4), 252. https://doi.org/10.1038/s41419-019-1479-3.
Zhao, L., Liu, Y., Zhang, J., Liu, Y., & Qi, Q. (2019). LncRNA SNHG14/miR-5590-3p/ZEB1 positive feedback loop promoted diffuse large B cell lymphoma progression and immune evasion through regulating PD-1/PD-L1 checkpoint. *Cell Death & Disease, 10*(10), 731. https://doi.org/10.1038/s41419-019-1886-5.
Zhou, J., Li, X., Wu, X., Zhang, T., Zhu, Q., Wang, X., et al. (2018). Exosomes released from tumor-associated macrophages transfer miRNAs that induce a Treg/Th17 cell imbalance in epithelial ovarian cancer. *Cancer Immunology Research, 6*(12), 1578–1592. https://doi.org/10.1158/2326-6066.CIR-17-0479.
Zhou, J., Tang, Z., Gao, S., Li, C., Feng, Y., & Zhou, X. (2020). Tumor-associated macrophages: Recent insights and therapies. *Frontiers in Oncology, 10*, 188. https://doi.org/10.3389/fonc.2020.00188.
Zhu, Q., Li, Y., Guo, Y., Hu, L., Xiao, Z., Liu, X., et al. (2019). Long non-coding RNA SNHG16 promotes proliferation and inhibits apoptosis of diffuse large B-cell lymphoma cells by targeting miR-497-5p/PIM1 axis. *Journal of Cellular and Molecular Medicine, 23*(11), 7395–7405. https://doi.org/10.1111/jcmm.14601.
Zhu, X., Shen, H., Yin, X., Yang, M., Wei, H., Chen, Q., et al. (2019). Macrophages derived exosomes deliver miR-223 to epithelial ovarian cancer cells to elicit a chemoresistant phenotype. *Journal of Experimental & Clinical Cancer Research, 38*(1), 81. https://doi.org/10.1186/s13046-019-1095-1.
Zhu, F. Q., Zeng, L., Tang, N., Tang, Y. P., Zhou, B. P., Li, F. F., et al. (2016). MicroRNA-155 downregulation promotes cell cycle arrest and apoptosis in diffuse large B-cell lymphoma. *Oncology Research, 24*(6), 415–427. https://doi.org/10.3727/096504016X14685034103473.

CHAPTER FIVE

Immunological insights of selectins in human disease mechanism

Chandrabose Selvaraj[a,†,*], Rajaram Abhirami[a], Rajendran Vijayakumar[b], Faiz Abdulaziz Alfaiz[b], and Sanjeev Kumar Singh[a,†,*]

[a]Computer Aided Drug Design and Molecular Modelling Lab, Department of Bioinformatics, Science Block, Alagappa University, Karaikudi, India
[b]Department of Biology, College of Science in Zulfi, Majmaah University, Majmaah, Saudi Arabia
[*]Corresponding authors: e-mail address: selnikraj@bioclues.org; sksingh@alagappauniversity.ac.in

Contents

[†] The authors Chandrabose Selvaraj and Sanjeev Kumar Singh have equally contributed as corresponding authors of this work.

Advances in Protein Chemistry and Structural Biology, Volume 129
ISSN 1876-1623
https://doi.org/10.1016/bs.apcsb.2021.11.005

Abstract

Selectin enzymes are glycoproteins and are an important adhesion molecule in the mammalian immune system, especially in the inflammatory response and the healing process of tissues. Selectins play an important role in a variety of biological processes, including the rolling of leukocytes in endothelial cells, a process known as the adhesion cascade. It has recently been discovered and reported that the selectin mechanism plays a role in cancer and thrombosis disease. This process begins with non-covalent interactions-based selectin-ligand binding and the glycans play a role as a connector between cancer cells and the endothelium in this process. The selectin mechanism is critical for the immune system, but it is also involved in disease mechanisms, earning the selectins the nickname "Selectins-The Two Dr. Jekyll and Mr. Hyde Faces". As a result, the drug for selectins should have a multifaceted role and be a dynamic molecule that targets the disease mechanism specifically. This chapter explores the role of selectins in the disease mechanism at the mechanism level that provides the impact for identifying the selectin inhibitors. Overall, this chapter provides the molecular level insights on selectins, their ligands, involvement in normal and disease mechanisms.

1. Introduction

The human health system is highly depending on the immune system for the survival, in accordance to various disease mechanisms (Nicholson, 2016). The present condition of COVID-19 and allied disease has raised the need for understanding and improving the knowledge of human immune system, and for that, the immunological scientist is working on investigating the human immune status. The immune system has the multiple types of immune cells and that are in communicate with cytokines through cell-cell interactions (Garcia, 2020). These are highly recognized using the computational system biological approaches, that also delivers the changes happens in immune response mechanism (Dhillon, Smith, Baghela, Lee, & Hancock, 2020). Basically, the infection occurs through the foreign particle that may be virus, bacteria or in common as pathogen invades the human body cells and this trigger the immune response (Sattler, 2017). Immediate response from the immune system helps in eradicating the infection in the initial level itself, and if the immune response if week, that leads to strong infections. The strong infections are initiated through an illness and symptoms in the human body, due to weak immune system and pathogenic infection rate is high. In strengthening the human immune system, the white blood cells (WBC) are the key component and generated from bone marrows and those are part of the lymphatic system (Florindo et al., 2020). WBC have the tendency to flow through the blood and available all over the human system, that hunts for foreign particles to initiate the immune response mechanisms. Immune cells such as B- and T-cells, natural killer cells, and others make up the human immune system's WBC (Srikiatkhachorn, Mathew, & Rothman, 2017). Through these immune systems the antibodies are developed in order to fight the pathogenic microbes producing toxins. The foreign particles are recognized with the substance called antigens, most probably found in microbial surfaces and toxic chemical substance they produce (ElTanbouly, Schaafsma, Noelle, & Lines, 2020).

Most probably, the disorders in the human immune system leads to certain diseases like allergic disease and autoimmune diseases. In this, the allergic diseases are commonly caused due to allergen induced unfavorable immune response initiated by variety of symptoms. For allergic reactions, the immunological mechanism is represented in two phases.

(i) Sensitization and production of T-cell and B-Cell response, (ii) IgE production and effector production (Okada, Kuhn, Feillet, & Bach, 2010). In cellular level mechanism, it originates through allergen interactions with allergen specific IgE on the mast cell surface and basophils, which results in chemical mediators exit and influx the inflammatory cells. Among the allergies, the food allergies are the most common, and apart from that, Asthma, allergic rhinitis, atopic dermatitis are other allergic diseases (Galli & Tsai, 2012). Another immune oriented disease mechanism is the autoimmune disease, which are developed due to the defects in immune systems, that causes the auto immune response to self-tissue. In simple terms, the autoimmune disease is caused by T-Cells and developed antibodies against human cells itself, in the event of immune system malfunctions or it fails (Recher & Lang, 2006). These autoimmune diseases are initiated through various pathogens like virus, bacteria, or it may be through drugs, chemicals and other factors that affects the immune systems in various levels. Genetic and environmental factors combine to cause autoimmune disease, which is a complex genetic trait (Starshinova et al., 2019). This is because of having high genetic risk effect with the initiation of human leukocyte antigen (HLA) region and specifically the HLA allele associations that leads genetically for the autoimmune diseases. In these mechanisms, several susceptibility genes are cause for T-cell activation, innate immunity, and immune cell signaling role in the immunological pathway (van Drongelen & Holoshitz, 2017). The common immune system consists of complicated complex network of cell-cell interactions with both innate immunity (non-specific antigen) and adaptive immunity (antigen specific).

Thymus-derived lymphocytes (T cells) and bone marrow-derived lymphocytes initiate antigen-specific adaptive immunity in this model (B cells). Both T cells and B cells are having the antigen specific recognition surface receptors (Taefehshokr, Taefehshokr, Hemmat, & Heit, 2020). For B cell, the surface immunoglobulins take the role and for the T cells, the T-cell receptor (TCR) takes the role. The non-specific antigen is induced by pattern recognition receptors (PRRs) in terms of innate immunity, which also includes nucleotide-binding oligomerization (NLR) and domain-like receptors (TLRs). In the event of immunity, the cells involved are mainly WBC or leukocytes, which are classified in to polymorphonuclear leukocytes and mononuclear leukocytes (Mogensen, 2009). In this, the polymorphonuclear leukocytes are sub-categorized into basophils, eosinophils, and neutrophils. These eosinophils and neutrophils function to rupture the dead cell and infectious particles. The other leucocyte namely mononuclear leukocytes are categorized into lymphocytes and monocytes (Chan, Tsai, Huang, Zheng, & Hung, 2010). Here monocytes are segregated into macrophages cells, which are functionally eliminating the infectious agents and dead/damaged cells. The notable specialized cells are sub-categorized from the Lymphocytes cells, and from this, the T lymphocytes are having the tendency to infect the cell directly and destroy them (Murray & Wynn, 2011). Along with this, the T-cells are helping in identifying the infected cells, label them to highlight the damaged cells to WBCs to kill the infected cells. Here B-cells are other lymphocytes get triggered with infectious agents, variate those cells and synthesize the antigen specific antibodies. Since there are multiple mechanism involved in the immune system and human disease, this chapter will elucidate the structural aspects of proteins involved, especially leucocytes involved selectin enzymes for the human disease mechanisms (Waldman, Fritz, & Lenardo, 2020).

2. Adhesion mechanism of leucocytes

The adhesion mechanism is initiated by encompassing the leukocyte rolling to firm adhesion and this mechanism is reported in 19th century by the pathologist namely Rudolf Wagner and Rudolf Virchow (Wagner, 1972). Arfors et al. was the first researcher to report the first experimental reports for the availability of adhesion molecules that mediates the rolling and stabilize the adhesion mechanism. In this, the leukocyte adhesion mechanism initiates from the tethering process, by the communication with endothelium and rolling for the activation process, that results to the firm adhesion. In this, several events like integrin clustering and signaling are function and with proper endothelial and leukocyte stimuli, the transmission occurs and that leads to migration in the tissue (Ivetic, Hoskins Green, & Hart, 2019). Here the strength for leukocytes is required in the post adhesion mechanism, else it will detach and flow in the bloodstream.

Fig. 1 depicts the schematic representation of endothelial cell rolling and adhesion mechanisms in endothelial cells. The steps initiate from rolling of leukocytes and that contributes from the adhesion molecules such as selectin proteins on the leukocytes or endothelial cells with their ligands on the cellular receptors. Here the selectins and mucins are mediating the tethering mechanism, followed by rolling mechanism are generating the signal transduction pathways that leads to integrin activation (Wedepohl et al., 2012). The members of immunoglobulin superfamily along with integrins are functioning in the rolling, arrest and transmigration mechanism. In other terms, the firm adhesion mechanism is mediated with immobilized chemo attractants on the endothelium cells and the integrins on the leukocytes and CAMs on the surface of the endothelium (Wang & Springer, 1998). Arrest process initiated after the attachment process by mediating the activated integrins and with immunoglobulin superfamily counter ligands. Final step involved with the transmigration and exit process to reach the tissue and process for the interactions with neural cells with tendency of leukocytes crossing the endothelial basement membranes (Muller, 2013).

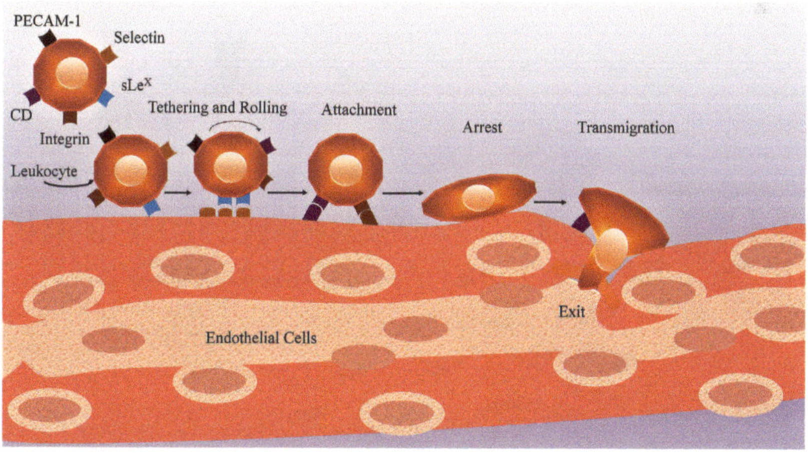

Fig. 1 Pictorial representation of steps involved in the leukocyte trans endothelial rolling and adhesion mechanism.

3. Carbohydrate recognition by selectins

A wide range of biological and pathological events are mediated by carbohydrates, an essential biological macromolecule. Carbohydrates can be recognized by both small and large (Macro) molecules. Cell adhesion during inflammation, differentiation, and the regulation of various intra-cellular communication and signal transduction are all processes in which biomolecules play an important role (Reisz, Bansal, Qian, Zhao, & Furdui, 2014). Biomolecules also serve as structural components and provide energy. Selectin molecules, in contrast to other adhesive molecules, bind to their ligand on basis of protein-protein interaction. Selectins use a protein scaffold or lipid carrier molecule modified by specific carbohydrates for this interaction. Selectins are a new class of glycoprotein found on the surface of cells that were discovered to be cell adhesion molecules (Smith & Bertozzi, 2021). Fig. 2 shows that endothelial cells, leukocytes, and platelets all have selectins, which are carbohydrate-binding transmembrane molecules. The selectins responsible for the interaction of sugars linked to glycoproteins and glycolipids on leukocyte surfaces have a typical structure with Ca^{2+}-dependent carbohydrate recognition domains (C-type CRD). The amino-terminal calcium-dependent lectin domain is found on all three types of selectins (Bouyain, Rushton, & Drickamer, 2001). In both cases, Sialylated and fucosylated tetra saccharides recognize all three types of selectins by a common

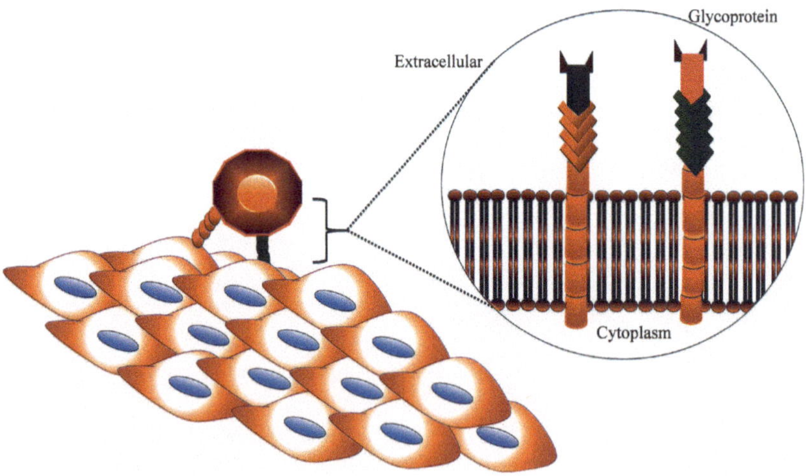

Fig. 2 Represents the carbohydrate moiety (glycans) on the cell surface proteins with N and O-linked glycosylation on cell surface proteins.

motif; both galacto, fucose, and sialic acid-linked oligosaccharides differ in their linking of galactose and fucose to *N*-acetylglucosamine, but they share similar spatial orientation (Li et al., 2008). The most important functional groups that interact directly with the selectin binding pocket are sialic acid, fucose, and galactose, with *N*-acetylglucosamine primarily serving as a scaffold. During the early stages of inflammation or metastasis, selectins and carbohydrates interacted (Impellizzeri & Cuzzocrea, 2014). The binding of selectins and glycoprotein moiety as illustrated in Fig. 2. Tissue injury results in the production of cytokines, which signal endothelial cells, and are responsible for the production of *E*-selectins, which attract neutrophils to the site of injury. The multivalent interaction of sLeX and selectin molecules on the cell surface initiates the tight interaction between integrins on neutrophils to the site of injury (Rossi & Constantin, 2008). Scientists synthesized selectin ligand molecules like sLeX to better understand the structure-function relationship between selectin molecules and carbohydrate ligands. Besides confirming the structure and function of ligands, this research revealed the interaction between essential ligand groups that are responsible for the interaction with carbohydrate ligands (Trinchera, Aronica, & Dall'Olio, 2017). This interaction has shown that the three hydroxyl groups of fucose, the 2-6hydroxy groups of galactoses, and the carboxylate of neuraminic acid play an important role in the binding mechanism of the *E*- and L-selectin preferentially interacts with sLeX. The functional groups that play a key role in P-selectin are similar to those found in other selectin molecules, with the exception of the 2- or 4-hydroxyl group of fucose. On endothelial venules, the sulfated form of SLeX may recognize sites such as Gal-6-SO^4 or GlcNAc-6-SO^4 (Bowman, Cook, de Graffenried, & Bertozzi, 2001). Several studies have also shown that optimal interaction between L-selectin and endothelial cells requires sialation, fucolation and carbohydrate sulfation It not only binds to PAGL-1, which is expressed on leukocytes, but also to proteins that are heavily glycosylated, such as GlyCAM-1, CD^{34}, MAdCAM-1, and Sgp200 (Kappelmayer & Nagy Jr., 2017). To recognize ligands, selectin molecules have lectin domains that act as a Ca^{2+}-dependent mechanism for cell-cell interaction and to recognize carbohydrate, the small lectin molecule region of *E*-selectin is critical, and another domain, the EGF-like domain, also helps promote recognition of ligands and cell-cell interactions by stabilizing the lectin domain's conformation or by indirect interaction with ligand (Crockett-Torabi, 1998). There are three types of lectins in the mammal family: Ca^{2+}-dependent (C-type) lectins, galectins, mannose-6-phosphate-binding lectins (P-type), and immunoglobulin-like lectins (I-type).

Understanding the molecular basis of a protein's carbohydrate recognition is commonly accomplished using NMR, X-ray studies, molecular modeling, and site-directed mutagenesis (Raposo, Canelas, & Barros, 2021). The lectin domains recognize the sLeX and these are clearly understandable through site-directed mutagenesis. The study on P- and *E*-selectins revealed that the fucose residues of SLeX are coordinated to a Ca^{2+} ion via C_2OH and C_3-OH groups. Several glycoproteins have been identified as carbohydrate-containing counter ligands for the selectins in a significant number of studies (Kovalchuk et al., 2018). P-selectins have the highest ligand specificity of the three selectins, with a well-known single ligand called PSGL-1. Selectins have the ability to recognize ligands that have been modified with specific carbohydrate motifs, such as protein or lipid scaffolds bearing sLeX and sLeA, which both have termini of O-glycans, N-glycans, or neolactosphingolipds expressed on the surface of leukocytes (John et al., 2003). Two key proteins are involved in cell adhesion and endothelial activation, and they can be blocked with liposomes containing glycolipids from the sLeX carbohydrate. Many studies have been conducted to investigate the role of a fucosylated carbohydrate in the adhesion mediated by E-selectin. The isomeric carbohydrate (NeuNaca2-3GalI-3(Fucl-4)GlcNac) can adopt a related conformation in solution with both the carboxylate and the fucose pointing in the general direction, indicating that both of these oligosaccharides may act in selectin-mediated carbohydrate recognition, according to NMR analysis of selectin ligands (Kannagi, 1997). The mechanism for carbohydrate recognition is more complicated; on protein backbones, several types of oligosaccharide ligands are present that appear to contribute to the binding activity and specificity of selectin molecules (Tvaroska, Selvaraj, & Koca, 2020).

4. Structure of selectin

Selectins are lectin cell adhesion molecules (Lec-CAMs) that are highly expressed on endothelial cells and regulate leukocyte circulation. Selectins (E, L, and P) have a similar structure and are stored in platelet secretory granules, particularly the P-selectin. Human P-selectin has 120 amino acids and a molecular weight of 140 kDa. It extends 40 nm from the endothelial surface and has a 120 amino acid lectin or carbohydrate binding domain, an epidermal growth factor-like domain, a membrane-spanning domain, nine consensus repeats, and a 35 amino acid cytoplasmic tail (Tedder, Steeber, Chen, & Engel, 1995). The lectin domain of P-selectin is strikingly similar to a variety of calcium-dependent carbohydrate binding domains (C-type lectin). The important members of the C-type lectin family include the low-affinity IgE receptor (CD23), the asialoglycoprotein receptor, and the pulmonary surfactant apoprotein (Drouin, Saenz, & Chiffoleau, 2020). Each member's lectin domain has a similar amino acid sequence, with 14 invariant and 17 highly conserved residues mainly involved in the formation of the hydrophobic core. A 60–70% identity is shared by all three types of selectins, and they all effectively bind with carbohydrates (Xie, 2012).

Even though the mass of P-selectin when expressed on platelets is around 149 kDa, it contains 28.8% carbohydrates by weight of 120 kDa under reducing conditions. P-selectins' soluble form is created by splicing mRNA without exon 14 and deleting 40 amino acids, including the transmembrane domain (Cummings & Smith, 1992). The expression of stimulated endothelial cells and platelets cells is influenced by both the soluble and membrane form. Glycoprotein *E*-selectin has a molecular weight of 116 kDa and 610 amino acids, and it's found in high concentrations on the outer membrane of vascular endothelial cells, where it helps regulate the expression of cytokines like tumor necrosis factor (Vallien, Langley, Jennings, Specian, & Granger, 2000). Fig. 3 depicts the typical structure of selectins, which includes the lectin domain, EGF domain, and cytoplasmic tail. It has a characteristic lectin domain, an EGF domain, and two consensus repeats of

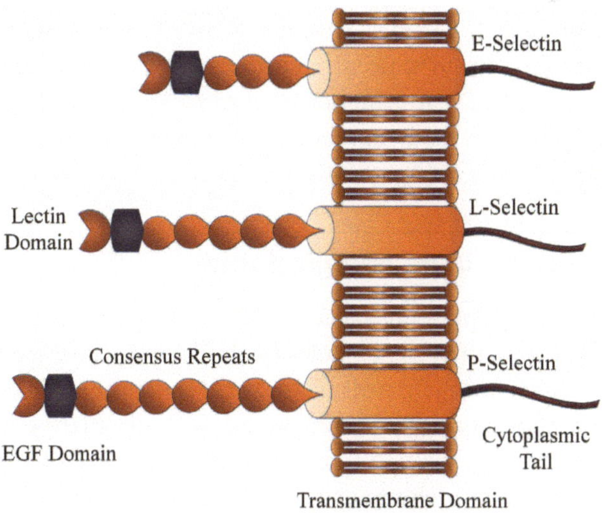

Fig. 3 Represents the typical structure of selectin adhesion molecule (L); L-selectin (E); E-selectin (P); P-selectin.

the complement-regulatory protein domain. About six cysteine-rich repeats can be found in the *E*-selectin amino acid sequence, and an N-terminal lectin domain of 119 residues is responsible for oligosaccharide binding (Aleisa et al., 2020). The EGF-like domain of *E*-selectin connects this domain with the stem of six cysteine-rich repeats and the bottom of the stem through single transmembrane α-helices to a small C-terminal cytoplasmic domain NMR and X-ray analysis of E-selectin found the homologous C-type animal lectin domain in the rat mannose-binding protein, which provided enough data to construct a structural model of the E-selectin lectin domain (Ohmori et al., 2000). The presence of the carbohydrate-recognition domain in C-type lectin defines its structure. This domain contains about 120 amino acids, 14 of which are invariant and 18 of which are characterized by conservation. Once E-selectin is activated, it is responsible for drawing neutrophils to the area of inflammation and then shedding them from the cell's surface. Proteolysis and increased levels of solubilized selectins are both caused by cleavage at the membrane's proximity (Iozzo & Schaefer, 2015). It recognizes sialylated LewisX and LewisA, and binds to sulfatides, such as fucoidan and heparin, like other lectin molecules. As a lymphocyte homing receptor, L-selectin was first discovered to be highly expressed in most peripheral blood leukocytes and plays an important role in leukocyte trafficking in the systemic microcirculation. L-selectin has an extracellular domain organization similar to that of P- and E-selectin and two CRs of the complement regulatory protein domain with molecular weights of approximately 75 and 110 kDa (Rosen, 2004). The nine amino acids at the NH2 end of the ligand molecule's chain are critical to the binding mechanism with L-selectin. The lymphocytes had to attach to the HEV (High endothelial venules) in the lymph nodes and the neutrophils had to invade the area of inflammation (Ager, 2017).

5. Selectin's immunological role

Selectins are commonly called as the cell surface lectins function to mediate the adhesion of WBC to endothelial cells and platelets. Selectin proteins can recognize the sialylated, fucosylated, and sulfated ligands that are expressed on scaffold glycoproteins or counter receptors. In the transcriptional level, the selectins are regulated through proteolytic processing accessed via cellular sorting and through upregulated expression of glycosyl-transferases, which are cause for active functional ligands (McEver, 2015). Functionally, these selectins are playing the role in inflammation, lymphocyte homing, immune response and homing for bone marrow stem cells. In addition, they play the role in inflammatory diseases and metastatic spreading of few cancers. Immunologically, the leukocyte migration is the important function, that has to occur for lymphocyte homing, surveillance and immune invasion sites of inflammations (Pals, de Gorter, & Spaargaren, 2007). The event of intracellular signaling pathways and regulation of cell-cell interactions involves monocytes, lymphocytes, platelets and endothelial cells with the dependency of selectin mediated cell-cell interactions. The leukocytes adhesion role is initiated with multistep process that functions during constitutive homing of naive lymph nodes, and this endorse the leukocytes and endothelial cell interactions (Ala, Dhillon, & Hodgson, 2003). This clarifies the role of selectin, that initiates the multicellular adhesive and signaling events in the cell. Platelets along with PSGL-1 on myeloid cells initiates the signaling mechanism through selectin molecule. Platelets produce chemokines that activate integrins and stimulate the production of other mediators in myeloid cells. CTH is secreted by activated cells and is responsible for the induction of protease-activated receptors, which leads to selectin-induced adhesion and bidirectional signaling in platelets and myeloid cells. The interaction of neutrophil L-selectin with sulfatides significantly increases the expression of TNF and interleukin-8, which are important in the host defense mechanism (McEver, 2015). TNF-α expression induces *E*-selectin and P-selectin dependent rolling in arterioles that require E-selection. The P and E-selectin have been studied in the activation signaling transduction of leukocytes in various studies. Adhesion mediated by P-selectin results in the production of O_2, which is dependent on the carbohydrate ligand sLeX interacting with P-selectin and this O_2-generations influence neutrophil transmigration. To regulate the interaction between neutrophils and endothelial cells, L-selectin communicates with the cells by both directly and indirectly stimulates neutrophil production of reactive oxygen metabolites (Kappelmayer & Nagy Jr., 2017). In leukocyte signal transduction, the interaction between P-selectin and the-r counter-receptors is critical. For example, the carbohydrate ligand molecule (CHO) Ag CD15 for P- and L-selectin is important for leukocyte signal transduction. Leukocyte carbohydrate ligand expression on monocyte and neutrophil mediates cell adhesion to activated platelets and induces cytoplasmic calcium fluxes, activation of the respiratory burst and expression of granule-associated proteins. *E*-selectin's cytoplasmic domain contains 32 amino acids, including 2 tyrosine residues, has been heavily implicated in transmembrane signaling (Zhang et al., 2013). Tyrosine phosphorylation in this domain is important for receptor-mediated transmembrane signal transmission, particularly for activation of extracellular signal-regulated kinase ½ (ERK1/2) (Mebratu & Tesfaigzi, 2009). The selectin molecules, as is natural, are involved in the process of lymphocyte homing and migration. Monocytes constantly leave the blood during this process to replenish the tissue-macrophage population at the site of inflammation. The presence of L-selectin has been demonstrated in numerous studies to be a critical receptor for lymphocytes attempting to enter the peripheral lymphatic system. Use of L-selectin inhibitor monoclonal antibodies prevented lymphocyte adhesion in vitro and prevented their localization in peripheral lymph nodes in vivo to lymph node high endothelial venules (HEV). Monoclonal antibodies that block P-selectin and PSGL-1 are associated with reduced platelet activation (Arbones et al., 1994). In mice, this blocking mechanism reduced the arterial thrombosis and the size of infracts. The soluble *E*-Selectin mutant attracts neutrophils by acting as a chemoattractant. L-selectin has been found to be a mediator of 2-integrin-mediated neutrophil aggregation in other research studies. It has been proven in numerous studies that selectin molecules effectively initiate signals that lead to increased leukocyte adhesion. Selectin is involved in ischemia-reperfusion, shock, and various inflammatory reactions. Thus, a better understanding of the adhesion cascade mechanism is required to develop new therapeutic interventions (Woollard & Chin-Dusting, 2010).

6. Interaction of selectins with their ligands

Calcium-dependent transmembrane glycoprotein selectins play an important role in pathological processes by mediating the interaction of leukocytes and the endothelium as shown in the Table 1. During acute inflammation, this interaction primarily influences and initiates the adhesion and signaling cascade mechanism, which leads to leukocyte tethering to and rolling on the surface of endothelial cells. PSGL-1 is a ligand molecule with a high affinity for all three types of selectins as shown in Fig. 4, as well as other molecules. Small sialylated, fucosylated oligosaccharides have a weak affinity for selectin, whereas glycoproteins or proteoglycans have a strong interaction (Park et al., 2020). Selectin-mediated leukocyte adhesion during inflammation is dependent on these high-affinity ligand interactions.

In a lectin-like interaction, glycoproteins serve as the primary counter-receptor for selectins. They have a molecular weight of about 120 kDa and contain 402 amino acids. Post-translational modifications are made to this native protein after it is expressed in order to make it functional (Tvaroska et al., 2020). PSGL-1, the P-selectin ligand, was post-translationally modified by sialic acid and fucose exclusively on O-glycans and to sulfated on tyrosine

Table 1 Represents the selectins, ligand molecules, and expression site.

S. no.	Selectins	Respective ligands	Site of expression
1.	P-Selectin	P-selectin glycoprotein ligand-1 (PSGL-1), CD24 (Heat Stable Antigen or nectadrin)	Endothelial cells, platelets
2	*E*-Selectin	P-selectin glycoprotein ligand-1 (PSGL-1), E-selectin ligand-1 (ESL-1), L-Selectin, Podocalyxin (Sialo Glycoprotein)	Activated endothelial cells
3	L-Selectin	P-selectin glycoprotein ligand-1 (PSGL-1) Glycosylation-dependent Cell adhesion molecule-1 (GlyCAM-1), Mucosal Addressin Cell Adhesion molecule 1 (MAdCAM-1), CD34 (Transmembrane Phospho glycoprotein protein), Podocalyxin (Sialo Glycoprotein)	Leukocytes

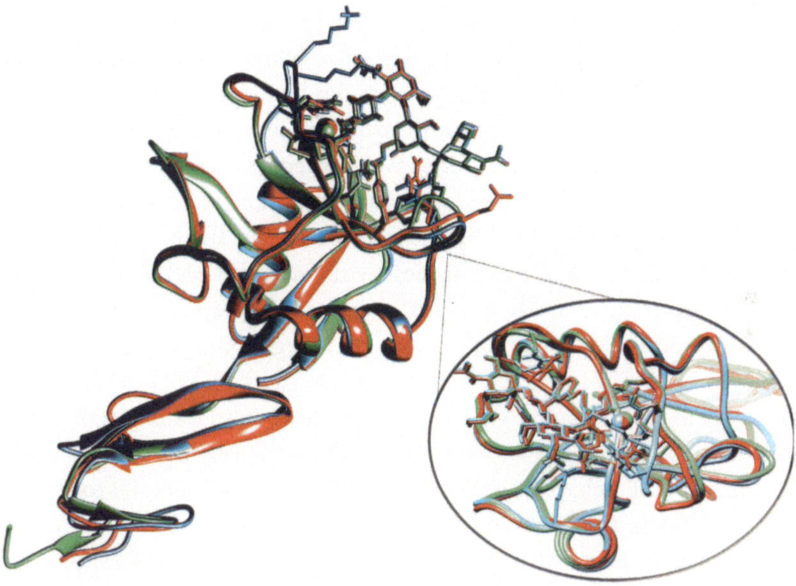

Fig. 4 Interaction of sLeX with three types of Selectins showing unique binding of interactions in the 3D conformation. Here, the red shows P-selectin, blue shows E-selectin and green shows the L-selectin.

residues to become functional. T cell migration is mediated by selectin and its ligand interaction in a multistep cascade reaction. The selectin-ligand interaction initiates the localization of T cells in all tissues, and the synthesis of sialyl Lewis X is critical for selectin ligand binding, which is closely related to a cutaneous lymphocyte-associated antigen (CLA). High-affinity ligands for both P- and E-selectins bind to the amino terminus of PSGL-1, particularly the extracellular domain, which contains mucin and is rich in serine, threonine, and proline, as well as 14–16 decamer repeats (Epperson, Patel, McEver, & Cummings, 2000). Along with this, three clustered tyrosine sulfate residues, sialic acid residues and fucose are also required. Several crystallographic studies revealed that P-selectins have a positive electrostatic potential surface. The initial step in the adhesion signaling cascade mechanism is mediated by the binding of selectin and its corresponding glycoproteins, and this leads to the recruitment of circulating leukocytes to the injury site.

7. Selectin ligands

7.1 PSGL-1

In leukocytes, PSGL-1 is a disulfide-linked homodimer protein consisting of 412 amino acids, including the signal sequence and a consensus tetrapeptide cleavage site. These have two subunits each contain two to three N-glycans, which are critical for P-selectin binding as shown in Fig. 5. While the PSG-1 sequence encodes a signal peptide in mammals like humans, mice and rats, the sequence for bovine and equine PSG-1 encodes a propeptide, predicted to be cleaved by paired basic amino acids (Tvaroska et al., 2020). There are three selectin molecules that PSGL-1 can contact via the N-terminus of the extracellular domain, and the affinities are different for each. The serine and threonine-rich extracellular domain contains a single Cys residue at 320, which is responsible for the prediction of a single transmembrane domain spanning residues 321–341 and the cytoplasmic domain (341–412). A major role is played by P-selectin, which regulates neutrophil rolling/tethering on endothelial cells and promotes targeted extravasation into tissues (Baisse, Galisson, Giraud, Schapira, & Spertini, 2007). It is also essential for neutrophil recruitment and host defense against *Salmonella typhimurium*. Recent research has revealed that the sialic acid-binding toxin staphylococcal superantigen-like 5 (SSL5) plays an important role in the binding of sLeX expressed on PSGL-1 by neutrophils, initiating neutrophil activation and recruitment.

7.2 E-selectin ligands

E-selectin interacts with two major glycoprotein ligands, E-selectin ligand-1 (ESL-1) and PSGL-1, which are both glycoprotein ligands. Among the

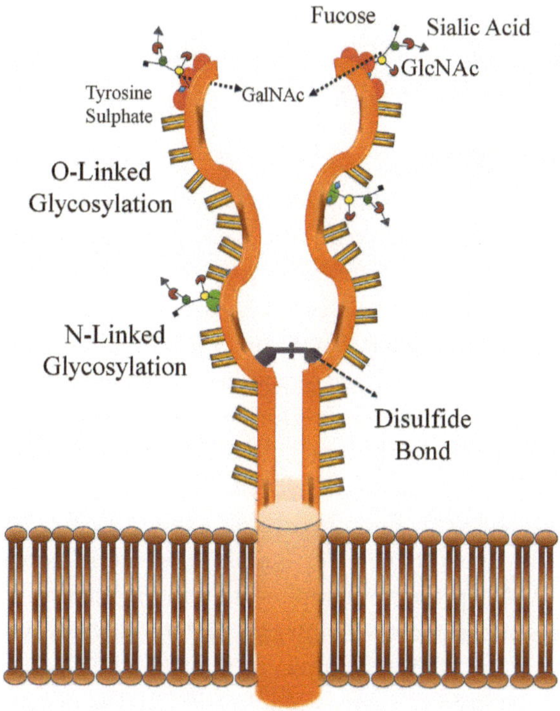

Fig. 5 Represents the structure of P-selectin ligand molecule PSGL-1.

ligands, ESL-1 is a specific receptor for E-selectin. A glutamine-rich NH_2-terminal segment of 70 amino acids and cysteine-rich repeats make up the protein's type-I transmembrane domain, which has a molecular weight of about 150kDa. These five potential N-glycosylation sites have O-linked carbohydrate side chains that alter their structure. Fucosylated and sialylated glycolipids and proteins recognized by E-selectins can serve as high affinity ligands, while P-selectin and L-selectin significantly recognize sugars that are highly expressed with sulfated tyrosine residue on ligands (Steegmaier et al., 1995). Human P and E-selectin crystal structures revealed that the lectin and EGF (LE) domains complexed with sLeX, as well as the NH2-terminal domain of human PSGL-1. ESL-1 is found in a wide variety of cells, primarily in the Golgi apparatus, with a small amount also found in the plasma membrane and also, the ESL-1 requires fucosyltransferase IV modifications during the interaction (Steegmaier, Borges, Berger, Schwarz, & Vestweber, 1997). The short hairpin RNA strategy used in the knockdown study revealed that the binding of a recombinant soluble form of E-selectin was reduced when ESL-1 was absent, and it also noticed that the expression of soluble E-selectin was also reduced when PSGL-1 was absent (Chase, Magnani, & Simon, 2012).

7.3 L-selectin ligands

In the selectin family, L-selectin is the third member. L-selectin is a receptor for HEV (high endothelial veins) via sLeX epitopes present on various glycoproteins O-glycans. In peripheral lymph nodes, four L-selectin ligands have been found, including one called MAdCAM-1 and three called sulfated L-selectin ligands (Sgp50, Sgp90 and Spg200). HEV peripheral lymph node GlyCAM-1 mucin secreted ligand has a soluble secreted moiety and may primarily transduce signals into leukocytes rather than support leukocyte adhesion (Mitoma et al., 2007). GlyCAM-1 undergoes modification with core 2 O-glycans that are sialylated, fucosylated, and sulfated when it binds to the L-selectin. Endothelial cells express Sp90, a surface marker like CD34, and Sgp200 mediates lymphocytes initial loose binding to HEV. MAdCAM-1 promotes lymphocyte tethering and rolling via interactions with L-selectin and α4-β7 integrin, and it is also involved in lymphocyte tethering on the HEV (Uchimura et al., 2005). PSGL-1 and L-selectin interact during neutrophil aggregation, but PSGL-1 is not recognized as a primary L-selectin ligand. L-Selectin recognizes the glycoprotein's main structural components, such as sLeX and sLeA, as well as their sulfate derivatives.

8. Disease mechanism through selectin signaling

The emigration of leukocytes is an important event in lymphocyte homing and surveillance as well as in the invasion of inflammatory sites. Selectin-dependent cell-cell interactions help activate intracellular signaling pathways and regulate cell-cell interactions involving monocytes, lymphocytes, platelets, and endothelial cells. During physiological or pathological inflammation and hemostasis, selectins initiate multicellular adhesive and signaling events, which emphasize the interactions between leukocytes and endothelial cells. As a result, the initiation of the leucocyte adhesion cascade is a multi-step process that occurs during constitutive homing of naive lymph nodes (Patel, Cuvelier, & Wiehler, 2002). The platelet-PSGL-1 interaction on myeloid cells starts the signaling by the selectin molecule. By coming into close contact with glycoprotein and other ligands on platelets, αMβ2 activates chemokine expression, which in turn activates integrins and other myeloid cell mediators. Stabilization necessitates activation of SFKs as well. Platelets release chemokines that activate integrins and cause myeloid cells to produce additional mediators. Platelets and myeloid cells with activated adhesion secrete CTH, which leads to the induction of protease-activated receptors and, as a result, bidirectional signaling (Tvaroska et al., 2020). Attaching neutrophil L-selectin to sulfatides significantly increases the expression of TNF- and interleukin-8, both of which are important in the host defense mechanism. TNF-α and interleukin-8 act on neutrophils, as well as inflammatory cells, by binding with their receptors. P- and *E*-selectin have been studied in the activation signaling transduction of leukocytes in various studies. O_2 was generated by P-selectin-dependent adhesion, and this was due to the interaction of the carbohydrate ligand sLeX with P-selectin, also this O2 function in regulation of neutrophil transmigration (Kubes, 2002). Thus, L-selectin regulates neutrophil-endothelial cell interactions via direct cell-to-cell communication as well as indirectly through neutrophil production of reactive oxygen metabolites. Leukocyte carbohydrate ligand expression is higher on monocytes and neutrophils, which helps these cells adhere to activated platelets and it leads to the activation of the respiratory burst and expression of granule-associated proteins. Many studies have linked the 32 amino acid E-selectin cytoplasmic domain to transmembrane signaling, including the presence of two tyrosine residues (Witz, 2006). Monocytes constantly leave the blood to replenish the tissue-macrophage population at the site of inflammation while selectin molecules are involved in lymphocyte homing. According to numerous studies, selectin molecules are effective in triggering the signals that lead to leukocyte adhesion enhancement. Additionally, selectin plays a significant role in the processes of ischemia-reperfusion injury, shock, as well as various inflammatory reactions. Therefore, new approaches to the development of therapeutic interventions require a better understanding of the adhesion cascade mechanism of leukocyte interaction (Chia, Teo, & Yeo, 2020).

8.1 Role in metastasis and cancer

Adhesion between cells is critical in cancer metastasis, and tumors that have already spread to other organs or even the bloodstream interact with host cells like platelets, leukocytes, and endothelial cells to modulate their extravasation from the bloodstream. These cancer cells are known as circulatory tumor cells because of their interaction with these host cells. Epithelial cells are normally covered in mucins, either on the cell's surface or in the soluble epithelium layers they construct. Malignant transformation has been linked significantly to the presence of mucin, which is a glycan-attached molecule that increases expression of sLeX and sLeA. This increased level of expression is frequently linked to a poor prognosis in various types of cancer. Tumor cells can spread through the bloodstream and block the microvasculature with selectins because they mediate specific interactions between the selectin-expressing host cell and the ligand on tumor cells (Micalizzi, Maheswaran, & Haber, 2017). Several In vivo experiments have clearly demonstrated the involvement of selectins in cancer metastasis. In a colon carcinoma cell model, p-selectin-deficient mice showed significant inhibition of metastasis compared to wild-type controls, indicating that P-selectins are directly involved in the metastasis process. P-selectin-deficient mice have reduced cell arrest in their lungs, as shown by the experiments described above. P-selectin was found to play a critical role in colon carcinoma-platelet adhesion and to facilitate metastasis, according to the research. Because of the high level of P-selectin expression on the surface of endothelial cells, the metastatic process is triggered. NK cells regulate tumor cells when platelet-tumor cell interactions are lacking. Breast cancer, lung cancer, prostate cancer, and colorectal cancer all express high levels of *E*-selectin during progression and metastasis (Chistiakov & Chekhonin, 2018). E-selectin is highly expressed in cancer cells, because of inflammatory cytokines produced by cancer cells, which stimulate the endothelium to recruit immune components early in cancer development. In this process, E-selectins are acting as a mediator of the adhesion cascade. Both cancer cells and tumor stroma produce the cytokines like TNF-α and IL-1β which responsible for the expression of Ap-1 and subsequently activate E-selectin transcription. Oncogenic E-selectin overexpression promotes the expression of immune infiltrates like sLeX and sLeA, two key tumor stromal components that aid in cancer cell growth, invasion, and angiogenesis. E-selectin dependent adhesion mechanisms regulate tissue migration at distal organs in cancer cells that have been disseminated by expressing E-selectin ligands on their surface to make first contact with the vascular endothelium (Garcia, 2020). In metastatic breast cancer, CD44v4 is an *E*-selectin ligand that regulates the interaction between cancer cells and endothelial cells via E-selectin, resulting in bidirectional signaling that aids cancer cell migration through the endothelium. Another E-selectin ligand, α-2,3-sialyltransferase, is highly expressed in pancreatic tumor cells and promotes pancreatic cancer cell transmigration in an E-selectin-dependent manner, like CD44. A wide variety of carbohydrate molecules, such as sialyl Lewis X (sLeX) and sialyl Lewis A (sLeA), are heavily incorporated with sialic acids and both the structures have α 2–3 linked sialic acid on the GlcNAc backbone and represent the minimal recognition motif for ligands of selectins effectively involved leukocyte trafficking. In lymphocyte homing, these glycoproteins and glycolipids in carbohydrate molecules interact with selectin molecules, causing leukocytes to slowly tether and roll on the endothelial cell surface, which is the first step in leukocyte extravasation (Ala et al., 2003). Although it is found on only a small percentage of immune cells (neutrophils), L-selectin is highly expressed on all these cell types and plays an important role in lymphocyte recycling, as well as aiding polymorphonuclear neutrophil adhesion to vascular endothelial cells in inflamed tissues. Several studies have suggested that L-selectin is also involved in leukocyte aggregation and facilitates white blood cells in adhering to vascular surfaces via P-selectin and cooperating during an inflammatory response. L-selectin abundantly express in regulatory T cells (T_{reg}) which having a $CD4^+$, $CD25^+$ phenotype and are involved in regulation of inflammatory responses, tolerance, T cell homeostasis and utilize these adhesion molecules expression for migration and maintenance of normal tissues distribution (Mitoma et al., 2007). The permissive microenvironment and premetastatic niche are critical for metastasis establishment and are responsible for circulating cancer cells colonizing distant organs. A premetastatic niche is formed by several factors, including primary tumor-derived secreting factors, exosomes, bone marrow-derived cells, and stromal cells, and in this process the selectins and their ligands also help to keep the metastatic niche in shape. EGFR and other tumor-derived secretin factors promote *E*-selectin expression in lung metastases by activating focal adhesion kinase-FAK dependent expression (Rossi & Constantin, 2008).

8.2 Role in thrombosis

Over the last few decades, researchers have focused increasingly on the connection between inflammation and blood clotting, despite selectins' primary role in inflammation as adhesion molecules. Thrombosis and a pro-inflammatory event are now widely accepted as being influenced by leukocyte recruitment directly. When P-selectin is shed, it turns into a soluble form known as soluble P-selectin (sP-Selectin), which is responsible for activating platelets and endothelial cells. P-selectin was found to play a critical role in thrombosis in several studies, and the links between inflammation, thrombosis, and coagulation were also demonstrated (Baisse et al., 2007). Palabrica et al. (1992) used a baboon thrombosis model with an arteriovenous shunt to prove the role of P-selectin in blood coagulation and thrombosis. Using an anti-P-selectin antibody, thrombus formation in a thrombogenic shunt is significantly slowed in the development of fibrin. A mechanism involving P-selectin binding to its receptor PSGL-1 activates the elaboration of cytokines after it upregulates tissues factor generation in monocytes and initiates signaling pathways in leukocytes (Palabrica et al., 1992). By binding to vascular endothelial cells, P-selectin stimulates platelet activation and leukocyte adhesion and due to this, it was found that having this expression increased the risk of venous thrombosis by 1.7 times. Most of the time, it interacts with leukocytes via the PSGL-1 receptor, which then triggers neutrophil and macrophage recruitment, resulting in the formation of procoagulant microparticles. Platelets, leukocytes, and endothelial cells all excrete microparticles that contribute to thrombosis amplification in a calcium-dependent mechanism (Palabrica et al., 1992). Genetically modified animals with reduced levels of P- and *E*-selectin were used in several studies to examine the possible role of these proteins in the pathogenesis of thrombosis and other blood vessel diseases. In genetically modified animals, thrombus weights were significantly lower, whereas in wild types, high circulating P-selectin concentrations were associated with larger thrombi (Mebratu & Tesfaigzi, 2009). The clotting system is an important part of thrombosis, and changes or abnormalities in the clotting system in either the anticoagulant or the procoagulant system increase the risk of deep vein thrombosis. In this process, the cell adhesion molecules soluble P-selectin mediates the binding of specific carbohydrate-containing ligands, by which it supports initial tethering of leukocytes to activated endothelial cells and platelets. P-selectin expression is also important in the interaction between deep vein thrombosis and the inflammatory response of the vein. Initially, leukocytes attach to the vessel wall and roll along it, before becoming activated and migrating into the vein wall while still firmly attached to the surface (Tedder et al., 1995).

9. Selectin as drug target

Several pathological conditions benefit from the multifunctionality of selectins, which are adhesion molecules said to play various functions. Selectins also serve to initiate the interaction between circulating leukocytes and the endothelial cells. Leukocyte recruitment depends on endothelial interaction, making it an appealing therapeutic target for chronic inflammation (Tvaroska et al., 2020). Endothelial cells are activated by targeting anti-inflammatory drugs, which reduces the overall side effects by targeting only them. For controlling multiple physiological expressions including inflammation and cancer metastasis, selectin molecules are suitable for targeting, because they strictly control the adhesion of leukocytes or cancer cells on endothelial cells to limit the many physiological expressions. Recently Šmak et al., has also successfully reported as selectin as drug target for COVID-19, and also some other researchers have reported the selectins for viral infections that initiates the cytokine storm (Smak, Chandrabose, Tvaroska, & Koca, 2021).

10. Selectins for drug delivery

P-selectin and *E*-selectins are significantly expressed on the surface of the endothelial cells and this mechanism is explored for drug delivery approaches with the understanding of selectin molecules are conjugated with antibodies that facilitate drug delivery. The role of selection molecules and their receptors in a variety of pathological disorders, including immune cell trafficking, has been revealed in numerous advanced molecular and cell biology studies. Inflammatory diseases like lung injury, psoriasis, and rheumatoid arthritis, P-selectins play an important role (Hua, 2013). A number of pathological conditions, including inflammation, cardiovascular disease, and cancer, are strongly linked to E-selectin, while HIV, Diabetes, Meningeal leukemia, multiple sclerosis, and sepsis is linked to L-selectin. This has sparked interest in the selectin family and its receptors among scientists in recent decades, with the goal of using them as pharmacological targets in the treatment of various diseases and as molecular targets for the delivery of therapeutic and diagnostic agents (Tvaroska et al., 2020). These cell adhesion molecules are studied for drug delivery to cytokine-activated endothelium in both cell culture and animal models of inflammation has provided the fantastic insights for the selectin mechanism and drug delivery approaches.

11. Conclusion and future prospects

The role of selectins molecules and their interacting ligands are summarized in the current chapter, which highlights the critical role of selectins in cancer metastasis and thrombosis. Several steps are involved in cancer cell metastatic spread, and each one is crucial for the development of a metastatic focus and here the selectins are involved in hematogenous metastasis. Due to the fact that selectin molecules act as powerful receptors and contribute to pathological processes, they serve as diagnostic biomarkers and are a major target for the invasion of cancer metastasis and thrombosis. However, there are only a few drug molecules that target selectin receptors, so further research into the mechanism underlying the selectin molecule's interaction with its ligands, as well as the diverse roles played by this versatile molecule, is needed. Selectins and their ligands, such as sLe X, sLeA, and CD^{44} isoforms, interact to facilitate various physiological processes, including leukocyte recruitment. Multiple sclerosis, ischemic stroke, cancer, and inflammation all have high rates of leukocyte recruitment, making selectins an appealing strategy for developing new therapeutics for treatment. However, there are still many unanswered questions about the structural and precious mechanisms that trigger selectin ligand interactions and how they are translated into cellular interactions. More research into how selectin molecules are involved in physiological processes such as inflammatory trafficking, tethering, leukocyte rolling, metastasis, and CNS demyelination will help develop new treatment and diagnostic strategies for human disorders. In addition, selectins will be future drug targets for virus that introduce the cytokinin storm and thus versatile selectin inhibitors has to be identified and studied in future.

Acknowledgments

The authors C.S. and S.K.S. thankfully acknowledge the Tamil Nadu State Council for Higher Education (TANSCHE) for the research grant (Au/S.o. (P&D): TANSCHE Projects: 117/2021). The authors would like to thank Deanship of Scientific Research at Majmaah University for supporting this work under project number R-2021-296.

Declaration of interest

The Authors declares there is nothing to declare with this chapter.

References

Ager, A. (2017). High endothelial venules and other blood vessels: Critical regulators of lymphoid organ development and function. *Frontiers in Immunology*, *8*, 45.

Ala, A., Dhillon, A. P., & Hodgson, H. J. (2003). Role of cell adhesion molecules in leukocyte recruitment in the liver and gut. *International Journal of Experimental Pathology*, *84*(1), 1–16.

Aleisa, F. A., Sakashita, K., Lee, J. M., AbuSamra, D. B., Al Alwan, B., Nozue, S., et al. (2020). Functional binding of E-selectin to its ligands is enhanced by structural features beyond its lectin domain. *The Journal of Biological Chemistry*, *295*(11), 3719–3733.

Arbones, M. L., Ord, D. C., Ley, K., Ratech, H., Maynard-Curry, C., Otten, G., et al. (1994). Lymphocyte homing and leukocyte rolling and migration are impaired in L-selectin-deficient mice. *Immunity*, *1*(4), 247–260.

Baisse, B., Galisson, F., Giraud, S., Schapira, M., & Spertini, O. (2007). Evolutionary conservation of P-selectin glycoprotein ligand-1 primary structure and function. *BMC Evolutionary Biology*, 7, 166.

Bouyain, S., Rushton, S., & Drickamer, K. (2001). Minimal requirements for the binding of selectin ligands to a C-type carbohydrate-recognition domain. *Glycobiology*, *11*(11), 989–996.

Bowman, K. G., Cook, B. N., de Graffenried, C. L., & Bertozzi, C. R. (2001). Biosynthesis of L-selectin ligands: Sulfation of sialyl Lewis x-related oligosaccharides by a family of GlcNAc-6-sulfotransferases. *Biochemistry*, *40*(18), 5382–5391.

Chan, Y. K., Tsai, M. H., Huang, D. C., Zheng, Z. H., & Hung, K. D. (2010). Leukocyte nucleus segmentation and nucleus lobe counting. *BMC Bioinformatics*, *11*, 558.

Chase, S. D., Magnani, J. L., & Simon, S. I. (2012). E-selectin ligands as mechanosensitive receptors on neutrophils in health and disease. *Annals of Biomedical Engineering*, *40*(4), 849–859.

Chia, P. Y., Teo, A., & Yeo, T. W. (2020). Overview of the assessment of endothelial function in humans. *Frontiers in Mededicine (Lausanne)*, *7*, 542567.

Chistiakov, D. A., & Chekhonin, V. P. (2018). Circulating tumor cells and their advances to promote cancer metastasis and relapse, with focus on glioblastoma multiforme. *Experimental and Molecular Pathology*, *105*(2), 166–174.

Crockett-Torabi, E. (1998). Selectins and mechanisms of signal transduction. *Journal of Leukocyte Biology*, *63*(1), 1–14.

Cummings, R. D., & Smith, D. F. (1992). The selectin family of carbohydrate-binding proteins: Structure and importance of carbohydrate ligands for cell adhesion. *Bioessays*, *14*(12), 849–856.

Dhillon, B. K., Smith, M., Baghela, A., Lee, A. H. Y., & Hancock, R. E. W. (2020). Systems biology approaches to understanding the human immune system. *Frontiers in Immunology*, *11*, 1683.

Drouin, M., Saenz, J., & Chiffoleau, E. (2020). C-type lectin-like receptors: Head or tail in cell death immunity. *Frontiers in Immunology*, *11*, 251.

ElTanbouly, M. A., Schaafsma, E., Noelle, R. J., & Lines, J. L. (2020). VISTA: Coming of age as a multi-lineage immune checkpoint. *Clinical and Experimental Immunology*, *200*(2), 120–130.

Epperson, T. K., Patel, K. D., McEver, R. P., & Cummings, R. D. (2000). Noncovalent association of P-selectin glycoprotein ligand-1 and minimal determinants for binding to P-selectin. *The Journal of Biological Chemistry*, *275*(11), 7839–7853.

Florindo, H. F., Kleiner, R., Vaskovich-Koubi, D., Acurcio, R. C., Carreira, B., Yeini, E., et al. (2020). Immune-mediated approaches against COVID-19. *Nature Nanotechnology*, *15*(8), 630–645.

Galli, S. J., & Tsai, M. (2012). IgE and mast cells in allergic disease. *Nature Medicine*, *18*(5), 693–704.

Garcia, L. F. (2020). Immune response, inflammation, and the clinical spectrum of COVID-19. *Frontiers in Immunology*, *11*, 1441.

Hua, S. (2013). Targeting sites of inflammation: Intercellular adhesion molecule-1 as a target for novel inflammatory therapies. *Frontiers in Pharmacology*, *4*, 127.

Impellizzeri, D., & Cuzzocrea, S. (2014). Targeting selectins for the treatment of inflammatory diseases. *Expert Opinion on Therapeutic Targets*, *18*(1), 55–67.

Iozzo, R. V., & Schaefer, L. (2015). Proteoglycan form and function: A comprehensive nomenclature of proteoglycans. *Matrix Biology*, *42*, 11–55.

Ivetic, A., Hoskins Green, H. L., & Hart, S. J. (2019). L-selectin: A major regulator of leukocyte adhesion, migration and signaling. *Frontiers in Immunology*, *10*, 1068.

John, A. E., Lukacs, N. W., Berlin, A. A., Palecanda, A., Bargatze, R. F., Stoolman, L. M., et al. (2003). Discovery of a potent nanoparticle P-selectin antagonist with anti-inflammatory effects in allergic airway disease. *The FASEB Journal*, *17*(15), 2296–2298.

Kannagi, R. (1997). Carbohydrate-mediated cell adhesion involved in hematogenous metastasis of cancer. *Glycoconjugate Journal*, *14*(5), 577–584.

Kappelmayer, J., & Nagy, B., Jr. (2017). The interaction of selectins and PSGL-1 as a key component in Thrombus formation and cancer progression. *BioMed Research International*, *2017*, 6138145.

Kovalchuk, S. N., Buinovskaya, N. S., Likhatskaya, G. N., Rasskazov, V. A., Son, O. M., Tekutyeva, L. A., et al. (2018). Mutagenesis studies and structure-function relationships for GalNAc/gal-specific lectin from the sea mussel Crenomytilus grayanus. *Marine Drugs*, *16*(12), 471. https://doi.org/10.3390/md16120471.

Kubes, P. (2002). The complexities of leukocyte recruitment. *Seminars in Immunology*, *14*(2), 65–72.

Li, Y. L., Wu, G. Z., Dawe, G. S., Zeng, L., Cui, S. S., Loers, G., et al. (2008). Cell surface sialylation and fucosylation are regulated by L1 via phospholipase Cgamma and cooperate to modulate neurite outgrowth, cell survival and migration. *PLoS One*, *3*(12), e3841.

McEver, R. P. (2015). Selectins: Initiators of leucocyte adhesion and signalling at the vascular wall. *Cardiovascular Research*, *107*(3), 331–339.

Mebratu, Y., & Tesfaigzi, Y. (2009). How ERK1/2 activation controls cell proliferation and cell death: Is subcellular localization the answer? *Cell Cycle*, *8*(8), 1168–1175.

Micalizzi, D. S., Maheswaran, S., & Haber, D. A. (2017). A conduit to metastasis: Circulating tumor cell biology. *Genes & Development*, *31*(18), 1827–1840.

Mitoma, J., Bao, X., Petryanik, B., Schaerli, P., Gauguet, J. M., Yu, S. Y., et al. (2007). Critical functions of N-glycans in L-selectin-mediated lymphocyte homing and recruitment. *Nature Immunology*, *8*(4), 409–418.

Mogensen, T. H. (2009). Pathogen recognition and inflammatory signaling in innate immune defenses. *Clinical Microbiology Reviews*, *22*(2), 240–273. Table of Contents.

Muller, W. A. (2013). Getting leukocytes to the site of inflammation. *Veterinary Pathology*, *50*(1), 7–22.

Murray, P. J., & Wynn, T. A. (2011). Protective and pathogenic functions of macrophage subsets. *Nature Reviews. Immunology*, *11*(11), 723–737.

Nicholson, L. B. (2016). The immune system. *Essays in Biochemistry*, *60*(3), 275–301.

Ohmori, K., Kanda, K., Mitsuoka, C., Kanamori, A., Kurata-Miura, K., Sasaki, K., et al. (2000). P- and E-selectins recognize sialyl 6-sulfo Lewis X, the recently identified L-selectin ligand. *Biochemical and Biophysical Research Communications*, *278*(1), 90–96.

Okada, H., Kuhn, C., Feillet, H., & Bach, J. F. (2010). The 'hygiene hypothesis' for autoimmune and allergic diseases: An update. *Clinical and Experimental Immunology*, *160*(1), 1–9.

Palabrica, T., Lobb, R., Furie, B. C., Aronovitz, M., Benjamin, C., Hsu, Y. M., et al. (1992). Leukocyte accumulation promoting fibrin deposition is mediated in vivo by P-selectin on adherent platelets. *Nature*, *359*(6398), 848–851.

Pals, S. T., de Gorter, D. J., & Spaargaren, M. (2007). Lymphoma dissemination: The other face of lymphocyte homing. *Blood*, *110*(9), 3102–3111.

Park, E. J., Myint, P. K., Ito, A., Appiah, M. G., Darkwah, S., Kawamoto, E., et al. (2020). Integrin-ligand interactions in inflammation, cancer, and metabolic disease: Insights into the multifaceted roles of an emerging ligand irisin. *Frontiers in Cell and Development Biology*, *8*, 588066.

Patel, K. D., Cuvelier, S. L., & Wiehler, S. (2002). Selectins: Critical mediators of leukocyte recruitment. *Seminars in Immunology*, *14*(2), 73–81.

Raposo, C. D., Canelas, A. B., & Barros, M. T. (2021). Human lectins, their carbohydrate affinities and where to find them. *Biomolecules*, *11*(2), 188. https://doi.org/10.3390/biom11020188.

Recher, M., & Lang, K. S. (2006). Innate (over)immunity and adaptive autoimmune disease. *Current Topics in Microbiology and Immunology*, *305*, 89–104.

Reisz, J. A., Bansal, N., Qian, J., Zhao, W., & Furdui, C. M. (2014). Effects of ionizing radiation on biological molecules—Mechanisms of damage and emerging methods of detection. *Antioxidants & Redox Signaling*, *21*(2), 260–292.

Rosen, S. D. (2004). Ligands for L-selectin: Homing, inflammation, and beyond. *Annual Review of Immunology*, *22*, 129–156.

Rossi, B., & Constantin, G. (2008). Anti-selectin therapy for the treatment of inflammatory diseases. *Inflammation & Allergy Drug Targets*, 7(2), 85–93.

Sattler, S. (2017). The role of the immune system beyond the fight against infection. *Advances in Experimental Medicine and Biology*, *1003*, 3–14.

Smak, P., Chandrabose, S., Tvaroska, I., & Koca, J. (2021). Pan-selectin inhibitors as potential therapeutics for COVID-19 treatment: In silico screening study. *Glycobiology*, *31*(8), 975–987.

Smith, B. A. H., & Bertozzi, C. R. (2021). The clinical impact of glycobiology: Targeting selectins, Siglecs and mammalian glycans. *Nature Reviews. Drug Discovery*, *20*(3), 217–243.

Srikiatkhachorn, A., Mathew, A., & Rothman, A. L. (2017). Immune-mediated cytokine storm and its role in severe dengue. *Seminars in Immunopathology*, *39*(5), 563–574.

Starshinova, A. A., Malkova, A. M., Basantsova, N. Y., Zinchenko, Y. S., Kudryavtsev, I. V., Ershov, G. A., et al. (2019). Sarcoidosis as an autoimmune disease. *Frontiers in Immunology*, *10*, 2933.

Steegmaier, M., Borges, E., Berger, J., Schwarz, H., & Vestweber, D. (1997). The E-selectin-ligand ESL-1 is located in the Golgi as well as on microvilli on the cell surface. *Journal of Cell Science*, *110*(Pt 6), 687–694.

Steegmaier, M., Levinovitz, A., Isenmann, S., Borges, E., Lenter, M., Kocher, H. P., et al. (1995). The E-selectin-ligand ESL-1 is a variant of a receptor for fibroblast growth factor. *Nature*, *373*(6515), 615–620.

Taefehshokr, N., Taefehshokr, S., Hemmat, N., & Heit, B. (2020). Covid-19: Perspectives on innate immune evasion. *Frontiers in Immunology*, *11*, 580641.

Tedder, T. F., Steeber, D. A., Chen, A., & Engel, P. (1995). The selectins: Vascular adhesion molecules. *The FASEB Journal*, *9*(10), 866–873.

Trinchera, M., Aronica, A., & Dall'Olio, F. (2017). Selectin ligands Sialyl-Lewis a and Sialyl-Lewis x in gastrointestinal cancers. *Biology (Basel)*, *6*(1), 16. https://doi.org/10.3390/biology6010016.

Tvaroska, I., Selvaraj, C., & Koca, J. (2020). Selectins—The two Dr. Jekyll and Mr. Hyde faces of adhesion molecules—A review. *Molecules*, *25*(12), 2835. https://doi.org/10.3390/molecules25122835.

Uchimura, K., Gauguet, J. M., Singer, M. S., Tsay, D., Kannagi, R., Muramatsu, T., et al. (2005). A major class of L-selectin ligands is eliminated in mice deficient in two sulfotransferases expressed in high endothelial venules. *Nature Immunology*, *6*(11), 1105–1113.

Vallien, G., Langley, R., Jennings, S., Specian, R., & Granger, D. N. (2000). Expression of endothelial cell adhesion molecules in neovascularized tissue. *Microcirculation*, 7(4), 249–258.

van Drongelen, V., & Holoshitz, J. (2017). Human leukocyte antigen-disease associations in rheumatoid arthritis. *Rheumatic Diseases Clinics of North America*, *43*(3), 363–376.

Wagner, D. (1972). Rudolf Virchow as a medical historian. *Z Gesamte Hyg*, *18*(6), 442–447.

Waldman, A. D., Fritz, J. M., & Lenardo, M. J. (2020). A guide to cancer immunotherapy: From T cell basic science to clinical practice. *Nature Reviews. Immunology*, *20*(11), 651–668.

Wang, J., & Springer, T. A. (1998). Structural specializations of immunoglobulin superfamily members for adhesion to integrins and viruses. *Immunological Reviews*, *163*, 197–215.

Wedepohl, S., Beceren-Braun, F., Riese, S., Buscher, K., Enders, S., Bernhard, G., et al. (2012). L-selectin—A dynamic regulator of leukocyte migration. *European Journal of Cell Biology*, *91*(4), 257–264.

Witz, I. P. (2006). The involvement of selectins and their ligands in tumor-progression. *Immunology Letters*, *104*(1–2), 89–93.

Woollard, K. J., & Chin-Dusting, J. (2010). P-selectin antagonism in inflammatory disease. *Current Pharmaceutical Design*, *16*(37), 4113–4118.

Xie, J. (2012). The C-type lectin-like receptors of Dectin-1 cluster in natural killer gene complex. *Glycoconjugate Journal*, *29*(5–6), 273–284.

Zhang, Y., Jiang, N., Zarnitsyna, V. I., Klopocki, A. G., McEver, R. P., & Zhu, C. (2013). P-selectin glycoprotein ligand-1 forms dimeric interactions with E-selectin but monomeric interactions with L-selectin on cell surfaces. *PLoS One*, *8*(2), e57202.

CHAPTER SIX

CoVaccine HT™ adjuvant is superior to Freund's in eliciting ovine polyclonal antibodies against human tumor necrosis factor-alpha

Owen R. Griffiths[a,b,*], John Landon[a], R. Keith Morris[b], Philip E. James[b], and Rachel A. Adams[b]

[a]Micropharm Ltd, Carmarthenshire, United Kingdom
[b]Department of Biomedical Sciences, Cardiff Metropolitan University, Cardiff, United Kingdom
*Corresponding author: e-mail address: owen.rees.griffiths@outlook.com

Contents

Abstract

Introduction: Passive immunotherapy using polyclonal antibodies plays an important role in preventing and treating antigenic and pathogenic diseases. Polyclonal antibodies are used for therapeutic, diagnostic and investigational purposes, with adjuvants employed to enhance the immune response against proteins that are poorly antigenic or self-antigens. This study aimed to optimize current immunization methods by evaluating the novel adjuvant CoVaccine HT™ against the established Freund's at

ISSN 1876-1623
https://doi.org/10.1016/bs.apcsb.2021.11.009

producing ovine polyclonal antibodies against pro-inflammatory cytokine human recombinant tumor necrosis factor alpha (TNF-α).

Methods: Castrated male Aberfield cross sheep were immunized with TNF-α in CoVaccine HT™ or Freund's adjuvant. The binding titer of antibodies for TNF-α and neutralization titer were determined *in vitro,* as well as the strength of antibody binding by a simple small scale affinity chromatography elution experiment. Animal welfare was monitored through inspection of immunization site reactions at regular time points and graded according to reaction size. The second part of the study looked at re-immunization using Freund's adjuvant alone every 4- or 8-weeks.

Results: Freund's generated significantly higher antibody binding titers than CoVaccine HT™ but were less effective at neutralizing TNF-alpha which is a better indicator of functional potency. CoVaccine HT™ also caused fewer immunization site reactions, while no statistical difference was observed in the binding strength of antibodies. Re-immunization every 4- and 8-weeks showed no statistical difference.

Conclusion: This study provides evidence that CoVaccine HT™ is superior to Freund's adjuvant for the production of antibodies to TNF-α, and supports the use of this alternative adjuvant for clinical and experimental use. The outcomes gained through this study are applicable to passive and active immunotherapy for the generation of polyclonal antibodies in human and veterinary medicine.

1. Introduction

The immune system's remarkable ability to generate immunoglobulin antibodies that can bind with considerable strength and specificity in response to antigen exposure is one of nature's remarkable abilities. Immunoglobulins present in blood are heterogeneous and are produced by a range of B cell clones, each with unique specificity towards a protein. Immunoglobulins can be expressed either on the cellular surface or as soluble secretions. Consequently, antigen-specific polyclonal antibodies (PcAbs) have been used for decades for investigational, diagnostic and therapeutic purposes.

Passive immunotherapy involves the transfer of antibodies or immune components in a bid to neutralize target proteins, which plays on nature's own lines of utilizing PcAbs to prevent or treat antigenic and pathogenic conditions by binding to and neutralizing a recipient antigen. On similar lines active immunotherapy exists, where the body produces its own PcAbs, such as following vaccination, to combat disease. Passive immunotherapy using PcAbs has revolutionized the way we approach and treat medical emergencies such as drug overdose, snake envenoming and other clinical indications. However, it could also play an important role in the future by preventing and treating viral conditions, and would be most beneficial to the immunocompromised, pregnant, infants, the elderly and those with chronic conditions.

The benefits of using sheep to produce antibodies are manifold; as well as being an established means of producing PcAbs, benefits include their ease of care, relative affordability to keep, consistent biological function, and reduced immunogenicity compared to antibodies derived from horse or human when injected into mice (Redwan, Fahmy, El Hanafy, Abd El-Baky, & Sallam, 2009). Large volumes of hyperimmunized serum can be obtained with lower immune boosting needs and lowered maintenance costs in comparison to horses. The downstream processing of ovine derived antibody purification from serum has already been investigated and optimized (Al-Abdulla, Casewell, & Landon, 2014), but little work has been focused on fine-tuning and optimizing the generation of high-quality PcAbs, particularly toward TNF-α.

Adjuvants, taken from the Latin word, "*adjuvare*," meaning, "to help," are immunopotentiators used to enhance an antigen's immunogenicity by increasing the intensity and duration of the humoral immune response, similarly used during the process of vaccination. The choice of adjuvant is critical for poorly immunogenic and self-antigens. Their exact mechanism is likely to vary but several mechanisms are postulated, including acting like a depot to ensure the gradual and continuous release of the immunogen, recruitment of immune cells, induction of cytokines and chemokines, the promotion of antigen transport to drainage lymph nodes, stimulating phagocytosis through non-specific activation of lymphocytes and causing aggregation of the immunogen to increase its molecular weight (Awate, Babiuk, & Mutwiri, 2013). When selecting an adjuvant, formulations with minimal reactogenicity consistent with good antibody response are favored, with particular consideration given to minimal adjuvant-related pain, distress or adverse reactions.

It is known that antibodies present in animal serum taken at different times following immunization have greater affinity, a process known as affinity maturation (Eisen & Siskind, 1964; Victora & Nussenzweig, 2012), and that antibody responses develop slowly and vary between adjuvants (Lakhan, Stevens, Diener, & Hayball, 2016; Stevens et al., 2013), with each adjuvant giving rise to a unique antibody response. Understanding immune kinetics is important when generating durable PcAbs for passive immunotherapy, which has been demonstrated to vary between adjuvants (Pedersen, Wørzner, Andersen, & Christensen, 2020).

Several groups have investigated the use of different adjuvants to obtain the highest levels of the highest affinity antibodies (Fraser, Diener, Brown, & Hayball, 2007; Jennings, 1995; Johnston, Eisen, & Fry, 1991; Lakhan et al.,

2016; Leenaars, Hendriksen, Angulo, Koedam, & Claassen, 1994; Stevens et al., 2013). For years, Freund's has been considered the gold standard adjuvant in animals and, despite evidence of adjuvant-related pain and distress (Amyx, 1987; Stevens et al., 2013; Stills, 2005), it is believed that the present day Freund's adjuvant is much less toxic than original formulations (Stills, 2005). The standard adjuvant employed for antibody generation is Freund's Complete Adjuvant (FCA) for the primary immunization and Freund's Incomplete Adjuvant (FIA) for all subsequent immunizations. CoVaccine HT™ has been suggested as an alternative because, when compared with Freund's, it produced significantly higher titers of ovine PcAb capable of neutralizing the highly immunogenic influenza agglutin both *in vitro* and *in vivo* (Stevens et al., 2013), as well as the poorly immunogenic endogenous protein High Mobility Group Box-1 (HMGB1) (Lakhan et al., 2016). Both studies looked at CoVaccine HT™ over a limited time period of 70 and 80 days, respectively.

CoVaccine HT™ comprises sucrose fatty acid sulphate esters immobilized on the oil droplets of a submicron emulsion of squalene-in-water (Hilgers & Blom, 2006). It forms an oil-in-water emulsion when mixed gently with an aqueous solution of the immunogen. In contrast, FCA and FIA must be mixed vigorously with an aqueous solution of the immunogen until a stable water-in-oil emulsion form. CoVaccine HT™ has been used in ferrets (Bodewes et al., 2010; Heldens et al., 2010), rhesus macaques (Heldens et al., 2010; Kusi et al., 2011), rabbits (Heldens et al., 2010; Younis, Faber, Kocken, & Remarque, 2019), pony stallions (Turkstra et al., 2005) and pigs (Turkstra et al., 2011), but only two groups have compared the use of CoVaccine HT™ to Freund's at eliciting antibodies in sheep (Lakhan et al., 2016; Stevens et al., 2013).

Freund's adjuvant is an established adjuvant, but adverse reactions are well-documented. These include granuloma formation, tissue swelling, necrosis, sloughing, abscess formation and other pathological lesions, which have led some to suggest that pain and distress are invariable (Stills, 2005; Wolfensohn & Lloyd, 1994). FCA consists of heat-killed *Mycobacterium tuberculosis*, which acts as an immunostimulant, in non-metabolizable oils (paraffin oil and mannide monooleate) (Freund, Casals, & Hosmer, 1937; Freund & McDermott, 2016). Injection of an antigen in FCA induces a humoral antibody response. This is essential for the delayed-type hypersensitivity reactions (Raffel, 1948) witnessed with FCA but not with FIA. FCA causes a Th1-dominated response, while injections which lack mycobacterial components induce a Th2-dominated response. An essential component of FCA-mediated response is an intense inflammatory reaction at the site of antigen deposition. For most applications, FCA is usually only necessary for the initial immunization, while FIA is the adjuvant of choice for subsequent immunizations. Inoculant is prepared by emulsifying the antigen in an aqueous solution with the oil, producing a water-in-oil emulsion. A stable water-in-oil emulsion is critical for the effectiveness of FCA or FIA.

This study sets out to generate antibodies toward TNF-α for the oral treatment of IBD, a chronic and often life-long condition that demands significant healthcare time and expense. Optimizing antibody production carries commercial benefits for the generation of diagnostic, therapeutic or scientific investigational purposes. To produce large volumes of serum with high titer specific PcAbs, adjuvants are essential by enhancing an immune response with higher levels of neutralizing antibodies, thus keeping the number of animals required to a minimum.

In addition, we aimed to optimize the production of large volumes of specific PcAbs toward TNF-α without compromising animal health or welfare. The study is divided into two arms: the initial arm comparing two adjuvants, Freund's and CoVaccine HT™, while the second phase looks at re-immunization with Freund's and Brij35 every 4- or 8-weeks.

2. Materials and methods

2.1 Chemicals and reagents

Dulbecco's Modified Eagle's Medium (DMEM), Fetal Bovine Serum (FBS), 2 mM L-Glutamine, trypsin/EDTA, HEPES, Brij35, Tween®20, 1% Penicillin-streptomycin, Donkey Anti-Sheep IgG Antibody HRP conjugate, Neutral Red Solution (0.33%) and Phosphate buffered saline (PBS) were obtained from Sigma (St. Louis, Missouri, USA). 0.22 μm filter, PBS, Immulon^{R} 4HBX flat bottom microfilter plates, Freund's Complete Adjuvant and Freund's Incomplete Adjuvant (Thermo Fischer Scientific), Recombinant human TNF-α (Peprotech, London, UK) and CellTiter-Blue® reagent from Promega (Madison, Wisconsin, USA), BioFX® TMB from Surmodics, and CoVaccine HT™ from BTG (UK).

2.2 Methodology

2.2.1 Ethical and home office approval for research and animal studies

Ethical approval was granted by Cardiff Metropolitan University (Project reference number: 9871). The production of hyperimmune ovine antisera, directed toward human recombinant TNF-α, was contracted out to

Ig-Innovations (Llandysul, UK) and was performed under full Home Office approval according to strict ethical guidelines for animal welfare under Home Office Animals (Scientific Procedures) Act 1986, Project License No PF0D06B6B.

2.2.2 Sheep immunization and hyperimmunized serum collection

Phase 1: Ten male Aberfield cross sheep were immunized at six different injection sites and assessed for reactions at injection sites. The immunogen for the primary immunization comprised Freund's complete adjuvant with Brij35 or CoVaccine HT™ adjuvant and 125 μg of human recombinant TNF-α per sheep. The protein-adjuvant mixture was injected subcutaneously and equally into 6 injection sites situated bilaterally, chosen for their proximity to the axillary, inguinal, and prescapular drainage lymph nodes. Each sheep was re-immunized at 28-day intervals with TNF-α mixed in Freund's incomplete adjuvant + Brij35 or CoVaccine HT™, with blood samples collected 14 days later. This method was used for the duration of the study. A total of 10 mL of blood per kg of body weight was collected from the external jugular vein without detriment to the animal. Sheep were not terminally bled.

Phase 2 followed on from the initial 22-week adjuvant comparison study. Sheep continued to be immunized every 4 weeks. Sheep previously in the CoVaccine HT™ group were converted to and immunized with Freund's + Brij35 for a minimum of 3 months, then on the fourth month, the study commenced. Sheep previously on Freund's and Brij35 continued with the previous immunization schedule for the 3 months bridging interval. At 3 months, 10 sheep were randomly assigned into two groups of 5 and re-immunized every 4- or 8-weeks with blood taken at 4-weekly intervals from 2-weeks after immunization. During the course of the study, a veterinary surgeon assessed and graded the size of injection site reactions.

2.2.3 Purification of antibody as control

Human TNF-α hyperimmunized ovine serum (Martindale Holdings Pty Ltd., Mintaro, South Australia, Australia) was raised by BTG, as previously described (Abbott et al., 2013), and the IgG fraction purified by adding caprylic acid (6% v/v) and further diluted with 2 parts saline to precipitate non-IgG proteins. The product was dialyzed and formulated at a concentration of 133 g/L in 20 mM citrate buffer, containing 153 mM NaCl (pH 6.0 ± 0.2) before storing at −20 °C. Antibody concentration was measured by absorbance at 280 nm using extinction coefficients of 1.5 (1 g/L, 1 cm path length) for ovine IgG (Dowall et al., 2016).

2.2.4 Determination of antibody titer by enzyme immunoassay (EIA)

The binding titers of ovine polyclonal anti-TNF-α antibodies in serum was determined by EIA. Briefly, EIA plates were coated with human recombinant TNF-α at a concentration of 1 μg/mL (100 μL/well) using coating buffer (phosphate-buffered saline (PBS)) and incubated at 4 °C overnight. Plates were washed (3 ×) using PBS containing 0.1% Tween 20 (PBST) and blocked with 2.5% fetal bovine serum in PBS (150 μL/well) for 2 h (hrs) at 37 °C. Plates were subsequently washed three times with PBST and incubated with antisera for TNF-α (100 μL/well) at appropriate dilutions. Serum from sheep which hadn't been immunized (pre-immunized serum) served as negative control. Wells were washed (3 ×) with PBST and 100 μL of diluted donkey anti-sheep antibody coupled to Horse Radish Peroxidase was added and incubated for 1 h at 37 °C. Then the plates were washed with PBST three times and 100 μL of substrate solution 3,3′,5,5′-tetramethylbenzidine (TMB). Plates were allowed to stand at room temperature for 5 min. The reaction was stopped by adding 50 μL/well of 1 M HCl and plates were read in a PolarStar plate reader at 450 and 690 nm. All samples were tested in either duplicate or triplicate. The dilution required to cause 50% maximum binding was estimated using GraphPad Prism V8.4.0. This value is defined as the EC_{50}, or binding titer.

2.2.5 Small-scale affinity chromatography (SSAC) to isolate specific anti-TNF-α PcAb

Protein G columns (NAbTM Protein G Spin Kit #89949, Thermo Scientific) were used to extract IgG from serum according to manufacturer's specifications. The extracted IgG was added to TNF-α affinity columns (AminoLinkTM Protein Coupling Kit, 0.1 mL, #SK259455, Thermo Scientific) prepared as per manufacturer's specifications to produce PcAbs specific for TNF-α. A minimum of three washes and seven elutions were performed for each run, and columns used for a maximum of 10 times.

2.2.6 Determination of specific antibody as a percentage of total IgG in serum by EIA

EIA was used to determine the concentration of specific anti-TNF-α PcAb by plotting a standard curve using specific PcAb of known concentration generated by small scale affinity chromatography. Each sample of unknown concentration was determined by interpolating from the standard curve. Based on in-house studies with the same sheep breed, sheep serum was determined to contain 25 g/L total IgG.

Three samples of known concentration (prepared independently of the standard curve samples) were assayed per plate and the values interpolated from the standard curve. Failure to fall within 15% of the intended concentration, the plate was rejected and repeated.

2.2.7 Cell culture

2.2.7.1 Maintenance of L929 cells

Cells were cultured in 75 cm^3 flasks using DMEM supplemented with 10% FBS, 2 mM L-Glutamine and 1% Penicillin-streptomycin. The L929 cells were seeded at a rate of 2×10^6 cells per flask in 30 mL of the culture medium and incubated at 37 °C in a 5% CO_2 humidified atmosphere.

2.2.7.2 Determination of antibody neutralization titer by cytotoxicity assay

L929 cells (Sigma, UK) are mouse derived fibroblast cells which are used as a target in TNF-α detection assays (Humphreys & Wilson, 1999). Cells were cultured in DMEM supplemented with 10% FBS, 2 mM L-Glutamine and 1% Penicillin-streptomycin. L929 cells were grown at 7.5×10^3 cells/well (100 μL) for 24 h (±1 h) at 37 °C in a 5% CO_2 humidified atmosphere. For each serum sample, serial dilutions were performed in a 96-well dilution plate, followed by media containing TNF-α (12 ng/mL). As a control, a serial dilution curve was included using purified anti-TNF-α PcAb of known concentration and known EC_{50} to account for inter-assay variability and standardize results. Serum in media acted as negative control, and TNF-α challenging dose (12 ng/mL) alone as positive control. The dilutions were left to rest for 1 h (±15 min) before transferring over to the indicator plate and left for 48 h (±15 min). Neutral Red Uptake assay was performed to assess cellular viability (Repetto, del Peso, & Zurita, 2008). The dilution required to protect 50% of the cell monolayer was estimated using GraphPad Prism V8.4.0. This value is defined as the EC_{50}, or neutralizing titer.

2.2.8 Determination of antibody binding strength using small scale affinity chromatography

Hyperimmunized serum was added to affinity columns (AminoLink Plus, 2 mL, #1856083, Thermo Scientific) coupled with 4.5 mg TNF-α as per manufacturer's instructions. 1 mL of 0.22 μm sterile filtered serum mixed in 1 mL of PBS was added and allowed to rotate for 1 h. Four washes with

2 mL PBS were performed and the protein concentration of the fourth wash measured to ensure all non-specific antibodies and residual serum protein had been washed through. Eight elutions were performed using glycine-HCl pH 3.7–2.3 in increasing order of acidity. Post elution, three washes with 2 mL glycine-HCl pH 2.6 were performed to ensure the removal of all residual antibodies before being washed (3 ×) with PBS. The columns were re-used a maximum of 10 times. The concentration of each elution was determined using a spectrophotometer at wavelength 280 nm.

2.2.9 L929 cell viability assessment using CellTiter Blue (CTB)

Viable cells reduce resazurin to resorufin, which emits fluorescence at 590 nm; the greater the number of viable cells, the higher the fluorescence reading. L929 cells were grown to confluence in 96-well plates with serum and antibody added, and left for 48 h. 20 μL of CTB was added per 100 μL media, left for 3 h then read in a PolarStar plate reader at Fluorescence 544_{Ex} 590_{Em}.

2.2.10 Statistical analysis

Data is presented as mean (±standard deviation) using GraphPad Prism v8 (San Diego, USA) and statistical analysis performed using Minitab Statistical Software 20.2.0.0 (State College, Pennsylvania, USA). Data was analyzed using Analysis of Variance (ANOVA) with the General Linear Model used to analyze with treatment effect as the response and time as factor. Significance was assumed with $P < 0.05$ throughout with precise values presented to $P < 0.0005$. For all data sets, when performing statistical analysis, the zero-time points were excluded. The determination of pre-immunization titers was undertaken to confirm that baseline serum samples were absent from specific anti-TNF-α antibodies.

3. Results

3.1 Phase 1 study—A comparison between sheep immunized with 125 μg TNF-α in CoVaccine HT™ or Freund's adjuvant with Brij 35 detergent every 4 weeks

3.1.1 Effect of adjuvant on sheep serum specific anti-TNF-α antibody concentration

Antisera from sheep immunized with 125 μg TNF-α in CoVaccine HT™ or Freund's were assessed for specific antibody concentration (g/L) using EIA by interpolating from a standard curve, generated using specific anti-TNF-α IgG PcAb of known concentration. Sheep serum contained approximately 25 g/L total IgG (data not shown) and the concentration of specific IgG was determined as a percentage of total IgG.

At week 0 (pre-immunization), serum from both CoVaccine HT™ and Freund's groups contained 0 g/L specific anti-TNF-α antibodies (Fig. 1). From week 6 onward, the percentage of specific antibodies increased, with a significantly higher concentration of specific antibodies in the Freund's group. However, time had no statistical effect on the concentration of specific antibodies in serum.

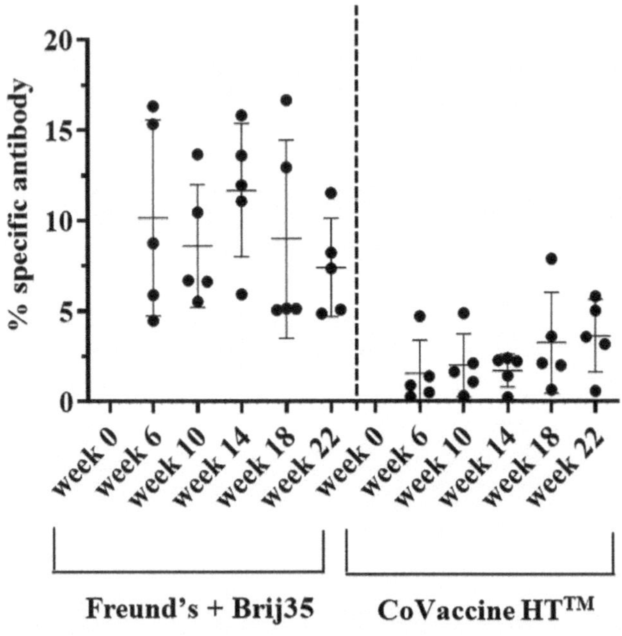

Fig. 1 Quantification of specific anti-TNF-α antibodies present in serum of sheep immunized with 125 μg TNF-α in CoVaccine HT™ or Freund's + Brij 35 adjuvant, determined by EIA. A standard curve generated using specific anti-TNF-α was generated and the concentration of anti-TNF-α antibodies in serum determined. The percentage of specific IgG was determined by dividing the previosuly quantified mean concentration of total IgG present in sheep serum. Data is displayed as mean ± SD with each data point representing individual sheep analyzed in duplicate ($N = 5$, $n = 10$). All control samples were within the 15% acceptance limit. The General Linear Model method of ANOVA with factor analysis confirmed that levels of TNF-α specific IgG did not change significantly between weeks 6 and 22 ($P = 0.0903$). However, Freund's treatment resulted in significantly higher specific antibody levels than the CoVaccine HT™ group ($P < 0.0005$).

3.1.2 Effect of adjuvant on sheep serum anti-TNF-α binding titer

Data in Fig. 2 shows antisera samples from individual sheep assessed by EIA to determine the 50% binding titer (EC_{50}). The mean binding titer for each adjuvant was plotted at 4-weekly time points from weeks 6 to 22 post-primary immunization. Pre-immunization samples contained near zero binding antibodies, while samples from week 6 onward contained antibodies against TNF-α.

Over time, the EC_{50} dilution binding titers increased, meaning that a lower concentration of serum was required to cause 50% maximum binding. Sheep immunized with Freund's generated better binding antibodies, which matches the data in Fig. 1 where higher concentrations of specific anti-TNF-α antibodies were generated by Freund's.

Data presented in Fig. 2 demonstrates a highly significant difference in binding titer of the anti-TNF-α between the Freund's adjuvant and CoVaccine HT™ treatment. Data presented in Fig. 2 also confirms higher binding of antibodies produced using the Freund's adjuvant compared to CoVaccine HT™. Binding titers of antibodies when Freund's was used reached their maximum within first 6 weeks of the experiment. By contrast, a slight trend of increase was observed with CoVaccine HT™ between weeks 10 and 14. This trend was not significant; but may have become

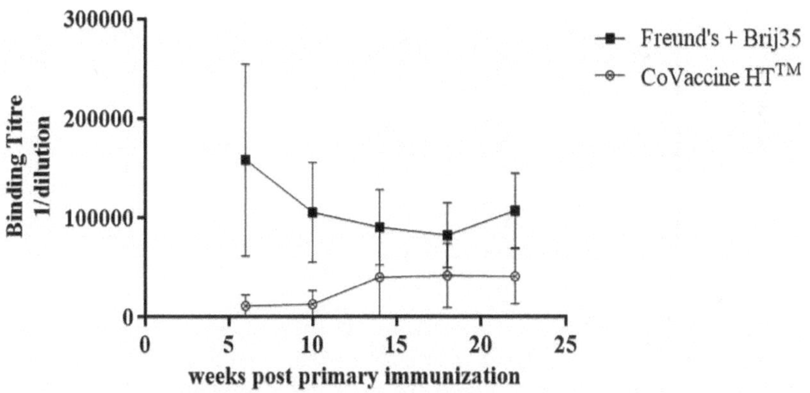

Fig. 2 Comparison of antibody binding titer (EC_{50}) for anti-TNF-α antibodies in serum of sheep immunized with 125 μg TNF-α in CoVaccine HT™ or Freund's + Brij 35, determined by EIA. Each serum sample was serial diluted across an EIA plate coated with 1 μg/mL TNF-α. The EC_{50} value was determined computationally by GraphPad. Each EC_{50} value plotted represents the mean of 5 sheep performed in duplicate ± SD ($N=5$, $n=10$). The General Linear Model method of ANOVA with factor analysis confirmed that the effect of time was non-significant ($P=0.07603$). However, Freund's treatment resulted in significantly higher antibody binding titers ($P<0.0005$).

significant if the experiment had continued for longer. This correlates with our previous observations using a different antigen where CoVaccine HT™ required a considerably longer time than Freund's + Brij 35 to reach a plateau in binding titers.

The zero-time point was not included as part of the analysis because this would have given a statistically significant result based on pre- and post-immunization instead of the effect of adjuvant; however, no significant time effect was observed over the remaining time period.

3.1.3 Effect of adjuvant on sheep serum anti-TNF-α neutralization titer

Antisera samples from individual sheep were assessed for functionality using a cell-based immunocytotoxicity assay to determine the 50% neutralization titer (Fig. 3). The mean neutralization titer for each adjuvant group was plotted at 4 weekly time points from weeks 6 to 22 post primary immunization. Pre-immunization samples were also assessed and were found to contain no neutralizing antibodies, while samples from week 6 onward contained antibodies capable of neutralizing TNF-α. Viability assessments were performed on the L929 cell line which confirmed that antibody and serum alone did not reduce viability by more than 10% (data not shown).

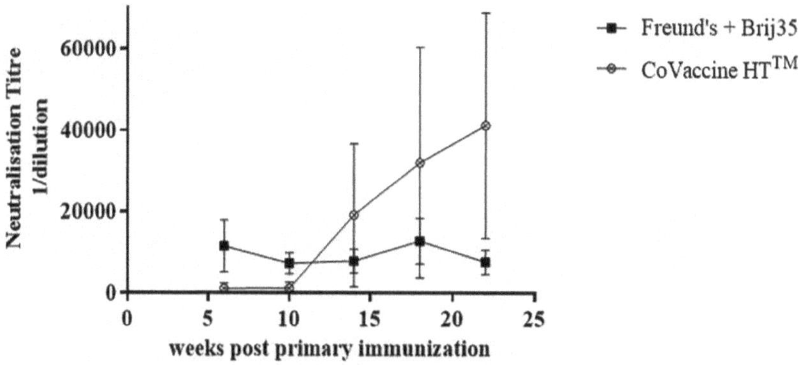

Fig. 3 Comparison of antibody neutralization titer (EC_{50}) for anti-TNF-α antibodies in serum of sheep immunized with 125 μg TNF-α in CoVaccine or Freund's + Brij 35 adjuvant, determined by a cell-based immunocytotoxicity assay. Hyperimmunized sheep serum containing anti-TNF-α antibodies were serially diluted across a cell plate containing L929 cells, and challenged with 12 ng/mL TNF-α for 48 h. EC_{50} was determined computationally using GraphPad. Each data point represents a mean of 5 sheep performed in duplicate ± SD ($N=5$, $n=10$). The General Linear Model method of ANOVA with factor analysis confirmed that the effect of time was significant ($P=0.041$) and a statistically significant difference seen in specific antibody binding between CoVaccine HT™ and Freund's adjuvant ($P=0.020$).

Following subsequent immunizations, the EC_{50} dilution neutralization titer increased for CoVaccine HT™, which meant that a lower concentration of serum was required to cause 50% neutralization. For weeks 6 and 10, Freund's contained better neutralization antibodies than CoVaccine HT™, but from week 14 onward, CoVaccine HT™ surpassed Freund's. The neutralization capacity of antibodies generated by Freund's reached a maximum at 6 weeks. However, neutralizing antibodies were not generated until at least 10 weeks post-immunization (3 immunizations) when CoVaccine HT™ was used as an adjuvant. Despite the delayed response to immunization using CoVaccine HT™, the neutralizing capacity of antibodies collected at week 14 post-immunization was greater than that with Freund's and continued to increase substantially until the end of the study at 22 weeks.

3.1.4 Effect of adjuvant on antibody avidity (binding strength)

To assess the binding strength of specific anti-TNF-α antibodies in serum, after extracting total IgG from serum then added to affinity columns and eluted using different pH in a stepwise manner from pH 3.8 to 2.3 (Fig. 4). Theoretically, the stronger bound antibodies would require a lower pH elution buffer in order to be removed from the column, thus meaning that the antibodies had stronger binding abilities.

Elution buffer of lower pH was required to dissociate a higher percentage of PcAbs generated using CoVaccine HT™ adjuvant. The need for lower pH suggests that the antibodies are tighter binding or have higher avidity. To check the robustness of the SSAC's, the total concentration of specific antibody eluted from the columns were compared to the concentration of specific IgG determined by enzyme immunoassay (EIA) and were found to be of similar values. Statistical analysis revealed that there were no differences observed in anti-TNF-α binding strength in the CoVaccine HT™ or Freund's group.

3.1.5 Effect of-immunization with 125 μg TNF-α in CoVaccine HT™ or Freund's + Brij35 every 4-weeks on immunization site reactions

Sheep administered CoVaccine HT™ adjuvant displayed few injection site reactions, whereas sheep in the Freund's group all developed reactions by week 8 which increased through to week 20 and remained the same at week 24 (Fig. 5). All reactions were minor to small in size and no concerns were raised over the welfare of individual sheep. A statistically higher number of sheep in the Freund's group experienced reactions which were greater over time.

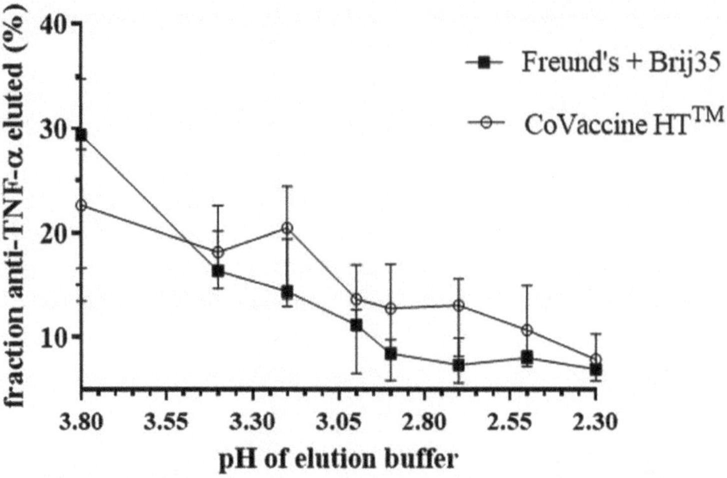

Fig. 4 Comparison of antibody binding strength of anti-TNF-α antibodies in serum of sheep immunized with 125 μg TNF-α in CoVaccine or Freund's + Brij 35 adjuvant, determined by small scale affinity chromatography. IgG was extracted from serum using protein G columns followed by TNF-α affinity purification using affinity columns and the concentration of anti-TNF-α eluted during each pH converted to percentage. The fraction of anti-TNF-α removed during each elution at different pH compared antibodies raised using Freund's + Brij35 or CoVaccine HT™ adjuvant at 22 weeks post initial immunization. Pre-immunization serum was used as negative control to determine non-specific binding. Each data point represents the mean of 5 sheep in duplicate ±SD. The General Linear Model method of ANOVA with factor analysis confirmed that pH had a significant effect on the fraction eluted ($P < 0.005$), with no significant difference associated with Freund's and CoVaccine HT™.

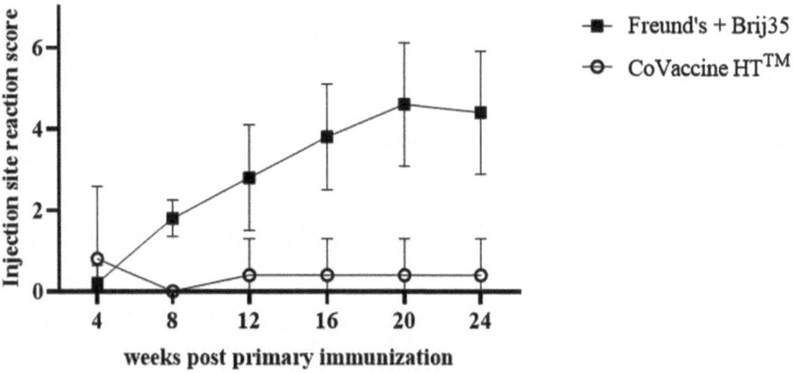

Fig. 5 Immunization site reaction scoring for sheep immunized with 125 μg TNF-α in Freund's + Brij 35 and CoVaccine HT™. Each of the six immunization sites were assessed every 4 weeks for immunization related reactions at 6 different sites and graded between 0 and 4 based on reaction size using pre-defined scoring criteria. The score was combined to give a cumulative score (displayed) for each sheep, with larger scores indicative of more severe or multi-site reactions. Each data point represents the sum of reaction scores for 5 sheep ($n = 5$). The General Linear Model method of ANOVA with factor analysis confirmed that Freund's gave rise to significantly bigger reactions ($P < 0.005$), which were significantly greater over time ($P = 0.007$).

3.2 Phase 2—Comparison of titers and site reaction from re-immunization every 4- or 8-weeks

Sheep in the CoVaccine HT™ group were converted to Freund's for a minimum of 3 months before commencing phase 2 of the study, being the comparison of re-immunization every 4- or 8-weeks after a maximum titer had been achieved. The purpose of the second phase was to determine whether 4- or 8-weekly re-immunization had any effect on the concentration of specific antibody generated.

3.2.1 Effect of re-immunization every 4- or 8-weeks on sheep serum specific anti-TNF-α antibody concentration

Data in Fig. 6 shows that the concentration of specific anti-TNF-α antibodies did not show any statistical difference within each group (either 4- or 8-weeks re-immunization intervals) over the time of the experiment. However, a significant difference was observed in the concentration of specific antibodies between the two groups (Fig. 5).

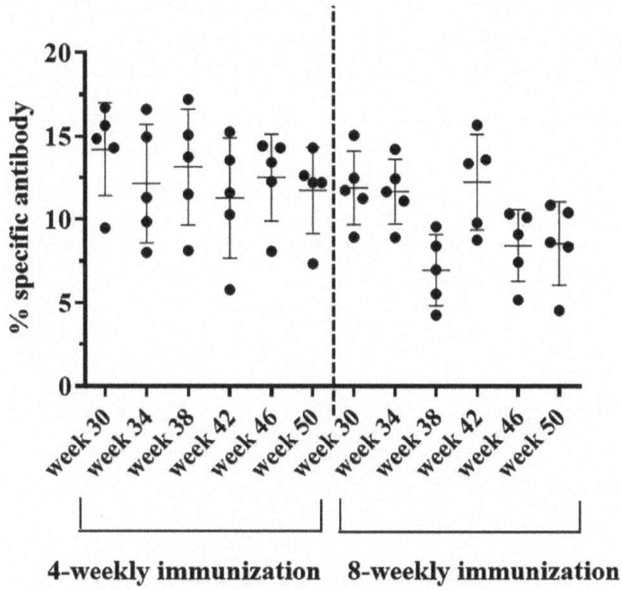

Fig. 6 Quantification of specific anti-TNF-α antibodies in serum of sheep immunized with 125 μg TNF-α in Freund's + Brij 35 adjuvant at 4- or 8-weekly intervals, determined by EIA. A standard curve generated using specific anti-TNF-α ovine PcAb was plotted, and the concentration of anti-TNF-α antibodies in serum determined. The percentage of specific IgG was determined by dividing the previosuly determined mean concentration of total IgG present in sheep serum. Data is displayed as mean ± SD with each data point representing each individual sheep performed in duplicate ($N = 5$, $n = 10$). All control samples were within the 15% acceptance limit. The General Linear Model method of ANOVA with factor analysis confirmed that time was non-significant ($P = 0.149$). However, a significant difference was seen between immunization every 4- or 8-weeks ($P = 0.001$).

3.2.2 Effect of re-immunization with 125 μg TNF-α every 4- or 8-weeks on sheep serum anti-TNF-α binding titer

Data in Fig. 7 shows no obvious correlation between immunizing every 4- or 8-weeks, with antibody binding titers seen to fluctuate unpredictably for both groups. There is a peak in antibody titer at week 46 for both groups which could be attributed to the weather and could be responsible for statistical significance seen with time.

3.2.3 Effect of re-immunization every 4- or 8-weeks on sheep serum anti-TNF-α antibody neutralization titer

As with Fig. 7, Fig. 8 shows no difference in neutralization of TNF-α between immunizing every 4- and 8-weeks. Antibody neutralization titers fluctuate for both groups. The peak at week 46 is likely to be responsible for the statistical significance observed over time.

3.2.4 Effect of re-immunization with 125 μg TNF-α in Freund's + Brij35 every 4- or 8-weeks on immunization site reactions

In phase 2, all sheep had been converted to Freund's adjuvant for a minimum of 3 months before blood samples were taken. Sheep were graded for reactions at immunization sites (Fig. 9). All sheep developed mild reactions

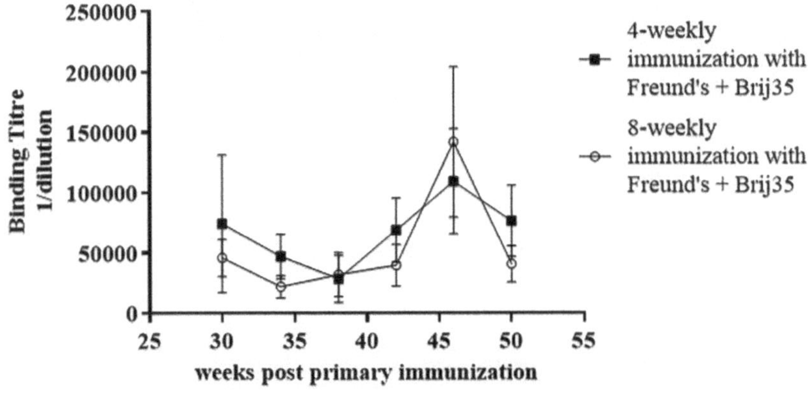

Fig. 7 Comparison of antibody binding titer (EC_{50}) for anti-TNF-α antibodies in serum of sheep immunized with 125 μg TNF-α in Freund's + Brij 35 adjuvant every 4- or 8-weeks, determined by EIA. Each hyperimmunized sheep serum sample was serially diluted across an EIA plate coated with 1 μg/mL TNF-α. The EC_{50} was determined computationally by GraphPad. Each EC_{50} value plotted represents the mean of 5 sheep performed in duplicate ± SD ($N=5$, $n=10$). The General Linear Model method of ANOVA with factor analysis confirmed that the effect of time was significant ($P=0.0005$), however, whether immunized every 4- or 8- weeks showed no statistical significance ($P=0.118$).

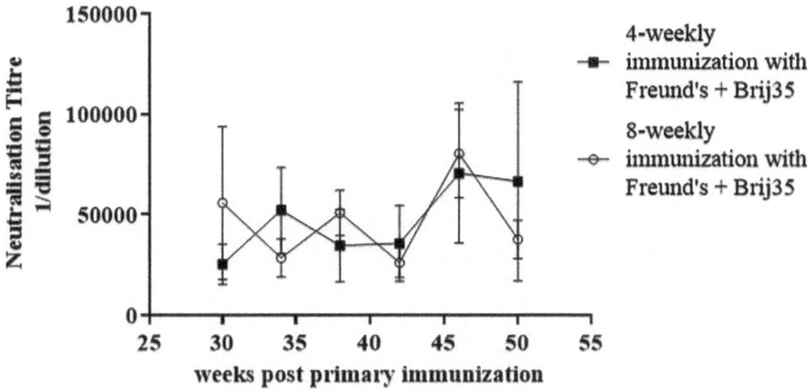

Fig. 8 Comparison of anti-TNF-α antibody neutralization titer (EC_{50}) in serum of sheep immunized with 125 μg TNF-α in Freund's + Brij 35 adjuvant every 4- or 8-weeks, determined using a cell-based immunocytotoxity assay. Serum containing anti-TNF-α antibodies were serial diluted across a cell plate containing L929 cells and challenged with 12 ng/mL TNF-α for 48 h. The EC_{50} was determined computationally by GraphPad. Each data point represents the mean of 5 sheep performed in duplicate ± SD ($N=5$, $n=10$). The General Linear Model method of ANOVA with factor analysis confirmed that the effect of time was significant ($P=0.007$). However, immunization every 4- or 8-weeks was non-significant ($P=0.886$).

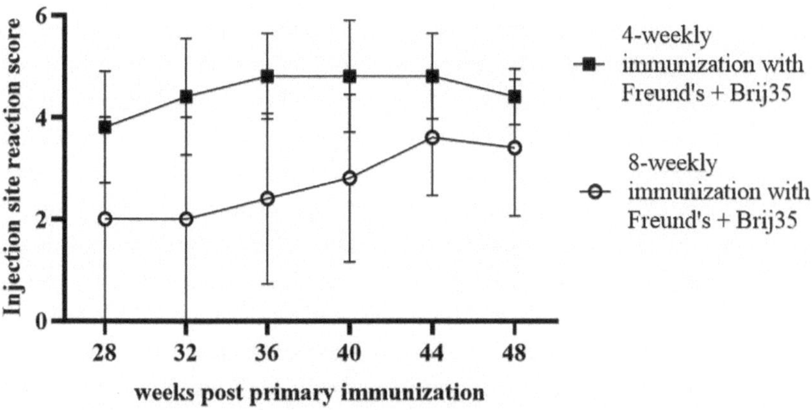

Fig. 9 Immunization site reaction score in sheep immunized with 125 μg TNF-α in either Freund's + Brij 35 adjuvant at 4- or 8-weeks interval. Each of the six immunization sites were assessed every 4 weeks for immunization related reactions and graded between 0 and 4 based on reaction size, using pre-defined scoring criteria. The score was combined to give a cumulative score for each sheep, with larger scores indicative of more severe or multi-site reactions. Each data point represents the sum of reaction scores for 5 sheep *($n = 5$)*. The General Linear Model method of ANOVA with factor analysis confirmed that there was a statistically significant difference between the two groups ($P < 0.005$). There was no statistical significance in the site reaction score within each group during the time of the experiment ($P=0.275$).

confirming that the reactions were adjuvant related, or more specifically, related to Freund's, with a significantly greater number of reactions observed in the 4-weekly immunization group.

4. Discussion

The success of passive immunotherapy relies on the production of high volumes of high neutralizing antibodies. The function of an antibody relies on its ability to recognize a structure and bind tightly with the aim of blocking its antigenic effects. This study aimed to optimize antibody production in sheep by comparing the efficacy of two adjuvants, CoVaccine HT™ and Freund's + Brij 35. In addition, to optimize immunization schedules, a comparison of re-immunization at 4- or 8-week intervals was performed using Freund's + Brij 35. Our ultimate aim was to generate high volumes of TNF-α neutralizing antibodies without compromising animal welfare while saving significantly on antigen use and expense.

During an adaptive immune response, a B-cell can select one of two pathways. An extrafollicular response is a T-cell independent pathway involving rapid production of large volumes of short-lived neutralizing antibodies of modest affinity. These provide short term immunity, but do not afford long term protection. By contrast, a germline center response (or follicular) is a T-cell dependent pathway, where the production of antibodies is delayed, but generates a broad spectrum of highly specific, high affinity antibodies that are longer lasting, with memory and long-lived plasma cells.

In some animal models it is known what type of response a particular immunization or antigen elicits; however, it remains unclear why some favor the extrafollicular pathway over germline center response (Elsner & Shlomchik, 2020). Some bacteria have been shown to follow the extrafollicular pathway with suppression of germline centers, as observed with *B. burgdoferi*, *E. muris* and *STm* (Elsner & Shlomchik, 2020; Klasse, 2016). Freund's, being heat-killed *M. tuberculosis*, may favor the extrafollicular pathway, which may explain the high levels of modest affinity antibodies compared with CoVaccine HT™'s slower generation of antibodies with improved neutralizing ability, as seen in Figs. 1–3.

Another factor influencing the pathway that an antigen-activated B cell follows is the initial affinity between B-cell receptor and antigen. Having high affinity for the antigen usually adopts the extrafollicular route (Palm & Henry, 2019), while reducing B-cell receptor affinity or increasing antigen density can lower the extrafollicular plasma cell response which promotes the alternative germline center response (Paus et al., 2006). It has been proposed that the stability of antigen in CoVaccine HT™ adjuvant is better than in emulsified Freund's and maintains TNF-α's native structure. Freund's adjuvant requires vigorous mixing which could distort the native TNF-α structure.

Emulsion based adjuvants (MF59TM and GLA-SE) generate a rapid increase in antibody titers while depot forming adjuvants (CAF01 and AH) gave initially poor antibody responses (Pedersen et al., 2020). Depot-forming adjuvants, such as aluminum hydroxide or cationic liposome-based adjuvant release antigen slowly by constantly feeding germinal centers with high-affinity, long-term antibody response (Pedersen et al., 2020), and could account for CoVaccine HT™'s superiority in generating such antibodies. Slow release of antigen may be beneficial by constantly feeding antigen to germinal centers promoting affinity maturation (Cirelli & Crotty, 2017; Pedersen et al., 2020; Tam et al., 2016; Turner, Benet, & Grigorova, 2017). Slow antigen release can result in low presentation to B-cells leading to delayed B-cell activation and germinal center formation (Pedersen et al., 2020). Based on reactions at immunization sites, Freund's is more reactive and elicits a greater immune response, whereas CoVaccine HT™ generates a slower paced response which favors the germline route.

The higher neutralization titers seen with CoVaccine HT™ (Fig. 3) may also be due to the stability and integrity of the antigen after mixing with adjuvant, which was interestingly coupled with a lower incidence of reactions at immunization sites. It may be that, in Freund's, the antigen (TNF-α) undergoes partial or complete denaturation in the water-in-oil emulsion during its preparation which requires vigorous mixing, possibly deforming the native epitopes of TNF-α. This leads to the generation of poorly fitting antibodies which recognize disrupted secondary and tertiary protein structures (Lakhan et al., 2016; Leenaars et al., 1994). This is highly likely for antigens such as TNF-α where the native structure consists of three monomers held together by non-covalent interactions. A breakdown or derangement in TNF-α structure would generate poor affinity antibodies through poorly recognized native structure. Contrariwise, CoVaccine HT™ is an oil-in water emulsion of hydrophobic and negatively-charged sucrose fatty acid sulphate esters formulated with submicron emulsions of squalene that is combined with aqueous antigen and thus would result in less protein denaturation than with Freund's adjuvant (Hilgers & Blom, 2006; Stevens et al., 2013). However, this was not supported by the antibody binding study (Fig. 4) where antibodies raised using CoVaccine HT™ or

Freund's showed no statistical difference when eluted from small affinity columns using various elution buffer pH.

The statistically higher incidence of reactions at immunization sites with Freund's implies that the adjuvant is more reactive and better at eliciting an immune response, potentially adopting the extrafollicular pathway. These antibodies may still bind to native TNF-α but the strength of the association may be weaker due to it recognizing distorted or partially denatured epitopes. Furthermore, the blocking of TNF-α function might be due to binding of those antibodies in the proximity of the antigen active site, causing a steric hindrance in binding to the receptor.

Factors known to influence whether an antibody response is *via* the extrafollicular or germline route help to explain the differences observed between antibody binding and neutralization. CoVaccine HT™ produces lower levels of antibodies with improved neutralization capacity which is a result of antibody generation *via* germline centers. The success of an antibody response, whether for vaccinations or PcAb production, is measured through antibody titers, but determining if a response is *via* the germline or extrafollicular pathway could present a better indicator of antibody functionality and long-term protection. Future studies should focus on determining the pathway in which an antibody is generated and whether those pathways would depend on the adjuvant characteristics alone, or if the physicochemical features of the antigen are important.

The timing of booster immunizations has been shown to be critical, being most effective when administered after antibody responses have peaked (Pedersen et al., 2020; Siegrist, 2018). It has been suggested that higher antibody responses are seen when the first booster immunization is given later than 4 weeks post-priming (Pedersen et al., 2020), after the contraction of germinal centers (Siegrist, 2018). Comparing immunization every 4 *versus* 8 weeks, with the same adjuvant, there were no statistical differences observed for antibody binding, neutralization and total specific IgG levels. As this study was commenced at 30 weeks, it shows that once titer max has been achieved, re-immunization can be performed less frequently. Future studies should focus on re-immunization at intervals greater than 8 weeks. It has been observed that re-immunization at 4 (first booster) and 16 weeks (second booster) resulted in different responses with an increase affinity and class-switching which promotes high affinity antibody responses observed at 16 weeks (De Silva & Klein, 2015; Pedersen et al., 2020).

With the viral protein hemagglutinin, very few inflammatory site reactions were observed with CoVaccine HT™ compared to Freund's

(Stevens et al., 2013). These findings correspond with our findings where sheep immunized with CoVaccine HT™ displayed minimal reactions. One sheep developed small reactions bilaterally in the fore flank region at weeks 16 and 20, and a second developed small reactions at four sites by week 4 which had fully resolved by week 8. In the Freund's group, all sheep developed reactions by the third immunization at week 8. These ranged from minor to plum sized reactions. It was also observed that CoVaccine HT™ gave rise to significantly higher antibody tires with stronger affinity and more functional potency than antibodies induced with Freund's adjuvants (Stevens et al., 2013) which supports our phase 1 findings. For phase 2, all sheep had been converted to Freund's adjuvant and those previously on CoVaccine HT™ all developed few reactions at injection sites, which remained the case throughout the study. These observations highlight CoVaccine HT™ as a safer and highly effective potential alternative to Freund's for adjuvanting antigens, particularly when factoring in the time-consuming and often hazardous emulsification periods needed with Freund's adjuvant and other water-in-oil adjuvants prior to administration (Lakhan et al., 2016).

In a study comparing CoVaccine HT™ to Freund's adjuvant at eliciting antibodies against the endogenous alarmin HMGB1, CoVaccine HT™ produced higher titers with increased affinity and better functional potency when compared to Freund's (Lakhan et al., 2016). However, with only 2 sheep in the study, the statistical strength is weak. Binding was assessed using endpoint and indirect EIA, with neutralization activity determined using the murine RAW 264.7 macrophage cell line, and confirmed with an *in vivo* murine model of endotoxemia. They also observed fewer and less severe adverse site reactions in the CoVaccine HT™ group which matches the findings of this study and those of Stevens et al. (2013). A 10-fold reduction of antigen dose (20 μg over 200 μg) could be used without compromising antibody titers, with the lower antigen dose group giving rise to a lower surface temperature at sites of injection. Although this study did not compare antigen dosing it is particularly pertinent when the cost of antigen is a factor, as with TNF-α. Additionally, no significant differences were observed in the size of reactions at injection sites between the two antigen dose groups.

It was believed that antibodies generated using CoVaccine HT™ would have increased avidity. This would be evidenced by the need for lower pH elution buffer to remove antibodies from the affinity column and imply that PcAbs raised using CoVaccine HT™ adjuvant bind tighter, possibly by having better recognition of the antigenic binding region. However, no statistical difference was observed between the binding of antibodies raised using CoVaccine HT™ and Freund's.

The limitations of this study include small group sizes ($n = 5$), although in some previously published studies, even smaller group sizes have been used (8). Future studies should aim to address this by increasing the sheep numbers per group. The second phase of this study looked at re-immunization every 4- or 8-weeks using Freund's. This decision was based on preliminary binding data which revealed that Freund's was superior to CoVaccine HT™. However, neutralization data, which were generated later, provided contrary outcomes. Future studies should focus on CoVaccine HT™ and different re-immunization time points.

For both studies, 10 sheep were divided into two groups of 5. A large variation in PcAb percentage, binding and neutralizing ability exists between each sheep and within groups, which is not uncommon. This can only be explained by natural variation and the varying ability each sheep has in mounting a humoral immune response, which is expected with live subjects. The low group numbers, however, weaken the statistical strength of the results. The breed, age and bodyweight of sheep was kept constant throughout to minimize potential genetic variability. The percentage of specific antibodies, binding and neutralization undergoes an increase which correlates with B-cell maturation with repeated immunizations, with some sheep less responsive to the antigen-adjuvant than others.

In conclusion, CoVaccine HT™ is superior to Freund's when used as an adjuvant for the production of ovine anti-TNF-α antibodies. In particular, antibodies generated with CoVaccine HT™ had a greater ability to neutralize TNF-α and were therefore functionally more potent. While Freund's adjuvant enhanced antibody production leading to high antibody titers, the antibodies had lower neutralizing ability. Furthermore, use of CoVaccine HT™ minimized adverse immunization site reactions in immunized sheep. Re-immunization of sheep more frequently at 4-weekly rather than 8-weekly intervals did not elicit a greater immunological response. An immunization regime using CoVaccine HT™ as adjuvant with an 8-weekly re-immunization schedule is therefore likely to be most effective for production of high concentration and quality ovine anti-TNF-α polyclonal antibody. This would be a focus for future studies.

Competing interests

No competing interests to declare.

Author contributions

This work was undertaken by Owen Griffiths and the manuscript reviewed by PhD supervisors.

Acknowledgments

Owen Griffiths' work was supported by a PhD grant from Knowledge Economy Skills Scholarships (KESS 2), a major pan-Wales operation supported by European Social Funds (ESF) through the Welsh Government.

References

Abbott, W. M., Snow, M., Eckersley, S., Renshaw, J., Davies, G., Norman, R. A., et al. (2013). Characterization of the complex formed between a potent neutralizing ovine-derived polyclonal anti-TNFα Fab fragment and human TNFα. *Bioscience Reports*, *33*(4). https://doi.org/10.1042/BSR20130044.

Al-Abdulla, I., Casewell, N. R., & Landon, J. (2014). Single-reagent one-step procedures for the purification of ovine IgG, F(ab')2 and Fab antivenoms by caprylic acid. *Journal of Immunological Methods*, *402*(1), 15–22. https://doi.org/10.1016/j.jim.2013.11.001.

Amyx, H. L. (1987). Control of animal pain and distress in antibody production and infectious disease studies. *Journal of the American Veterinary Medical Association*, *191*(10), 1287–1289.

Awate, S., Babiuk, L. A., & Mutwiri, G. (2013). Mechanisms of action of adjuvants. *Frontiers in Immunology*, *4*. https://doi.org/10.3389/fimmu.2013.00114.

Bodewes, R., Kreijtz, J. H. C. M., van Amerongen, G., Geelhoed-Mieras, M. M., Verburgh, R. J., Heldens, J. G. M., et al. (2010). A single immunization with CoVaccine HT-adjuvanted H5N1 influenza virus vaccine induces protective cellular and humoral immune responses in ferrets. *Journal of Virology*, *84*(16), 7943–7952. https://doi.org/10.1128/JVI.00549-10.

Cirelli, K. M., & Crotty, S. (2017). Germinal center enhancement by extended antigen availability. *Current Opinion in Immunology*, *47*, 64–69. https://doi.org/10.1016/j.coi.2017.06.008.

De Silva, N. S., & Klein, U. (2015). Dynamics of B cells in germinal centres. *Nature Reviews. Immunology*, *15*(3), 137–148. https://doi.org/10.1038/nri3804.

Dowall, S. D., Callan, J., Zeltina, A., Al-Abdulla, I., Strecker, T., Fehling, S. K., et al. (2016). Development of a cost-effective ovine polyclonal antibody-based product, EBOTAb, to treat Ebola virus infection. *The Journal of Infectious Diseases*, *213*(7), 1124–1133. https://doi.org/10.1093/infdis/jiv565.

Eisen, H. N., & Siskind, G. W. (1964). Variations in affinities of antibodies during the immune response. *Biochemistry*, *3*, 996–1008. https://doi.org/10.1021/bi00895a027.

Elsner, R. A., & Shlomchik, M. J. (2020). Germinal center and extrafollicular B cell responses in vaccination, immunity, and autoimmunity. *Immunity*, *53*(6), 1136–1150. https://doi.org/10.1016/j.immuni.2020.11.006.

Fraser, C. K., Diener, K. R., Brown, M. P., & Hayball, J. D. (2007). Improving vaccines by incorporating immunological coadjuvants. *Expert Review of Vaccines*, *6*(4), 559–578. https://doi.org/10.1586/14760584.6.4.559.

Freund, J., Casals, J., & Hosmer, E. P. (1937). Sensitization and antibody formation after injection of tubercle bacilli and paraffin oil. *Proceedings of the Society for Experimental Biology and Medicine*, *37*(3), 509–513. https://doi.org/10.3181/00379727-37-9625.

Freund, J., & McDermott, K. (2016). Sensitization to horse serum by means of adjuvants. *Proceedings of the Society for Experimental Biology and Medicine*. https://doi.org/10.3181/00379727-49-13625.

Heldens, J. G. M., Glansbeek, H. L., Hilgers, L. A. T., Haenen, B., Stittelaar, K. J., Osterhaus, A. D. M. E., et al. (2010). Feasibility of single-shot H5N1 influenza vaccine in ferrets, macaques and rabbits. *Vaccine*, *28*(51), 8125–8131. https://doi.org/10.1016/j.vaccine.2010.09.097.

Hilgers, L. A. T., & Blom, A. G. (2006). Sucrose fatty acid sulphate esters as novel vaccine adjuvant. *Vaccine*, *24*, S81–S82. https://doi.org/10.1016/j.vaccine.2005.01.133.

Humphreys, D. T., & Wilson, M. R. (1999). Modes of L929 cell death induced by TNF-alpha and other cytotoxic agents. *Cytokine*, *11*(10), 773–782. https://doi.org/10.1006/cyto.1998.0492.

Jennings, V. M. (1995). Review of selected adjuvants used in antibody production. *ILAR Journal*, *37*(3), 119–125. https://doi.org/10.1093/ilar.37.3.119.

Johnston, B. A., Eisen, H., & Fry, D. (1991). An evaluation of several adjuvant emulsion regimens for the production of polyclonal antisera in rabbits. *Laboratory Animal Science*, *41*(1), 15–21.

Klasse, P. J. (2016). How to assess the binding strength of antibodies elicited by vaccination against HIV and other viruses. *Expert Review of Vaccines*, *15*(3), 295–311. https://doi.org/10.1586/14760584.2016.1128831.

Kusi, K. A., Remarque, E. J., Riasat, V., Walraven, V., Thomas, A. W., Faber, B. W., et al. (2011). Safety and immunogenicity of multi-antigen AMA1-based vaccines formulated with CoVaccine HT™ and Montanide ISA 51 in rhesus macaques. *Malaria Journal*, *10*, 182. https://doi.org/10.1186/1475-2875-10-182.

Lakhan, N., Stevens, N. E., Diener, K. R., & Hayball, J. D. (2016). CoVaccine HT™ adjuvant is superior to Freund's adjuvants in eliciting antibodies against the endogenous alarmin HMGB1. *Journal of Immunological Methods*, *439*, 37–43. https://doi.org/10.1016/j.jim.2016.09.008.

Leenaars, P. P., Hendriksen, C. F., Angulo, A. F., Koedam, M. A., & Claassen, E. (1994). Evaluation of several adjuvants as alternatives to the use of Freund's adjuvant in rabbits. *Veterinary Immunology and Immunopathology*, *40*(3), 225–241. https://doi.org/10.1016/0165-2427(94)90022-1.

Palm, A.-K. E., & Henry, C. (2019). Remembrance of things past: Long-term B cell memory after infection and vaccination. *Frontiers in Immunology*, *10*. https://doi.org/10.3389/fimmu.2019.01787.

Paus, D., Phan, T. G., Chan, T. D., Gardam, S., Basten, A., & Brink, R. (2006). Antigen recognition strength regulates the choice between extrafollicular plasma cell and germinal center B cell differentiation. *Journal of Experimental Medicine*, *203*(4), 1081–1091. https://doi.org/10.1084/jem.20060087.

Pedersen, G. K., Wørzner, K., Andersen, P., & Christensen, D. (2020). Vaccine adjuvants differentially affect kinetics of antibody and germinal center responses. *Frontiers in Immunology*, *11*. https://doi.org/10.3389/fimmu.2020.579761.

Raffel, S. (1948). The components of the tubercle bacillus responsible for the delayed type of "infectious" allergy. *The Journal of Infectious Diseases*, *82*(3), 267–293. https://doi.org/10.1093/infdis/82.3.267.

Redwan, E.-R. M., Fahmy, A., El Hanafy, A., Abd El-Baky, N., & Sallam, S. M. A. (2009). Ovine anti-rabies antibody production and evaluation. *Comparative Immunology, Microbiology and Infectious Diseases*, *32*(1), 9–19. https://doi.org/10.1016/j.cimid.2007.11.004.

Repetto, G., del Peso, A., & Zurita, J. L. (2008). Neutral red uptake assay for the estimation of cell viability/cytotoxicity. *Nature Protocols*, *3*(7), 1125–1131. https://doi.org/10.1038/nprot.2008.75.

Siegrist, C.-A. (2018). 2—Vaccine immunology. In S. A. Plotkin, W. A. Orenstein, P. A. Offit, & K. M. Edwards (Eds.), *Plotkin's vaccines* (7th ed., pp. 16–34.e7). Elsevier. https://doi.org/10.1016/B978-0-323-35761-6.00002-X.

Stevens, N. E., Fraser, C. K., Alsharifi, M., Brown, M. P., Diener, K. R., & Hayball, J. D. (2013). An empirical approach towards the efficient and optimal production of influenza-neutralizing ovine polyclonal antibodies demonstrates that the novel adjuvant CoVaccine HT™ is functionally superior to Freund's adjuvant. *PLoS One*, *8*(7), e68895. https://doi.org/10.1371/journal.pone.0068895.

Stills, H. F. (2005). Adjuvants and antibody production: Dispelling the myths associated with Freund's complete and other adjuvants. *ILAR Journal*, *46*(3), 280–293. https://doi.org/10.1093/ilar.46.3.280.

Tam, H. H., Melo, M. B., Kang, M., Pelet, J. M., Ruda, V. M., Foley, M. H., et al. (2016). Sustained antigen availability during germinal center initiation enhances antibody responses to vaccination. *Proceedings of the National Academy of Sciences of the United States of America*, *113*(43), E6639–E6648. https://doi.org/10.1073/pnas.1606050113.

Turkstra, J. A., van der Meer, F. J. U. M., Knaap, J., Rottier, P. J. M., Teerds, K. J., Colenbrander, B., et al. (2005). Effects of GnRH immunization in sexually mature pony stallions. *Animal Reproduction Science*, *86*(3–4), 247–259. https://doi.org/10.1016/j.anireprosci.2004.07.010.

Turkstra, J. A., van der Staay, F. J., Stockhofe-Zurwieden, N., Woelders, H., Meloen, R. H., & Schuurman, T. (2011). Pharmacological and toxicological assessment of a potential GnRH vaccine in young-adult male pigs. *Vaccine*, *29*(21), 3791–3801. https://doi.org/10.1016/j.vaccine.2011.03.023.

Turner, J. S., Benet, Z. L., & Grigorova, I. L. (2017). Antigen acquisition enables newly arriving B cells to enter ongoing immunization-induced germinal centers. *Journal of Immunology (Baltimore, Md.: 1950)*, *199*(4), 1301–1307. https://doi.org/10.4049/jimmunol.1700267.

Victora, G. D., & Nussenzweig, M. C. (2012). Germinal centers. *Annual Review of Immunology*, *30*, 429–457. https://doi.org/10.1146/annurev-immunol-020711-075032.

Wolfensohn, S., & Lloyd, M. (1994). *The handbook of laboratory animal management and welfare*. Oxford University Press.

Younis, S., Faber, B. W., Kocken, C. H. M., & Remarque, E. J. (2019). Identification of adjuvants for clinical trials performed with Plasmodium falciparum AMA1 in rabbits. *BMC Immunology*, *20*(1), 25. https://doi.org/10.1186/s12865-019-0307-y.

CHAPTER SEVEN

Clostridioides difficile: Current overview and future perspectives

Joanna Giles[a,*] and April Roberts[b]
[a]MicroPharm Ltd, Newcastle Emlyn, United Kingdom
[b]Toxins Group, National Infection Service, Public Health England, Porton Down, United Kingdom
[*]Corresponding author: e-mail address: joanna.giles@micropharm.co.uk

Contents

Abstract

The most common world-wide cause of antibiotic-associated infectious diarrhea and colitis is the toxin producing bacterium, *Clostridioides difficile* (*C. difficile*). Here we review the background and characteristics of the bacterium and the toxins produced together with the epidemiology and the complex pathogenesis that leads to a broad clinical spectrum of disease. The review describes the difficulties faced in obtaining a quick and accurate diagnosis despite the range of sensitive and specific diagnostic tools available. We also discuss the problem of disease recurrence and the importance of disease prevention. The high rates of infection recurrence mean that treatment strategies are constantly under review and we outline the diverse treatment options that are currently

Advances in Protein Chemistry and Structural Biology, Volume 129
ISSN 1876-1623
https://doi.org/10.1016/bs.apcsb.2021.11.003

in use and explore the emerging treatment options of pulsed antibiotic use, microbial replacement therapies and the use of monoclonal antibodies. We summarize the future direction of treatment strategies which include the development of novel antibiotics, the administration of oral polyclonal antibody formulations, the use of vaccines, the administration of competitive non-toxigenic spores and the neutralization of antibiotics at the microbiota level. Future successful treatments will likely involve a combination of therapies to provide the most effective and robust approach to *C. difficile* management.

1. Introduction

Clostridioides difficile, previously referred to as *Clostridium difficile* (*C. difficile*), is a highly prevalent human pathogen, responsible for a spectrum of debilitating symptoms. The symptoms range from mild diarrhea to life-threatening colitis with an approximately 5% mortality in patients (Czepiel et al., 2019; Rupnik, Wilcox, & Gerding, 2009). It has a world-wide prevalence with a high mortality rate in symptomatic patients, it is also present in some asymptomatic individuals with approximately 5% of all healthy adults colonized by *C. difficile* where 15%–70% of infants are also colonized. The prevalence is greatly increased in hospital settings (Leffler & Lamont, 2015). Worldwide, both the incidence and severity of the disease have dramatically increased over the last 15 years (Krutova et al., 2018) and in 2019 it was listed as one of the top three urgent antibiotic resistance threats by the Centre of Disease Control and Prevention (CDC). The illness associated with *C. difficile* infection typically occurs in adults after the use of antibiotics, which result in a disruption to the normal gut microbiota and subsequent colonization by *C. difficile*. *C. difficile* Infection (CDI) is the most common health-care associated infection (Del Prete, Ronga, & Addati, 2019); however, increasing rates of infection have also been recorded in the community in people previously considered as low-risk (young and healthy individuals) (Sammons & Toltzis, 2013).

In this review we provide an overview of the history of CDI and discuss the current understanding of its pathogenesis, diagnosis and treatment. We also look to the future and discuss prospective treatment options and the outlook in our battle with such antibiotic resistant bacteria.

2. The bacterium

The bacterium, *C. difficile,* is a Gram-positive anaerobic species of spore-forming, motile bacteria found in feces which was first described by Hall and O'Toole (1935). Initial findings suggested that this organism was part of the normal intestinal flora, as it was identified in healthy new-borns and it was observed to show a slow rate of replication. However, Hall and O'Toole also noted the release of soluble exotoxins by the bacterium which suggested that it could be disease causing. Two toxins were identified and were named toxin A (TcdA) and toxin B (TcdB). In 1987 a third, and unrelated toxin, *C. difficile* binary Toxin (CDT) was identified (Gerding, Johnson, Rupnik, & Aktories, 2014). The role of this toxin in the pathogenesis of *C. difficile* is unclear although it may act in concert with TcdA to enhance pathogenesis (Kuehne et al., 2014). Since it's early discovery it has become apparent that *C. difficile* is responsible for a spectrum of clinical conditions and the prevalence of disease has markedly increased worldwide since the 2000s.

Several strains of the bacterium exist, which have been isolated from almost all mammals and soil. Various typing methods for *C. difficile* are currently used, including PCR ribotyping, toxinotyping and pulsed field gel electrophoresis. PCR ribotyping is the preferred method in Europe and Australia. Toxinotyping gives a picture of the toxigenic status of the strain and matches well with ribotypes (Huber, Foster, Riley, & Paterson, 2013). More recently, the use of MALDI-TOF analysis is being tested in Brazil as a fast and reliable alternative to the methods already listed (Calderaro et al., 2021). Several ribotypes are associated with disease where severity of disease differs between different ribotypes.

The existence of different ribotypes was initially uncovered when a strain was discovered that was unable to produce TcdA (Borriello et al., 1992). Since then, at least 86 ribotypes have been identified (Herbert et al., 2019), each showing a different level of virulence. The increasing prevalence and mortality attributed to CDI has been linked to the emergence of multiple novel hypervirulent ribotypes of *C. difficile,* particularly RT027. RT027 was considered the cause of a *C. difficile* epidemic, which began in Europe in 2003. RT027 has been shown to produce a substantially higher concentration of toxins than other ribotypes which resulted in an increased severity of disease in patients and a threefold mortality (Barbut, Jones, & Eckert, 2011; Couturier, Davies, Gateau, & Barbut, 2018).

3. *C. difficile* toxins

CDI is a toxin mediated disease, where many strains of *C. difficile* are known to produce up to three different exotoxins. It is noteworthy, that some strains produce fewer toxins due to a deficit or truncation in the genes encoding for the toxins. Strains of C. *difficile* unable to produce toxins are termed nontoxigenic C. *difficile* strains (NTCD) and mostly present in asymptomatic patients, (Shim, Johnson, Samore, Bliss, & Gerding, 1998) although asymptomatic patients can also harbor toxigenic strains.

Two large toxins, TcdA and TcdB were first identified in 1981 by three separate groups (Banno et al., 1984; Sullivan, Pellett, & Wilkins, 1982; Taylor, Thorne, & Bartlett, 1981) and have long been considered the main virulence factors. The third toxin is CDT and was first discovered in 1987. The CDT producing strains are observed to be linked to a higher risk of re-infection and mortality and have shown an increasing prevalence (Kuehne et al., 2014; Spigaglia & Mastrantonio, 2004).

It is well established that both TcdA and TcdB are virulence factors which cause severe damage to colonic epithelial cells. Both TcdA and TcdB, when administered intraperitoneally in mice, demonstrate similar lethal potencies (Ball et al., 1993) but TcdB is observed to show up to 1000-fold greater potency in vitro in cultured cells (Carter et al., 2015; Voth & Ballard, 2005). Such studies suggest that TcdB is the potent and harmful toxin, responsible for cell damage, whereas TcdA primarily presents a role in expression of clinical symptoms. Wild type TcdA(+) TcdB(+) and mutant TcdA(−) TcdB(+) strains are fully virulent but TcdA(+) and TcdB(−) mutants are attenuated in virulence by comparison (Carter et al., 2015).

TcdA and TcdB have been studied intensively due to their primary role in pathogenesis and also as useful markers of disease and potential targets for treatment. Multiple toxinotypes have been identified (Rupnik & Janezic, 2016) especially of TcdB, with slight variability in the genes encoding TcdA and TcdB. Hypervirulent strains are noted to produce up to 20 times greater amounts of TcdA and TcdB (McDonald et al., 2005). TcdA and TcdB are among the largest bacterial toxins reported at 308 and 270 kDa respectively. Hamster studies have revealed that giving the animals TcdA and TcdB leads to a breakdown of the cytoskeleton within cells, providing the first insights into the mechanism of toxin action. TcdA and TcdB both belong to a family of large clostridial glucosylating toxins (Voth & Ballard, 2005) and have well defined roles as catalysts for the glucosylation and subsequent irreversible inactivation of Rho-GTPases, Rac and Cdc42. The Rho-GTPases normally perform a regulatory function on the G-actin of the eukaryotic cytoskeleton. Upon targeted Rho-GTPase inactivation, cell death is triggered through actin condensation, cell rounding and the loss of structure within the cytoskeleton (Just & Gerhard, 2005).

Between 6% and 12.5% of *C. difficile* strains produce the third toxin, CDT (Burnham & Carroll, 2013) which has been less well studied and understood compared to TcdA and TcdB and its precise role in disease is not completely certain. CDT (+) TcdA (−) TcdB (−) strains of *C. difficile* do not appear to cause disease (Gerding et al., 2014). Unlike TcdA and TcdB, CDT is a member of the ADP-ribosylating toxin family with an enzymatic subunit (CDTa) of 53 kDa and a binding subunit (CDTb) of 98.8 kDa (Gerding et al., 2014). The CDTb subunit mediates binding to host cells while the CDTa subunit shares a close resemblance to iota toxin (Vandekerckhove, Schering, Bärmann, & Aktories, 1987) which functions by translocating into the cytosol of the host cells and ribosylates the actin of the host cells to prevent polymerization (Gülke et al., 2001), with deconstruction of the actin filament and eventual breakdown of the actin cytoskeleton.

4. Symptoms and diagnosis

Historically, *C. difficile* had been considered a straightforward infection to diagnose and treat. However, due to an improved understanding of the disease and a wider availability of enhanced diagnostic tools, it is now recognized that the disease has historically been grossly under or misdiagnosed (Davies et al., 2014). Such incorrect and slow diagnosis of patients may have contributed to its observed rapid spread and nosocomial infections. There still remains to be a defined and established diagnostic procedure, and no single standalone test is currently suitable to establish disease in patients. Testing procedures vary greatly between regions and centers.

At present, the diagnosis of CDI is considered challenging, despite the wide range of diagnostic tools that are available. One reason that diagnosis is not straightforward is the fact that patients harboring toxigenic strains may not always have clinical disease (Viscidi, Willey, & Bartlett, 1981). Further difficulties in diagnosis can be attributed to the broad range of symptoms that may be displayed by patients, many of which overlap with the symptoms of irritable bowel syndrome (IBS) or inflammatory bowel disease (IBD). The symptoms these gastrointestinal disorders include debilitating diarrhea, abdominal pain with fever and weight loss. There is the added complication that an IBD flare may itself be triggered by *C. difficile* infection. Occurrence of CDI in IBD patients leads to poor prognosis.

C. difficile can present with a spectrum of clinical manifestations, as illustrated in Fig. 1. Thus, mild disease may present with new onset of

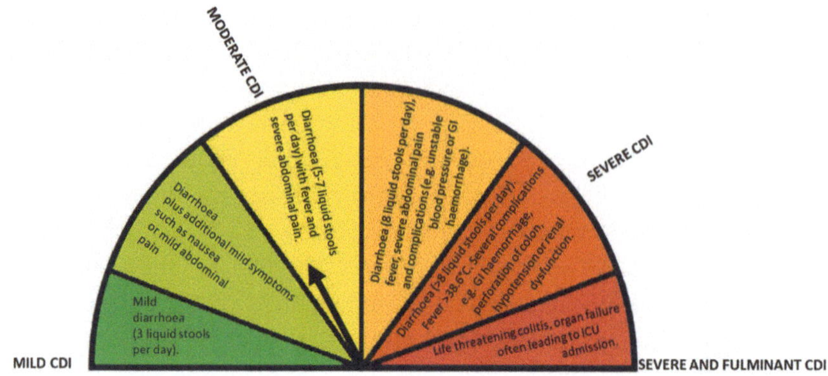

Fig. 1 *C. difficile* infection may present with a spectrum of clinical symptoms ranging from mild diarrhoea to life-threatening colitis. The severity of disease must be classified in order to establish the correct treatment strategy.

approximately 3 episodes of diarrhea per day, with mild abdominal discomfort and a white blood cell count (WBC) of <15,000 cells/mm^3. Moderate disease progresses to between 3 and 5 episodes of diarrhea per day, moderate abdominal discomfort, nausea and fever above 38 °C, and patients with severe and fulminant disease present with debilitating symptoms such as hypotension, shock, organ failure, small bowel disease, colitis and a WBC of >35,000 cells/mm^3 often leading to intensive care admission, colectomy or death (Cohen et al., 2010; McDonald et al., 2018).

A further complication with CDI is that it has an estimated 10%–30% likelihood of disease recurrence (Fekety et al., 1997; Johnson, 2009). Recurrence of disease is defined as a re-occurrence of CDI within 8 weeks after initial onset of a previous episode post resolution of initial symptoms (Debast, Bauer, & Kuijper, 2014). Such recurrences are difficult to treat due to a lack of data regarding treatment strategies and are also linked to a significantly higher risk of mortality. Prevention of recurrent disease is a very prominent clinical issue and multiple risk factors have been linked with recurrence (Song & Kim, 2019). Briefly, the pre-use of antibiotics has already been mentioned as a risk factor for the development of CDI, supported by meta-analysis of 33 studies including 18,530 patients (Deshpande et al., 2015). Advancing age is another commonly reported risk factor for recurrent CDI where the same meta study demonstrated an age of >65 years as being a strong independent risk factor (Deshpande et al., 2015).

Although there is no clear reason for this association between age and risk, recurrence is likely attributed to a weaker immune system and the presence of co-morbidities. Meta-analyses have also identified the suppression of gastric acid as a significant risk factor for recurrent CDI (Garey, Sethi, Yadav, & DuPont, 2008). The concomitant use of antacid medications, intended for treatment of acid-related diseases, was unexpectedly observed to be significantly associated with increased risk of recurrent CDI. The final factor to note that is linked to an increased recurrence rate is the strain of *C. difficile* that infects a patient. Hypervirulent strains have been observed to lead to a greater percentage of recurrence, particularly RT027 (Marsh et al., 2012). This is likely due to the increased TcdA and TcdB production as well as production of binary toxin. This subsequently leads to greater colonization of the gut and a greater risk of recurrence.

A quick, accurate and reliable diagnostic procedure for *C. difficile* patients would provide a most effective means to reduce the spread of infection and provide an improved opportunity for effective management and recovery.

Current diagnosis strategies are based upon the analysis of displayed symptoms in combination with positive testing for glutamate dehydrogenase (GDH) tests or nucleic acid amplification tests (NAAT) as well as a positive TcdB test. Microbiological evidence of *C. difficile* toxin in stool samples can also be obtained, but this is not performed routinely. It is widely accepted that patients should only be tested upon the display of symptoms suggestive of active CDI (primarily the production of 3 or more unexplainable unformed stools within 24 h). Sometimes colonoscopic or histopathological evidence demonstrating pseudomembranous colitis is also considered (Debast et al., 2014).

There are currently multiple methods for the laboratory testing to establish a diagnosis of CDI and typically more than one method is adopted to confirm such an infection. The methods display a range of sensitivities and specificities, which will be discussed below and are referred to in Figs. 2 and 3.

Clinical Practice Guidelines recommend that the most sensitive and best method of diagnosing CDI is the use of NAAT. Alternatively, multi-step testing for GDH and TcdA and/or TcdB may also be used (McDonald et al., 2018). A positive result from a NAAT or GDH test would warrant confirmation through further testing using a more specific assay, such as a toxin immunoassay.

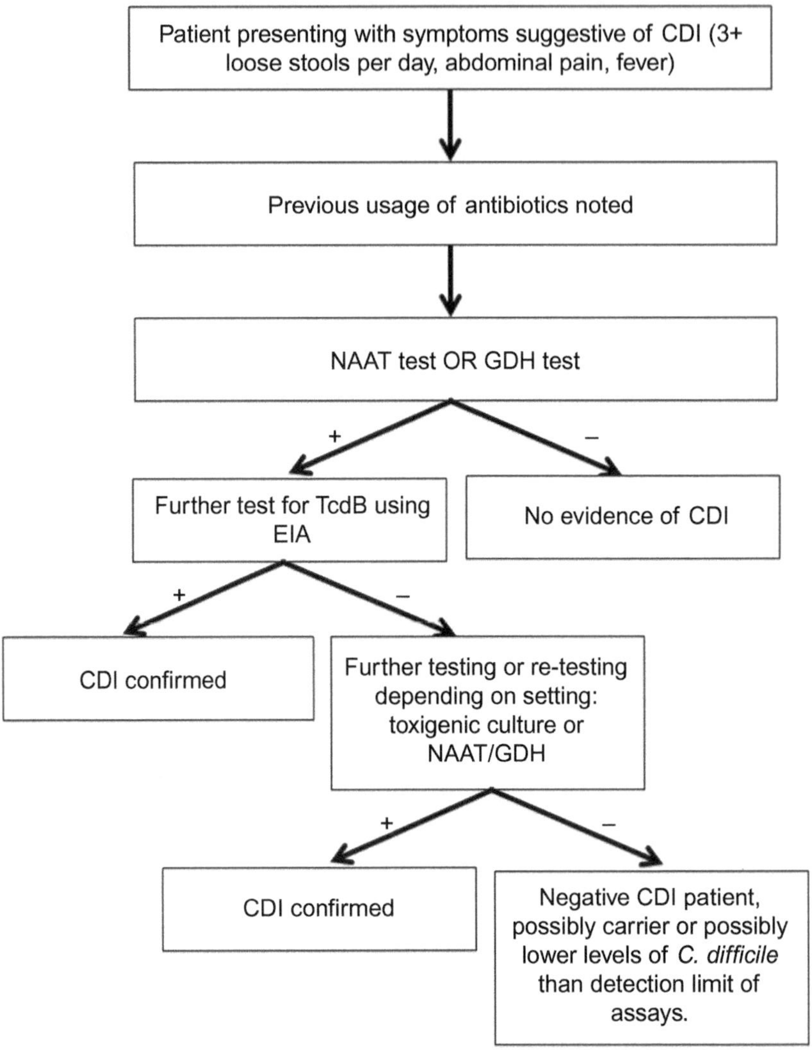

Fig. 2 The strategy used in the United Kingdom for the diagnosis of CDI is outlined above. A combination of 2 tests, the first of which is NAAT or GDH should be conducted, followed by a more sensitive toxin EIA assay. Two positive results in the flow diagram confirm CDI. If NAAT/GDH is positive but subsequent EIA is negative then CDI could be present and further testing may be warranted. No single test result is considered absolute and the clinical condition of the patient must be considered when treatment strategy is being decided. Patients that are thought to be positive for *C. difficile* will be isolated in order to prevent possible spread of infection.

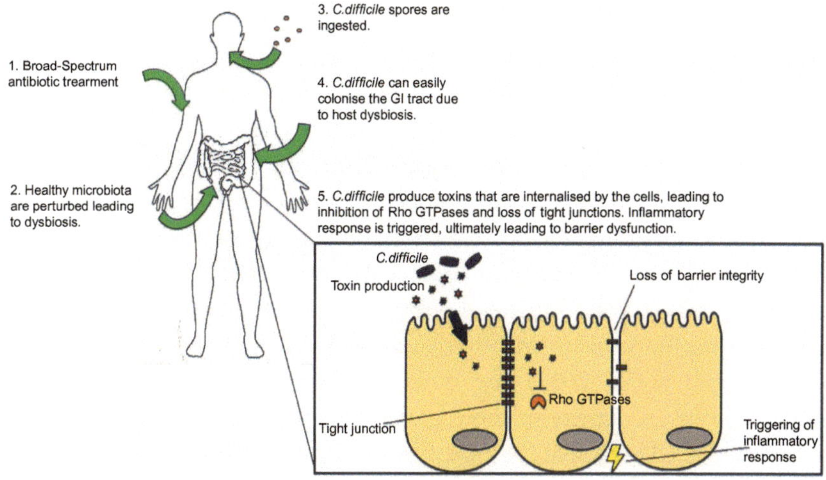

Fig. 3 *C. difficile* infection is commonly preceded by the use of broad-spectrum antibiotics, which alters the composition of the gut microbiota. *C. difficile* are able to rapidly colonise the host GI tract due to the microbiota dysbiosis. *C. difficile* pathogenesis is attributed to the toxins that are produced in the gut. TcdA and TcdB are endocytosed and inhibit the Rho GTPases, leading to breakdown of the actin cytoskeleton and loss of tight junctions. These cellular changes result in the loss of barrier integrity and trigger an inflammatory response, that leads to further damage of the lining.

4.1 NAAT tests

Although a more costly method than the other techniques, the use of NAAT to detect genes encoding TcdA and TcdB is a commonly used diagnostic assay due to its quick turnaround and very high sensitivity. Commercial testing kits are available which add to the convenience of the test when used in tandem with one of the other tests (Crobach, Duszenko, Terveer, Verduin, & Kuijper, 2017). Despite their high sensitivity and fast turnaround, they are still not capable of being used alone as due to their tendency to over diagnose patients that have *C. difficile* present in their faeces, but are not suffering with CDI. The purpose of using a second testing method is to improve the specificity of diagnosis.

4.2 GDH testing

The testing of stool samples for GDH by antigen detection is a more rapid technique than the toxigenic culture method, but unlike the toxigenic culture method, GDH is not able to differentiate between toxigenic and non-toxigenic strains of *C. difficile* as GDH is conserved in all *C. difficile* strains. The technique is therefore usually used in tandem with toxin immunoassays (Burnham & Carroll, 2013).

4.3 Toxin immunoassays

Many enzyme Immunoassays (EIA) are available commercially for the detection of *C. difficile* toxins. The preferred tests detect both TcdA and TcdB, rather than a single toxin, due to the existence of TcdA negative strains that are still able to cause disease (Kim et al., 2008). Therefore, the majority of commercial EIAs are designed to detect TcdB only. The specificity of these monoclonal and polyclonal antibody based assays is generally high, up to 100% (Chapin, Dickenson, Wu, & Andrea, 2011) but some assays lack specificity and are not adequate as positive predictors of disease. In addition, sensitivity can be variable with some assays displaying a sensitivity as low as 43% (Chapin et al., 2011) and others as high as 98.7% (Musher et al., 2007). The heterogeneity of these assays have been compared by Burnham and Carroll (2013) where they explain the inadequacy of these tests alone as a diagnostic tool, despite the low cost per assay. Rather, they should be used in tandem with the other methods discussed here.

4.4 Toxigenic culture

Traditionally, toxigenic culture has been the gold standard for the diagnosis of CDI. The process requires a sample of stool from an individual confirmed not to be using laxatives or any other medication that may explain their diarrhea. The sample is then cultured anaerobically in agar or in broth, in combination with agents to inhibit the growth of other members of the fecal flora for 48 h and then held for up to a week to determine if the samples may produce any colonies. The colonies are then further tested to determine whether *C. difficile* is present using methods such as Gram stain and toxin testing (Wren, 2010). Although relatively inexpensive, this process is slow and requires facilities for anaerobic cell culture (Delmée, Van Broeck, Simon, Janssens, & Avesani, 2005). Since it may take up to 7 days in order to achieve a result and this test is unsuitable to be used alone for routine diagnostic testing. Although highly sensitive, the process is less specific and may yield false-positive results from non-toxigenic strains which may lead to misdiagnosis of CDI (Lee, Plechot, Gohil, & Le, 2021).

4.5 Cell culture cytotoxicity neutralization assay

Several cell lines are appropriate for the testing of stool samples for cytotoxicity, including human diploid fibroblasts, VERO cells and Hep2 cells. They can be cultured as monolayers and then incubated with a sample of stool filtrate for up to 48 h. After this time achieved using an appropriate antiserum.

Like the toxigenic culture tests, this procedure is lengthy but is able to determine presence of toxin B producing strains (Lyerly, Krivan, & Wilkins, 1988). Although previously considered a gold-standard, its sensitivity and the time and labor intensity of the procedure means that it is now more appropriately used in tandem with other methods (Planche & Wilcox, 2011). Furthermore, this method alone is also liable to produce false negatives, which could lead to further spread of disease by environmental contamination by incorrectly diagnosed patients (Delmée et al., 2005).

5. The severity of disease

Upon making a diagnosis of CDI, the severity of the infection must be established through careful analysis of clinical features and laboratory results. In order to monitor the progress of the disease and to tailor the most appropriate course of treatment. Severity can range from mild to moderate, to severe and fulminant, with the added variable of whether the disease is recurring.

Much confusion exists around the definition of severe CDI and although scoring systems are used, these are not universal nor fully validated. A typical scoring system may take note of the mental status of the patient, abdominal pain, tachycardia ($\geq$110 beats/min), hypotension, fever, leukocytosis or leucopenia, ascites or colitis documented by imaging studies; hypoalbuminemia (<2.5 mg/dL) or admission to an intensive care unit (Velazquez-Gomez et al., 2008). Each feature present in the patient will be scored 1 point, and a score between 1 and 3 (post positive diagnosis) represents mild disease, with increasing scores representing more severe disease. CDI is classified as severe and fulminant when a patient is in shock, presents with sepsis, is admitted to an intensive care unit, is hypotensive or requires colectomy. While such classification systems are helpful in defining management of disease, the features do not allow for the identification of at-risk individuals and do not allow prediction of disease outcome.

Several studies have provided evidence for the role of cytokines in the pathogenesis of CDI, which means that in the future they may function as useful serum biomarkers of disease, predictors of disease recurrence and predictors of adverse outcomes. Such studies have indicated a plethora of cytokines that show a correlation between disease severity and cytokine level. These include granulocyte-macrophage colony-stimulating factor, interleukin (IL)-1β, IL-8, IL-12p70, interferon-α, IL-15, IL-2 (Hamo, Azrad, Nitzan, & Peretz, 2019), CCL-5 (Abhyankar et al., 2020), HGF, procalcitonin, IP-10 (Dieterle et al., 2020) and more recently IL-16 and BCA-1 (Gotshal, Azrad, Hamo, Nitzan, & Peretz, 2021). In the future, it is possible that a selection of these cytokines may be used as a serum-based biomarker panel for the prediction of disease outcome for CDI and therefore more rapid and directed treatment and amelioration of disease.

6. Epidemiology

C. difficile is recognized as one of the leading causes of hospital associated diarrhea and hospital outbreaks in adults, worldwide (Voss & Hopman, 2014). The incidence of such outbreaks have increased dramatically since the 2000s (Bauer et al., 2011) with the number of cases in Europe alone being up to 124,000 per year (Suetens et al., 2018).

The improved sensitivity of diagnostic techniques and an increase in awareness by clinicians may partly explain the increase in numbers. However, there are also an increased number of strains being identified, with hypervirulent ribotype 027 being strongly attributed to the marked increase in cases in the early 2000s (Loo et al., 2005; Warny et al., 2005), particularly across Europe, North America and Australia. The increased incidence of outbreaks and hospitalizations have had a knock-on effect on the heavy burden on healthcare facilities. Further adding to this burden is the 10%–30% likelihood of recurrent disease (Fekety et al., 1997; Song & Kim, 2019) and the 17% of cases that lead to further health complications or death. The common recurrence of infection is problematic in itself, as it leads to difficulties in assessing treatment regimens and lengthens the recovery process.

At present 125 distinct *C. difficile* clones, or ribotypes have been identified in 19 European countries (Davies et al., 2016). Although RT027 is still prevalent, cases associated with this ribotype are decreasing in Europe (Bauer et al., 2011) and ribotypes other than 027 are widespread in Europe. Standardization of diagnostic methods and molecular typing are needed to more carefully, closely and comparatively assess the epidemiology and global burden of CDI.

7. Pathogenesis

C. difficile is a prevalent anaerobic, Gram-positive, spore-forming bacterium found at low levels in soil, air and water that spreads easily through ingested spores. The spores are highly stable in the environment. Infection with *C. difficile* usually occurs when a susceptible individual comes into contact with spores in the environment e.g. hospital surfaces. There is also evidence of *C. difficile* colonization upon food, particularly meats but also vegetables. Vegetables could be contaminated through the spread of infected manure upon crops or irrigation with infected water supplies (Eckert, Burghoffer, & Barbut, 2013).

Infection in humans is commonly preceded by the use of broad-spectrum antibiotics, specifically clindamycin (Bartlett, Chang, Gurwith, Gorbach, & Onderdonk, 1978), cephalosporins, quinolones, carbapenems and combination penicillins (Hensgens, Goorhuis, Dekkers, & Kuijper, 2012; Slimings & Riley, 2014). This perturbs the healthy microbiota of these patients, which allows the *C. difficile* to easily colonize their digestive tract upon ingestion of spores of *C. difficile* (Borriello, 1998). Furthermore, resistance to these antibiotics plays an important role in the emergence of new *C. difficile* clones (Baines & Wilcox, 2015).

Colonization of the gut by *C. difficile* is thought to be facilitated by the production of CDT. This toxin has been observed to produce microtubule protrusions which capture additional *C. difficile* cells and encourage colonization. However, CDT is not produced by all strains of *C. difficile,* therefore other colonization mechanisms must exist. There is a role for non-toxic *C. difficile* cell surface proteins (S-layer proteins), such as Cwp84 (Bradshaw, Kirby, Roberts, Shone, & Acharya, 2017) that assist in adhesion and internalization of the *C. difficile*, and boost colonization in the gut (Bradshaw, Roberts, Shone, & Acharya, 2018). These S-layer proteins are also linked to an increased formation of biofilm which is associated with increased survival of bacteria and confers protection of *C. difficile* from antibiotics (Tijerina-Rodríguez, Villarreal-Treviño, Morfín-Otero, Camacho-Ortíz, & Garza-González, 2019).

The pathogenesis of *C. difficile* is attributed to the secretion of two toxins, TcdA and TcdB, which have been shown in animal models to be primarily responsible for the symptoms associated with CDI (Kelly & Kyne, 2011). Furthermore, passive immunotherapy using avian antibodies directed to recombinant TcdA and TcdB has shown to rescue animals that were anticipated to die from CDI in the hamster model (Kink & Williams, 1998).

TcdA and TcdB bind to the host cells and then endocytosed (Carter, Rood, & Lyras, 2010) and function to monoglucosylate the Rho family of GTPases via the N-terminal glucosyltransferase domains. This covalent modification results in the inhibition of Rho activity, which in turn leads to the degradation of the actin cytoskeleton, cell rounding, breakdown of the tight junctions and loss of function in the intestinal epithelial barrier.

Furthermore, the release of IL-1ß, TNF-α and IL-8 from intestinal epithelial cells are triggered, leading to neutrophil influx and breakdown of the intestinal lining to increase vascular permeability (Pothoulakis, 2006). IL-8 has been linked to neutrophil recruitment and activation of lymphoid cells (Abt, McKenney, & Pamer, 2016). Fecal IL-8 levels are also elevated in CDI patients with a positive correlation between IL-8 levels and disease severity (Steiner, Flores, Pizarro, & Guerrant, 1997). It is the combination of these two pathways: loss of the actin cytoskeleton and the release of proinflammatory cytokines that ultimately lead to destruction of the epithelial cell layer and the symptoms of diarrhea and colitis.

8. Prevention

Guidance for the prevention of CDI in acute healthcare settings are continually reviewed and have provided comprehensive recommendations (Tschudin-Sutter et al., 2018). The first point of action is accurate and fast diagnosis, which has already been discussed. Further recommendations include effective hand hygiene techniques, surveillance of CDI symptoms in healthcare settings and subsequent isolation of patients, use of appropriate personal protective equipment (PPE) and daily sporicidal disinfection.

It is well established that probiotics are able to provide many health benefits, particularly in diseases with impaired gut microbiota (La Fata, Weber, & Mohajeri, 2018). Probiotics can improve human immune function and prevent gastrointestinal infection in animals and humans (Zheng et al., 2017). Various strains of probiotics have been considered as preventative therapies for CDI, by boosting the impaired gut microbiota that is observed after antibiotic treatment and before CDI colonisation. The most common strains that have been studied include the *Bifidobacterium* and *Lactobacillus* genera. In addition to these strains a probiotic medicinal yeast culture of *Saccharomyces* has been tested as a preventative treatment for CDI due to the success of *Saccharomyces boulardii* CNCM I-745 in the treatment and prevention of various types of diarrhea (Moré & Vandenplas, 2018). Treatment using this yeast strain has shown anti-toxin effects, a boost in microbiota, several immunological effects and enhanced action of digestive enzymes. However, assessing these strains for the treatment of CDI in clinical trials is particularly problematic due to the number of patients required, timing of dosing, achieving control across treatment programs and due to difficulties in achieving randomization (McFarland, 2020). Despite some data showing a promising result with the early administration of *S. boulardii* in the prevention of CDI (Wombwell, Patterson, Bransteitter, & Gillen, 2020), at present the use of probiotic treatments as a preventative measure for CDI is not recommended. However, there are many studies ongoing and the use of probiotics has been shown to be safe for patients with CDI and associated diarrhea (Goldenberg et al., 2017).

9. Current treatment strategies

9.1 Antibiotics

Before any treatment of CDI is decided, the severity of disease must be established as the treatment pathways differ for non-severe and severe disease. Overall, there have been advances in the management and treatment of initial and recurrent CDI due to improvements in diagnosis and increase in treatment options.

Treatment regimens differ between countries. The Infectious Diseases Society of America/ Society for Healthcare Epidemiology of America clinical practice guidelines (IDSE/SHEA-CPG) now recommend oral fidaxomicin, as the first-choice treatment for patients with their first and non-severe case of CDI, with vancomycin providing a suitable alternative (Johnson et al., 2021). However, the European Society of Clinical Microbiology and Infectious Diseases clinical practice guidelines (ESCMID-CPG) and the UK Department of Health recommend metronidazole as the first line drug in non-severe CDI and vancomycin for severe episodes (Debast et al., 2014). Doses and treatment durations will differ depending upon the severity of disease and the risk of recurrence.

Treatment of recurrent CDI includes repeat administration of oral antibiotics with repeat metronidazole is recommended if the first episode was treated with metronidazole. This strategy has been shown to achieve a 50% cure rate in patients (Leffler & Lamont, 2009).

9.1.1 Metronidazole

Metronidazole is an FDA approved oral antibiotic that may be used for the treatment of mild initial infections due to *C. difficile*. It is cytotoxic to obligate anaerobic microorganisms and protozoans through the inhibition of protein synthesis. It's mechanism of action begins with rapid uptake of the drug into the bacterial cell and subsequent interaction with DNA leading to strand breakage and loss of DNA integrity, which inhibits protein synthesis (Tocher & Edwards, 1992). Although metronidazole is no longer the first-line antibiotic of choice, it is still an option for an initial non-severe

CDI episode and is prescribed to be taken orally. The risks associated with metronidazole include a low risk that it may be carcinogenic and that prolonged use is linked with neurological disturbances due to neurotoxicity.

9.1.2 Vancomycin

Vancomycin is a tricyclic glycopeptide antibiotic with bactericidal activity against *C. difficile* and other Gram-positive bacteria. Vancomycin functions by inhibiting the synthesis of the peptidoglycan cell wall of the bacterium so that the cell walls become weak with leakage of intracellular components which leads to bacterial cell death (Lee, Pang, Abraham, & Coombs, 2019).

Treatment using vancomycin is FDA-approved by intravenous injection or by oral administration. A high-dose of oral vancomycin is one of the preferred treatment strategies for the first instance of severe *C. difficile* infection. Oral vancomycin has a low systemic absorption and is therefore not appropriate for systemic infections. Duration of treatments are dependent on multiple factors including severity and presentation of disease. For patients with severe and fulminant disease including a paralytic ileus, colonic diversion or dilated colon, rectal vancomycin may be offered. Although considered a fundamental antibiotic in the management of CDI, it is also linked with nephrotoxicity and ototoxicity and should not be prescribed inappropriately (Monteiro, Hahn, Gonçalves, & Fresco, 2018).

9.1.3 Fidaxomicin

Fidaxomicin is a narrow-spectrum antibiotic with minimal systemic absorption. It is a macrocyclic antibacterial approved by the FDA for treatment of CDI. It functions as a bactericidal agent where the mechanism of action relates to inhibition of an RNA polymerase leading to inhibited protein synthesis. It is particularly active against the genus *Clostridioides*. Significantly fewer patients treated with fidaxomicin have a recurrence of the infection compared to patients treated with vancomycin (13.3% vs. 24.0%) (Louie et al., 2011). The adverse-event profile for the treatment regimens with fidaxomicin and vancomycin have been found to be similar.

9.1.4 Ridinilazole

Ridinilazole is a relatively new small-molecule, narrow-spectrum antibiotic that is currently being considered as a potential treatment for CDI (Vickers et al., 2016). It is considered a highly selective and potent antibiotic toward *C. difficile* that is well tolerated in patients. Treatment with ridinilazole has shown to result in diminished production of *C. difficile* toxins while preserving host gut microbiota compared to patients receiving vancomycin in Phase II clinical trials (Cho, 2018; Vickers et al., 2017). Ridinilazole functions through the inhibition of bacterial DNA synthesis, to ultimately inhibit sporulation (Petrosillo, Granata, & Cataldo, 2018). The specificity of this antibiotic toward *C. difficile* while sparing other clostridia offers a great advantage in maintaining the integrity of the host microbiota during treatment of CDI. Studies to date suggest that ridinilazole has the potential to be an effective treatment option for CDI patients, and ongoing clinical trials will provide further clarity.

9.1.5 Pulse dosing

New dosing regimens of currently approved antibiotics is another area of study that is aiming to improve the success of antibiotic treatment. Such studies are analyzing the use of pulse dosing and tapered pulse methods for administration of currently available antibiotics. Tapering or pulse dosing vancomycin has been shown to improve the cure rate in recurrent disease (McFarland, Elmer, & Surawicz, 2002) and such studies have been extended to trials of recurrent CDI patients using vancomycin and fidaxomicin (Guery et al., 2018; Sirbu et al., 2017). It is hoped the optimization of dosing strategies will provide a much-improved cure rate and a lower recurrence rate.

9.2 Immunotherapy

In addition to antibiotics, particularly with more severe CDI, immunotherapy offers a promising approach to the safe and effective treatment of disease. In 2016 bezlotoxumab was approved as the first monoclonal antibody-based treatment delivered as an intravenous infusion for the prevention of recurrent CDI in patients at risk of relapse, to be used in conjunction with antibiotics. This therapy is marketed as Zinplava by Merck, Sharp and Dohme in the United Kingdom. Bezlotoxumab is a fully humanized monoclonal antibody (Bartlett, 2017) that targets the TcdB molecule and was shown to result in a 11% lower rate of recurrence compared to placebo when administered systemically alongside standard of care therapy (Wilcox et al., 2017). Given that many strains of *C. difficile* exist with multiple toxinotypes of TcdB it is not surprising that a uniform response to treatment was not observed for this monoclonal antibody in these trials. The use of a second monoclonal antibody targeting TcdA, actoxumab in combination with bezlotoxumab did not improve the efficacy of treatment (Wilcox et al., 2017), despite pre-clinical animal studies showing efficacy (Ray, 2017). In vitro studies assessing the neutralizing potency of bezlotoxumab and actoxumab demonstrate a range in efficacy with a significantly lower potency for RT036 and the hypervirulent strains RT027 and RT078 (Hernandez et al., 2015). The authors describe the binding to bezlotoxumab epitopes within TcdB and the positive or negative impact of key residues. Future CDI immunotherapy treatments may prove more successful if multiple epitopes of TcdB are targeted, in order to capture multiple toxinotypes and dampen the damaging effect of TcdB. Nonetheless, bezlotoxumab is currently an attractive treatment option for patients with severe CDI with risk factors for poor outcome, such as the elderly or those who are immunocompromised. It is worth noting that the use of Bezlotoxumab should be avoided in patients with a history of heart failure, due to an increased risk of heart failure in these patients.

9.3 Microbial replacement treatments

Despite several options for antibiotic treatment and the increasing use of immunotherapy, recurrences of CDI with antibiotic treatment alone occur in 25% of patients (Eyre et al., 2012). Therefore, there is an urgent need for additional treatment options, particularly for recurrent disease. The most recent option that has shown great efficacy is the use of fecal microbial transplant (FMT). Since the disruption to the diversity and overall health of the gut microbiota is one of the driving factors for *C. difficile* colonisation, restoration of the normal and healthy gut microbiota is a logical strategy and the principle of FMT. The first studies using FMT were conducted in 1958 (Eiseman, Silen, Bascom, & Kauvar, 1958), and many uncontrolled studies have taken place over the last few decades. It is only more recently that FMT has proven to provide a safe and efficacious treatment for recurrent CDI (Quraishi et al., 2017) with better efficacy observed in patients with recurrent CDI rather than refractory CDI or compared to placebo (Tariq, Pardi, Bartlett, & Khanna, 2019).

FMT relies on the donation of fresh stool suspension from pre-screened suitable donors who do not harbor any transmissible pathogens, do not use illegal drugs and have not previously received a tissue/organ transplant among other criteria (Cammarota et al., 2017). Donor selection has been shown to be key in the success rate of the FMT, with some specific microbiological features coming to light for effective treatment of CDI (Bibbò et al., 2020). Although standardized combinations of bacteria have been identified, it is likely that the success of treatment will be greater with transfer of a complete fecal microbiota, rather than through strain selection.

Donated stool samples have been tested and administered as fresh samples, frozen samples and lyophilized samples, where lyophilized samples have shown the lowest cure rates (Jiang et al., 2017) and microbial diversity in the recipient was increased at similar rates in the fresh and frozen samples. Delivery of fecal microbiota may be via the upper gastrointestinal tract using a naso-enteric tube, or an esophagogastroduodenoscopy or by the ingestion of capsules. However, delivery to the lower tract has yielded better results and cure rates (Quraishi et al., 2017). This method of delivery may be considered more invasive and more difficult to perform, and may be achieved by colonoscopy, flexible sigmoidoscopy or by enema. The treatment method depends greatly on institutional practices and patient preference.

At present, studies have proven that FMT is considered a highly effective form of treatment with up to 90% cure rates. It is also mostly considered a safe form of treatment but is not yet approved by the FDA. The long-term safety of FMT is not completely understood and despite the good safety profile that is growing, long term safety is still being studied. There is a potential risk of infection transmission, development of autoimmunological disorders, and other adverse events (Kelly et al., 2021).

10. Future directions

At present there are 16 novel antimicrobial agents under study for *C. difficile* (Petrosillo et al., 2018) of which and benefits of one such molecule, ridinilazole, have already been discussed. Many such antimicrobials show promise in minimizing destruction of the microbiota through their narrow spectrum of activity, inhibition of sporulation and reducing toxin production. As such, there may well be new antibiotic options available in the near future.

However, antibiotic overuse is one factor that drives the development of alternative therapies for potential treatments to CDI. There is a growing interest in the use of antibody mediated TcdA and TcdB neutralisation in the treatment of CDI. Expanding on the current use of immunotherapies are the studies involving the administration of polyclonal antibodies against CDI toxins as a means of ameliorating disease. Thus, several studies have shown success using this approach and it is logical to anticipate a more efficient response compared to monoclonals due to the multiple epitope targets that are neutralized using polyclonal antibodies and the 34 toxinotypes that have been shown to exist (Rupnik & Janezic, 2016). Synergistic effects have been demonstrated unambiguously by Cole and her colleagues.

When a combination of monoclonal antibodies, specific to major toxin domains are compared to equivalent concentrations of monoclonal antibodies against one or two domains in the hamster model of CDI (Cole et al., 2019). For CDI, the approach of using polyclonal ovine antibodies is particularly attractive due to the low cost of manufacture and good safety profile, as has been shown for snake envenomation treatments (Dart & McNally, 2001). Despite the shorter circulatory half-life of these ovine polyclonal antibodies and the fact that long term administration is not possible due to the probability of evoking host immune responses, polyclonal antibodies offer an attractive treatment for rapid toxin-neutralisation in cases of fulminant disease. Passive immunization of hamsters using antibodies directed against TcdA and TcdB has shown a protective effect upon challenge with these toxins (Roberts et al., 2012) and these studies have been further expanded to optimize the delivery of these antibodies directly to the gut rather than systemically as pre-clinical studies using a novel oral formulation, called OraCAb (Roberts et al., 2020). Through oral administration, the treatment minimizes any humoral immune response, making it more suited to mild or moderate disease. The formulation has been optimized to protect the integrity of the antibodies from digestion and allows for the specific targeting of the antibodies to the gut to allow for more rapid binding to the target toxins. Although a promising strategy, clinical trials are required to fully assess the safety and efficacy in humans.

Recently, the 3D structure of TcdB in complex with a human receptor, chondroitin sulfate proteoglycan 4 (CSPG4), has been determined (Chen et al., 2021). This is an exciting and elucidating development that reveals a unique but conserved binding site between TcdB and CSPG4 that offers a potential therapeutic target in terms of developing CSPG4 decoy molecules. The therapeutic features of such decoy molecules are being examined as an alternative and potentially superior treatment option to bezlotoxumab.

Vaccines also offer a possibility as an effective and cheap medical intervention for long-term prevention of CDI. The target groups for receiving such vaccines are the elderly, those that have been exposed to antibiotics and are subject to long-term stays at hospital settings. The efficacy of multiple vaccine candidates have been studied. Antigens including toxoids, recombinant toxin-based proteins and surface-associated proteins and carbohydrates have been assessed as vaccine epitopes.

Until October 2017, the toxoid-based vaccines were the most advanced in terms of human trials. The toxoid vaccine (Sanofi Pasteur toxoid vaccine) consists of formalin-detoxified TcdA and TcdB from a hyper-productive strain (VPI 10463) (Leuzzi, Adamo, & Scarselli, 2014). Phase II studies using the toxoid vaccine reported a high level of antibody production after 3 immunizations in 90% of individuals (aged 40–64) and manufacture at a global scale was considered achievable. However, the studies using this vaccine were discontinued as careful analysis of the Phase III data suggested that full prevention of CDI would not be achieved and that the study met the criteria for futility (Riley, Lyras, & Douce, 2019). It may be postulated that the failure in efficacy of this vaccine may be attributed to the intravenous route of administration that was utilized in this trial, and a more successful trial may have been achieved through targeted administration to the gut.

The efficacy of toxin-based vaccines relies on the production of strong neutralizing antibodies against TcdA and TcdB, and the administration of such vaccines has been shown to reduce the symptoms of CDI. However, studies do not yet show evidence for reducing persistence of *C. difficile* in the host (Riley et al., 2019) where administration in animal models has not been shown to prevent colonisation (Spencer et al., 2014). Several cell surface antigens have been investigated as potential vaccine epitopes, including Cwp84 and flagellar antigens (Péchiné et al., 2011). Although these have shown immunogenicity, none have provided high levels of protection from *C. difficile* colonisation (Bruxelle et al., 2016) and no vaccine of this type is currently at pre-clinical trial stage.

The experimental vaccines against *C. difficile* recently under clinical evaluation are a Pfizer genetically and chemically detoxified TcdA and TcdB candidate and a Valneva recombinant chimeric protein linking binding domains of TcdA and TcdB. Recently, the trials for the Pfizer vaccine have been concluded and results from this Phase III trial are currently pending. Valneva have completed their Phase II studies and could advance to Phase III trials in the future.

Another approach that has been explored to reduce recurrence of disease is the administration of nontoxigenic spores of *C. difficile* following antibiotic treatment for diagnosed CDI (Gerding et al., 2015). The study followed 157 patients to completion, where patients were assigned to receive oral liquid formulation of a nontoxigenic strain of *C. difficile* (M3, VP20621; NTCD-M3) as 10^4 spores/day for 7 days, 10^7 spores/day for 7 days, 10^7spores/day for 14 days or placebo for 14 days. The study demonstrated safety and tolerability of the formulations and NTCD-M3 was shown to colonize the gastrointestinal tract and demonstrated reduction in CDI recurrence. Although the mechanism of prevention of recurrence is unclear, it is attributed to the NTCD-M3 outcompeting the toxigenic strains of the host. Despite this small study size, these studies offer exciting preliminary evidence of this approach for the development of another safe and potential treatment for CDI.

Neutralisation of deleterious effect of an antibiotic on the commensal microbiota in the gastrointestinal tract is another innovative approach currently being explored. The aim is to allow antibiotic absorption in the upper gastrointestinal tract to target *C. difficile*, but to prevent the action of antibiotic on the microbiota in the lower gastrointestinal tract and colon to preserve the protective mechanism of the microbiota. At present, there are two techniques under evaluation. One is to use oral beta-lactamase to target and inactivate beta-lactam antibiotics. One such beta-lactamase, SYN-004 (or ribaxamase), is well tolerated when co-administered with ceftriaxone and has demonstrated a reduction in CDI cases in preliminary reports (Kokai-Kun et al., 2019). The second approach is the use of non-specific activated charcoal in an oral capsule, intended for release in the distal ileum as a means of absorption of antibiotics. A phase I clinical study using an oral charcoal capsule, termed DAV132, has been tested with administration of moxifloxacin and has been shown to be safely tolerated and to effectively reduce the exposure of the intestinal microbiota to the antibiotic (de Gunzburg et al., 2018). Both of these antibiotic neutralization strategies show promise and are being advanced in clinical studies as a means of preventing *C. difficile* associated disease.

11. Summary and outlook

The threat of CDI as one of the most serious and immediate antibiotic-resistant infections and means for the prevention and treatment are urgent health concerns.

Currently, early antibiotic treatment is the best means to resolve symptoms and minimize spread of infection. Antibiotics are the gold standard treatment and have been the mainstay of CDI treatment for over four decades, with fidaxomicin, vancomycin and metronidazole predominating worldwide as the preferred treatments. Ridinilazole is the most recent experimental oral antibiotic that has generated promising results in a small phase II clinical trial. Ridinilazole offers a promising future treatment option and a hope for minimizing recurrent disease.

Fortunately, there are a number of other approaches that focus on preventative strategies under clinical development. These include immunotherapy, vaccines, gut microbiota restoration, antibiotic neutralisation in the lower gut and administration of nontoxigenic spores to outcompete toxigenic *C. difficile*. This broad range of approaches show promise in the ultimate goal of preventing CDI and recurrent CDI and it is likely that future effective strategies for prevention and treatment of CDI will require a combination of approaches that will be weighed against cost and efficacy.

References

Abhyankar, M. M., Ma, J. Z., Scully, K. W., Nafziger, A. J., Frisbee, A. L., Saleh, M. M., et al. (2020). Immune profiling to predict outcome of *Clostridioides difficile* infection. *MBio, 11*(3). https://doi.org/10.1128/mBio.00905-20.

Abt, M. C., McKenney, P. T., & Pamer, E. G. (2016). Clostridium difficile colitis: Pathogenesis and host defence. *Nature Reviews Microbiology, 14*(10), 609–620. https://doi.org/10.1038/nrmicro.2016.108.

Baines, S., & Wilcox, M. (2015). Antimicrobial resistance and reduced susceptibility in *Clostridium difficile*: Potential consequences for induction, treatment, and recurrence of *C. difficile* infection. *Antibiotics, 4*(3), 267–298. https://doi.org/10.3390/antibiotics4030267.

Ball, D. W., Van Tassell, R. L., Roberts, M. D., Hahn, P. E., Lyerly, D. M., & Wilkins, T. D. (1993). Purification and characterization of alpha-toxin produced by *Clostridium novyi* type A. *Infection and Immunity, 61*(7), 2912–2918. https://doi.org/10.1128/iai.61.7.2912-2918.1993.

Banno, Y., Kobayashi, T., Kono, H., Watanabe, K., Ueno, K., & Nozawa, Y. (1984). Biochemical characterization and biologic actions of two toxins (D-1 and D-2) from Clostridium difficile. *Reviews of Infectious Diseases, 6*(Suppl. 1), S11–S20. https://doi.org/10.1093/clinids/6.supplement_1.s11.

Barbut, F., Jones, G., & Eckert, C. (2011). Epidemiology and control of *Clostridium difficile* infections in healthcare settings: An update. *Current Opinion in Infectious Diseases, 24*(4), 370–376. https://doi.org/10.1097/QCO.0b013e32834748e5.

Bartlett, J. G. (2017). Bezlotoxumab—A new agent for *Clostridium difficile* infection. *New England Journal of Medicine, 376*(4), 381–382. https://doi.org/10.1056/NEJMe1614726.

Bartlett, J. G., Chang, T. W., Gurwith, M., Gorbach, S. L., & Onderdonk, A. B. (1978). Antibiotic-associated pseudomembranous colitis due to toxin-producing clostridia. *New England Journal of Medicine, 298*(10), 531–534. https://doi.org/10.1056/NEJM197803092981003.

Bauer, M. P., Notermans, D. W., van Benthem, B. H. B., Brazier, J. S., Wilcox, M. H., Rupnik, M., et al. (2011). *Clostridium difficile* infection in Europe: A hospital-based survey. *Lancet (London, England), 377*(9759), 63–73. https://doi.org/10.1016/S0140-6736(10)61266-4.

Bibbò, S., Settanni, C. R., Porcari, S., Bocchino, E., Ianiro, G., Cammarota, G., et al. (2020). Fecal microbiota transplantation: Screening and selection to choose the optimal donor. *Journal of Clinical Medicine, 9*(6), 1757. https://doi.org/10.3390/jcm9061757.

Borriello, S. P. (1998). Pathogenesis of *Clostridium difficile* infection. *The Journal of Antimicrobial Chemotherapy, 41*(Suppl. C), 13–19. https://doi.org/10.1093/jac/41.suppl_3.13.

Borriello, S. P., Wren, B. W., Hyde, S., Seddon, S. V., Sibbons, P., Krishna, M. M., et al. (1992). Molecular, immunological, and biological characterization of a toxin A-negative, toxin B-positive strain of Clostridium difficile. *Infection and Immunity, 60*(10), 4192.

Bradshaw, W. J., Kirby, J. M., Roberts, A. K., Shone, C. C., & Acharya, K. R. (2017). Cwp2 from *Clostridium difficile* exhibits an extended three domain fold and cell adhesion *in vitro*. *The FEBS Journal, 284*(17), 2886–2898. https://doi.org/10.1111/febs.14157.

Bradshaw, W. J., Roberts, A. K., Shone, C. C., & Acharya, K. R. (2018). The structure of the S-layer of *Clostridium difficile*. *Journal of Cell Communication and Signaling, 12*(1), 319–331. https://doi.org/10.1007/s12079-017-0429-z.

Bruxelle, J.-F., Mizrahi, A., Hoys, S., Collignon, A., Janoir, C., & Péchiné, S. (2016). Immunogenic properties of the surface layer precursor of *Clostridium difficile* and vaccination assays in animal models. *Anaerobe, 37*, 78–84. https://doi.org/10.1016/j.anaerobe.2015.10.010.

Burnham, C.-A. D., & Carroll, K. C. (2013). Diagnosis of *Clostridium difficile* infection: An ongoing conundrum for clinicians and for clinical laboratories. *Clinical Microbiology Reviews, 26*(3), 604–630. https://doi.org/10.1128/CMR.00016-13.

Calderaro, A., Buttrini, M., Martinelli, M., Farina, B., Moro, T., Montecchini, S., et al. (2021). Rapid classification of *Clostridioides difficile* strains using MALDI-TOF MS peak-based assay in comparison with PCR-ribotyping. *Microorganisms, 9*(3). https://doi.org/10.3390/microorganisms9030661.

Cammarota, G., Ianiro, G., Tilg, H., Rajilić-Stojanović, M., Kump, P., Satokari, R., et al. (2017). European consensus conference on faecal microbiota transplantation in clinical practice. *Gut, 66*(4), 569–580. https://doi.org/10.1136/gutjnl-2016-313017.

Carter, G. P., Chakravorty, A., Pham Nguyen, T. A., Mileto, S., Schreiber, F., Li, L., et al. (2015). Defining the roles of TcdA and TcdB in localized gastrointestinal disease, systemic organ damage, and the host response during *Clostridium difficile* infections. *MBio, 6*(3). https://doi.org/10.1128/mBio.00551-15.

Carter, G. P., Rood, J. I., & Lyras, D. (2010). The role of toxin A and toxin B in Clostridium difficile-associated disease: Past and present perspectives. *Gut Microbes, 1*(1), 58–64. https://doi.org/10.4161/gmic.1.1.10768.

Chapin, K. C., Dickenson, R. A., Wu, F., & Andrea, S. B. (2011). Comparison of five assays for detection of *Clostridium difficile* toxin. *The Journal of Molecular Diagnostics, 13*(4), 395–400. https://doi.org/10.1016/j.jmoldx.2011.03.004.

Chen, P., Zeng, J., Liu, Z., Thaker, H., Wang, S., Tian, S., et al. (2021). Structural basis for CSPG4 as a receptor for TcdB and a therapeutic target in *Clostridioides difficile* infection. *Nature Communications, 12*(1), 3748. https://doi.org/10.1038/s41467-021-23878-3.

Cho, J. (2018). Ridinilazole: A novel antimicrobial for *Clostridium difficile* infection. *Annals of Gastroenterology*. https://doi.org/10.20524/aog.2018.0336.

Cohen, S. H., Gerding, D. N., Johnson, S., Kelly, C. P., Loo, V. G., McDonald, L. C., et al. (2010). Clinical practice guidelines for *Clostridium difficile* infection in adults: 2010 update by the Society for Healthcare Epidemiology of America (SHEA) and the Infectious Diseases Society of America (IDSA). *Infection Control & Hospital Epidemiology, 31*(05), 431–455. https://doi.org/10.1086/651706.

Cole, L. E., Li, L., Jetley, U., Zhang, J., Pacheco, K., Ma, F., et al. (2019). Deciphering the domain specificity of *C. difficile* toxin neutralizing antibodies. *Vaccine, 37*(29), 3892–3901. https://doi.org/10.1016/j.vaccine.2019.05.040.

Couturier, J., Davies, K., Gateau, C., & Barbut, F. (2018). Ribotypes and new virulent strains across Europe. *Advances in Experimental Medicine and Biology, 1050*, 45–58. https://doi.org/10.1007/978-3-319-72799-8_4.

Crobach, M. J. T., Duszenko, N., Terveer, E. M., Verduin, C. M., & Kuijper, E. J. (2017). Nucleic acid amplification test quantitation as predictor of toxin presence in *Clostridium difficile* infection. *Journal of Clinical Microbiology, 56*(3) e01316-17, /jcm/56/3/e01316-17.atom. https://doi.org/10.1128/JCM.01316-17.

Czepiel, J., Dróżdż, M., Pituch, H., Kuijper, E. J., Perucki, W., Mielimonka, A., et al. (2019). Clostridium difficile infection: Review. *European Journal of Clinical Microbiology & Infectious Diseases, 38*(7), 1211–1221. https://doi.org/10.1007/s10096-019-03539-6.

Dart, R. C., & McNally, J. (2001). Efficacy, safety, and use of snake antivenoms in the United States. *Annals of Emergency Medicine*, *37*(2), 181–188. https://doi.org/10.1067/mem.2001.113372.

Davies, K. A., Ashwin, H., Longshaw, C. M., Burns, D. A., Davis, G. L., Wilcox, M. H., et al. (2016). Diversity of *Clostridium difficile* PCR ribotypes in Europe: Results from the European, multicentre, prospective, biannual, point-prevalence study of *Clostridium difficile* infection in hospitalised patients with diarrhoea (EUCLID), 2012 and 2013. *Euro Surveillance: Bulletin Europeen Sur Les Maladies Transmissibles = European Communicable Disease Bulletin*, *21*(29). https://doi.org/10.2807/1560-7917.ES.2016.21.29.30294.

Davies, K. A., Longshaw, C. M., Davis, G. L., Bouza, E., Barbut, F., Barna, Z., et al. (2014). Underdiagnosis of *Clostridium difficile* across Europe: The European, multicentre, prospective, biannual, point-prevalence study of *Clostridium difficile* infection in hospitalised patients with diarrhoea (EUCLID). *The Lancet Infectious Diseases*, *14*(12), 1208–1219. https://doi.org/10.1016/S1473-3099(14)70991-0.

de Gunzburg, J., Ghozlane, A., Ducher, A., Le Chatelier, E., Duval, X., Ruppé, E., et al. (2018). Protection of the human gut microbiome from antibiotics. *The Journal of Infectious Diseases*, *217*(4), 628–636. https://doi.org/10.1093/infdis/jix604.

Debast, S. B., Bauer, M. P., & Kuijper, E. J. (2014). European Society of Clinical Microbiology and Infectious Diseases: Update of the treatment guidance document for *Clostridium difficile* infection. *Clinical Microbiology and Infection*, *20*, 1–26. https://doi.org/10.1111/1469-0691.12418.

Del Prete, R., Ronga, L., & Addati, G. (2019). *Clostridium difficile*. A review on an emerging infection. *La Clinica Terapeutica*, *1*, 41–47. https://doi.org/10.7417/CT.2019.2106.

Delmée, M., Van Broeck, J., Simon, A., Janssens, M., & Avesani, V. (2005). Laboratory diagnosis of *Clostridium difficile*-associated diarrhoea: A plea for culture. *Journal of Medical Microbiology*, *54*(2), 187–191. https://doi.org/10.1099/jmm.0.45844-0.

Deshpande, A., Pasupuleti, V., Thota, P., Pant, C., Rolston, D. D. K., Hernandez, A. V., et al. (2015). Risk factors for recurrent *Clostridium difficile* infection: A systematic review and meta-analysis. *Infection Control & Hospital Epidemiology*, *36*(4), 452–460. https://doi.org/10.1017/ice.2014.88.

Dieterle, M. G., Putler, R., Perry, D. A., Menon, A., Abernathy-Close, L., Perlman, N. S., et al. (2020). Systemic inflammatory mediators are effective biomarkers for predicting adverse outcomes in *Clostridioides difficile* infection. *MBio*, *11*(3). https://doi.org/10.1128/mBio.00180-20.

Eckert, C., Burghoffer, B., & Barbut, F. (2013). Contamination of ready-to-eat raw vegetables with *Clostridium difficile* in France. *Journal of Medical Microbiology*, *62*(9), 1435–1438. https://doi.org/10.1099/jmm.0.056358-0.

Eiseman, B., Silen, W., Bascom, G. S., & Kauvar, A. J. (1958). Fecal enema as an adjunct in the treatment of pseudomembranous enterocolitis. *Surgery*, *44*(5), 854–859.

Eyre, D. W., Walker, A. S., Wyllie, D., Dingle, K. E., Griffiths, D., Finney, J., et al. (2012). Predictors of first recurrence of *Clostridium difficile* infection: Implications for initial management. *Clinical Infectious Diseases*, *55*(Suppl. 2), S77–S87. https://doi.org/10.1093/cid/cis356.

Fekety, R., McFarland, L. V., Surawicz, C. M., Greenberg, R. N., Elmer, G. W., & Mulligan, M. E. (1997). Recurrent *Clostridium difficile* diarrhea: Characteristics of and risk factors for patients enrolled in a prospective, randomized, double-blinded trial. *Clinical Infectious Diseases*, *24*(3), 324–333. https://doi.org/10.1093/clinids/24.3.324.

Garey, K. W., Sethi, S., Yadav, Y., & DuPont, H. L. (2008). Meta-analysis to assess risk factors for recurrent *Clostridium difficile* infection. *Journal of Hospital Infection*, *70*(4), 298–304. https://doi.org/10.1016/j.jhin.2008.08.012.

Gerding, D. N., Johnson, S., Rupnik, M., & Aktories, K. (2014). *Clostridium difficile* binary toxin CDT: Mechanism, epidemiology, and potential clinical importance. *Gut Microbes*, *5*(1), 15–27. https://doi.org/10.4161/gmic.26854.

Gerding, D. N., Meyer, T., Lee, C., Cohen, S. H., Murthy, U. K., Poirier, A., et al. (2015). Administration of spores of nontoxigenic *Clostridium difficile* strain M3 for prevention of recurrent *C difficile* infection: A randomized clinical trial. *JAMA*, *313*(17), 1719. https://doi.org/10.1001/jama.2015.3725.

Goldenberg, J. Z., Yap, C., Lytvyn, L., Lo, C. K.-F., Beardsley, J., Mertz, D., et al. (2017). Probiotics for the prevention of *Clostridium difficile*-associated diarrhea in adults and children. *Cochrane Database of Systematic Reviews*. https://doi.org/10.1002/14651858.CD006095.pub4.

Gotshal, D., Azrad, M., Hamo, Z., Nitzan, O., & Peretz, A. (2021). IL-16 and BCA-1 serum levels are associated with disease severity of *C. difficile* infection. *Pathogens*, *10*(5), 631. https://doi.org/10.3390/pathogens10050631.

Guery, B., Menichetti, F., Anttila, V.-J., Adomakoh, N., Aguado, J. M., Bisnauthsing, K., et al. (2018). Extended-pulsed fidaxomicin versus vancomycin for *Clostridium difficile* infection in patients 60 years and older (EXTEND): A randomised, controlled, open-label, phase 3b/4 trial. *The Lancet Infectious Diseases*, *18*(3), 296–307. https://doi.org/10.1016/S1473-3099(17)30751-X.

Gülke, I., Pfeifer, G., Liese, J., Fritz, M., Hofmann, F., Aktories, K., et al. (2001). Characterization of the enzymatic component of the ADP-ribosyltransferase toxin CDTa from *Clostridium difficile*. *Infection and Immunity*, *69*(10), 6004–6011. https://doi.org/10.1128/IAI.69.10.6004-6011.2001.

Hall, I. C., & O'Toole, E. (1935). Intestinal flora in new-born infants: With a description of a new pathogenic anaerobe, *Bacillus difficilis*. *American Journal of Diseases of Children*, *49*(2), 390–402. https://doi.org/10.1001/archpedi.1935.01970020105010.

Hamo, Z., Azrad, M., Nitzan, O., & Peretz, A. (2019). Characterization of the immune response during infection caused by *Clostridioides difficile*. *Microorganisms*, 7(10), 435. https://doi.org/10.3390/microorganisms7100435.

Hensgens, M. P. M., Goorhuis, A., Dekkers, O. M., & Kuijper, E. J. (2012). Time interval of increased risk for *Clostridium difficile* infection after exposure to antibiotics. *The Journal of Antimicrobial Chemotherapy*, *67*(3), 742–748. https://doi.org/10.1093/jac/dkr508.

Herbert, R., Hatcher, J., Jauneikaite, E., Gharbi, M., d'Arc, S., Obaray, N., et al. (2019). Two-year analysis of *Clostridium difficile* ribotypes associated with increased severity. *Journal of Hospital Infection*, *103*(4), 388–394. https://doi.org/10.1016/j.jhin.2019.06.003.

Hernandez, L. D., Racine, F., Xiao, L., DiNunzio, E., Hairston, N., Sheth, P. R., et al. (2015). Broad coverage of genetically diverse strains of *Clostridium difficile* by Actoxumab and Bezlotoxumab predicted by *in vitro* neutralization and epitope modeling. *Antimicrobial Agents and Chemotherapy*, *59*(2), 1052–1060. https://doi.org/10.1128/AAC.04433-14.

Huber, C. A., Foster, N. F., Riley, T. V., & Paterson, D. L. (2013). Challenges for standardization of *Clostridium difficile* typing methods. *Journal of Clinical Microbiology*, *51*(9), 2810–2814. https://doi.org/10.1128/JCM.00143-13.

Jiang, Z. D., Ajami, N. J., Petrosino, J. F., Jun, G., Hanis, C. L., Shah, M., et al. (2017). Randomised clinical trial: Faecal microbiota transplantation for recurrent *Clostridum difficile* infection—Fresh, or frozen, or lyophilised microbiota from a small pool of healthy donors delivered by colonoscopy. *Alimentary Pharmacology & Therapeutics*, *45*(7), 899–908. https://doi.org/10.1111/apt.13969.

Johnson, S. (2009). Recurrent *Clostridium difficile* infection: A review of risk factors, treatments, and outcomes. *Journal of Infection*, *58*(6), 403–410. https://doi.org/10.1016/j.jinf.2009.03.010.

Johnson, S., Lavergne, V., Skinner, A. M., Gonzales-Luna, A. J., Garey, K. W., Kelly, C. P., et al. (2021). Clinical practice guideline by the Infectious Diseases Society of America (IDSA) and Society for Healthcare Epidemiology of America (SHEA): 2021 Focused update guidelines on management of *Clostridioides difficile* infection in adults. *Clinical Infectious Diseases*, ciab549. https://doi.org/10.1093/cid/ciab549.

Just, I., & Gerhard, R. (2005). *Large clostridial cytotoxins. In* Vol. 152, *Reviews of physiology, biochemistry and pharmacology* (pp. 23–47). Berlin Heidelberg: Springer. https://doi.org/10.1007/s10254-004-0033-5.

Kelly, C. P., & Kyne, L. (2011). The host immune response to *Clostridium difficile. Journal of Medical Microbiology*, *60*(8), 1070–1079. https://doi.org/10.1099/jmm.0.030015-0.

Kelly, C. R., Yen, E. F., Grinspan, A. M., Kahn, S. A., Atreja, A., Lewis, J. D., et al. (2021). Fecal microbiota transplantation is highly effective in real-world practice: Initial results from the FMT National Registry. *Gastroenterology*, *160*(1), 183–192.e3. https://doi.org/10.1053/j.gastro.2020.09.038.

Kim, H., Riley, T. V., Kim, M., Kim, C. K., Yong, D., Lee, K., et al. (2008). Increasing prevalence of toxin A-negative, toxin B-positive isolates of *Clostridium difficile* in Korea: Impact on laboratory diagnosis. *Journal of Clinical Microbiology*, *46*(3), 1116–1117. https://doi.org/10.1128/JCM.01188-07.

Kink, J. A., & Williams, J. A. (1998). Antibodies to recombinant *Clostridium difficile* toxins A and B are an effective treatment and prevent relapse of *C. difficile*-associated disease in a hamster model of infection. *Infection and Immunity*, *66*(5), 2018–2025. https://doi.org/10.1128/IAI.66.5.2018-2025.1998.

Kokai-Kun, J. F., Roberts, T., Coughlin, O., Le, C., Whalen, H., Stevenson, R., et al. (2019). Use of ribaxamase (SYN-004), a β-lactamase, to prevent *Clostridium difficile* infection in β-lactam-treated patients: A double-blind, phase 2b, randomised placebo-controlled trial. *The Lancet Infectious Diseases*, *19*(5), 487–496. https://doi.org/10.1016/S1473-3099(18)30731-X.

Krutova, M., Kinross, P., Barbut, F., Hajdu, A., Wilcox, M. H., Kuijper, E. J., et al. (2018). How to: Surveillance of *Clostridium difficile* infections. *Clinical Microbiology and Infection*, *24*(5), 469–475. https://doi.org/10.1016/j.cmi.2017.12.008.

Kuehne, S. A., Collery, M. M., Kelly, M. L., Cartman, S. T., Cockayne, A., & Minton, N. P. (2014). Importance of toxin A, toxin B, and CDT in virulence of an epidemic *Clostridium difficile* strain. *The Journal of Infectious Diseases*, *209*(1), 83–86. https://doi.org/10.1093/infdis/jit426.

La Fata, G., Weber, P., & Mohajeri, M. H. (2018). Probiotics and the gut immune system: Indirect regulation. *Probiotics and Antimicrobial Proteins*, *10*(1), 11–21. https://doi.org/10.1007/s12602-017-9322-6.

Lee, T., Pang, S., Abraham, S., & Coombs, G. W. (2019). Antimicrobial-resistant CC17 *Enterococcus faecium*: The past, the present and the future. *Journal of Global Antimicrobial Resistance*, *16*, 36–47. https://doi.org/10.1016/j.jgar.2018.08.016.

Lee, H. S., Plechot, K., Gohil, S., & Le, J. (2021). *Clostridium difficile*: Diagnosis and the consequence of over diagnosis. *Infectious Diseases and Therapy*, *10*(2), 687–697. https://doi.org/10.1007/s40121-021-00417-7.

Leffler, D. A., & Lamont, J. T. (2009). Treatment of *Clostridium difficile*-associated disease. *Gastroenterology*, *136*(6), 1899–1912. https://doi.org/10.1053/j.gastro.2008.12.070.

Leffler, D. A., & Lamont, J. T. (2015). *Clostridium difficile* infection. *The New England Journal of Medicine*, *372*(16), 1539–1548. https://doi.org/10.1056/NEJMra1403772.

Leuzzi, R., Adamo, R., & Scarselli, M. (2014). Vaccines against *Clostridium difficile. Human Vaccines & Immunotherapeutics*, *10*(6), 1466–1477. https://doi.org/10.4161/hv.28428.

Loo, V. G., Poirier, L., Miller, M. A., Oughton, M., Libman, M. D., Michaud, S., et al. (2005). A predominantly clonal multi-institutional outbreak of *Clostridium difficile*-associated diarrhea with high morbidity and mortality. *New England Journal of Medicine*, *353*(23), 2442–2449. https://doi.org/10.1056/NEJMoa051639.

Louie, T. J., Miller, M. A., Mullane, K. M., Weiss, K., Lentnek, A., Golan, Y., et al. (2011). Fidaxomicin versus vancomycin for *Clostridium difficile* infection. *New England Journal of Medicine*, *364*(5), 422–431. https://doi.org/10.1056/NEJMoa0910812.

Lyerly, D. M., Krivan, H. C., & Wilkins, T. D. (1988). *Clostridium difficile*: Its disease and toxins. *Clinical Microbiology Reviews*, *1*(1), 1–18. https://doi.org/10.1128/CMR.1.1.1.

Marsh, J. W., Arora, R., Schlackman, J. L., Shutt, K. A., Curry, S. R., & Harrison, L. H. (2012). Association of relapse of *Clostridium difficile* disease with BI/NAP1/027. *Journal of Clinical Microbiology*, *50*(12), 4078–4082. https://doi.org/10.1128/JCM.02291-12.

McDonald, L. C., Gerding, D. N., Johnson, S., Bakken, J. S., Carroll, K. C., Coffin, S. E., et al. (2018). Clinical practice guidelines for *Clostridium difficile* infection in adults and children: 2017 Update by the Infectious Diseases Society of America (IDSA) and Society for Healthcare Epidemiology of America (SHEA). *Clinical Infectious Diseases*, *66*(7), e1–e48. https://doi.org/10.1093/cid/cix1085.

McDonald, L. C., Killgore, G. E., Thompson, A., Owens, R. C., Kazakova, S. V., Sambol, S. P., et al. (2005). An epidemic, toxin gene-variant strain of *Clostridium difficile*. *The New England Journal of Medicine*, *353*(23), 2433–2441. https://doi.org/10.1056/NEJMoa051590.

McFarland, L. V. (2020). When It's too trying for trials: A commentary on a different approach for testing a probiotic, *Saccharomyces boulardii*, to prevent healthcare-associated *Clostridioides difficile* infections. *Clinical Infectious Diseases*, ciaa798. https://doi.org/10.1093/cid/ciaa798.

McFarland, L. V., Elmer, G. W., & Surawicz, C. M. (2002). Breaking the cycle: Treatment strategies for 163 cases of recurrent *Clostridium difficile* disease. *American Journal of Gastroenterology*, *97*(7), 1769–1775. https://doi.org/10.1111/j.1572-0241.2002.05839.x.

Monteiro, J. F., Hahn, S. R., Gonçalves, J., & Fresco, P. (2018). Vancomycin therapeutic drug monitoring and population pharmacokinetic models in special patient subpopulations. *Pharmacology Research & Perspectives*, *6*(4)e00420 https://doi.org/10.1002/prp2.420.

Moré, M. I., & Vandenplas, Y. (2018). *Saccharomyces boulardii* CNCM I-745 improves intestinal enzyme function: A trophic effects review. *Clinical Medicine Insights: Gastroenterology*, *11*, 117955221775267. https://doi.org/10.1177/1179552217752679.

Musher, D. M., Manhas, A., Jain, P., Nuila, F., Waqar, A., Logan, N., et al. (2007). Detection of *Clostridium difficile* toxin: Comparison of enzyme immunoassay results with results obtained by cytotoxicity assay. *Journal of Clinical Microbiology*, *45*(8), 2737–2739. https://doi.org/10.1128/JCM.00686-07.

Péchiné, S., Denève, C., Le Monnier, A., Hoys, S., Janoir, C., & Collignon, A. (2011). Immunization of hamsters against *Clostridium difficile* infection using the Cwp84 protease as an antigen. *FEMS Immunology & Medical Microbiology*, *63*(1), 73–81. https://doi.org/10.1111/j.1574-695X.2011.00832.x.

Petrosillo, N., Granata, G., & Cataldo, M. A. (2018). Novel antimicrobials for the treatment of *Clostridium difficile* infection. *Frontiers in Medicine*, *5*, 96. https://doi.org/10.3389/fmed.2018.00096.

Planche, T., & Wilcox, M. (2011). Reference assays for *Clostridium difficile* infection: One or two gold standards? *Journal of Clinical Pathology*, *64*(1), 1–5. https://doi.org/10.1136/jcp.2010.080135.

Pothoulakis, C. (2006). Effects of *Clostridium difficile* toxins on epithelial cell barrier. *Annals of the New York Academy of Sciences*, *915*(1), 347–356. https://doi.org/10.1111/j.1749-6632.2000.tb05263.x.

Quraishi, M. N., Widlak, M., Bhala, N., Moore, D., Price, M., Sharma, N., et al. (2017). Systematic review with meta-analysis: The efficacy of faecal microbiota transplantation for the treatment of recurrent and refractory *Clostridium difficile* infection. *Alimentary Pharmacology & Therapeutics*, *46*(5), 479–493. https://doi.org/10.1111/apt.14201.

Ray, K. (2017). Modifying recurrence of *Clostridium difficile* infection. *Nature Reviews Gastroenterology & Hepatology*, *14*(3), 137. https://doi.org/10.1038/nrgastro.2017.17.

Riley, T. V., Lyras, D., & Douce, G. R. (2019). Status of vaccine research and development for *Clostridium difficile*. *Vaccine*, *37*(50), 7300–7306. https://doi.org/10.1016/j.vaccine.2019.02.052.

Roberts, A. K., Harris, H. C., Smith, M., Giles, J., Polak, O., Buckley, A. M., et al. (2020). A novel, orally delivered antibody therapy and its potential to prevent *Clostridioides difficile* infection in pre-clinical models. *Frontiers in Microbiology, 11*, 578903. https://doi.org/10.3389/fmicb.2020.578903.

Roberts, A., McGlashan, J., Al-Abdulla, I., Ling, R., Denton, H., Green, S., et al. (2012). Development and evaluation of an ovine antibody-based platform for treatment of *Clostridium difficile* infection. *Infection and Immunity, 80*(2), 875–882. https://doi.org/10.1128/IAI.05684-11.

Rupnik, M., & Janezic, S. (2016). An update on *Clostridium difficile* toxinotyping. *Journal of Clinical Microbiology, 54*(1), 13–18. https://doi.org/10.1128/JCM.02083-15.

Rupnik, M., Wilcox, M. H., & Gerding, D. N. (2009). *Clostridium difficile* infection: New developments in epidemiology and pathogenesis. *Nature Reviews Microbiology*, 7(7), 526–536. https://doi.org/10.1038/nrmicro2164.

Sammons, J. S., & Toltzis, P. (2013). Recent trends in the epidemiology and treatment of *C. difficile* infection in children. *Current Opinion in Pediatrics, 25*(1), 116–121. https://doi.org/10.1097/MOP.0b013e32835bf6c0.

Shim, J. K., Johnson, S., Samore, M. H., Bliss, D. Z., & Gerding, D. N. (1998). Primary symptomless colonisation by *Clostridium difficile* and decreased risk of subsequent diarrhoea. *The Lancet, 351*(9103), 633–636. https://doi.org/10.1016/S0140-6736(97)08062-8.

Sirbu, B. D., Soriano, M. M., Manzo, C., Lum, J., Gerding, D. N., & Johnson, S. (2017). Vancomycin taper and pulse regimen with careful follow-up for patients with recurrent *Clostridium difficile* infection. *Clinical Infectious Diseases, 65*(8), 1396–1399. https://doi.org/10.1093/cid/cix529.

Slimings, C., & Riley, T. V. (2014). Antibiotics and hospital-acquired *Clostridium difficile* infection: Update of systematic review and meta-analysis. *Journal of Antimicrobial Chemotherapy, 69*(4), 881–891. https://doi.org/10.1093/jac/dkt477.

Song, J. H., & Kim, Y. S. (2019). Recurrent *Clostridium difficile* infection: Risk factors, treatment, and prevention. *Gut and Liver, 13*(1), 16–24. https://doi.org/10.5009/gnl18071.

Spencer, J., Leuzzi, R., Buckley, A., Irvine, J., Candlish, D., Scarselli, M., et al. (2014). Vaccination against *Clostridium difficile* using toxin fragments: Observations and analysis in animal models. *Gut Microbes, 5*(2), 225–232. https://doi.org/10.4161/gmic.27712.

Spigaglia, P., & Mastrantonio, P. (2004). Comparative analysis of *Clostridium difficile* clinical isolates belonging to different genetic lineages and time periods. *Journal of Medical Microbiology, 53*(11), 1129–1136. https://doi.org/10.1099/jmm.0.45682-0.

Steiner, T. S., Flores, C. A., Pizarro, T. T., & Guerrant, R. L. (1997). Fecal lactoferrin, interleukin-1beta, and interleukin-8 are elevated in patients with severe *Clostridium difficile* colitis. *Clinical Diagnostic Laboratory Immunology*, 4(6), 719–722. https://doi.org/10.1128/cdli.4.6.719-722.1997.

Suetens, C., Latour, K., Kärki, T., Ricchizzi, E., Kinross, P., Moro, M. L., et al. (2018). Prevalence of healthcare-associated infections, estimated incidence and composite antimicrobial resistance index in acute care hospitals and long-term care facilities: Results from two European point prevalence surveys, 2016 to 2017. *Euro Surveillance: Bulletin Europeen Sur les maladies Transmissibles = European communicable disease bulletin, 23*(46). https://doi.org/10.2807/1560-7917.ES.2018.23.46.1800516.

Sullivan, N. M., Pellett, S., & Wilkins, T. D. (1982). Purification and characterization of toxins A and B of *Clostridium difficile. Infection and Immunity, 35*(3), 1032–1040. https://doi.org/10.1128/IAI.35.3.1032-1040.1982.

Tariq, R., Pardi, D. S., Bartlett, M. G., & Khanna, S. (2019). Low cure rates in controlled trials of fecal microbiota transplantation for recurrent *Clostridium difficile* infection: A systematic review and meta-analysis. *Clinical Infectious Diseases, 68*(8), 1351–1358. https://doi.org/10.1093/cid/ciy721.

Taylor, N. S., Thorne, G. M., & Bartlett, J. G. (1981). Comparison of two toxins produced by *Clostridium difficile*. *Infection and Immunity*, *34*(3), 1036–1043. https://doi.org/10.1128/IAI.34.3.1036-1043.1981.

Tijerina-Rodríguez, L., Villarreal-Treviño, L., Morfín-Otero, R., Camacho-Ortíz, A., & Garza-González, E. (2019). Virulence factors of *Clostridioides* (*Clostridium*) *difficile* linked to recurrent infections. *Canadian Journal of Infectious Diseases and Medical Microbiology*, *2019*, 1–7. https://doi.org/10.1155/2019/7127850.

Tocher, J. H., & Edwards, D. I. (1992). The interaction of reduced metronidazole with DNA bases and nucleosides. *International Journal of Radiation Oncology Biology Physics*, *22*(4), 661–663. https://doi.org/10.1016/0360-3016(92)90498-7.

Tschudin-Sutter, S., Kuijper, E. J., Durovic, A., Vehreschild, M. J. G. T., Barbut, F., Eckert, C., et al. (2018). Guidance document for prevention of *Clostridium difficile* infection in acute healthcare settings. *Clinical Microbiology and Infection*, *24*(10), 1051–1054. https://doi.org/10.1016/j.cmi.2018.02.020.

Vandekerckhove, J., Schering, B., Bärmann, M., & Aktories, K. (1987). *Clostridium perfringens* iota toxin ADP-ribosylates skeletal muscle actin in Arg-177. *FEBS Letters*, *225*(1–2), 48–52. https://doi.org/10.1016/0014-5793(87)81129-8.

Velazquez-Gomez, I., Rocha-Rodriguez, R., Toro, D. H., Gutierrez-Nuñez, J. J., Gonzalez, G., & Saavedra, S. (2008). A severity score index for *Clostridium difficile* infection. *Infectious Diseases in Clinical Practice*, *16*(6), 376–378. https://doi.org/10.1097/IPC.0b013e318182557f.

Vickers, R. J., Tillotson, G., Goldstein, E. J. C., Citron, D. M., Garey, K. W., & Wilcox, M. H. (2016). Ridinilazole: A novel therapy for *Clostridium difficile* infection. *International Journal of Antimicrobial Agents*, *48*(2), 137 143. https://doi.org/10.1016/j.ijantimicag.2016.04.026.

Vickers, R. J., Tillotson, G. S., Nathan, R., Hazan, S., Pullman, J., Lucasti, C., et al. (2017). Efficacy and safety of ridinilazole compared with vancomycin for the treatment of *Clostridium difficile* infection: A phase 2, randomised, double-blind, active-controlled, non-inferiority study. *The Lancet Infectious Diseases*, *17*(7), 735–744. https://doi.org/10.1016/S1473-3099(17)30235-9.

Viscidi, R., Willey, S., & Bartlett, J. G. (1981). Isolation rates and toxigenic potential of *Clostridium difficile* isolates from various patient populations. *Gastroenterology*, *81*(1), 5–9. https://doi.org/10.1016/0016-5085(81)90644-2.

Voss, A., & Hopman, J. (2014). Survey of health care-associated infections. *The New England Journal of Medicine*, *370*(26), 2542. https://doi.org/10.1056/NEJMc1405194.

Voth, D. E., & Ballard, J. D. (2005). *Clostridium difficile* toxins: Mechanism of action and role in disease. *Clinical Microbiology Reviews*, *18*(2), 247–263. https://doi.org/10.1128/CMR.18.2.247-263.2005.

Warny, M., Pepin, J., Fang, A., Killgore, G., Thompson, A., Brazier, J., et al. (2005). Toxin production by an emerging strain of *Clostridium difficile* associated with outbreaks of severe disease in North America and Europe. *The Lancet*, *366*(9491), 1079–1084. https://doi.org/10.1016/S0140-6736(05)67420-X.

Wilcox, M. H., Gerding, D. N., Poxton, I. R., Kelly, C., Nathan, R., Birch, T., et al. (2017). Bezlotoxumab for prevention of recurrent *Clostridium difficile* infection. *New England Journal of Medicine*, *376*(4), 305–317. https://doi.org/10.1056/NEJMoa1602615.

Wombwell, E., Patterson, M. E., Bransteitter, B., & Gillen, L. R. (2020). The effect of *Saccharomyces boulardii* primary prevention on risk of hospital-onset *Clostridioides difficile* infection in hospitalized patients administered antibiotics frequently associated with *C. difficile* infection. *Clinical Infectious Diseases*, ciaa808. https://doi.org/10.1093/cid/ciaa808.

Wren, M. (2010). Clostridium difficile *isolation and culture techniques*. In P. Mullany & A. P. Roberts (Eds.), Vol. 646. *Clostridium difficile* (pp. 39–52). Humana Press. https://doi.org/10.1007/978-1-60327-365-7_3.

Zheng, M., Zhang, R., Tian, X., Zhou, X., Pan, X., & Wong, A. (2017). Assessing the risk of probiotic dietary supplements in the context of antibiotic resistance. *Frontiers in Microbiology*, *8*, 908. https://doi.org/10.3389/fmicb.2017.00908.

CHAPTER EIGHT

A computational model revealing the immune-related hub genes and key pathways involved in rheumatoid arthritis (RA)

Ambritha Balasundaram, S. Udhaya Kumar, and C. George Priya Doss*

School of Biosciences and Technology, Vellore Institute of Technology, Vellore, TN, India
*Corresponding author: e-mail address: georgepriyadoss@vit.ac.in

Contents

Abstract

Rheumatoid arthritis (RA) has one of the highest disability rates among inflammatory joint disorders. However, the reason and possible molecular events are still unclear. There are various treatment options available, but no complete cure. To obtain early

Advances in Protein Chemistry and Structural Biology, Volume 129
ISSN 1876-1623
https://doi.org/10.1016/bs.apcsb.2021.11.006

diagnosis and successful medication in RA, it is necessary to explore gene susceptibility and pathogenic factors. The main intend of our work is to explore the immune-related hub genes with similar functions that are differentially expressed in RA patients. Three datasets such as GSE21959, GSE55457, and GSE77298, were taken to analyze the differently expressed genes (DEGs) among 55 RA and 33 control samples. We obtained 331 upregulated and 275 downregulated DEGs from three Gene Expression Omnibus (GEO) datasets using the R package. Furthermore, a protein–protein interaction network was built for upregulated and downregulated DEGs using Cytoscape. Subsequently, MCODE analysis was performed and obtained the top two modules in each DEG's upregulated and downregulated protein–protein interactions (PPIs) network. CytoNCA and cytoHubba were performed and identified overlapping DEGs. In addition, we narrowed down DEGs by filtering with immune-related genes and identified DE-IRGs. Gene ontology (GO) and KEGG pathway analysis in upregulated and downregulated DEGs were executed with the DAVID platform. Our study obtained the nine most significant DE-IRGs in RA such as *CXCR4, CDK1, BUB1, BIRC5, AGTR1, EGFR, EDNRB, KALRN,* and *GHSR*. Among them, *CXCR4, CDK1, BUB1,* and *BIRC5* are overexpressed in RA and may contribute to the pathophysiology of the disease. Similarly, *AGTR1, EGFR, EDNRB, KALRN*, and *GHSR* are all low expressed in RA and may have a contribution to pathogenesis. GO, KEGG functional enrichment, and GeneMANIA showed that the dysregulated process of DE-IRGs causes RA development and progression. These findings may be helpful in future studies in RA diagnosis and therapy.

Abbreviations

ACPAs	anti-citrullinated protein antibodies
BP	biological process
DEGs	differently expressed genes
DE-IRGs	differently expressed immune related genes
EULAR	European Alliance of Associations for Rheumatology
FC	fold-change
FDR	false discovery rate
FLS	fibroblast-like synoviocytes
GEO	Gene Expression Omnibus
GO	Gene Ontology
GWAS	Genome-Wide Association Study
IRGs	immune-related genes
KEGG	Kyoto Encyclopedia of Genes and Genomes
MCODE	Molecular Complex Detection
MF	molecular functions
MLS	macrophage-like synoviocytes
NCBI	National Center for Biotechnology Information
PAD	peptidyl arginine deiminases
PPIs	protein–protein interactions
RA	rheumatoid arthritis
STRING	Search Tool for the Retrieval of Interacting Genes/Proteins
TNF	tumor necrosis factor

1. Introduction

RA is considered as an autoimmune disease that forms chronic synovium inflammation, leading to damage in the joints & bone, and rises the mortality rate (Alamanos & Drosos, 2005). Synovitis was categorized as synovium inflammation, which causes tissue tethering and decreases the surface of the joint (Hitchon & El-Gabalawy, 2011). Fibroblast-like synoviocytes (FLS) and macrophage-like synoviocytes (MLS) are two main types of synovial cells. Upregulation of inflammatory cytokines in the rheumatoid synovium leads to MLS formation. FLS may contribute to local joint injury (Xiong et al., 2019; Yoshitomi, 2019). RA originates from genetic and epigenetic components, but the environment also plays a crucial part in it, such as cigarette smoke, exposure to dust, and the microbiome (internal environment). The components of the adaptive and innate immune responses interacted significantly (Scherer, Häupl, & Burmester, 2020).

Pathological changes are mainly caused by persistent, symmetrical, and multiple joint inflammatory lesions in the peripheral region. As the RA progresses, other body components like the lungs, eyes, heart, kidneys, and other internal organs can also be affected (Zhou, Zou, Chen, & Huang, 2019). RA has two main subgroups based on the presence or lack of biologic anti-citrullinated protein antibodies (ACPAs) (Wu, Yang, & Lai, 2020). Peptidyl arginine deiminases (PAD) triggers citrullination that converts a positively charged arginine to polar but leads to neutral citrulline through post-translational alteration. Compared to RA's negative ACPA fraction, the positive ACPA subset exhibits a more aggressive clinical profile. The negative ACPA fraction of RA exhibits distinct genetic association patterns and immune reactions to citrulline antigens than the positive ACPA population. Response to therapy of rituximab/methotrexate (MTX) was found to be less success rate in negative ACPA RA patients (Guo et al., 2018). With an estimated global prevalence of three cases for every 10,000 population, RA is the more familiar type of rheumatic disorder. RA usually begins around 35 to 50 years of age. The prevalence ratio of women to men varies from 3:1, but gender differences diminish in older age groups. According to the EULAR references, RA therapy contains numerous therapeutic alternatives but no cure (Fakhouri et al., 2018). Therapy primarily focuses on pain relief, preventing/limiting joint deterioration, improving/preserving joint function, and optimizing the quality of life.

Immune cells such as thymus cells, bone marrow cells, and macrophages have an essential play in RA development. The immune cells usually persist in the synovium or revolve in the blood. Bone marrow cells release functionally significant proteins include rheumatoid factors, proinflammatory cytokines, and ACPA. Thymus cell activation is well intermediated by bone marrow cells through costimulatory molecules. Thymus cells' primary purpose in RA is to stimulate macrophages & fibroblasts, converting them into devastating tissue cells. Similarly, activated macrophages release many cytokines & chemokines to support joint inflammation (Yap et al., 2018). So far, several genes have been linked to the etiology of RA. Large Genome-Wide Association Studies (GWAS) reported several loci related to pathogenesis in RA, in addition to human leukocyte antigen. Protein tyrosine phosphatase, peptidyl arginine deiminase type IV, nonreceptor type 22, signal transducer & activator of transcription 4, tumor necrosis factor, cytotoxic T-lymphocyte-associated protein 4, tumor necrosis factor-receptor associated factor 1, and others are among the genes that have been identified (Korczowska, 2014). According to a previous study published in the International Journal of Rheumatology, roughly 50% of RA patients in Europe discontinued their TNF inhibitor biological therapy during the first 5 years of treatment due to ineffectiveness or adverse medication responses (Arora, Mahajan, Spurden, Boyd, & Porter, 2013). Though previously available drugs are ineffective in many patients, there is a need to find or assess immune-specific genes and pathways of importance for the progression of RA. These findings will improve the prevention, diagnosis, and treatment of RA. Microarray techniques can look at the expression levels of thousands of genes in a single mRNA sample all at once. Microarrays have been used to find diagnostic or prognostic biomarkers, classify diseases, track treatment responses, and learn about the mechanisms involved in disease genesis (Tarca, Romero, & Draghici, 2006; Udhaya Kumar et al., 2020a, 2020b). This research aims to find new RA susceptibility genes and see if they can be used as novel biomarkers for RA prognosis, diagnosis, and treatments. For this, we selected three GEO datasets and performed mRNA expression profiles to detect the DEGs among RA patients & the controls. Furthermore, significant modules and immune-related hub DEGs linked with RA were analyzed via computational approaches.

2. Methods

2.1 Retrieval of microarray datasets

The RA-related data sets were obtained from the GEO database hosted by the National Center for Biotechnology Information (NCBI). Based on the keywords "Rheumatoid arthritis" and "*Homo sapiens,*" we selected three microarray datasets GSE21959, GSE55457, and GSE77298, containing RA and control samples. The microarray dataset GSE21959 expression profiling was executed in the platform Agilent-014850 Whole Human Genome Microarray 4x44K G4112F (GPL4133), consists of six synovial fibroblasts tissue from RA patients and six from sex and age-matched healthy donors (Del Rey et al., 2010). The microarray data set GSE55457 was constructed on the platform (GPL96) Affymetrix Human Genome U133A Array [HG-U133A], containing 33 RA patients and 20 controls (Woetzel et al., 2014). The microarray data set GSE77298 was executed in the platform (GPL570) Affymetrix Human Genome U133 Plus 2.0 Array [HG-U133_Plus_2], contained 16 RA synovial biopsies and seven synovial biopsies from individuals without a joint disease (control) (Broeren et al., 2016). A total of 88 samples, including 55 RA and 33 control samples, were used in this study.

2.2 Microarray data pre-processing and DEGs screening

The microarray dataset was analyzed by utilizing the online tool GEO2R to detect DEGs in a GEO dataset by comparing RA and control groups (Davis & Meltzer, 2007). DEGs selection criteria were as follows: to decrease the FDR, utilized Benjamini and Hochberg method with adjusted P-value <0.05 (Reiner-Benaim, 2007) and log fold-change (FC) should be greater than 1 or less than 1. In RStudio software, the DEGs analysis results of the GSE21959, GSE55457, and GSE77298 datasets were displayed in a volcano plot map. Using a web application offered in VIB and Ghent University, the overlap of the three datasets DEGs were represented in a Venn diagram.

2.3 PPIs network and submodule construction

Overlapping DEGs was imported into the online database "Search Tool for the Retrieval of Interacting Genes/Proteins (STRING)" to visualize the PPIs (Szklarczyk et al., 2015). To show the PPIs between the statistically significant DEG encoded proteins with the cutoff ≥ 0.9 (interaction score with high confidence) was used. We then generate results of the PPIs into a TSV format and used Cytoscape software for module analysis (Shannon, 2003). Cytoscape's Molecular Complex Detection (MCODE) plugin is utilized in a complicated topology-based network to locate closely linked nodes. Therefore, we used MCODE to obtain significant modules in upregulated and downregulated PPIs networks with default parameters.

2.4 Identification of significant differentially expressed immune-related genes (DE-IRGs)

CytoNCA is a multiple centrality analysis that integrates biologic data with topologic data to distinguish definite nodes. The biologic data of both nodes & edges in the PPIs networks was imported in the CytoNCA plugin of Cytoscape software to measure betweenness, Local Average Connectivity-based Centrality (LAC), closeness Centrality, network Centrality, and eigen-vector Centrality with "Without weight" parameter (Tang, Li, Wang, Pan, & Wu, 2015). Using Cytohubba centrality analysis, Genes and their regulators were subjected to five different statistical methods (Maximal Clique Centrality (MCC), Maximum Neighborhood Component (MNC), Stress, Density of Maximum Neighborhood Component (DMNC), and Degree) to identify their connectivity (Chin et al., 2014). DE-centrality hub genes were identified based on the top 50 genes in each centrality method of CytoNCA and cytoHubba, respectively. The overlapping genes were then chosen as the DE-centrality hub genes. Furthermore, the DE-centrality hub genes and immune-related genes (IRGs) overlapped, which are downloaded from the Immport database to identify DE-IRGs. To detect the significant DE-IRGs, the selected DE-IRGs were further filtered by overlapping with the most significant modules of the PPIs network.

2.5 Functional enrichment analysis for DEGs in RA

DAVID is a web-based annotation tool designed for the functional enrichment of gene lists using an automated method (Huang et al., 2007). We performed DAVID to identify the KEGG pathways (Kanehisa, Goto, Furumichi, Tanabe, & Hirakawa, 2010) and GO-BP and GO-MF involved in DEGs of RA. Statistically significant functional annotation of DEGs selected with a P -value of <0.05 and a threshold count of two.

3. Results

3.1 DEGs on RA

DEGs were identified with $\text{logFC} \geq 1$ and P -value ≤ 0.05 for each dataset. They were visualized using a volcano plot analysis (Fig. 1). After screening the DEGs with its significant criteria, a set of 1720 DEGs were found from the GSE21959 dataset, which included 792 upregulated DEGs and 928

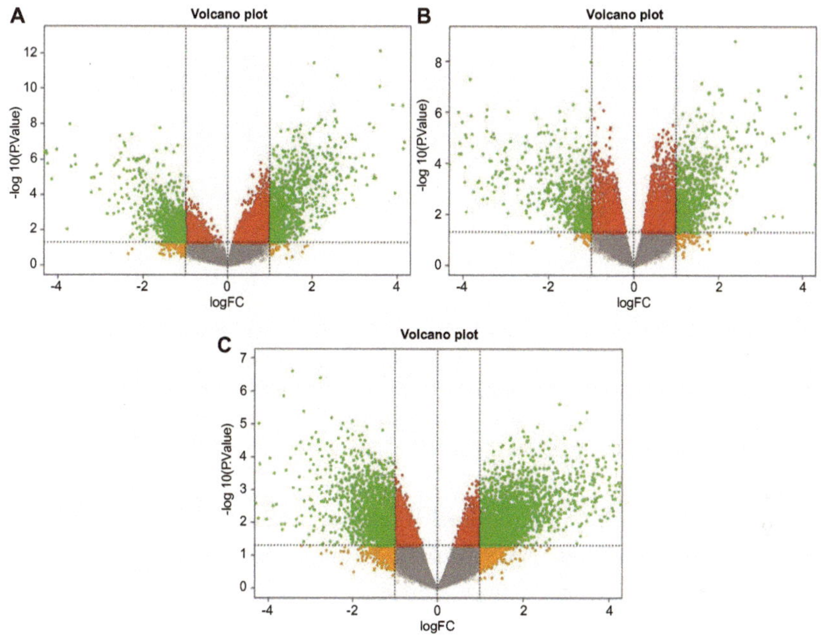

Fig. 1 Volcano plot of DEGs in RA samples from (A) GSE21959, (B) GSE55457, (C) GSE77298 mRNA expression profiling datasets. DEGs (Green dots) were selected in each dataset by $P < 0.05$ and $\log FC > 1$ or < -1. The X-axis shows the fold change in gene expression between RA samples and controls (logFC), where the Y-axis shows the statistical significance of the differences (*P*-value). Dots in the plot represent genes, each color dot represents different genes. Green dots represent significantly DEGs in RA including upregulated ($\log FC > 1$; $P < 0.05$) and downregulated ($\log FC < -1$; $P < 0.05$). Red dots represent no or less significant DEGs.

downregulated DEGs between RA samples and Normal samples. The GSE55457 dataset contained 1331 DEGs, including 820 upregulated DEGs and 511 downregulated DEGs. Furthermore, 3828 DEGs were identified from the GSE77298 dataset, including 2379 upregulated genes and 1449 downregulated genes. These genes shared 32 DEGs among the three datasets, with 19 upregulated and 13 downregulated genes. As an outcome, 331 upregulated and 275 downregulated DEGs were identified in at least two of the three data sets (Fig. 2).

3.2 PPIs network establishment and cluster analysis of RA

RA-specific 331 upregulated DEGs were uploaded to the STRING database, a total of 99 nodes & 135 edges with a combined score of greater than

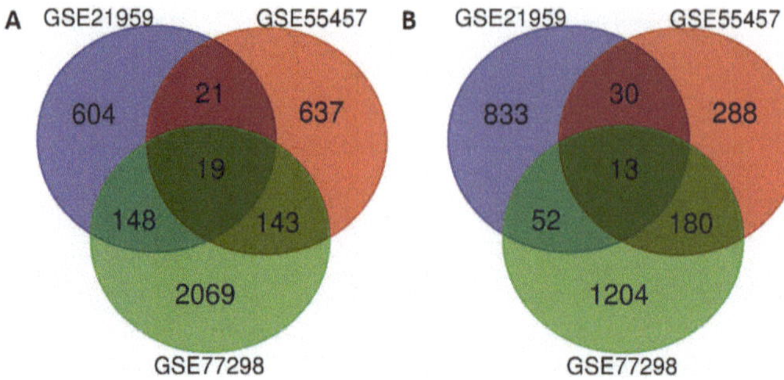

Fig. 2 A Venn diagram of (A) upregulated DEGs (B) downregulated DEGs with the cut-off criteria ($P < 0.05$ and logFC > 1 or < −1) in three datasets, respectively. The blue, red, and green circles of the Venn diagram contain DEGs from the GSE21959, GSE55457, and GSE77298 datasets. In total, 19 upregulated and 13 downregulated DEGs are common in three datasets.

or equal to 0.9. The PPI network of upregulated DEGs acquired from STRING was displayed in Cytoscape. Node widths were positively connected with connectivity degree, and edges width and transparency were positively correlated with the combined score (Fig. 3A). Similarly, 275 downregulated DEGs specific to RA were uploaded to the STRING database, yielding a total of 119 nodes & 380 edges with a combined score of greater than 0.9. Visualized the downregulated DEGs PPI network in Cytoscape, in which the node size correlates to the degree and width, and transparency of edges associated with the combined score (Fig. 3B). The PPI network did not represent the isolated nodes. We carried out the MCODE analysis on both the upregulated and downregulated PPI network to determine the key modules from the complex network. In each PPI network, we selected the top two modules. Module 1 (Fig. 4A) of upregulated PPI network containing 8 genes and 28 edges with the MCODE score of 8. Module 2 (Fig. 4B) of upregulated PPI network containing five genes and nine edges with the MCODE score 4.5. Module 1 (Fig. 4C) in the downregulated PPI contains 16 genes and 114 edges with an MCODE score of 15.2, whereas Module 2 (Fig. 4D) contains 9 genes and 114 edges with an MCODE score of 9. These four modules are interpreted as the most important based on their significant MCODE scores and obtained nodes.

3.3 Immune-related hub gene identification

We further curated the hub genes from an overlapped analysis for the five distinct CytoNCA and cytoHubba plugin approaches in the upregulated and

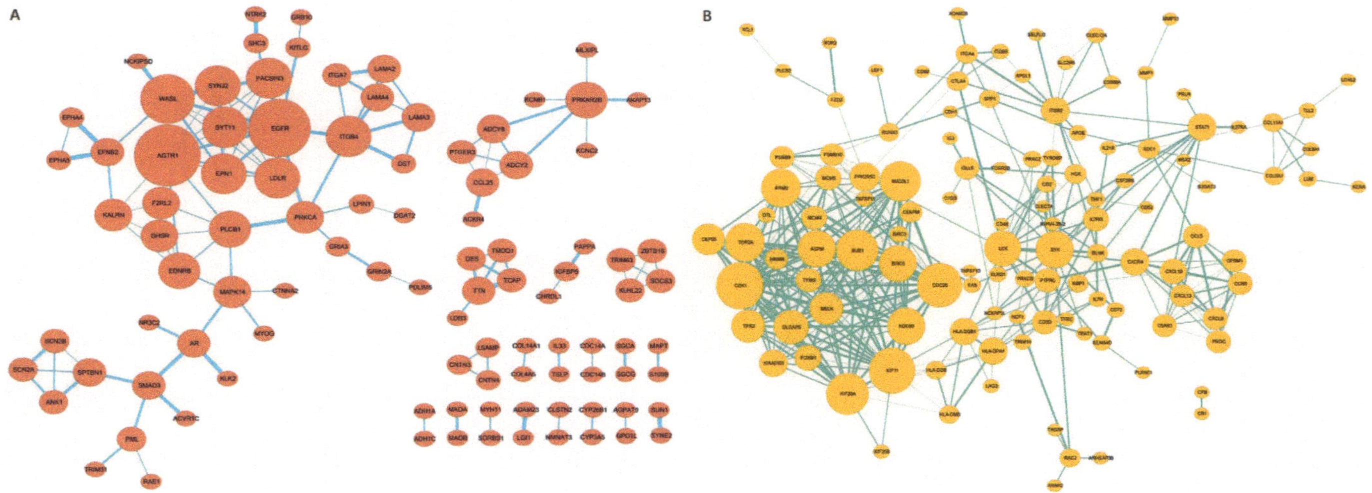

Fig. 3 The protein–protein interaction (PPI) network of **(A)** upregulated DEGs and (B) downregulated DEGs in RA from at least 2 of those 3 data sets were determined using STRING and Cytoscape. The nodes and edges are represented by spheres and lines, respectively. Spheres contain gene symbols, and the size of the spheres represents the degree score of the particular gene (a larger size indicates a higher degree). Lines width and transparency represent the combined score of DEGs (a wider or thick line indicates a higher combined score).

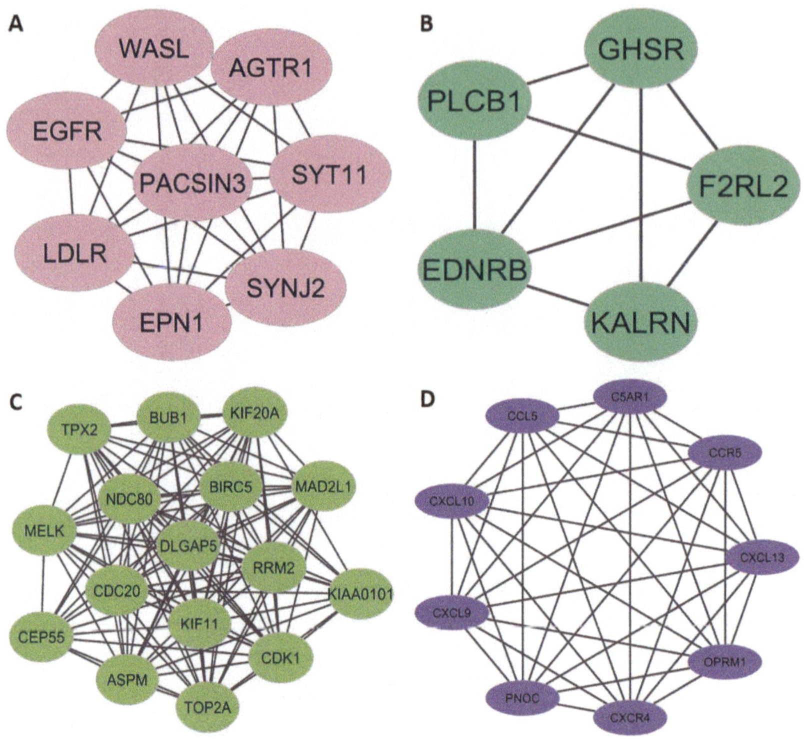

Fig. 4 The two most significant modules (module 1(A), module 2(B)) of DEGs in the upregulated PPI network. (A) Module 1 (MCODE score 8) upregulated genes denoted in pink nodes; (B) Module 2 (MCODE score 4.5) upregulated genes denoted in green nodes and their edges denoted in gray lines. The two most important DEG modules (module 1 (C), module 2 (D)) in the downregulated PPI network. (C) Module 1 (MCODE score 15.2) downregulated genes denoted in light green nodes; (D) **M**odule 2 (MCODE score 9) downregulated genes denoted in purple nodes and their edges denoted in gray lines.

downregulated PPI network. In the upregulated PPI network, we first selected the top 50 DEGs from five CytoNCA approaches (Table 1 in Supplementary Material in the online version at https://doi.org/10.1016/bs.apcsb.2021.11.006) and discovered that 26 genes were common. Second, we chose the top 50 DEGs in upregulated PPI network with 5 approaches of cytoHubba (Table 2 in Supplementary Material in the online version at https://doi.org/10.1016/bs.apcsb.2021.11.006) and determined that 21 DEGs are common. Furthermore, the genes from the CytoNCA and cytoHubba approaches overlapped, resulting in 16 DE-centrality hub genes of upregulated PPI network (Table 3 in Supplementary Material in the online version at https://doi.org/10.1016/bs.apcsb.2021.11.006). Immune-related genes from *H. sapiens* were obtained from the Immport database. This database contains 4677 manually curated genes overlapped with 16 hub genes obtained from an upregulated PPI network, 12 of the 16 genes are immune-related. As module 1 and module 2 are considered to have the most significant module in upregulated DEGs of RA, we discovered five immune-associated final hub genes Angiotensin II receptor type 1 (*AGTR1*), epidermal growth factor receptor (*EGFR*), Endothelin Receptor Type B (*EDNRB*), *Kalirin* RhoGEF Kinase (*KALRN*), and Growth Hormone Secretagogue Receptor (*GHSR*) by overlapping the 16 differentially expressed immune-related genes (DE-IRGs) with module 1/module 2 of the upregulated PPI network. The overall selection methods of immune-related hub genes from upregulated genes are illustrated in Fig. 5A. Similarly, we curated the DE-centrality hub genes from the downregulated PPI network by choosing the top 50 genes of 5 distinct CytoNCA (Table 4 in Supplementary Material in the online version at https://doi.org/10.1016/bs.apcsb.2021.11.006), and cytoHubba (Table 5 in Supplementary Material in the online version at https://doi.org/10.1016/bs.apcsb.2021.11.006) approaches. Among the downregulated PPI network, we identified that 19 genes are identical in 5 CytoNCA methods and 16 identical in 5 cytoHubba methods. Further we overlapped the CytoNCA and cytoHubba of downregulated DEGs and found 12 common DEGs (Table 6 in Supplementary Material in the online version at https://doi.org/10.1016/bs.apcsb.2021.11.006). To identify DE-IRGs, we merged the immune-related gene list from the Immport database with the DE downregulated DEGs from the CytoNCA and cytoHubba methods. There are 7 DE-IRGs present in the downregulated PPI network of RA. Only four of the seven DE-IRGs are found in the two most important modules of the downregulated PPI network. These four DE-IRGs C-X-C chemokine receptor type 4 (*CXCR4*), Cyclin-dependent kinase 1 (*CDK1*), budding uninhibited by benzimidazole 1 (*BUB1*), and baculoviral inhibitor of apoptosis repeat-containing 5 (*BIRC5*) are considered to be the final hub genes of downregulated DEGs. Fig. 5B depicts the overall selection process of downregulated hub genes. We discovered nine hub genes (*AGTR1*, *EGFR*, *EDNRB*, *KALRN*, *GHSR*, *CXCR4*, *CDK1*, *BUB1*, and *BIRC5*) in total immune-related genes and also found in the topmost clusters of the RA network (Table 1).

3.4 GO analysis of DEGs in RA

GO enrichment analysis was executed to determine the significant functions of 331 upregulated and 275 downregulated DEGs. The top 15 GO

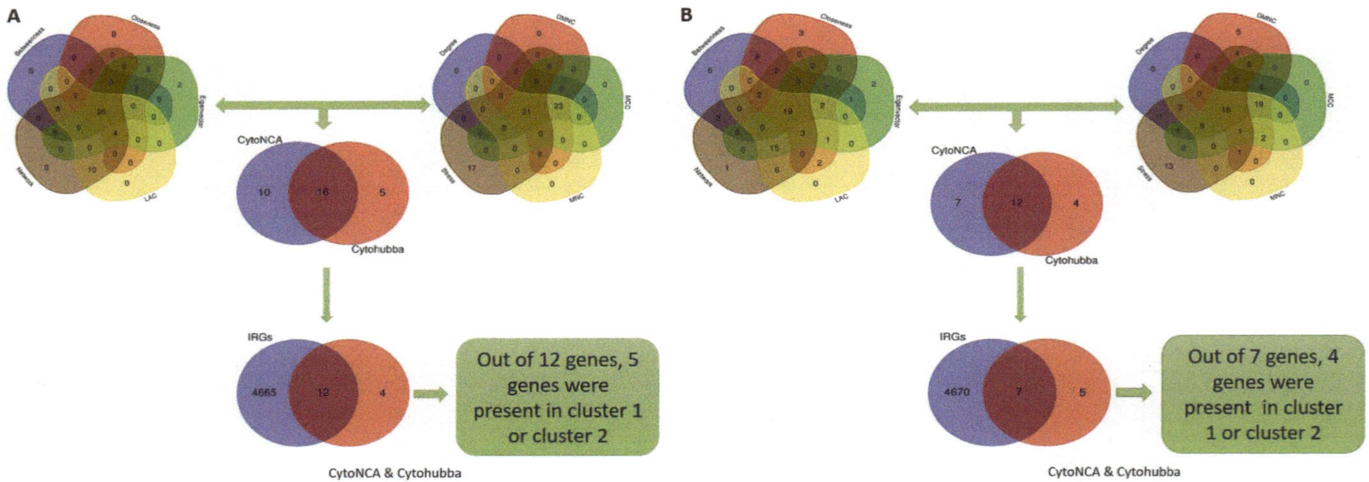

Fig. 5 (A) A right-side venn diagram displayed 21 overlapped genes from the top 50 upregulated DEGs of 5 distinct cytoHubba methods (stress, Degree, DMNC, MCC, MNC). A Venn diagram on the left side showed 26 overlapped genes from the top 50 DEGs of the upregulated PPI network using 5 different cytoNCA techniques (betweenness, closeness, eigenvector, LAC, network). Furthermore, by overlapping 21 CytoHubba and 26 cytoNCA genes and found 16 common genes (middle venn-diagram). These 16 genes were then overlapped with immune-related genes (last-venn diagram) and found 12 genes. (B) In downregulated PPI network, 5 different methods of cytoHubba intersecting 16 genes are displayed on the right-side venn diagram. Similarly, 5 different methods of cytoNCA intersecting 19 genes are displayed on the left-side venn diagram. A middle venn-diagram intersecting 12 genes from 10 different methods of CytoNCA and cytohubba. Furthermore, only 7 out of 12 genes intersected CytoNCA & cytoHubba, and immune-related genes (last venn-diagram). Among these 7 genes, selected only genes which are present in the module 1 and module 2 of downregulated genes.

Table 1 Centrality analysis results of hub genes by CytoNCA and cytoHubba.

	MCODE	Centrality analysis by CytoNCA					Centrality analysis by CytoHubba				
Gene	MCODE score	Betweenness	Closeness	Eigenvector	LAC	Network	Degree	DMNC	MCC	MNC	Stress
Upregulated hub genes											
AGTR1	8	460.7	0.018893387	0.39268887	5	10.5	12	0.768339	5088	7	926
EGFR[a]	8	642.4	0.018878829	0.35057327	4	6.766667	11	0.768339	5044	7	924
EDNRB[a]	4.5	324.1	0.018882466	0.12228116	3.333333	4.6	6	0.475492	50	6	590
KALRN	4.5	115.8	0.018813591	0.111419834	2.4	3	5	0.568394	25	4	138
GHSR	4.5	4.1	0.01882805	0.118020885	3.6	4.5	5	0.583436	48	5	20
Downregulated hub genes											
CXCR4	9	2133.3938	0.2030981	0.001704927	5.636364	8.288889	11	0.52601	40,326	11	11,774
CDK1	16	681.20807	0.19124797	0.2747297	12.4	23.40274	25	0.651379	6.75E+09	25	5274
BUB1	16	244.894	0.18730159	0.25792152	13.14286	19.40325	21	0.780021	6.75E+09	21	2442
BIRC5	16	131.1721	0.1864139	0.2375682	13.22222	16.38926	18	0.874145	6.75E+09	18	1726

[a]Denotes those genes are present in all three datasets and others are present in any of the two datasets.

biological processes in upregulated DEGs are "cell adhesion (GO:0007155)" "MAPK cascade (GO:0000165)" "chemical synaptic transmission (GO: 0007268)" "cytoskeleton organization (GO:0007010)" "muscle contraction (GO:0006936)" "axon guidance (GO:0007411)" "peptidyl-tyrosine phosphorylation (GO:0018108)" "response to wounding (GO:0009611)" "memory (GO:0007613)" "learning or memory (GO:0007611)" "negative regulation of osteoblast differentiation (GO:0045668)" "glutamate signaling pathway (GO:0007215)" "synapse assembly regulation (GO:0051963)" "cardiac muscle cell development (GO:0055013)" and "skeletal muscle myosin thick filament assembly (GO:0030241)" (Fig. 6D). The top 15 GO molecular functions of upregulated DEGs are "protein homodimerization activity (GO:0042803)" "actin-binding membrane (GO:0003779)" "structural molecule activity (GO:0005198)" "actin filament binding (GO:0051015)" "activity of Ras guanyl-nucleotide exchange factor (GO:0005085)" "ion channel binding (GO:0044325)" "structural constituent of the cytoskeleton (GO:0005200)" "activity of growth factor (GO:0008083)" "activity of protein tyrosine kinase (GO:0004713)" "activity of metallopeptidase (GO: 0008237)" "cytoskeletal protein binding (GO:0008092)" "activity of phos-phatidylinositol-4,5-bisphosphate 3-kinase (GO:0046934)" "transforming growth factor-beta receptor binding (GO:0005160)" "activity of GTPase regulator (GO:0030695)" and "activity of thiamine transmembrane transporter (GO:0015234)" (Fig. 6E).

The top 15 GO biological processes of downregulated DEGs with the lowest *P*-value are "signal transduction (GO:0007165)" "immune response (GO:0006955)" "innate immune response (GO:0045087)" "inflammatory response (GO:0006954)" "cell adhesion (GO:0007155)" "cell surface receptor signaling pathway(GO:0007166)" "immune response regulation (GO:0050776)" "adaptive immune response (GO:0002250)" "B cell receptor signaling pathway (GO:0050853)" "T cell receptor signaling pathway (GO: 0050852)" "complement activation, classical pathway (GO:0006958)" "chemokine-mediated signaling pathway (GO:0070098)" and "positive regulation of T cell proliferation (GO:0042102)" "phagocytosis, recognition (GO:0006910)" and "positive regulation of B cell activation (GO: 0050871)" were derived from DAVID GO annotation (Fig. 6A). The top 15 GO molecular functions of downregulated DEGs are "protein homodimerization activity (GO:0042803)" "binding of receptor (GO:0005102)" "antigen-binding activity (GO:0003823)" "protein kinase binding process (GO:0019901)" "receptor activity (GO:0004852)" "activity of transmembrane signaling receptor (GO:0004888)" "activity of serine-type endopeptidase (GO:0004252)" "immunoglobulin receptor binding activity

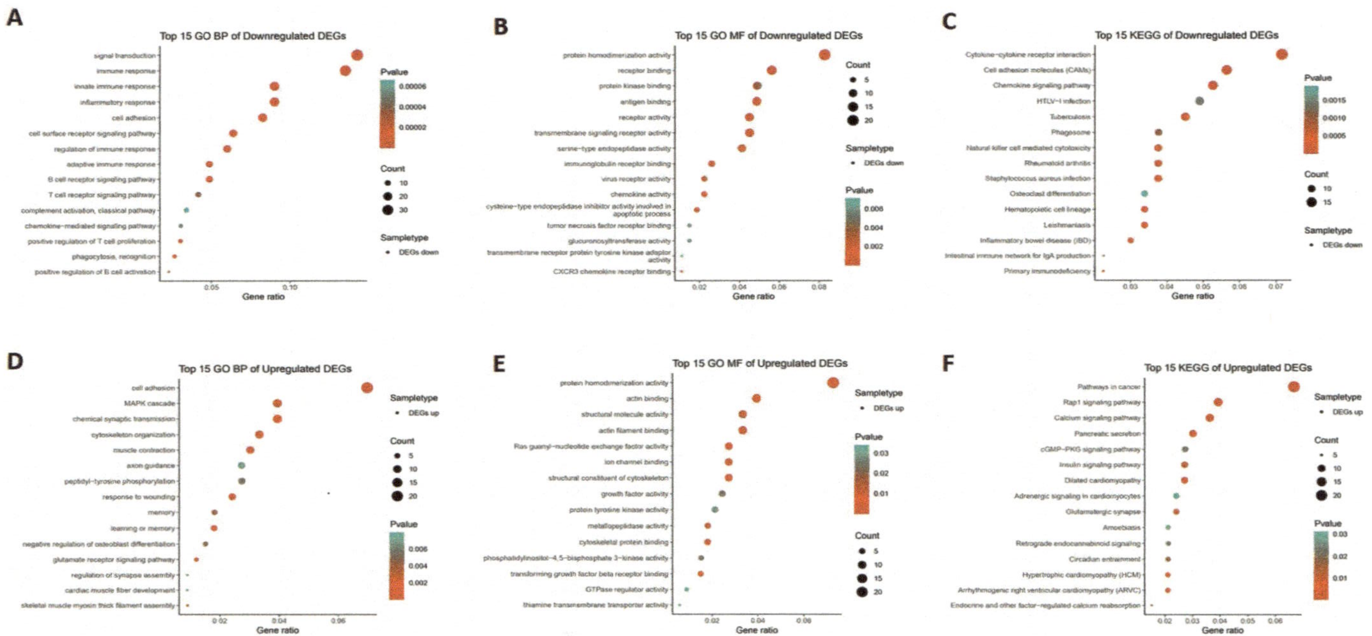

Fig. 6 Enrichment analysis revealed the biological process, molecular function, and KEGG pathways of DEGs (including upregulated and downregulated genes) from the RA dataset. Sphere (Count) size reflects the number of genes involved in the process and spheres color ranges from red to cyan reflects the *P*-value. (A) Top 15 GO-biological pathways of upregulated DEGs in RA; (B) Top 15 GO-molecular function of upregulated DEGs in RA; (C) Top 15 GO-KEGG pathways of upregulated DEGs in RA; (D) Top 15 GO-biological pathways of downregulated DEGs in RA; (E) Top 15 GO-molecular function of downregulated DEGs in RA; (F) Top 15 GO-KEGG pathways of downregulated DEGs in RA.

(GO:0034987)" "virus receptor activity (GO:0001618)" "chemokine activity (GO:0008009)" "the activity of cysteine-type endopeptidase inhibitor have participated in the apoptotic process (GO:0043027)" "tumor necrosis factor receptor binding (GO:0005164)" "glucuronosyltransferase activity (GO:0015020)" "activity of transmembrane receptor protein tyrosine kinase adaptor (GO:0005068)" and "*CXCR3* chemokine receptor binding (GO: 0048248)" (Fig. 6B).

3.5 KEGG analysis for DEGs in RA

In upregulated DEGs of RA, 15 most involved KEGG pathways analysis were as follows: pathways in cancer (has05200), Rap 1 signaling pathway (hsa04015), Dilated cardiomyopathy (hsa05414), Calcium signaling pathway (hsa04020), Pancreatic secretion (hsa04972), cGMP-PKG signaling pathway (hsa04022), Insulin signaling pathway (hsa04910), Adrenergic signaling in cardiomyocytes (hsa04261), Glutamatergic synapse (hsa04724), Amoebiasis (hsa05146), Retrograde endocannabinoid signaling (hsa04723), Arrhyth-mogenic right ventricular cardiomyopathy (ARVC) (hsa05412), Circadian entrainment (hsa04713), Hypertrophic cardiomyopathy (HCM) (hsa05410), and Endocrine & other factor-regulated calcium reabsorption (hsa04961) (Fig. 6F). The following are the top 15 KEGG pathways involved in downregulated DEGs in RA: Cytokine-cytokine receptor interaction (hsa04060), Cell adhesion molecules (CAMs) (hsa04514), Chemokine signaling pathway (hsa04062), HTLV-I infection (hsa05166), Tuberculosis (hsa05152), Phagosome (hsa04145), Natural killer cell-mediated cytotoxicity (hsa04650), *Staphylococcus aureus* infection (hsa05150), Osteoclast differentiation (hsa04380), Hematopoietic cell lineage pathway (hsa04640), Leishmaniasis (hsa05140), Inflammatory bowel disease (IBD) (hsa05321), Rheumatoid arthritis (hsa05323), Intestinal immune network for IgA production (hsa04672) and Primary immunodeficiency (hsa05340). The final upregulated hub genes mainly were engaged in calcium signaling pathways, neuroactive ligand-receptor interactions, cancer pathways, and the cGMP-PKG signaling pathway. The final downregulated hub genes were primarily associated with pathways like protein kinase & protein threonine/serine kinase activity (Fig. 6C).

4. Discussion

RA is an autoimmune joint disease that can damage cartilage and bones and cause disability. An early diagnosis is essential for successful treatment in individuals with well-defined risk factors for poor outcomes. So far, numerous RA susceptibility genes (such as *HLADRB1*, *IL2RA*, *DKK1*, *GRZB*, *MMP9*, *SPAG16*, and *FOXO3*) have also been linked to severity. However, no reliable and consistent genetic biomarkers have been discovered (Smolen et al., 2018). Molecular study of RA will enhance the diagnosis and therapy of the disease. Since high-throughput sequencing can give gene expression levels in the human genome simultaneously, it has been frequently utilized to identify treatment targets for RA (Chen, Chang, Hsu, Chen, & Kuo, 2017). Many studies have employed the DNA microarray method to find new biomarkers for early disease diagnosis, treatment monitoring, and disease progression. Microarray and gene expression profiles data have also aided us in better understanding the molecular pathways of biological systems (Gang et al., 2017; Udhaya Kumar, Madhana Priya, et al., 2021; Udhaya Kumar, Saleem et al., 2021). Several research groups have utilized computational strategies to explore the significant hub genes from RA (Cheng, Chen, Wu, & Du, 2021; Wu et al., 2021). Since no reliable immune-related biomarkers in RA. Our study aimed to assess significant immune-related hub genes with similar functions that are expressed aberrantly in RA compared to normal samples. We extracted the mRNA expression profiles of GSE21959, GSE55457, and GSE77298 downloaded from the GEO database and identified 331 upregulated and 275 down-regulated overlap DEGs in at least two or three datasets between RA and normal control. The DEGs upregulated PPI and downregulated PPI network, which consisting of 99 nodes with 135 edges and 119 nodes with 380 edges, were built using STRING and Cytoscape. We optimized the significant DEGs using network parameters and MCODE analysis. Furthermore, the DEGs upregulated PPI and downregulated PPI network was analyzed to identify DE-central hub genes using CytoNCA and cytoHubba. Then, overlapping genes of CytoNCA and cytoHubba were overlapped with IRGs and we identified the DE-IRGs. The obtained DE-IRGs are overlapped with the top two modules of MCODE analysis to identify the most significant DE-IRGs. Overall, we found the nine most significant DE-IRGs, including four of which are overexpressed in RA (*CXCR4*, *CDK1*, *BUB1*, and *BIRC5*) & five of which are lower expressed in RA (*AGTR1*, *EGFR*, *EDNRB*, *KALRN*, and *GHSR*) as compared with the normal group and, they are associated with RA progression. We also performed GO, BP, MF, and KEGG pathway analysis for upregulated and downregulated DEGs to understand DEG linkages and their functions. The BP of upregulated genes are primarily involved in cell adhesion, MAPK cascade, chemical synaptic transmission, cytoskeleton organization, and muscle contraction, where downregulated genes are involved in signal transduction, immune response, innate immune response, inflammatory response, and cell adhesion. The MF of upregulated genes is mainly involved in protein actin-binding, homodimerization activity, structural molecule activity, Ras guanyl-nucleotide exchange factor, and actin filament binding activity, where downregulated genes are involved in protein homodimerization activity, receptor binding, protein kinase binding, antigen-binding, and receptor activity. Dysregulated of these BP and MF may lead to the development of RA. According to the KEGG study, dysregulated pathways such as calcium signaling pathways, neuroactive ligand-receptor interactions, cancer pathways, the cGMP-PKG signaling, protein kinase, and protein threonine/serine kinase activity may involve in the progression of RA.

Our study showed that *CXCR4*, *CDK1*, *BUB1*, and *BIRC5* genes are overexpressed in RA when compared to control. *CXCR4* is a chemokine receptor expressed in various leukocytes and recognized through the CXC chemokine ligand 12 (*CXCL12*). *CXCR4* has a role in various immune processes, including cell apoptosis, proliferation, survival, differentiation, and chemotaxis. Overexpressed *CXCR4* in systemic lupus erythematosus (SLE) B cells was correlated with kidney and disease activity. Overactivity of the *JAK/STAT* and *PI3K*/Akt pathways might play a role in *CXCR4* dysregulation in SLE patients (Zhao et al., 2017). According to a previous report by Li et al. (2013), Celastrol reduced hypoxia-induced *CXCR4* expression in RA-FLSs on protein and mRNA levels. Celastrol prevents hypoxia-induced migration and invasion by suppressing HIF-1-mediated *CXCR4* expression in fibroblast-like synoviocytes (FLSs) under hypoxic circumstances. They suggested that celastrol may be used as an alternative for the treatment of RA (Li et al., 2013). Kanbe et al., 2004 reported that CXCR4 overexpression increased MMP-9 and MMP-13 levels by 3.1- and 2.1-fold, respectively. *CXCR4*, as well as *MMP-9* and *MMP-13*, were found to be involved in OA and RA. Using a retrovirus transfection technique and mutant *CXCR4*, they can prevent the release of *MMP-13* and *MMP-9* in chondrocytes despite the low transfection efficiency (Kanbe et al., 2004). Our findings suggest that overexpressed CXCR4 may be involved in the dysregulated chemokine-mediated signaling pathways that lead to RA development. *CDK1* gene encodes the *CDK1* protein, which regulates the cell cycle. In RA disease, *CDK1* contributes to the excess of synovial fibroblasts, resulting in a hyperplastic pannus. *CDK1* expression profiles in the Egyptian population were examined using quantitative PCR (qRT-PCR) and found to be 3.29-fold higher in RA patients than healthy controls (Fattah et al., 2020). Wu et al. (2015) reported *CDK1* overexpression might lead to excessively showed type I IFN signaling, where inhibition of *CDK1* may be utilized to downregulate type I of IFN pathway signal in SLE (Wu et al., 2015). As previously stated, RA patients had a high level of *CDK1* gene expression, which is consistent with the earlier findings. This suggests that *CDK1* overexpression is linked to the pathophysiology of RA. *BUB1*, a kinetochore-bound protein kinase, is required for the spindle checkpoint signaling as well as proper chromosomal alignment during mitosis (Grabsch et al., 2003). The aberrant *BUB1 gene* expression was detected in many cancers and cancer cell lines (Fujibayashi et al., 2020; Pinto et al., 2008; Wang et al., 2015). *BUB1* gene expression was found to be upregulated in RA-FLSs prior to Wantong Jingu Tablet (WJT) treatment. WJT employed its pro-apoptosis and anti-proliferation effects, perhaps through *BUB1* suppressing expression in RA-FLSs (Li, Qi, & Li, 2020). Therefore, the therapeutic target of the *BUB1* gene may be a promising strategy for RA recovery. BIRC5/Survivin is considered to have multifunctional protein in cellular and molecular networks, cell cycle progression, apoptosis, developmental morphogenesis, and homeostasis encoded by *BIRC5*. Survivin upregulation worsens autoimmune disorders (AIDs) such as myasthenia gravis and multiple sclerosis via lowering the stimulation of survivin-specific cytotoxic Thymus cells (Pahlavan et al., 2019). Sharief, Noori, Douglas, and Semra (2002) found that *ex vivo* survivin expression triggered T lymphocytes that correlate with cellular resistance to apoptosis and multiple sclerosis (MS) clinical activity. They reported that the diversity of apoptotic defects in MS suggests that survivin overexpression is unlikely to be a main pathogenic cause (Sharief et al., 2002). Survivin was overexpressed in nearly all cancer cells, and increase in survivin expression has also been linked to poor prognosis and treatment resistance among individuals with cancer (Soleimanpour & Babaei, 2015). Prior studies suggested that survivin might be formed from FLS, and this survivin could have a role in joint damage among RA patients (Ahn et al., 2010; Bokarewa, Lindblad, Bokarew, & Tarkowski, 2005). *AGTR1*, *EGFR*, *EDNRB*, *KALRN*, and *GHSR* genes are less expressed in RA than in control. These genes are negatively correlated with RA progression since they have decreased expression in RA as compared with normal. In tumors of the pancreas, ovary, prostate, and breast. *AGTR1* was associated with cell growth and proliferation through numerous intracellular pathways like Janus kinase/signal transducer & activator of transcription (*JAK/STAT3*), mitogen-activated protein kinase (*MAPK*), and *PI3K*/Akt. (Oh et al., 2016). *AGTR1* activation increases lipid desaturation via SCD1 activation, which reduces endoplasmic reticulum stress in multicellular spheroids (MCS). This mechanism explains the link between high *AGTR1* levels and poor clinical outcomes in epithelial ovarian cancer (EOC) patients (Zhang et al., 2019). *AGTR1* was found to be downregulated in thyroid cancer when compared to normal thyroid tissue, with benign tumors having higher relative expression than malignant tumors (Prasad et al., 2008). Similarly, our work revealed downregulated *AGTR1* in RA compared to normal, which is negatively associated with blood vessel regulation. *EGFR* is generally dysregulated in many human cancers and could be used as a therapeutic target. *EGFR* dysregulation has been linked to chemoresistance and a poor prognosis in various tumor types (Dai et al., 2005). Downregulation of biological marker expression levels, such as *Ki-67*, *MCM-2*, and *EGFR*, was linked to worsening of laryngeal squamous cell carcinoma (LSCC) (Vivatvakin et al., 2021). The downregulation of *EGFR* in high-grade malignant cells has been explained by various aspects. Hypoxia stimulates the activation of prolyl hydroxylase 3 (*PHD3*), which targets the hypoxia-inducible factors (HIFs), resulting in *EGFR* internalization through endocytosis which leads to *EGFR* downregulation (Garvalov et al., 2014). *EDNRB* was discovered to be downregulated in primary nasopharyngeal carcinoma (NPC) tissues and NPC cell lines, and *EDNRB* methylation levels in NPC biopsies were greater than in chronic nasopharyngitis biopsies. They also suggested that aberrant promoter methylation and gene deletion might impact *EDNRB* expression and have a role in NPC development (Zhou et al., 2007). Another study discovered that decreased *EDNRB* mRNA transcripts were associated with abnormal methylation in lung adenocarcinoma (Chen et al., 2006). MacLean (2018) found that EDNRB downregulation leads to B cell damage lymphopoiesis in Trisomy 21 pluripotent stem cells (MacLean, 2018). The protein kalirin is encoded by *KALRN*. Kalirin, a synaptic regulator, it is important in synaptic plasticity and the creation of dendritic arbors and spines. Autism spectrum disorder, Alzheimer's disease (AD), schizophrenia, addiction, and intellectual difficulties have all been associated with dysregulation of the *KALRN* gene. Downregulation of *KALRN* transcript expression was detected in a small cohort of Alzheimer's disease affected patients, revealing a link between disease progression and kalirin dysregulation (Parnell et al., 2021). Reduced kalirin levels have been linked to increased activity of inducible nitric oxide synthetase (iNOS) & polymorphisms of the iNOS promoter region, resulted in severity of allergic asthma. Downregulated *KALRN* linked with potentially functional variations in asthma patients (DeWan et al., 2012). In addition, *KALRN* was also found to be downregulated in both dilated cardiomyopathy (DCM) and ischemic cardiomyopathy (ICM) patients (Molina-Navarro et al., 2013). *GHSR* also recognized as the ghrelin receptor, regulates food intake and energy metabolism as well as release different types of hormones such as prolactin, growth hormone, cortisol and adrenocorticotropic hormone. *GHSR* also involved in other process such as glucose & lipid metabolism, immunological function control, neuronal & cardiovascular cell protection, gastrointestinal motility & secretion and pancreatic function (Yin, Li, & Zhang, 2014). Ghrelin selectively inhibits the expression of proinflammatory anorectic cytokines like IL-1, TNF and IL-6, through *GHSR* (Dixit et al., 2004) *GHSR* is abnormally expressed in a variety of malignant carcinomas and is implicated in a number of cancer development processes, particularly metastasis and proliferation (Lin & Hsiao, 2017). In our case, *GHSR* is downregulated in RA and plays a role in dysregulated inflammatory response function.

Moreover, we conducted an interrelationship investigation on the discovered nine final DE-IRGs. GeneMANIA was used to map the network, and Cytoscape displayed it (Fig. 7). The nine central nodes reflect the discovered hub genes that were highly linked with 20 other genes in terms of physical connections, co-expression, predictions, co-localization, and genetic interactions. *MKI67, TPX2, TOP2A, SPC25*, and *CCNA2* were the top five genes having the highest associations with hub genes. *TOP2A* has physically linked to *CDK1* and has co-localization with *CDK1* and *BIRC5*. Physical interactions of *SPC25* with *CDK1, BUB1*, and *BIRC5*. Further functional analysis indicated that these proteins correlated mostly with mitotic nuclear division (FDR=7.5e-15). *CDK1, BUB1, BIRC5, MKI67, TPX2, DLGAP5, UBE2C, KIF2C, KIF18B, NEK2, CDCA8, CENPF, KIF4A*, and *BUB1*, are involved in mitotic nuclear division. These hub genes were also linked to the chromosomal area, chromosome segregation, condensed chromosome, segregation of sister chromatid, kinetochore, segregation of nuclear chromosome, centromeric region, spindle, mitotic sister chromatid segregation, and microtubule. According to our findings, overexpressed genes in RA such as *CXCR4, CDK1, BUB1*, and *BIRC5* are mostly involved in mitotic processes. According to our findings, overexpressed genes in RA such as *CXCR4, CDK1, BUB1*, and *BIRC5* are primarily engaged in the mitotic nuclear division, whereas lower expressed genes in RA such as *AGTR1, EGFR, EDNRB, KALRN*, and *GHSR* are primarily engaged in the adverse regulation of blood vessel diameter. Our findings suggest that hub genes involved in the development of RA may play a role in mitotic nuclear division and the negative regulation of blood vessel diameter functions.

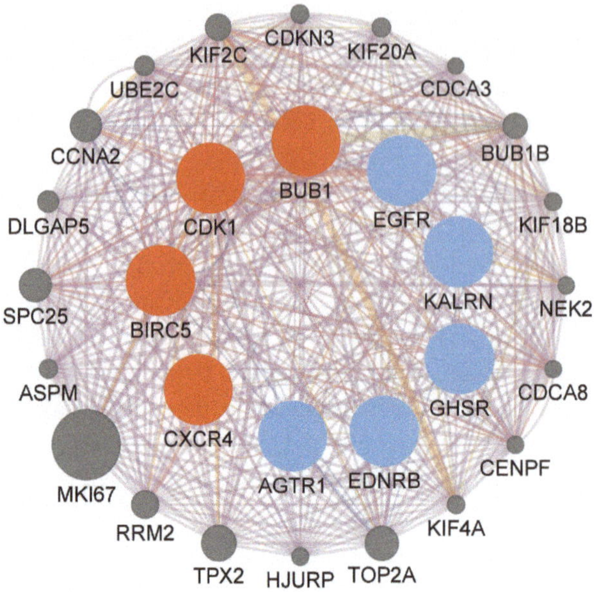

Fig. 7 Interrelationship analysis network constructed with GeneMANIA for final immune-related hub genes which were visualized in Cytoscape. The blue color node indicates low expressed gene in RA, while the orange color node indicates over-expressed genes in RA. The lavender color line depicts the coexpression, the red line depicts the physical interaction, the orange line depicts the predicted interaction, the blue line depicts the colocalization, yellow line depicts the shared protein. The genes were involved in the mitotic nuclear division and regulation of blood vessel diameter functions.

5. Conclusion

Numerous research has shown that these hub genes are linked to the advancement of various forms of cancer; however, relatively few studies have been published concerning these hub genes role in immune-related disease and RA. Our investigation discovered nine DE-IRGs (*CXCR4*, *CDK1*, *BUB1*, *BIRC5*, *AGTR1*, *EGFR*, *EDNRB*, *KALRN*, and *GHSR*) between RA patients and the control group using three different RA microarray datasets. Outcomes displayed that these nine DE-IRGs may have a significant impact on the development of RA. As a result, the findings of our study may give helpful knowledge and suggestions for future research in RA diagnosis and therapy. However, a series of tests should be conducted to verify these predictions in the future using molecular functional assays. In many cases, these findings may give insights into RA's possible biomarkers and molecular processes.

Acknowledgments

The authors would like to take this opportunity to thank the management of Vellore Institute of Technology (VIT), Vellore, India, for providing the necessary facilities and encouragement to carry out this work.

Conflict of interest

The authors have declared that no conflicts of interest exist.

Author contributions

A.B., S.U.K., and C.G.P.D. were involved in the design of the study. S.U.K. and A.B. were involved in the data collection and conducted the experiment. A.B., S.U.K., and C.G.P.D. were involved in the acquisition, analysis, and interpretation of the results. A.B. and S.U.K. drafted the manuscript. C.G.P.D. supervised the entire study and were involved in study design, the acquisition, analysis, understanding of the data, and critically reviewed the manuscript. All authors edited and approved the submitted version of the article.

Role of funding source

No funding agency was involved in the present study.

References

Ahn, J. K., Oh, J.-M., Lee, J., Bae, E.-K., Ahn, K.-S., Cha, H.-S., et al. (2010). Increased extracellular Survivin in the synovial fluid of rheumatoid arthritis patients: Fibroblast-like Synoviocytes as a potential source of extracellular Survivin. *Inflammation, 33*(6), 381–388. https://doi.org/10.1007/s10753-010-9196-1.

Alamanos, Y., & Drosos, A. (2005). Epidemiology of adult rheumatoid arthritis. *Autoimmunity Reviews, 4*(3), 130–136. https://doi.org/10.1016/j.autrev.2004.09.002.

Arora, A., Mahajan, A., Spurden, D., Boyd, H., & Porter, D. (2013). Long-term drug survival of TNF inhibitor therapy in RA patients: A systematic review of European National Drug Registers. *International Journal of Rheumatology, 2013*, 1–9. https://doi.org/10.1155/2013/764518.

Bokarewa, M., Lindblad, S., Bokarew, D., & Tarkowski, A. (2005). Balance between survivin, a key member of the apoptosis inhibitor family, and its specific antibodies determines erosivity in rheumatoid arthritis. *Arthritis Research & Therapy*, 7(2), R349. https://doi.org/10.1186/ar1498.

Broeren, M. G. A., de Vries, M., Bennink, M. B., Arntz, O. J., Blom, A. B., Koenders, M. I., et al. (2016). Disease-regulated gene therapy with anti-inflammatory Interleukin-10 under the control of the CXCL10 promoter for the treatment of rheumatoid arthritis. *Human Gene Therapy, 27*(3), 244–254. https://doi.org/10.1089/hum.2015.127.

Chen, Y.-J., Chang, W.-A., Hsu, Y.-L., Chen, C.-H., & Kuo, P.-L. (2017). Deduction of novel genes potentially involved in osteoblasts of rheumatoid arthritis using next-generation sequencing and Bioinformatic approaches. *International Journal of Molecular Sciences, 18*(11), 2396. https://doi.org/10.3390/ijms18112396.

Chen, S.-C., Lin, C.-Y., Chen, Y.-H., Fang, H.-Y., Cheng, C.-Y., Chang, C.-W., et al. (2006). Aberrant promoter methylation of EDNRB in lung cancer in Taiwan. *Oncology Reports.*. https://doi.org/10.3892/or.15.1.167.

Cheng, Q., Chen, X., Wu, H., & Du, Y. (2021). Three hematologic/immune system-specific expressed genes are considered as the potential biomarkers for the diagnosis of early rheumatoid arthritis through bioinformatics analysis. *Journal of Translational Medicine, 19*(1), 18. https://doi.org/10.1186/s12967-020-02689-y.

Chin, C.-H., Chen, S.-H., Wu, H.-H., Ho, C.-W., Ko, M.-T., & Lin, C.-Y. (2014). cytoHubba: Identifying hub objects and sub-networks from complex interactome. *BMC Systems Biology, 8*(Suppl. 4), S11. https://doi.org/10.1186/1752-0509-8-S4-S11.

Dai, Q., Ling, Y.-H., Lia, M., Zou, Y.-Y., Kroog, G., Iwata, K. K., et al. (2005). Enhanced sensitivity to the HER1/epidermal growth factor receptor tyrosine kinase inhibitor Erlotinib hydrochloride in chemotherapy-resistant tumor cell lines. *Clinical Cancer Research, 11*(4), 1572–1578. https://doi.org/10.1158/1078-0432.CCR-04-0993.

Davis, S., & Meltzer, P. S. (2007). GEOquery: A bridge between the gene expression omnibus (GEO) and BioConductor. *Bioinformatics (Oxford, England), 23*(14), 1846–1847. https://doi.org/10.1093/bioinformatics/btm254.

Del Rey, M. J., Izquierdo, E., Usategui, A., Gonzalo, E., Blanco, F. J., Acquadro, F., et al. (2010). The transcriptional response of normal and rheumatoid arthritis synovial fibroblasts to hypoxia: Gene expression in synovial fibroblasts in hypoxia. *Arthritis and Rheumatism, 62*(12), 3584–3594. https://doi.org/10.1002/art.27750.

DeWan, A. T., Egan, K. B., Hellenbrand, K., Sorrentino, K., Pizzoferrato, N., Walsh, K. M., et al. (2012). Whole-exome sequencing of a pedigree segregating asthma. *BMC Medical Genetics, 13*(1), 95. https://doi.org/10.1186/1471-2350-13-95.

Dixit, V. D., Schaffer, E. M., Pyle, R. S., Collins, G. D., Sakthivel, S. K., Palaniappan, R., et al. (2004). Ghrelin inhibits leptin- and activation-induced proinflammatory cytokine expression by human monocytes and T cells. *Journal of Clinical Investigation, 114*(1), 57–66. https://doi.org/10.1172/JCI200421134.

Fakhouri, W., Lopez-Romero, P., Antonelli, S., Losi, S., Rogai, V., Buda, S., et al. (2018). Treatment patterns, health care resource utilization and costs of rheumatoid arthritis patients in Italy: Findings from a retrospective administrative database analysis. *Open Access Rheumatology: Research and Reviews, 10*, 103–111. https://doi.org/10.2147/OARRR.S164738.

Fattah, S. A., Abdel Fattah, M. A., Mesbah, N. M., Saleh, S. M., Abo-Elmatty, D. M., & Mehanna, E. T. (2020). The expression of zinc finger 804a (ZNF804a) and cyclin-dependent kinase 1 (CDK1) genes is related to the pathogenesis of rheumatoid arthritis. *Archives of Physiology and Biochemistry, 1–6*. https://doi.org/10.1080/13813455.2020.1716810.

Fujibayashi, Y., Isa, R., Nishiyama, D., Sakamoto-Inada, N., Kawasumi, N., Yamaguchi, J., et al. (2020). Aberrant BUB1 overexpression promotes mitotic segregation errors and chromosomal instability in multiple myeloma. *Cancers, 12*(8), 2206. https://doi.org/10.3390/cancers12082206.

Gang, X., Sun, Y., Li, F., Yu, T., Jiang, Z., Zhu, X., et al. (2017). Identification of key genes associated with rheumatoid arthritis with bioinformatics approach. *Medicine, 96*(31), e7673. https://doi.org/10.1097/MD.0000000000007673.

Garvalov, B. K., Foss, F., Henze, A.-T., Bethani, I., Gräf-Höchst, S., Singh, D., et al. (2014). PHD3 regulates EGFR internalization and signalling in tumours. *Nature Communications, 5*(1), 5577. https://doi.org/10.1038/ncomms6577.

Grabsch, H., Takeno, S., Parsons, W. J., Pomjanski, N., Boecking, A., Gabbert, H. E., et al. (2003). Overexpression of the mitotic checkpoint genesBUB1, BUBR1, andBUB3 in gastric cancer? Association with tumour cell proliferation. *The Journal of Pathology, 200*(1), 16–22. https://doi.org/10.1002/path.1324.

Guo, Q., Wang, Y., Xu, D., Nossent, J., Pavlos, N. J., & Xu, J. (2018). Rheumatoid arthritis: Pathological mechanisms and modern pharmacologic therapies. *Bone Research, 6*(1), 15. https://doi.org/10.1038/s41413-018-0016-9.

Hitchon, C. A., & El-Gabalawy, H. S. (2011). The synovium in rheumatoid arthritis. *The Open Rheumatology Journal, 5*(1), 107–114. https://doi.org/10.2174/1874312901105010107.

Huang, D. W., Sherman, B. T., Tan, Q., Kir, J., Liu, D., Bryant, D., et al. (2007). DAVID bioinformatics resources: Expanded annotation database and novel algorithms to better extract biology from large gene lists. *Nucleic Acids Research, 35*(Suppl. 2), W169–W175. https://doi.org/10.1093/nar/gkm415.

Kanbe, K., Takemura, T., Takeuchi, K., Chen, Q., Takagishi, K., & Inoue, K. (2004). Synovectomy reduces stromal-cell-derived factor-1 (SDF-1) which is involved in the destruction of cartilage in osteoarthritis and rheumatoid arthritis. *The Journal of Bone and Joint Surgery. British, 86-B(2)*, 296–300. https://doi.org/10.1302/0301-620X.86B2.14474.

Kanehisa, M., Goto, S., Furumichi, M., Tanabe, M., & Hirakawa, M. (2010). KEGG for representation and analysis of molecular networks involving diseases and drugs. *Nucleic Acids Research, 38*(Suppl. 1), D355–D360. https://doi.org/10.1093/nar/gkp896.

Korczowska, I. (2014). Rheumatoid arthritis susceptibility genes: An overview. *World Journal of Orthopedics, 5*(4), 544–549. https://doi.org/10.5312/wjo.v5.i4.544.

Li, G., Liu, D., Zhang, Y., Qian, Y., Zhu, Y., Guo, S., et al. (2013). Anti-invasive effects of celastrol in hypoxia-induced fibroblast-like synoviocyte through suppressing of HIF-1α/CXCR4 signaling pathway. *International Immunopharmacology, 17*(4), 1028–1036. https://doi.org/10.1016/j.intimp.2013.10.006.

Li, Z., Qi, F., & Li, F. (2020). Identification of drug targets and potential molecular mechanisms for Wantong Jingu tablet extract in treatment of rheumatoid arthritis: Bioinformatics analysis of fibroblast-like synoviocytes. *Chinese Medicine, 15*(1), 59. https://doi.org/10.1186/s13020-020-00339-5.

Lin, T.-C., & Hsiao, M. (2017). Ghrelin and cancer progression. *Biochimica et Biophysica Acta (BBA) - Reviews on Cancer, 1868*(1), 51–57. https://doi.org/10.1016/j.bbcan.2017.02.002.

MacLean, G. A. (2018). Downregulation of endothelin receptor B contributes to defective B cell Lymphopoiesis in trisomy 21 pluripotent stem cells. *Scientific Reports, 10.*

Molina-Navarro, M. M., Roselló-Lletí, E., Tarazón, E., Ortega, A., Sánchez-Izquierdo, D., Lago, F., et al. (2013). Heart failure entails significant changes in human nucleocytoplasmic transport gene expression. *International Journal of Cardiology, 168*(3), 2837–2843. https://doi.org/10.1016/j.ijcard.2013.03.192.

Oh, E., Kim, J. Y., Cho, Y., An, H., Lee, N., Jo, H., et al. (2016). Overexpression of angiotensin II type 1 receptor in breast cancer cells induces epithelial–mesenchymal transition and promotes tumor growth and angiogenesis. *Biochimica et Biophysica Acta (BBA) – Molecular Cell Research, 1863*(6), 1071–1081. https://doi.org/10.1016/j.bbamcr.2016.03.010.

Pahlavan, Y., Kahroba, H., Samadi, N., Karimi, A., Ansarin, K., & Khabbazi, A. (2019). Survivin modulatory role in autoimmune and autoinflammatory diseases. *Journal of Cellular Physiology, 234*(11), 19440–19450. https://doi.org/10.1002/jcp.28725.

Parnell, E., Shapiro, L. P., Voorn, R. A., Forrest, M. P., Jalloul, H. A., Loizzo, D. D., et al. (2021). KALRN: A central regulator of synaptic function and synaptopathies. *Gene, 768*, 145306. https://doi.org/10.1016/j.gene.2020.145306.

Pinto, M., Vieira, J., Ribeiro, F. R., Soares, M. J., Henrique, R., Oliveira, J., et al. (2008). Overexpression of the mitotic checkpoint genes *BUB1* and *BUBR1* is associated with genomic complexity in clear cell kidney carcinomas. *Analytical Cellular Pathology, 30*(5), 389–395. https://doi.org/10.1155/2008/820256.

Prasad, N. B., Somervell, H., Tufano, R. P., Dackiw, A. P. B., Marohn, M. R., Califano, J. A., et al. (2008). Identification of genes differentially expressed in benign versus malignant thyroid tumors. *Clinical Cancer Research, 14*(11), 3327–3337. https://doi.org/10.1158/1078-0432.CCR-07-4495.

Reiner-Benaim, A. (2007). FDR control by the BH procedure for two-sided correlated tests with implications to gene expression data analysis. *Biometrical Journal, 49*(1), 107–126. https://doi.org/10.1002/bimj.200510313.

Scherer, H. U., Häupl, T., & Burmester, G. R. (2020). The etiology of rheumatoid arthritis. *Journal of Autoimmunity, 110*, 102400. https://doi.org/10.1016/j.jaut.2019.102400.

Shannon, P. (2003). Cytoscape: A software environment for integrated models of biomolecular interaction networks. *Genome Research, 13*(11), 2498–2504. https://doi.org/10.1101/gr.1239303.

Sharief, M. K., Noori, M. A., Douglas, M. R., & Semra, Y. K. (2002). Upregulated survivin expression in activated T lymphocytes correlates with disease activity in multiple sclerosis. *European Journal of Neurology*, *9*(5), 503–510. https://doi.org/10.1046/j.1468-1331.2002.00454.x.

Smolen, J. S., Aletaha, D., Barton, A., Burmester, G. R., Emery, P., Firestein, G. S., et al. (2018). *Rheumatoid arthritis. Nature Reviews Disease Primers*, *4*(1), 18001. https://doi.org/10.1038/nrdp.2018.1.

Soleimanpour, E., & Babaei, E. (2015). Survivin as a potential target for cancer therapy. *Asian Pacific Journal of Cancer Prevention*, *16*(15), 6187–6191. https://doi.org/10.7314/APJCP.2015.16.15.6187.

Szklarczyk, D., Franceschini, A., Wyder, S., Forslund, K., Heller, D., Huerta-Cepas, J., et al. (2015). STRING v10: Protein–protein interaction networks, integrated over the tree of life. *Nucleic Acids Research*, *43*(D1), D447–D452. https://doi.org/10.1093/nar/gku1003.

Tang, Y., Li, M., Wang, J., Pan, Y., & Wu, F.-X. (2015). CytoNCA: A cytoscape plugin for centrality analysis and evaluation of protein interaction networks. *Biosystems*, *127*, 67–72. https://doi.org/10.1016/j.biosystems.2014.11.005.

Tarca, A. L., Romero, R., & Draghici, S. (2006). Analysis of microarray experiments of gene expression profiling. *American Journal of Obstetrics and Gynecology*, *195*(2), 373–388.

Udhaya Kumar, S., Madhana Priya, N., Thirumal Kumar, D., Anu Preethi, V., Kumar, V., Nagarajan, D., et al. (2021). Chapter ten—An integrative analysis to distinguish between emphysema (EML) and alpha-1 antitrypsin deficiency-related emphysema (ADL)—A systems biology approach. In R. Donev, & T. Karabencheva-Christova (Eds.), *Advances in protein chemistry and structural biology* (Vol. 127, pp. 315–342). Academic Press. https://doi.org/10.1016/bs.apcsb.2021.02.001.

Udhaya Kumar, S., Saleem, A., Thirumal Kumar, D., Anu Preethi, V., Younes, S., Zayed, H., et al. (2021). Chapter eleven—A systemic approach to explore the mechanisms of drug resistance and altered signaling cascades in extensively drug-resistant tuberculosis. In R. Donev, & T. Karabencheva-Christova (Eds.), *Advances in protein chemistry and structural biology* (Vol. 127, pp. 343–364). Academic Press. https://doi.org/10.1016/bs.apcsb.2021.02.002.

Udhaya Kumar, S., Thirumal Kumar, D., Bithia, R., Sankar, S., Magesh, R., Sidenna, M., et al. (2020a). Analysis of differentially expressed genes and molecular pathways in familial hypercholesterolemia involved in atherosclerosis: A systematic and bioinformatics approach. *Frontiers in Genetics*, *11*, 734. https://doi.org/10.3389/fgene.2020.00734.

Udhaya Kumar, S., Thirumal Kumar, D., Siva, R., George Priya Doss, C., Younes, S., Younes, N., et al. (2020b). Dysregulation of signaling pathways due to differentially expressed genes from the B-cell transcriptomes of systemic lupus erythematosus patients – A bioinformatics approach. *Frontiers in Bioengineering and Biotechnology*, *8*, 276. https://doi.org/10.3389/fbioe.2020.00276.

Vivatvakin, S., Ratchataswan, T., Leesutipornchai, T., Ruangritchankul, K., Keelawat, S., Mahattanasakul, P., et al. (2021). MCM-2, Ki-67, and EGFR downregulated expression levels in advanced stage laryngeal squamous cell carcinoma. *Scientific Reports*, *11*(1), 14607. https://doi.org/10.1038/s41598-021-94077-9.

Wang, Z., Katsaros, D., Shen, Y., Fu, Y., Canuto, E. M., Benedetto, C., et al. (2015). Biological and clinical significance of MAD2L1 and BUB1, genes frequently appearing in expression signatures for breast cancer prognosis. *PLoS One*, *10*(8), e0136246. https://doi.org/10.1371/journal.pone.0136246.

Woetzel, D., Huber, R., Kupfer, P., Pohlers, D., Pfaff, M., Driesch, D., et al. (2014). Identification of rheumatoid arthritis and osteoarthritis patients by transcriptome-based rule set generation. *Arthritis Research & Therapy*, *16*(2), R84. https://doi.org/10.1186/ar4526.

Wu, R., Long, L., Zhou, Q., Su, J., Su, W., & Zhu, J. (2021). Identification of hub genes in rheumatoid arthritis through an integrated bioinformatics approach. *Journal of Orthopaedic Surgery and Research*, *16*(1), 458. https://doi.org/10.1186/s13018-021-02583-3.

Wu, L., Qin, Y., Xia, S., Dai, M., Han, X., Wu, Y., et al. (2015). Identification of cyclin-dependent kinase 1 as a novel regulator of type 1 interferon signaling in systemic lupus erythematosus: CDK1 regulates IFN1 signaling in SLE. *Arthritis & Rheumatology*. https://doi.org/10.1002/art.39543.

Wu, C.-Y., Yang, H.-Y., & Lai, J.-H. (2020). Anti-Citrullinated protein antibodies in patients with rheumatoid arthritis: Biological effects and mechanisms of Immunopathogenesis. *International Journal of Molecular Sciences*, *21*(11), 4015. https://doi.org/10.3390/ijms21114015.

Xiong, Y., Mi, B., Liu, M., Xue, H., Wu, Q., & Liu, G. (2019). Bioinformatics analysis and identification of genes and molecular pathways involved in synovial inflammation in rheumatoid arthritis. *Medical Science Monitor*, *25*, 2246–2256. https://doi.org/10.12659/MSM.915451.

Yap, H.-Y., Tee, S., Wong, M., Chow, S.-K., Peh, S.-C., & Teow, S.-Y. (2018). Pathogenic role of immune cells in rheumatoid arthritis: Implications in clinical treatment and biomarker development. *Cell*, 7(10), 161. https://doi.org/10.3390/cells7100161.

Yin, Y., Li, Y., & Zhang, W. (2014). The growth hormone Secretagogue receptor: Its intracellular signaling and regulation. *International Journal of Molecular Sciences*, *15*(3), 4837–4855. https://doi.org/10.3390/ijms15034837.

Yoshitomi, H. (2019). Regulation of immune responses and chronic inflammation by fibroblast-like Synoviocytes. *Frontiers in Immunology*, *10*, 1395. https://doi.org/10.3389/fimmu.2019.01395.

Zhang, Q., Yu, S., Lam, M. M. T., Poon, T. C. W., Sun, L., Jiao, Y., et al. (2019). Angiotensin II promotes ovarian cancer spheroid formation and metastasis by upregulation of lipid desaturation and suppression of endoplasmic reticulum stress. *Journal of Experimental & Clinical Cancer Research*, *38*(1), 116. https://doi.org/10.1186/s13046-019-1127-x.

Zhao, L., Liang, D., Wu, X., Li, Y., Niu, J., Zhou, C., et al. (2017). Contribution and underlying mechanisms of CXCR4 overexpression in patients with systemic lupus erythematosus. *Cellular & Molecular Immunology*, *14*(10), 842–849. https://doi.org/10.1038/cmi.2016.47.

Zhou, L., Feng, X., Shan, W., Zhou, W., Liu, W., Wang, L., et al. (2007). Epigenetic and genetic alterations of the *EDNRB* gene in nasopharyngeal carcinoma. *Oncology*, *72*(5–6), 357–363. https://doi.org/10.1159/000113146.

Zhou, S., Zou, H., Chen, G., & Huang, G. (2019). Synthesis and biological activities of chemical drugs for the treatment of rheumatoid arthritis. *Topics in Current Chemistry*, *377*(5), 28. https://doi.org/10.1007/s41061-019-0252-5.

CHAPTER NINE

Understanding the activating mechanism of the immune system against COVID-19 by Traditional Indian Medicine: Network pharmacology approach

D. Thirumal Kumar[a], M.S. Shree Devi[b], S. Udhaya Kumar[c], Annie Sherlin[c], Aishwarya Mathew[c], M. Lakshmipriya[c], P. Sathiyarajeswaran[b], R. Gnanasambandan[c], R. Siva[c], R. Magesh[d], and C. George Priya Doss[c,*]

[a]Meenakshi Academy of Higher Education and Research (Deemed to be University), Chennai, India
[b]Siddha Central Research Institute (CCRS), Chennai, India
[c]School of BioSciences and Technology, Vellore Institute of Technology, Vellore, TN, India
[d]Department of Biotechnology, Sri Ramachandra Faculty of Biomedical Sciences and Technology, Sri Ramachandra Institute of Higher Education and Research (DU), Chennai, India
*Corresponding author: e-mail address: georgepriyadoss@vit.ac.in

Contents

Advances in Protein Chemistry and Structural Biology, Volume 129
ISSN 1876-1623
https://doi.org/10.1016/bs.apcsb.2021.11.007

Abstract

Severe Acute Respiratory Syndrome Coronavirus 2 (SARS-CoV-2) transmissions are occurring rapidly; it is raising the alarm around the globe. Though vaccines are currently available, the evolution and mutations in the SARS-CoV-2 threaten available vaccines' significance. The drugs are still undergoing clinical trials, and certain medications are approved for "emergency use" or as an "off-label" drug during the pandemic. These drugs have been effective yet accommodating side effects, which also can be lethal. Complementary and alternative medicine is highly demanded since it embraces a holistic approach. Since ancient times, natural products have been used as drugs to treat various diseases in the medical field and are still widely practiced. Medicinal plants contain many active compounds that serve as the key to an effective drug design. The Kabasura kudineer and Nilavembu kudineer are the two most widely approved formulations to treat COVID-19. However, the mechanism of these formulations is not well known. The proposed study used a network pharmacology approach to understand the immune-boosting mechanism by the Kabasura kudineer, Nilavembu kudineer, and JACOM in treating COVID-19. The plants and phytochemical chemical compounds in the Kabasura kudineer, Nilavembu kudineer, and JACOM were obtained from the literature. The Swiss target prediction algorithm was used to predict the targets for these phytochemical compounds. The common genes for the COVID-19 infection and the drug targets were identified. The gene–gene interaction network was constructed to understand the interactions between these common genes and enrichment analyses to determine the biological process, molecular functions, cellular functions, pathways involved, *etc.* Finally, virtual screening and molecular docking studies were performed to identify the most potential targets and significant phytochemical compounds to treat the COVID-19. The present study identified potential targets as ACE, Cathepsin L, Cathepsin B, Cathepsin K, DPP4, EGFR, HDAC2, IL6, RIPK1, and VEGFA. Similarly, betulinic acid, 5″-(2‴′-Hydroxybenzyl) uvarinol, antofine, (*S*)-1′-methyloctyl caffeate, (*Z*)-3-phenyl-2-propenal, 7-oxo-10α-cucurbitadienol, and PLX-4720 collectively to be potential treatment agents for COVID-19.

Abbreviations

ACE	angiotensin-converting enzyme
COVID-19	coronavirus disease 2019
CTSB	Cathepsin B
CTSL	Cathepsin L
DPP4	dipeptidyl peptidase-4
EGFR	epidermal growth factor receptor

HDAC2	histone deacetylase 2
IL-6	interleukin 6
PABPC1	poly(A)-binding protein C1
POLA	polymerase I
RIPK1	receptor-interacting serine/threonine-protein kinase 1
SARS-CoV-2	Severe Acute Respiratory Syndrome Coronavirus 2
SIGMAR1	Sigma non-opioid intracellular receptor 1
TCM	Traditional Chinese Medicine
TIM	Traditional Indian Medicine
VEGFA	vascular endothelial growth factor A
WHO	World Health Organization

1. Introduction

SARS-CoV-2 is a genetic variant of coronavirus known to cause respiratory illness and was reported as a global pandemic on March 11th, 2020 (Cucinotta & Vanelli, 2020). RT-PCR performed detects the infection and predicts the presence of Coronavirus. The virus has a close genetic resemblance with bat coronavirus, implying that it might have arisen from a bat-borne virus. This protocol has been published by World Health Organization (WHO) and used as a guideline for detecting the disease (Corman et al., 2020). Rapidly the viral infection begun to spread, increasing the number of cases in China and various parts of the world resulting from human-to-human transmission through close contact and primarily *via* respiratory droplets produced *via* coughs and sneezes (Dhand & Li, 2020). Currently, India stands behind the USA in the second position in the total number of people infected by COVID-19 (Kumar, Kumar, Christopher, & Doss, 2020; Kumar, Kumar, Siva, & Doss, 2020). As of September 6th, 2021, the reported cases are more than 220,563,227, and at least 4,565,483 people have died. The infection begins with the SARS-CoV-2 attaching to the enzyme called ACE2, which again is expressed in cells of various tissues and organs but is particularly abundant in type 2 lung pneumocytes. The people involved are frequently asymptomatic throughout the early stages of incubation, as well as the type I interferon response slows viral replication. With active replication and dissemination of SARS-CoV-2, the disease advances from mild to moderate symptoms such as sore throat, tear, fiber, and muscle pains, leading by both virus-associated respiratory tissue damage and antiviral activity (Cain & Cidlowski, 2020).

Several variants of concern (VOC) have appeared worldwide and seem to pose a significant threat to public health due to enhanced transmissibility or infectivity (Kumar, Shaikh, Kumar, Doss, & Zayed, 2021). Despite efficient vaccines, the coronavirus pandemic elicits widespread concern about discovering critical new variants from California, South Africa, UK, and Brazil (Plante et al., 2021). SARS-CoV-2 variants raise concerns about higher transmission and escaping on both vaccine and natural infection protection, especially in light of concerns about a specific mink strain, which triggered a person's illnesses and the possibility for future alterations. While most mutations are insignificant, the virus occasionally obtains a mutation that offers it an edge over other strains. The virus uses the spike protein to penetrate living cells through the ACE2 receptor. The spike protein's receptor-binding domain (RBD) is the most changeable component of the coronavirus genome (Tai et al., 2020). Mutations in viruses may lead to immune suppression in the host that the host antibodies initiate, necessitating their identification and monitoring for antibody therapy efficacy. The N439K, N440K, Q493K, and E484K spike mutations were prone to immunological escape, and this recent discovery is drawing attention (Srivastava, Banu, Singh, Sowpati, & Mishra, 2021). These mutations have given rise to several lineages. In the absence of direct drugs for the treatment, many efforts have been directed towards developing vaccines against COVID-19. Live attenuated or inactivated viruses, virus-like particles (VLP), protein subunits, viral vector (non-replicating and replicating), DNA, RNA, and nanoparticles, among additional vaccinations, are used in the conduit, each with its own set of advantages and disadvantages. According to the WHO report, these vaccines could only be 50% effective (What Is COVID-19 Vaccine Efficacy? | WHO | Regional Office for Africa, 2021a).

Complementary and alternative medicine is in high demand since it embraces a holistic approach. Since ancient times, natural products have been used as drugs in medical treatment and are still widely practiced. Medicinal plants contain many active compounds, which serve as the key to an effective drug design. Herbal medicine is achieving attention because of its extensive therapeutics like potent antiviral, immunomodulatory, anti-inflammatory, and anti-oxidant properties (Ross, 2009). Traditional Indian Medicine (TIM)—AYUSH and Traditional Chinese medicine (TCM) are ancient yet living traditions. These two traditional medicines are very philosophical and based on experiments, and they seem to be very effective in combatting viral diseases. Using these drugs from their natural origin is the main root of therapy (Patwardhan, Warude, Pushpangadan, & Bhatt, 2005).

Pharmaceutical companies have re-established their policies in favor of natural product drug development and discovery. China has effectively endorsed its therapies over the world with a scientific approach. The emerging popularity of TCM can be verified by the rapid upsurge in licensed Chinese medicine providers in the United States. Constant efforts in advancing these therapies in China have set TCM creditable (Youns, Hoheisel, & Efferth, 2010). Worldwide, Siddha is recognized and emerging, and there has been an increase in demand for medicinal plants in India. The growing use of traditional therapies requires more scientific evidence for the principles behind treatments and the effectiveness of medicines. Latest developments in the biological sciences, genomics, and proteomics can justify these therapies (Rathinam et al., 2020).

Several medicines mentioned by Ministry AYUSH, India have been in practice for viral diseases like Chikungunya and Dengue for the past two decades. Herbal formulations such as Kabasura kudineer and Nilavembu kudineer are widely used to treat phlegmatic and hemorrhagic fevers and are approved by the Siddha medicine (Jain et al., 2020; Natarajan et al., 2020). Decoding using *in silico* studies in Nilavembu kudineer against SARS-CoV-2 spike protein showcases that these medicines can be recognized as a valuable drug to combat COVID-19. Nilavembu kudineer interacts with ACE2 receptor, which serves as the pathogen entrance, and in the outcome, the pathogen cannot enter into the host body. Kabasura kudineer or choornam possesses antiviral solid, anti-bacterial, and immunomodulatory properties. Numerous studies have unveiled the anti-inflammatory properties of Kabasura kudineer. This herbal formulation became reasonably recognized during times of flu due to its therapeutic qualities. *In silico* docking was performed using Kabasura kudineer against SARS-CoV-2 spike protein. Its rich active Phyto-constituents revealed a favorable outcome that prevented the merging of viral replication binding with viral proteins and inhibited host receptors' binding. This verifies that Kabasura kudineer could be a potential herbal formulation to combat COVID-19 if proven with further preclinical and clinical confirmatory studies (Natarajan et al., 2020).

In our study, medicinal plants used in herbal formulations of Nilavembu kudineer were our primary focus to understand their drug inhibitory potential against SARS-CoV-2. In comparison to modern medicine, Siddha medicine's approach is more holistic. Hence, investigating and ameliorating the effectiveness of Siddha medicine to obtain the solution with the most negligible side effects on immune-compromised patients and the patients with co-morbid conditions (Kiran et al., 2020).

2. Materials and methods

2.1 Genes responsible for COVID-19

The human genes responsible for the COVID-19 disease were collected from the GeneCards database (Rebhan, Chalifa-Caspi, Prilusky, & Lancet, 1997). This database summarizes the current accessible biomedical information, including the human genes, proteins, and relevant diseases. The term "COVID-19" was used as the search term.

2.2 Siddha formulation as the supplement

The standard Siddha formulations such as Kabasura kudineer, Nilavembu kudineer, and JACOM that showed promising treatment results were chosen for this study (Jain et al., 2020; Kiran et al., 2020; Natarajan et al., 2020). The plant sources for these formulations were taken from the previous literature. The list of phytochemical compounds in these plant sources was obtained from the Chemical Entities of Biological Interest (ChEBI) database. This database provides an ontology of molecular entities focused on 'small' chemical compounds (Hastings et al., 2013).

2.3 Druglikeness of phytochemical compounds

The drug-likeness of the phytochemical compounds was evaluated using the SwissADME server (Daina, Michielin, & Zoete, 2017). The canonical SMILES format of phytochemical compounds was given as the input in SwissADME. We have considered the number of Lipinski violations given for each compound. The compounds with one or less than one Lipinski violations were selected for further analysis.

2.4 Target prediction for the phytochemical compounds

The various human protein targets for each phytochemical compound were identified using the SwissTargetPrediction (Gfeller et al., 2014). This server predicts bioactive molecules (query molecule) targets based on a blend of 2D and 3D resemblance procedures with known ligands. The top-ranking targets obtained for each phytochemical compound, with a probability of more than 0 was chosen for further analysis. The canonical SMILES format of phytochemical compounds was given as the input in SwissTargetPrediction.

2.5 Common target identification

A Venn diagram was created to find the expected targets for the identified phytochemical compounds and COVID-19 (genes affected by COVID-19). This was achieved using an online tool named Bioinformatics & Evolutionary Genomics. Using the Venn diagram, we identified the targets of phytochemical compounds with the same target as COVID-19 infection. Thus, the typical targets were used for further studies.

2.6 Enrichment analysis

The enrichment analysis was executed using the FunRich (Pathan et al., 2015). FunRich is a standalone package primarily utilized for the enrichment and interaction network analysis of proteins and genes. Enrichment analysis can be performed for biological process (BP), cellular process (CC), biological pathways, and molecular function (MF).

2.7 Pathway analysis

The pathway analysis was performed using the Reactome. Reactome is a comprehensive and well-annotated library of human molecular pathways and reactions (Fabregat et al., 2016).

2.8 Gene–gene interaction analysis

The standard targets obtained from the Venn diagram were subjected to a network analysis study. This was performed using the STRING database with default parameters (Szklarczyk et al., 2019). The STRING database aims to gather, score, and integrate all openly accessible protein–protein interaction information sources and complement these with computational predictions. Thus, this network analysis explains how these genes are biologically linked or overlapped through different pathways and functions.

2.9 Virtual screening and molecular interaction analysis

Several identified targets mapped in the gene–gene interaction have been targeted for more than one phytochemical compound. The virtual screening was performed using the AutoDock Vina plugin of the PyRx with the default parameter (Dallakyan & Olson, 2015; Trott & Olson, 2010). The protein structures with PDB IDs 6H5W, 2XU1, 6AY2, 2QT9, 4WKQ, 6WBW, 1ALU, 4ITJ, and 3QTK were taken for the proteins ACE, Cathepsin L, Cathepsin B, Cathepsin K, DPP4, EGFR, HDAC2, IL6, RIPK1, and VEGFA respectively. The compound showing the best affinity was taken for docking using the AutoDock standalone package (Morris et al., 2009). Blind docking protocol was performed using the Lamarckian Genetic Algorithm (Morris et al., 1998). The amino acid interaction with the compounds was visualized using the Discovery Studio.

3. Results and discussions

3.1 Genes responsible for COVID-19

Human genes that are involved in response to any disease are numerous and highly diverse within the genome. A total of 339 genes were obtained. A set of genes were retrieved from the GeneCards database with the keyword COVID. The detailed table with external identifiers is provided in Table 1.

3.2 Siddha formulation as the supplement

The list of plants involved in the formulation of Kabasura kudineer, Nilavembu kudineer, and JACOM were identified from the literature source. A total of 25 plants were identified in the formulation (Table 2). Further, a list of phytochemical compounds from the identified 25 plants was obtained from the ChEBI database. Three hundred and fourteen phytochemical compounds were identified in these 25 plants overall (Table 3).

3.3 Drug likeliness of phytochemical compounds

For a compound to persuade as a drug, the compound must satisfy a set of parameters that will prove its compatibility, efficiency, and toxicity. Lipinski's "rule of five" highlights possible bioavailability problems if two or more properties were violated. The SwissADME server was used to identify the Lipinski rule of 5 violations. The overall results from the SwissADME are tabulated in Table 4. The list of 285 compounds that satisfy the Lipinski rule of 5 is tabulated in Table 5.

3.4 Target prediction for the phytochemical compounds

The SwissTagetPrediction was used to find the target of the compounds that satisfied the Lipinski rule of 5. The overall 24,843 targets were predicted for the 285 compounds (Supplementary Table 1 in the online version at https://doi.org/10.1016/bs.apcsb.2021.11.007). The probability score will be provided for each of the targets. Considering the confidence level, the targets with a probability score of more than 0 were considered for further research (Supplementary Table 2 in the online version at https://doi.org/10.1016/bs.apcsb.2021.11.007).

Table 1 The list of genes responsible for the COVID-19 with the identifiers obtained from GeneCards database.

Input term	Symbol	Ensembl	HGNC	NCBI Entrez gene	OMIM	UniProtKB/Swiss-Prot
AKAP8	AKAP8	ENSG00000105127	378	10270	604692	O43823
AKAP8L	AKAP8L	ENSG00000011243	29857	26993	609475	Q9ULX6
AKAP9	AKAP9	ENSG00000127914	379	10142	604001	Q99996
GGCX	GGCX	ENSG00000115486	4247	2677	137167	P38435
MDN1	MDN1	ENSG00000112159	18302	23195	618200	Q9NU22
MAT2B	MAT2B	ENSG00000038274	6905	27430	605527	Q9NZL9
GDF15	GDF15	ENSG00000130513	30142	9518	605312	Q99988
GCC1	GCC1	ENSG00000179562	19095	79571	607418	Q96CN9
GCC2	GCC2	ENSG00000135968	23218	9648	612711	Q8IWJ2
MIPOL1	MIPOL1	ENSG00000151338	21460	145282	606850	Q8TD10
MIB1	MIB1	ENSG00000101752	21086	57534	608677	Q86YT6
GOLGA2	GOLGA2	ENSG00000167110	4425	2801	602580	Q08379
GOLGA3	GOLGA3	ENSG00000090615	4426	2802	602581	Q08378
GNB1	GNB1	ENSG00000078369	4396	2782	139380	P62873
GNG5	GNG5	ENSG00000174021	4408	2787	600874	P63218
GOLGB1	GOLGB1	ENSG00000173230	4429	2804	602500	Q14789

Continued

Table 1 The list of genes responsible for the COVID-19 with the identifiers obtained from GeneCards database.—cont'd

Input term	Symbol	Ensembl	HGNC	NCBI Entrez gene	OMIM	UniProtKB/Swiss-Prot
GORASP1	GORASP1	ENSG00000114745	16769	64689	606867	Q9BQQ3
CD14	CD14	ENSG00000170458	1628	929	158120	P08571
CCDC86	CCDC86	ENSG00000110104	28359	79080	611293	Q9H6F5
MAP7D1	MAP7D1	ENSG00000116871	25514	55700	NA	Q3KQU3
FBXL12	FBXL12	ENSG00000127452	13611	54850	609079	Q9NXK8
ATE1	ATE1	ENSG00000107669	782	11101	607103	O95260
FAM162A	FAM162A	ENSG00000114023	17865	26355	608017	Q96A26
EXOSC2	EXOSC2	ENSG00000130713	17097	23404	602238	Q13868
EXOSC3	EXOSC3	ENSG00000107371	17944	51010	606489	Q9NQT5
EXOSC5	EXOSC5	ENSG00000077348	24662	56915	606492	Q9NQT4
EXOSC8	EXOSC8	ENSG00000120699	17035	11340	606019	Q96B26
EZH2	EZH2	ENSG00000106462	3527	2146	601573	Q15910
AP2A2	AP2A2	ENSG00000183020	562	161	607242	O94973
AP2M1	AP2M1	ENSG00000161203	564	1173	601024	Q96CW1
AP3B1	AP3B1	ENSG00000132842	566	8546	603401	O00203
ARL6IP6	ARL6IP6	ENSG00000177917	24048	151188	616495	Q8N6S5
ARF6	ARF6	ENSG00000165527	659	382	600464	P62330

ATP13A3	ATP13A3	ENSG00000133657	24113	79572	610232	Q9H7F0
ATP1B1	ATP1B1	ENSG00000143153	804	481	182330	P05026
ATP6AP1	ATP6AP1	ENSG00000071553	868	537	300197	Q15904
ATP6V1A	ATP6V1A	ENSG00000114573	851	523	607027	P38606
FBLN5	FBLN5	ENSG00000140092	3602	10516	604580	Q9UBX5
FBN1	FBN1	ENSG00000166147	3603	2200	134797	P35555
FBN2	FBN2	ENSG00000138829	3604	2201	612570	P35556
FAM8A1	FAM8A1	ENSG00000137414	16372	51439	618409	Q9UBU6
FAM98A	FAM98A	ENSG00000119812	24520	25940	NA	Q8NCA5
ERC1	ERC1	ENSG00000082805	17072	23085	607127	Q8IUD2
ALG11	ALG11	ENSG00000253710	32456	440138	613666	Q2TAA5
ALG5	ALG5	ENSG00000120697	20266	29880	604565	Q9Y673
AGTR2	AGTR2	ENSG00000180772	338	186	300034	P50052
ERGIC1	ERGIC1	ENSG00000113719	29205	57222	617946	Q969X5
LOC117134593	LOC117134593	NA	NA	117134593	NA	NA
LOC117134604	LOC117134604	NA	NA	117134604	NA	NA
LOC117134605	LOC117134605	NA	NA	117134605	NA	NA
LOC117134606	LOC117134606	NA	NA	117134606	NA	NA

Continued

Table 1 The list of genes responsible for the COVID-19 with the identifiers obtained from GeneCards database.—cont'd

Input term	Symbol	Ensembl	HGNC	NCBI Entrez gene	OMIM	UniProtKB/Swiss-Prot
LOC117134607	LOC117134607	NA	NA	117134607	NA	NA
LOC117134608	LOC117134608	NA	NA	117134608	NA	NA
LOC117134611	LOC117134611	NA	NA	117134611	NA	NA
LOC117135104	LOC117135104	NA	NA	117135104	NA	NA
LOC117135105	LOC117135105	NA	NA	117135105	NA	NA
LOC117135106	LOC117135106	NA	NA	117135106	NA	NA
ERLEC1	ERLEC1	ENSG00000068912	25222	27248	611229	Q96DZ1
LOC117152610	LOC117152610	NA	NA	117152610	NA	NA
LOC117152611	LOC117152611	NA	NA	117152611	NA	NA
LOC117204000	LOC117204000	NA	NA	117204000	NA	NA
LOC117204001	LOC117204001	NA	NA	117204001	NA	NA
LOC117600004	LOC117600004	NA	NA	117600004	NA	NA
LOC117693187	LOC117693187	NA	NA	117693187	NA	NA
ERO1B	ERO1B	ENSG00000086619	14355	56605	615437	Q86YB8
ERP44	ERP44	ENSG00000023318	18311	23071	609170	Q9BS26
AGPS	AGPS	ENSG00000018510	327	8540	603051	O00116
FKBP10	FKBP10	ENSG00000141756	18169	60681	607063	Q96AY3

FKBP15	FKBP15	ENSG00000119321	23397	23307	617398	Q5T1M5
MEPCE	MEPCE	ENSG00000146834	20247	56257	611478	Q7L2J0
FKBP7	FKBP7	ENSG00000079150	3723	51661	607062	Q9Y680
BCKDK	BCKDK	ENSG00000103507	16902	10295	614901	O14874
ACE	ACE	ENSG00000159640	2707	1636	106180	P12821
ACE2	ACE2	ENSG00000130234	13557	59272	300335	Q9BYF1
ADAM9	ADAM9	ENSG00000168615	216	8754	602713	Q13443
ADAMTS1	ADAMTS1	ENSG00000154734	217	9510	605174	Q9UHI8
AAR2	AAR2	ENSG00000131043	15886	25980	617365	Q9Y312
AASS	AASS	ENSG00000008311	17366	10157	605113	Q9UDR5
AATF	AATF	ENSG00000275700	19235	26574	608463	Q9NY61
ACSL3	ACSL3	ENSG00000123983	3570	2181	602371	O95573
ACADM	ACADM	ENSG00000117054	89	34	607008	P11310
BSG	BSG	ENSG00000172270	1116	682	109480	P35613
FYCO1	FYCO1	ENSG00000163820	14673	79443	607182	Q9BQS8
G3BP1	G3BP1	ENSG00000145907	30292	10146	608431	Q13283
G3BP2	G3BP2	ENSG00000138757	30291	9908	NA	Q9UN86
BRD2	BRD2	ENSG00000204256	1103	6046	601540	P25440

Continued

Table 1 The list of genes responsible for the COVID-19 with the identifiers obtained from GeneCards database.—cont'd

Input term	Symbol	Ensembl	HGNC	NCBI Entrez gene	OMIM	UniProtKB/Swiss-Prot
BRD4	BRD4	ENSG00000141867	13575	23476	608749	O60885
MFGE8	MFGE8	ENSG00000140545	7036	4240	602281	Q08431
BZW2	BZW2	ENSG00000136261	18808	28969	NA	Q9Y6E2
FOXRED2	FOXRED2	ENSG00000100350	26264	80020	613777	Q8IWF2
FURIN	FURIN	ENSG00000140564	8568	5045	136950	P09958
C1ORF50	C1orf50	ENSG00000164008	28795	79078	NA	Q9BV19
LOX	LOX	ENSG00000113083	6664	4015	153455	P28300
FASTKD5	FASTKD5	ENSG00000215251	25790	60493	614272	Q7L8L6
ATP5MG	ATP5MG	ENSG00000167283	14247	10632	617473	O75964
CWC27	CWC27	ENSG00000153015	10664	10283	617170	Q6UX04
CYB5B	CYB5B	ENSG00000103018	24374	80777	611964	O43169
CYB5R3	CYB5R3	ENSG00000100243	2873	1727	613213	P00387
HS2ST1	HS2ST1	ENSG00000153936	5193	9653	604844	Q7LGA3
HS6ST2	HS6ST2	ENSG00000171004	19133	90161	300545	Q96MM7
CTSB	CTSB	ENSG00000164733	2527	1508	116810	P07858
CTSL	CTSL	ENSG00000135047	2537	1514	116880	P07711
CRP	CRP	ENSG00000132693	2367	1401	123260	P02741

HOOK1	HOOK1	ENSG00000134709	19884	51361	607820	Q9UJC3
HEATR3	HEATR3	ENSG00000155393	26087	55027	614951	Q7Z4Q2
HECTD1	HECTD1	ENSG00000092148	20157	25831	618649	Q9ULT8
GLA	GLA	ENSG00000102393	4296	2717	300644	P06280
GIGYF2	GIGYF2	ENSG00000204120	11960	26058	612003	Q6Y7W6
GFER	GFER	ENSG00000127554	4236	2671	600924	P55789
ETFA	ETFA	ENSG00000140374	3481	2108	608053	P13804
ANO6	ANO6	ENSG00000177119	25240	196527	608663	Q4KMQ2
GRPEL1	GRPEL1	ENSG00000109519	19696	80273	606173	Q9HAV7
CDK5RAP2	CDK5RAP2	ENSG00000136861	18672	55755	608201	Q96SN8
GTF2F2	GTF2F2	ENSG00000188342	4653	2963	189969	P13984
CEP112	CEP112	ENSG00000154240	28514	201134	NA	Q8N8E3
CEP135	CEP135	ENSG00000174799	29086	9662	611423	Q66GS9
CEP250	CEP250	ENSG00000126001	1859	11190	609689	Q9BV73
CEP350	CEP350	ENSG00000135837	24238	9857	617870	Q5VT06
CEP43	CEP43	ENSG00000213066	17012	11116	605392	O95684
CEP68	CEP68	ENSG00000011523	29076	23177	616889	Q76N32
CENPF	CENPF	ENSG00000117724	1857	1063	600236	P49454

Continued

Table 1 The list of genes responsible for the COVID-19 with the identifiers obtained from GeneCards database.—cont'd

Input term	Symbol	Ensembl	HGNC	NCBI Entrez gene	OMIM	UniProtKB/Swiss-Prot
GOLGA7	GOLGA7	ENSG00000147533	24876	51125	609453	Q7Z5G4
HLA-A	HLA-A	ENSG00000206503	4931	3105	142800	P04439
RHOA	RHOA	ENSG00000067560	667	387	165390	P61586
HMOX1	HMOX1	ENSG00000100292	5013	3162	141250	P09601
COQ8B	COQ8B	ENSG00000123815	19041	79934	615567	Q96D53
CLIP4	CLIP4	ENSG00000115295	26108	79745	NA	Q8N3C7
HLA-DRB1	HLA-DRB1	ENSG00000196126	4948	3123	142857	P01911
CISD3	CISD3	ENSG00000277972	27578	284106	611933	P0C7P0
CIT	CIT	ENSG00000122966	1985	11113	605629	O14578
CLCC1	CLCC1	ENSG00000121940	29675	23155	617539	Q96S66
RIPK1	RIPK1	ENSG00000137275	10019	8737	603453	Q13546
COLGALT1	COLGALT1	ENSG00000130309	26182	79709	617531	Q8NBJ5
COMT	COMT	ENSG00000093010	2228	1312	116790	P21964
HDAC2	HDAC2	ENSG00000196591	4853	3066	605164	Q92769
CHPF	CHPF	ENSG00000123989	24291	79586	610405	Q8IZ52
CHPF2	CHPF2	ENSG00000033100	29270	54480	608037	Q9P2E5
DNAJC11	DNAJC11	ENSG00000007923	25570	55735	614827	Q9NVH1

DNAJC19	DNAJC19	ENSG00000205981	30528	131118	608977	Q96DA6
DNMT1	DNMT1	ENSG00000130816	2976	1786	126375	P26358
MRPS2	MRPS2	ENSG00000122140	14495	51116	611971	Q9Y399
MRPS25	MRPS25	ENSG00000131368	14511	64432	611987	P82663
MRPS27	MRPS27	ENSG00000113048	14512	23107	611989	Q92552
MRPS5	MRPS5	ENSG00000144029	14498	64969	611972	P82675
DPP4	DPP4	ENSG00000197635	3009	1803	102720	P27487
MTCH1	MTCH1	ENSG00000137409	17586	23787	610449	Q9NZJ7
MTARC1	MTARC1	ENSG00000186205	26189	64757	614126	Q5VT66
EDEM3	EDEM3	ENSG00000116406	16787	80267	610214	Q9BZQ6
CD8A	CD8A	ENSG00000153563	1706	925	186910	P01732
GPT	GPT	ENSG00000167701	4552	2875	138200	P24298
GPX1	GPX1	ENSG00000233276	4553	2876	138320	P07203
CD209	CD209	ENSG00000090659	1641	30835	604672	Q9NNX6
GRIPAP1	GRIPAP1	ENSG00000068400	18706	56850	300408	Q4V328
GGH	GGH	ENSG00000137563	4248	8836	601509	Q92820
MOV10	MOV10	ENSG00000155363	7200	4343	610742	Q9HCE1
MPHOSPH10	MPHOSPH10	ENSG00000124383	7213	10199	605503	O00566

Continued

Table 1 The list of genes responsible for the COVID-19 with the identifiers obtained from GeneCards database.—cont'd

Input term	Symbol	Ensembl	HGNC	NCBI Entrez gene	OMIM	UniProtKB/Swiss-Prot
DDX10	DDX10	ENSG00000178105	2735	1662	601235	Q13206
DDX21	DDX21	ENSG00000165732	2744	9188	606357	Q9NR30
RNF41	RNF41	ENSG00000181852	18,401	10193	NA	Q9H4P4
COL6A1	COL6A1	ENSG00000142156	2211	1291	120220	P12109
DCAF7	DCAF7	ENSG00000136485	30915	10238	605973	P61962
DCAKD	DCAKD	ENSG00000172992	26238	79877	NA	Q8WVC6
NDUFAF2	NDUFAF2	ENSG00000164182	28086	91942	609653	Q8N183
EIF4E2	EIF4E2	ENSG00000135930	3293	9470	605895	O60573
EIF4H	EIF4H	ENSG00000106682	12741	7458	603431	Q15056
ELOB	ELOB	ENSG00000103363	11619	6923	600787	Q15370
ELOC	ELOC	ENSG00000154582	11617	6921	600788	Q15369
EMC1	EMC1	ENSG00000127463	28957	23065	616846	Q8N766
NEU1	NEU1	ENSG00000204386	7758	4758	608272	Q99519
NEK9	NEK9	ENSG00000119638	18591	91754	609798	Q8TD19
CSNK2A2	CSNK2A2	ENSG00000070770	2459	1459	115442	P19784
CSNK2B	CSNK2B	ENSG00000204435	2460	1460	115441	P67870
CRTC3	CRTC3	ENSG00000140577	26148	64784	608986	Q6UUV7

PCSK6	PCSK6	ENSG00000140479	8569	5046	167405	P29122
IL17A	IL17A	ENSG00000112115	5981	3605	603149	Q16552
IL17RA	IL17RA	ENSG00000177663	5985	23765	605461	Q96F46
IL10	IL10	ENSG00000136634	5962	3586	124092	P22301
IL2RA	IL2RA	ENSG00000134460	6008	3559	147730	P01589
SPART	SPART	ENSG00000133104	18514	23111	607111	Q8N0X7
PDZD11	PDZD11	ENSG00000120509	28034	51248	300632	Q5EBL8
PRRC2B	PRRC2B	ENSG00000130723	28121	84726	NA	Q5JSZ5
NPTX1	NPTX1	ENSG00000171246	7952	4884	602367	Q15818
SELENOS	SELENOS	ENSG00000131871	30396	55829	607918	Q9BQE4
NPC2	NPC2	ENSG00000119655	14537	10577	601015	P61916
PABPC1	PABPC1	ENSG00000070756	8554	26986	604679	P11940
PABPC4	PABPC4	ENSG00000090621	8557	8761	603407	Q13310
IDE	IDE	ENSG00000119912	5381	3416	146680	P14735
PRKACA	PRKACA	ENSG00000072062	9380	5566	601639	P17612
PCNT	PCNT	ENSG00000160299	16068	5116	605925	O95613
PRKAR2A	PRKAR2A	ENSG00000114302	9391	5576	176910	P13861
PDE4DIP	PDE4DIP	ENSG00000178104	15580	9659	608117	Q5VU43

Continued

Table 1 The list of genes responsible for the COVID-19 with the identifiers obtained from GeneCards database.—cont'd

Input term	Symbol	Ensembl	HGNC	NCBI Entrez gene	OMIM	UniProtKB/Swiss-Prot
SAAL1	SAAL1	ENSG00000166788	25158	113174	NA	Q96ER3
SCARB1	SCARB1	ENSG00000073060	1664	949	601040	Q8WTV0
SCCPDH	SCCPDH	ENSG00000143653	24275	51097	NA	Q8NBX0
SDF2	SDF2	ENSG00000132581	10675	6388	602934	Q99470
RRP9	RRP9	ENSG00000114767	16829	9136	NA	O43818
SBNO1	SBNO1	ENSG00000139697	22973	55206	614274	A3KN83
RTN4	RTN4	ENSG00000115310	14085	57142	604475	Q9NQC3
EGFR	EGFR	ENSG00000146648	3236	1956	131550	P00533
RPL36	RPL36	ENSG00000130255	13631	25873	617893	Q9Y3U8
ENPEP	ENPEP	ENSG00000138792	3355	2028	138297	Q07075
NGDN	NGDN	ENSG00000129460	20271	25983	610777	Q8NEJ9
NGLY1	NGLY1	ENSG00000151092	17646	55768	610661	Q96IV0
NIN	NIN	ENSG00000100503	14906	51199	608684	Q8N4C6
NINL	NINL	ENSG00000101004	29163	22981	609580	Q9Y2I6
SLC44A2	SLC44A2	ENSG00000129353	17292	57153	606106	Q8IWA5
NUP210	NUP210	ENSG00000132182	30052	23225	607703	Q8TEM1
NUP214	NUP214	ENSG00000126883	8064	8021	114350	P35658

SLC6A19	SLC6A19	ENSG00000174358	27960	340024	608893	Q695T7
SMOC1	SMOC1	ENSG00000198732	20318	64093	608488	Q9H4F8
PKP2	PKP2	ENSG00000057294	9024	5318	602861	Q99959
SLU7	SLU7	ENSG00000164609	16939	10569	605974	O95391
PITRM1	PITRM1	ENSG00000107959	17663	10531	618211	Q5JRX3
HSBP1	HSBP1	ENSG00000230989	5203	3281	604553	O75506
OS9	OS9	ENSG00000135506	16994	10956	609677	Q13438
POFUT1	POFUT1	ENSG00000101346	14988	23509	607491	Q9H488
POGLUT2	POGLUT2	ENSG00000134901	19350	79070	611613	Q6UW63
POGLUT3	POGLUT3	ENSG00000178202	28496	143888	618503	Q7Z4H8
POLA1	POLA1	ENSG00000101868	9173	5422	312040	P09884
POLA2	POLA2	ENSG00000014138	30073	23649	NA	Q14181
PLOD2	PLOD2	ENSG00000152952	9082	5352	601865	O00469
SNIP1	SNIP1	ENSG00000163877	30587	79753	608241	Q8TAD8
PLD3	PLD3	ENSG00000105223	17158	23646	615698	Q8IV08
PLEKHA5	PLEKHA5	ENSG00000052126	30036	54477	607770	Q9HAU0
PLEKHF2	PLEKHF2	ENSG00000175895	20757	79666	615208	Q9H8W4
SIRT5	SIRT5	ENSG00000124523	14933	23408	604483	Q9NXA8

Continued

Table 1 The list of genes responsible for the COVID-19 with the identifiers obtained from GeneCards database.—cont'd

Input term	Symbol	Ensembl	HGNC	NCBI Entrez gene	OMIM	UniProtKB/Swiss-Prot
SLC27A2	SLC27A2	ENSG00000140284	10996	11001	603247	O14975
SLC30A7	SLC30A7	ENSG00000162695	19306	148867	611149	Q8NEW0
SLC30A9	SLC30A9	ENSG00000014824	1329	10463	604604	Q6PML9
NUP54	NUP54	ENSG00000138750	17359	53371	607607	Q7Z3B4
NUP58	NUP58	ENSG00000139496	20261	9818	607615	Q9BVL2
NUP62	NUP62	ENSG00000213024	8066	23636	605815	P37198
NUP88	NUP88	ENSG00000108559	8067	4927	602552	Q99567
NUP98	NUP98	ENSG00000110713	8068	4928	601021	P52948
NUTF2	NUTF2	ENSG00000102898	13722	10204	605813	P61970
SLC25A21	SLC25A21	ENSG00000183032	14411	89874	607571	Q9BQT8
NSD2	NSD2	ENSG00000109685	12766	7468	602952	O96028
SEPSECS	SEPSECS	ENSG00000109618	30605	51091	613009	Q9HD40
SIGMAR1	SIGMAR1	ENSG00000147955	8157	10280	601978	Q99720
SIL1	SIL1	ENSG00000120725	24624	64374	608005	Q9H173
SRP19	SRP19	ENSG00000153037	11300	6728	182175	P09132
SRP54	SRP54	ENSG00000100883	11301	6729	604857	P61011
INS	INS	ENSG00000254647	6081	3630	176730	P01308

SRP72	SRP72	ENSG00000174780	11303	6731	602122	O76094
IMPDH2	IMPDH2	ENSG00000178035	6053	3615	146691	P12268
PTGES2	PTGES2	ENSG00000148334	17822	80142	608152	Q9H7Z7
INHBE	INHBE	ENSG00000139269	24029	83729	612031	P58166
PSMD8	PSMD8	ENSG00000099341	9566	5714	617844	P48556
INTS4	INTS4	ENSG00000149262	25048	92105	611348	Q96HW7
TOR1A	TOR1A	ENSG00000136827	3098	1861	605204	O14656
TOR1AIP1	TOR1AIP1	ENSG00000143337	29456	26092	614512	Q5JTV8
LARP1	LARP1	ENSG00000155506	29531	23367	612059	Q6PKG0
LARP4B	LARP4B	ENSG00000107929	28987	23185	616513	Q92615
LARP7	LARP7	ENSG00000174720	24912	51574	612026	Q4G0J3
TUBGCP2	TUBGCP2	ENSG00000130640	18599	10844	617817	Q9BSJ2
TUBGCP3	TUBGCP3	ENSG00000126216	18598	10426	617818	Q96CW5
UGGT2	UGGT2	ENSG00000102595	15664	55757	605898	Q9NYU1
TRIM59	TRIM59	ENSG00000213186	30834	286827	616148	Q8IWR1
UBAP2	UBAP2	ENSG00000137073	14185	55833	NA	Q5T6F2
TYSND1	TYSND1	ENSG00000156521	28531	219743	611017	Q2T9J0
UBAP2L	UBAP2L	ENSG00000143569	29877	9898	616472	Q14157

Continued

Table 1 The list of genes responsible for the COVID-19 with the identifiers obtained from GeneCards database.—cont'd

Input term	Symbol	Ensembl	HGNC	NCBI Entrez gene	OMIM	UniProtKB/Swiss-Prot
UPF1	UPF1	ENSG00000005007	9962	5976	601430	Q92900
ITGB1	ITGB1	ENSG00000150093	6153	3688	135630	P05556
PUSL1	PUSL1	ENSG00000169972	26914	126789	NA	Q8N0Z8
PVR	PVR	ENSG00000073008	9705	5817	173850	P15151
RAB2A	RAB2A	ENSG00000104388	9763	5862	179509	P61019
RAP1GDS1	RAP1GDS1	ENSG00000138698	9859	5910	179502	P52306
KNG1	KNG1	ENSG00000113889	6383	3827	612358	P01042
TCF12	TCF12	ENSG00000140262	11623	6938	600480	Q99081
TARS2	TARS2	ENSG00000143374	30740	80222	612805	Q9BW92
REEP5	REEP5	ENSG00000129625	30077	7905	125265	Q00765
REEP6	REEP6	ENSG00000115255	30078	92840	609346	Q96HR9
TBK1	TBK1	ENSG00000183735	11584	29110	604834	Q9UHD2
TBKBP1	TBKBP1	ENSG00000198933	30140	9755	608476	A7MCY6
RBX1	RBX1	ENSG00000100387	9928	9978	603814	P62877
RBM41	RBM41	ENSG00000089682	25617	55285	NA	Q96IZ5
RDX	RDX	ENSG00000137710	9944	5962	179410	P35241
RAB10	RAB10	ENSG00000084733	9759	10890	612672	P61026

JAKMIP1	JAKMIP1	ENSG00000152969	26460	152789	611195	Q96N16
STOM	STOM	ENSG00000148175	3383	2040	133090	P27105
RAB14	RAB14	ENSG00000119396	16524	51552	612673	P61106
RAB18	RAB18	ENSG00000099246	14244	22931	602207	Q9NP72
RAB1A	RAB1A	ENSG00000138069	9758	5861	179508	P62820
STC2	STC2	ENSG00000113739	11374	8614	603665	O76061
RALA	RALA	ENSG00000006451	9839	5898	179550	P11233
RAB5C	RAB5C	ENSG00000108774	9785	5878	604037	P51148
RAB7A	RAB7A	ENSG00000075785	9788	7879	602298	P51149
RAB8A	RAB8A	ENSG00000167461	7007	4218	165040	P61006
QSOX2	QSOX2	ENSG00000165661	30249	169714	612860	Q6ZRP7
RAE1	RAE1	ENSG00000101146	9828	8480	603343	P78406
SUN2	SUN2	ENSG00000100242	14210	25777	613569	Q9UH99
TLE1	TLE1	ENSG00000196781	11837	7088	600189	Q04724
TLE3	TLE3	ENSG00000140332	11839	7090	600190	Q04726
TLE5	TLE5	ENSG00000104964	307	166	600188	Q08117
TM2D3	TM2D3	ENSG00000184277	24128	80213	610014	Q9BRN9
TMPRSS2	TMPRSS2	ENSG00000184012	11876	7113	602060	O15393

Continued

Table 1 The list of genes responsible for the COVID-19 with the identifiers obtained from GeneCards database.—cont'd

Input term	Symbol	Ensembl	HGNC	NCBI Entrez gene	OMIM	UniProtKB/Swiss-Prot
TMPRSS4	TMPRSS4	ENSG00000137648	11878	56649	606565	Q9NRS4
IL6R	IL6R	ENSG00000160712	6019	3570	147880	P08887
PRKAR2B	PRKAR2B	ENSG00000005249	9392	5577	176912	P31323
PRIM1	PRIM1	ENSG00000198056	9369	5557	176635	P49642
PRIM2	PRIM2	ENSG00000146143	9370	5558	176636	P49643
PPT1	PPT1	ENSG00000131238	9325	5538	600722	P50897
PPIL3	PPIL3	ENSG00000240344	9262	53938	615811	Q9H2H8
TRMT1	TRMT1	ENSG00000104907	25980	55621	611669	Q9NXH9
MYCBP2	MYCBP2	ENSG00000005810	23386	23077	610392	O75592
NARS2	NARS2	ENSG00000137513	26274	79731	612803	Q96I59
CNTRL	CNTRL	ENSG00000119397	1858	11064	605496	Q7Z7A1
NAT14	NAT14	ENSG00000090971	28918	57106	NA	Q8WUY8
NOL10	NOL10	ENSG00000115761	25862	79954	616197	Q9BSC4
CUL2	CUL2	ENSG00000108094	2552	8453	603135	Q13617
MOGS	MOGS	ENSG00000115275	24862	7841	601336	Q13724
PLAT	PLAT	ENSG00000104368	9051	5327	173370	P00750
PLAUR	PLAUR	ENSG00000011422	9053	5329	173391	Q03405

POR	POR	ENSG00000127948	9208	5447	124015	P16435
HYOU1	HYOU1	ENSG00000149428	16931	10525	601746	Q9Y4L1
PMPCA	PMPCA	ENSG00000165688	18667	23203	613036	Q10713
PMPCB	PMPCB	ENSG00000105819	9119	9512	603131	O75439
IL6	IL6	ENSG00000136244	6018	3569	147620	P05231
RBM28	RBM28	ENSG00000106344	21863	55131	612074	Q9NW13
THTPA	THTPA	ENSG00000259431	18987	79178	611612	Q9BU02
TIMM10	TIMM10	ENSG00000134809	11814	26519	602251	P62072
TIMM10B	TIMM10B	ENSG00000132286	4022	26515	607388	Q9Y5J6
TIMM29	TIMM29	ENSG00000142444	25152	90580	617380	Q9BSF4
TIMM8B	TIMM8B	ENSG00000150779	11818	26521	606659	Q9Y5J9
TIMM9	TIMM9	ENSG00000100575	11819	26520	607384	Q9Y5J7
USP54	USP54	ENSG00000166348	23513	159195	NA	Q70EL1
ZC3H7A	ZC3H7A	ENSG00000122299	30959	29066	NA	Q8IWR0
ZDHHC5	ZDHHC5	ENSG00000156599	18472	25921	614586	Q9C0B5
ZC3H18	ZC3H18	ENSG00000158545	25091	124245	NA	Q86VM9
ZNF503	ZNF503	ENSG00000165655	23589	84858	613902	Q96F45
ZNF318	ZNF318	ENSG00000171467	13578	24149	617512	Q5VUA4

Continued

Table 1 The list of genes responsible for the COVID-19 with the identifiers obtained from GeneCards database.—cont'd

Input term	Symbol	Ensembl	HGNC	NCBI Entrez gene	OMIM	UniProtKB/Swiss-Prot
ZYG11B	ZYG11B	ENSG00000162378	25820	79699	618673	Q9C0D3
VPS11	VPS11	ENSG00000160695	14583	55823	608549	Q9H270
VPS39	VPS39	ENSG00000166887	20593	23339	612188	Q96JC1
WASHC4	WASHC4	ENSG00000136051	29174	23325	615748	Q2M389
USP13	USP13	ENSG00000058056	12611	8975	603591	Q92995
VEGFA	VEGFA	ENSG00000112715	12680	7422	192240	P15692
YIF1A	YIF1A	ENSG00000174851	16688	10897	611484	O95070
LMAN2	LMAN2	ENSG00000169223	16986	10960	609551	Q12907

Table 2 The list of plant involved in the formulation of Nilavembu kudineer, Kabasura kudineer, and JACOM.

S.·no	Plants name
1	*Zingiber officinale*
2	*Piper longum*
3	*Syzygium aromaticum*
4	*Tragia involucrata*
5	*Anacyclus pyrethrum*
6	*Andrographis paniculata*
7	*Hygrophila auriculata*
8	*Terminalia chebula*
9	*Justicia adhatoda*
10	*Plectranthus amboinicus*
11	*Saussurea lappa*
12	*Tinospora cordifolia*
13	*Rotheca serrata*
14	*Cypreus rotundus*
15	*Sida acuta* Burm.f.L
16	Adeloda serrata raf
17	*Carica Papaya*
18	*A. paniculata*
19	*Ocimum tenuiflorum*
20	*Vetiveria zizanioides*
21	*Santalum album*
22	*Piper nigrum*
23	*Hedyotis corymbosa*
24	*Plectranthus vettiveroides*
25	*Trichosanthes cucumerina*

Table 3 The list of phytochemical compounds found in the plant involved in the formulation of Nilavembu kudineer, Kabasura kudineer, and JACOM with ChEBI ID.

S.·no	Name of the plant	ChEBI ID	Name of the compound
1	*Zingiber officinale*	CHEBI:142262	Gingerenone B
2		CHEBI:64361	Beta-sesquiphellandrene
3		CHEBI:69294	3-(3,4-Dimethoxyphenyl)-4-[(Z)-3,4-dimethoxystyryl]cyclohex-1-ene
4		CHEBI:69295	3-(3,4-Dimethoxyphenyl)-4-[(E)-3,4-dimethoxystyryl]cyclohex-1-ene
5		CHEBI:26137	Pinocarveol
6		CHEBI:66502	Zerumboneoxide
7		CHEBI:68065	Ramonanin A, (rel)-
8		CHEBI:68066	Ramonanin B, (rel)-
9		CHEBI:68067	Ramonanin C, (rel)-
10		CHEBI:68068	Ramonanin D, (rel)-
11		CHEBI:63892	Zerumbone
12		CHEBI:66050	5-Hydroxyzerumbone
13		CHEBI:138043	(2E,6E)-hedycaryol
14		CHEBI:28817	Dodecane
15		CHEBI:146145	7,4′-Dimethylkaempferol
16		CHEBI:10115	Zingiberene
17		CHEBI:5414	Glucoputranjivin(1-)
18		CHEBI:79331	Glucoputranjivin
19		CHEBI:32389	All-cis-octadeca-6,9,12,15-tetraenoic acid
20	*Piper longum*	CHEBI:132658	Pipataline
21		CHEBI:66757	Pipercyclobutanamide A(rel)
22		CHEBI:69686	Pellitorine
23		CHEBI:67582	Gaudichaudianic acid, (−rac)
24		CHEBI:69675	Sarmentosumin A
25		CHEBI:69676	Sarmentosumin B

Table 3 The list of phytochemical compounds found in the plant involved in the formulation of Nilavembu kudineer, Kabasura kudineer, and JACOM with ChEBI ID.—cont'd

S.·no	Name of the plant	ChEBI ID	Name of the compound
26		CHEBI:69677	Sarmentosumin C
27		CHEBI:69678	Sarmentosumin D
28		CHEBI:69679	Isochamanetin
29		CHEBI:69680	7-Methoxychamanetin
30		CHEBI:69681	Dichamanetin
31		CHEBI:69682	7-Methoxydichamanetin
32		CHEBI:69683	5″-(2″″-Hydroxybenzyl)uvarinol
33		CHEBI:69687	2,4-Dodecadienamide
34		CHEBI:69689	7-Methoxyisochamanetin
35		CHEBI:70083	(2E,4E)-N-[2-(methylsulfinyl)ethyl]-2,4-decadienamide
36		CHEBI:70084	(2E,4E)-N-[(4-hydroxy-3-methoxyphenyl)ethyl]-2,4-decadienamide
37		CHEBI:70085	3-(4-Hydroxy-3,5-dimethoxyphenyl) propanoylpyrrole
38		CHEBI:70086	3-(3,4,5-Timethoxyphenyl) propanoylpyrrole
39		CHEBI:70087	1-[(2E,4E,6E)-2,4,6-dodecatrienoyl] pyrrolidine
40		CHEBI:70088	1-[(2E,4Z,8E)-9-(3,4-methylenedioxyphenyl)-2,4,8-nonatrienoyl]pyrrolidine
41		CHEBI:70089	1-[(2E,4E,10E)-10-(3,4-methylenedioxyphenyl)-2,4,10-undecatrienoyl]pyrrolidine
42		CHEBI:70090	1-[(4E,10E)-11-(3,4-methylenedioxyphenyl)-4,10-undecadienoyl]pyrrolidine
43		CHEBI:70091	1-[(9E)-10-(3,4-methylenedioxyphenyl)-9-decenoyl]pyrrolidine

Continued

Table 3 The list of phytochemical compounds found in the plant involved in the formulation of Nilavembu kudineer, Kabasura kudineer, and JACOM with ChEBI ID.—cont'd

S.·no	Name of the plant	ChEBI ID	Name of the compound
44		CHEBI:70092	1-[(2E,4E)-2,4-decadienoyl]pyrrolidine
45		CHEBI:70093	1-[(2E,4E)-2,4-dodecadienoyl]pyrrolidine
46		CHEBI:70094	1-[(2E)-7-(3,4-methylenedioxyphenyl)-2-heptenoyl]pyrrolidine
47		CHEBI:70095	1-[(2E,4E)-7-(3,4-methylenedioxyphenyl)-2,4-heptadienoyl]pyrrolidine
48		CHEBI:70096	1-[(2E,8E)-9-(3,4-methylenedioxyphenyl)-2,8-nonadienoyl]pyrrolidine
49		CHEBI:70097	1-[(8E)-9-(3,4-methylenedioxyphenyl)-8-nonenoyl]pyrrolidine
50		CHEBI:70098	1-[(2E,4E,8E)-9-(3,4-methylenedioxyphenyl)-2,4,8-nonatrienoyl]pyrrolidine
51		CHEBI:70099	1-[(2E,4E)-11-(3,4-methylenedioxyphenyl)-2,4-undecadienoyl]pyrrolidine
52		CHEBI:70100	1-[(2E,10E)-11-(3,4-methylenedioxyphenyl)-2,10-undecadienoyl]pyrrolidine
53		CHEBI:70101	(2E,4E)-N-isobutyl-2,4-dodecadienamide
54		CHEBI:70102	(2E,4E)-N-isobutyl-7-(3,4-methylenedioxyphenyl)-hepta-2,4-dienamide
55		CHEBI:70103	(8E)-N-isobutyl-9-(3,4-methylenedioxyphenyl)nona-8-enamide
56		CHEBI:70104	(2E,4E,8E)-N-isobutyl-11-(3,4-methylenedioxyphenyl)undeca-2,4,8-trienamide
57		CHEBI:70105	N-trans-sinapoyltyramine

Table 3 The list of phytochemical compounds found in the plant involved in the formulation of Nilavembu kudineer, Kabasura kudineer, and JACOM with ChEBI ID.—cont'd

S. no	Name of the plant	ChEBI ID	Name of the compound
58		CHEBI:70490	Dihydrocubebin, rel-
59		CHEBI:70491	Justiflorinol
60		CHEBI:70485	(−)-Sanguinolignan A
61		CHEBI:70486	(−)-Sanguinolignan B
62		CHEBI:70487	(−)-Sanguinolignan C
63		CHEBI:70488	(−)-Sanguinolignan D
64		CHEBI:70489	(7′*S*)-parabenzlactone
65		CHEBI:65899	Flavokawain B
66		CHEBI:66709	Methyl 3,4-dihydroxy-5-(3′-methyl-2′-butenyl)benzoate
67		CHEBI:69688	Pipercallosidine
68		CHEBI:132651	Kadsurenin C
69		CHEBI:132652	Kadsurenin K
70		CHEBI:132653	Kadsurenin L
71		CHEBI:69685	Pipercallosine
72		CHEBI:70483	(S)-1′-methylhexyl caffeate
73		CHEBI:132647	Futoenone
74		CHEBI:65684	(−)-Cubebin
75		CHEBI:65685	(−)-3,4-Dimethoxy-3,4-desmethylenedioxycubebin
76		CHEBI:70482	(S)-1′-methylbutyl caffeate
77		CHEBI:70484	(S)-1′-methyloctyl caffeate
78		CHEBI:65937	Futokadsurin B
79		CHEBI:65938	Futokadsurin C
80		CHEBI:28821	Piperine
81		CHEBI:65936	Futokadsurin A
82		CHEBI:132650	Burchellin
83		CHEBI:35697	Trans-cinnamic acid

Continued

Table 3 The list of phytochemical compounds found in the plant involved in the formulation of Nilavembu kudineer, Kabasura kudineer, and JACOM with ChEBI ID.—cont'd

S.·no	Name of the plant	ChEBI ID	Name of the compound
84		CHEBI:80484	Pinocembrin chalcone
85		CHEBI:132657	Piperlactam S
86		CHEBI:2871	Asebogenin
87		CHEBI:66470	(+)-Sesamin
88		CHEBI:30746	Benzoic acid
89		CHEBI:17818	N-feruloyltyramine
90		CHEBI:132654	Kadsurenin M
91		CHEBI:156227	(−)-Epicubenol
92		CHEBI:70626	Acacetin-8-C-neohesperidoside
93		CHEBI:70148	Monocerin
94		CHEBI:70149	(2S,3aR,9bR)-6,7-dihydroxy-8-methoxy-2-propyl-3,3a-dihydro-2H-furo[3,2-c]isochromen-5(9bH)-one
95		CHEBI:70150	Fusarentin 6,7-dimethyl ether
96		CHEBI:70151	Fusarentin 6-methyl ether
97		CHEBI:70152	(3*R*,4*R*)-4,8-dihydroxy-3-((*R*)-2-hydroxypentyl)-6,7-dimethoxyisochroman-1-one
98		CHEBI:70153	Colletotrialide, (+)-
99		CHEBI:132648	Galgravin
100		CHEBI:8240	Piperonal
101		CHEBI:28157	Pinocembrin
102		CHEBI:80788	1,4-Cineole
103		CHEBI:113455	Sodium benzoate
104		CHEBI:156224	(−)-Cubenol
105		CHEBI:37316	(E,E)-piperic acid
106		CHEBI:6116	Kavapyrone
107		CHEBI:10224	Alpha-cubebene
108		CHEBI:132649	Acuminatin

Table 3 The list of phytochemical compounds found in the plant involved in the formulation of Nilavembu kudineer, Kabasura kudineer, and JACOM with ChEBI ID.—cont'd

S.·no	Name of the plant	ChEBI ID	Name of the compound
109	*Syzygium aromaticum*	CHEBI:65486	Betulin di(3-carboxybutanoate)
110		CHEBI:65485	Dihydrobetulinic acid
111		CHEBI:65484	Bevirimat
112		CHEBI:65487	Platanic acid
113		CHEBI:132345	Canophyllal
114		CHEBI:27386	Cinnamic acid
115		CHEBI:69305	Cinnamtannin D-1
116		CHEBI:69307	Cassiatannin A
117		CHEBI:69304	Cinnamtannin B-1
118		CHEBI:69306	Parameritannin A-1
119		CHEBI:133634	Methyl linolenate
120		CHEBI:63892	Zerumbone
121		CHEBI:89729	(Z)-3-phenyl-2-propenal
122		CHEBI:136676	(E)-2-methoxycinnamic acid
123		CHEBI:3087	Betulinic acid
124	*Tragia involucrata*	CHEBI:133981	Heptacosan-1-ol
125	*Anacyclus pyrethrum*	CHEBI:27815	Pyrethrin I
126		CHEBI:27474	Pyrethrin II
127	*Andrographis paniculata*	CHEBI:69808	14-Deoxy-11,12-didehydroandrographolide
128		CHEBI:65408	Andrographolide
129		CHEBI:86612	Dihydroferulic acid
130		CHEBI:132373	Mesembryanthemoidigenic acid
131		CHEBI:142267	Methyl *N*-methylanthranilate
132		CHEBI:132830	Delta-elemene
133		CHEBI:65732	Decussatin
134	*Hygrophila auriculata*	CHEBI:3367	Capensinidin

Continued

Table 3 The list of phytochemical compounds found in the plant involved in the formulation of Nilavembu kudineer, Kabasura kudineer, and JACOM with ChEBI ID.—cont'd

S.·no	Name of the plant	ChEBI ID	Name of the compound
135	*Terminalia chebula*	CHEBI:66202	Termilignan B
136		CHEBI:69692	(Z)-9-hydroxybenzo[*c*]oxepin-3(1H)-one
137		CHEBI:69693	Cyclosordariolone, (rac)-
138		CHEBI:69694	(R)-3-Hydroxy-1-[(R)-4-hydroxy-1,3-dihydroisobenzofuran-1-yl] butan-2-one
139		CHEBI:69695	(R)-3-Hydroxy-1-[(S)-4-hydroxy-1,3-dihydroisobenzofuran-1-yl] butan-2-one
140		CHEBI:69696	(E)-2-(Hydroxymethyl)-3-(4-hydroxypent-1-enyl)phenol
141		CHEBI:69697	1-(3,9-Dihydroxy-1,3-dihydrobenzo[*c*] oxepin-3-yl)ethanone, (rac)-
142		CHEBI:69698	Pestalospirane A
143		CHEBI:69699	Pestalospirane B
144		CHEBI:145828	Methyl 3,4,5-trihydroxybenzoate
145		CHEBI:9908	Ursolic acid
146	*Justicia adhatoda*	CHEBI:2814	Arecoline
147		CHEBI:156072	Propyl benzoate
148	*Plectranthus amboinicus*	CHEBI:66763	Plectranthol A
149		CHEBI:66764	Plectranthol B
150		CHEBI:138963	11,20-Dihydroxysugiol
151		CHEBI:138962	11-Hydroxysugiol
152		CHEBI:86062	Abietatriene
153	*Saussurea lappa*	CHEBI:66024	1beta-hydroxy arbusculin A
154		CHEBI:2540	Alantolactone
155		CHEBI:244418	Dehydrocostus lactone
156		CHEBI:3900	Costunolide

Table 3 The list of phytochemical compounds found in the plant involved in the formulation of Nilavembu kudineer, Kabasura kudineer, and JACOM with ChEBI ID.—cont'd

S.·no	Name of the plant	ChEBI ID	Name of the compound
157		CHEBI:132820	Matairesinoside
158		CHEBI:138251	3-Hydroxyhexane-2,5-dione
159		CHEBI:133981	Heptacosan-1-ol
160	*Tinospora cordifolia*	CHEBI:66046	6-Hydroxyluteolin 7-O-laminaribioside
161		CHEBI:89715	Cyclotetradecane
162		CHEBI:142915	Cycloeucalenone
163		CHEBI:141063	Rubimaillin
164		CHEBI:132345	Canophyllal
165		CHEBI:132718	Stepharanine
166		CHEBI:134479	(Z)-icos-13-enoic acid
167		CHEBI:145828	Methyl 3,4,5-trihydroxybenzoate
168	*Rotheca serrata*	CHEBI:156193	Serratol
169		CHEBI:78330	Huperzine A
170		CHEBI:6701	Maytansine
171	*Cypreus rotundus*	CHEBI:66416	Mustakone
172		CHEBI:81377	(+)-Nootkatone
173	*Sida acuta* Burm.f.L	CHEBI:142397	Pectenotoxin-11
174		CHEBI:145981	Dinophysistoxin 2
175		CHEBI:142495	1-Icosanoylglycerol
176		CHEBI:131838	Swertisin
177		CHEBI:156193	Serratol
178		CHEBI:16634	Raffinose
179		CHEBI:90295	PLX-4720
180		CHEBI:78330	Huperzine A
181		CHEBI:6701	Maytansine
182		CHEBI:143119	N-(2-methoxyethyl)-4-{[6-(pyridin-4-yl) quinazolin-2-yl]amino}benzamide

Continued

Table 3 The list of phytochemical compounds found in the plant involved in the formulation of Nilavembu kudineer, Kabasura kudineer, and JACOM with ChEBI ID.—cont'd

S.·no	Name of the plant	ChEBI ID	Name of the compound
183	*Carica Papaya*	CHEBI:17127	Glucotropeolin
184		CHEBI:58021	Glucotropeolin(1-)
185		CHEBI:142658	Methyl 12-methyltetradecanoate
186		CHEBI:91143	(3*S*,5*R*,6*S*)-beta-cryptoxanthin 5,6-epoxide
187		CHEBI:3433	Carpaine
188		CHEBI:10362	Beta-cryptoxanthin
189		CHEBI:1307	24-Methylenecycloartanol
190		CHEBI:141360	Helvolic acid methyl ester
191		CHEBI:4316	Danielone
192		CHEBI:133683	2-Isobutylthiazole
193	*A. paniculata*	CHEBI:69808	14-Deoxy-11,12-didehydroandrographolide
194		CHEBI:65408	Andrographolide
195		CHEBI:86612	Dihydroferulic acid
196		CHEBI:132373	Mesembryanthemoidigenic acid
197		CHEBI:142267	Methyl *N*-methylanthranilate
198		CHEBI:132830	Delta-elemene
199		CHEBI:65732	Decussatin
200	*Ocimum tenuiflorum*	CHEBI:17580	Linalool
201		CHEBI:7545	Nevadensin
202		CHEBI:63710	7-Epi-sesquithujene
203	*Vetiveria zizanioides*	CHEBI:138051	Selina-4(15),7(11)-diene
204	*Santalum album*	CHEBI:16714	Codeine
205		CHEBI:45373	Sulfanilamide
206		CHEBI:65460	Avicularin

Table 3 The list of phytochemical compounds found in the plant involved in the formulation of Nilavembu kudineer, Kabasura kudineer, and JACOM with ChEBI ID.—cont'd

S.·no	Name of the plant	ChEBI ID	Name of the compound
207	*Piper nigrum*	CHEBI:66757	Pipercyclobutanamide A(rel)
208		CHEBI:65684	(−)-cubebin
209		CHEBI:65685	(−)-3,4-dimethoxy-3,4-desmethylenedioxycubebin
210		CHEBI:28821	Piperine
211		CHEBI:69686	Pellitorine
212		CHEBI:67582	Gaudichaudianic acid, (−rac)
213		CHEBI:69675	Sarmentosumin A
214		CHEBI:69676	Sarmentosumin B
215		CHEBI:69677	Sarmentosumin C
216		CHEBI:69678	Sarmentosumin D
217		CHEBI:69679	Isochamanetin
218		CHEBI:69680	7-Methoxychamanetin
219		CHEBI:69681	Dichamanetin
220		CHEBI:69682	7-Methoxydichamanetin
221		CHEBI:69683	5″-(2‴′-Hydroxybenzyl)uvarinol
222		CHEBI:69687	2,4-Dodecadienamide
223		CHEBI:69689	7-Methoxyisochamanetin
224		CHEBI:70083	(2E,4E)-N-[2-(methylsulfinyl)ethyl]-2,4-decadienamide
225		CHEBI:70084	(2E,4E)-N-[(4-hydroxy-3-methoxyphenyl)ethyl]-2,4-decadienamide
226		CHEBI:70085	3-(4-hydroxy-3,5-dimethoxyphenyl) propanoylpyrrole
227		CHEBI:70086	3-(3,4,5-timethoxyphenyl) propanoylpyrrole
228		CHEBI:70087	1-[(2E,4E,6E)-2,4,6-dodecatrienoyl] pyrrolidine

Continued

Table 3 The list of phytochemical compounds found in the plant involved in the formulation of Nilavembu kudineer, Kabasura kudineer, and JACOM with ChEBI ID.—cont'd

S.no	Name of the plant	ChEBI ID	Name of the compound
229		CHEBI:70088	1-[(2E,4Z,8E)-9-(3,4-methylenedioxyphenyl)-2,4,8-nonatrienoyl]pyrrolidine
230		CHEBI:70089	1-[(2E,4E,10E)-10-(3,4-methylenedioxyphenyl)-2,4,10-undecatrienoyl]pyrrolidine
231		CHEBI:70090	1-[(4E,10E)-11-(3,4-methylenedioxyphenyl)-4,10-undecadienoyl]pyrrolidine
232		CHEBI:70091	1-[(9E)-10-(3,4-methylenedioxyphenyl)-9-decenoyl]pyrrolidine
233		CHEBI:70092	1-[(2E,4E)-2,4-decadienoyl]pyrrolidine
234		CHEBI:70093	1-[(2E,4E)-2,4-dodecadienoyl]pyrrolidine
235		CHEBI:70094	1-[(2E)-7-(3,4-methylenedioxyphenyl)-2-heptenoyl]pyrrolidine
236		CHEBI:70095	1-[(2E,4E)-7-(3,4-methylenedioxyphenyl)-2,4-heptadienoyl]pyrrolidine
237		CHEBI:70096	1-[(2E,8E)-9-(3,4-methylenedioxyphenyl)-2,8-nonadienoyl]pyrrolidine
238		CHEBI:70097	1-[(8E)-9-(3,4-methylenedioxyphenyl)-8-nonenoyl]pyrrolidine
239		CHEBI:70098	1-[(2E,4E,8E)-9-(3,4-methylenedioxyphenyl)-2,4,8-nonatrienoyl]pyrrolidine
240		CHEBI:70099	1-[(2E,4E)-11-(3,4-methylenedioxyphenyl)-2,4-undecadienoyl]pyrrolidine
241		CHEBI:70100	1-[(2E,10E)-11-(3,4-methylenedioxyphenyl)-2,10-undecadienoyl]pyrrolidine

Table 3 The list of phytochemical compounds found in the plant involved in the formulation of Nilavembu kudineer, Kabasura kudineer, and JACOM with ChEBI ID.—cont'd

S. no	Name of the plant	ChEBI ID	Name of the compound
242		CHEBI:70101	(2E,4E)-N-isobutyl-2,4-dodecadienamide
243		CHEBI:70102	(2E,4E)-N-isobutyl-7-(3,4-methylenedioxyphenyl)-hepta-2,4-dienamide
244		CHEBI:70103	(8E)-N-isobutyl-9-(3,4-methylenedioxyphenyl)nona-8-enamide
245		CHEBI:70104	(2E,4E,8E)-N-isobutyl-11-(3,4-methylenedioxyphenyl)undeca-2,4,8-trienamide
246		CHEBI:70105	N-trans-sinapoyltyramine
247		CHEBI:70490	Dihydrocubebin, rel-
248		CHEBI:70491	Justiflorinol
249		CHEBI:70485	(−)-Sanguinolignan A
250		CHEBI:70486	(−)-Sanguinolignan B
251		CHEBI:70487	(−)-Sanguinolignan C
252		CHEBI:70488	(−)-Sanguinolignan D
253		CHEBI:70489	(7′S)-parabenzlactone
254		CHEBI:65899	Flavokawain B
255		CHEBI:66709	Methyl 3,4-dihydroxy-5-(3′-methyl-2′-butenyl)benzoate
256		CHEBI:69688	Pipercallosidine
257		CHEBI:132651	Kadsurenin C
258		CHEBI:132652	Kadsurenin K
259		CHEBI:132653	Kadsurenin L
260		CHEBI:69685	Pipercallosine
261		CHEBI:70483	(S)-1′-methylhexyl caffeate
262		CHEBI:132647	Futoenone
263		CHEBI:70482	(S)-1′-methylbutyl caffeate

Continued

Table 3 The list of phytochemical compounds found in the plant involved in the formulation of Nilavembu kudineer, Kabasura kudineer, and JACOM with ChEBI ID.—cont'd

S.·no	Name of the plant	ChEBI ID	Name of the compound
264		CHEBI:70484	(S)-1′-methyloctyl caffeate
265		CHEBI:65937	Futokadsurin B
266		CHEBI:65938	Futokadsurin C
267		CHEBI:65936	Futokadsurin A
268		CHEBI:132650	Burchellin
269		CHEBI:8240	Piperonal
270		CHEBI:156224	(−)-Cubenol
271		CHEBI:37316	(E,E)-piperic acid
272		CHEBI:35697	Trans-cinnamic acid
273		CHEBI:80484	Pinocembrin chalcone
274		CHEBI:132657	Piperlactam S
275		CHEBI:2871	Asebogenin
276		CHEBI:51226	Epicocconone
277		CHEBI:66470	(+)-Sesamin
278		CHEBI:132658	Pipataline
279		CHEBI:143911	(−)-Antofine
280		CHEBI:30746	Benzoic acid
281		CHEBI:88764	Ethyl butyrate
282		CHEBI:17818	N-feruloyltyramine
283		CHEBI:132654	Kadsurenin M
284		CHEBI:156227	(−)-Epicubenol
285		CHEBI:70626	Acacetin-8-C-neohesperidoside
286		CHEBI:70148	Monocerin
287		CHEBI:70149	(2S,3aR,9bR)-6,7-dihydroxy-8-methoxy-2-propyl-3,3a-dihydro-2H-furo [3,2-c]isochromen-5(9bH)-one
288		CHEBI:70150	Fusarentin 6,7-dimethyl ether
289		CHEBI:70151	Fusarentin 6-methyl ether

Table 3 The list of phytochemical compounds found in the plant involved in the formulation of Nilavembu kudineer, Kabasura kudineer, and JACOM with ChEBI ID.—cont'd

S.·no	Name of the plant	ChEBI ID	Name of the compound
290		CHEBI:70152	(3R,4R)-4,8-dihydroxy-3-((R)-2-hydroxypentyl)-6,7-dimethoxyisochroman-1-one
291		CHEBI:70153	Colletotrialide, (+)-
292		CHEBI:132648	Galgravin
293		CHEBI:133381	9,10-Epoxy-18-hydroxyoctadecanoic acid
294		CHEBI:28157	Pinocembrin
295		CHEBI:80788	1,4-Cineole
296		CHEBI:113455	Sodium benzoate
297		CHEBI:133325	9,10,18-Trihydroxyoctadecanoic acid
298		CHEBI:6116	Kavapyrone
299		CHEBI:10224	Alpha-cubebene
300		CHEBI:132649	Acuminatin
301	*Hedyotis corymbosa*	CHEBI:66123	Jerantinine E
302		CHEBI:66124	Jerantinine F
303		CHEBI:66121	Jerantinine C
304		CHEBI:66120	Jerantinine B
305		CHEBI:66122	Jerantinine D
306		CHEBI:66119	Jerantinine A
307		CHEBI:142075	Tabernaemontanine
308		CHEBI:6682	Mangiferin
309	*Plectranthus vettiveroides*	CHEBI:66763	Plectranthol A
310		CHEBI:66764	Plectranthol B
311		CHEBI:138963	11,20-Dihydroxysugiol
312		CHEBI:138962	11-Hydroxysugiol
313		CHEBI:86062	Abietatriene
314	*Trichosanthes cucumerina*	CHEBI:66838	7-Oxo-10α-cucurbitadienol

Table 4 The Lipinski rule calculation for the phytochemical compounds of the plant involved in the formulation of Nilavembu kudineer, Kabasura kudineer, and JACOM.

S.·no	Compounds	MW	#H-bond acceptors	#H-bond donors	iLOGP	Lipinski #violations
1	Gingerenone B	386.44	6	2	3.7	0
2	Beta-sesquiphellandrene	204.35	0	0	3.65	1
3	3-(3,4-Dimethoxyphenyl)-4-[(Z)-3, 4-dimethoxystyryl]cyclohex-1-ene	380.48	4	0	4.43	0
4	3-(3,4-Dimethoxyphenyl)-4-[(E)-3,4-dimethoxystyryl]cyclohex-1-ene	380.48	4	0	4.43	0
5	Pinocarveol	152.23	1	1	2.12	0
6	Zerumboneoxide	234.33	2	0	2.71	0
7	Ramonanin A, (rel)-	680.74	10	4	4.14	1
8	Ramonanin B, (rel)-	680.74	10	4	4.51	1
9	Ramonanin C, (rel)-	680.74	10	4	4.5	1
10	Ramonanin D, (rel)-	680.74	10	4	5.29	1
11	Zerumbone	218.33	1	0	2.72	0
12	5-Hydroxyzerumbone	234.33	2	1	2.14	0
13	(2E,6E)-hedycaryol	222.37	1	1	3.05	0
14	Dodecane	170.33	0	0	3.82	1
15	7,4′-Dimethylkaempferol	314.29	6	2	2.94	0

16	Zingiberene	204.35	0	0	3.63	1
17	Glucoputranjivin(1-)	360.38	10	4	0.86	0
18	Glucoputranjivin	361.39	10	5	0.14	0
19	All-cis-octadeca-6,9,12,15-tetraenoic acid	276.41	2	1	3.29	1
20	Pipataline	288.42	2	0	4.61	1
21	Pipercyclobutanamide A(rel)	570.68	6	0	5.26	1
22	pellitorine	223.35	1	1	3.61	0
23	Gaudichaudianic acid, (−rac)	340.46	3	1	3.85	1
24	Sarmentosumin A	680.74	8	6	3.7	2
25	Sarmentosumin B	680.74	8	6	3.97	2
26	Sarmentosumin C	786.86	9	7	4.13	3
27	Sarmentosumin D	786.86	9	7	4.06	3
28	Isochamanetin	362.38	5	3	2.68	0
29	7-Methoxychamanetin	376.4	5	2	3.2	0
30	Dichamanetin	468.5	6	4	3.01	0
31	7-Methoxydichamanetin	482.52	6	3	3.58	0
32	5″-(2‴′-Hydroxybenzyl)uvarinol	694.77	8	5	4.16	1
33	2,4-Dodecadienamide	195.3	1	1	2.87	0
34	7-Methoxyisochamanetin	376.4	5	2	3.5	0

Continued

Table 4 The Lipinski rule calculation for the phytochemical compounds of the plant involved in the formulation of Nilavembu kudineer, Kabasura kudineer, and JACOM.—cont'd

S.·no	Compounds	MW	#H-bond acceptors	#H-bond donors	iLOGP	Lipinski #violations
35	(2E,4E)-N-[2-(methylsulfinyl)ethyl]-2,4-decadienamide	257.39	2	1	3.05	0
36	(2E,4E)-N-[(4-hydroxy-3-methoxyphenyl)ethyl]-2,4-decadienamide	317.42	3	2	3.76	0
37	3-(4-Hydroxy-3,5-dimethoxyphenyl)propanoylpyrrole	275.3	4	1	2.63	0
38	3-(3,4,5-Timethoxyphenyl)propanoylpyrrole	289.33	4	0	3.21	0
39	1-[(2E,4E,6E)-2,4,6-dodecatrienoyl]pyrrolidine	247.38	1	0	3.86	0
40	1-[(2E,4Z,8E)-9-(3,4-methylenedioxyphenyl)-2,4,8-nonatrienoyl] pyrrolidine	325.4	3	0	3.88	0
41	1-[(2E,4E,10E)-10-(3,4-methylenedioxyphenyl)-2,4,10-undecatrienoyl]pyrrolidine	353.45	3	0	4.53	0
42	1-[(4E,10E)-11-(3,4-methylenedioxyphenyl)-4,10-undecadienoyl] pyrrolidine	355.47	3	0	4.6	0
43	1-[(9E)-10-(3,4-methylenedioxyphenyl)-9-decenoyl]pyrrolidine	343.46	3	0	4.43	0
44	1-[(2E,4E)-2,4-decadienoyl]pyrrolidine	221.34	1	0	3.45	0
45	1-[(2E,4E)-2,4-dodecadienoyl]pyrrolidine	249.39	1	0	3.95	0
46	1-[(2E)-7-(3,4-methylenedioxyphenyl)-2-heptenoyl]pyrrolidine	301.38	3	0	3.64	0
47	1-[(2E,4E)-7-(3,4-methylenedioxyphenyl)-2,4-heptadienoyl] pyrrolidine	299.36	3	0	3.65	0

48	1-[(2E,8E)-9-(3,4-methylenedioxyphenyl)-2,8-nonadienoyl] pyrrolidine	327.42	3	0	4.04	0
49	1-[(8E)-9-(3,4-methylenedioxyphenyl)-8-nonenoyl]pyrrolidine	329.43	3	0	4.16	0
50	1-[(2E,4E,8E)-9-(3,4-methylenedioxyphenyl)-2,4,8-nonatrienoyl] pyrrolidine	325.4	3	0	3.88	0
51	1-[(2E,4E)-11-(3,4-methylenedioxyphenyl)-2,4-undecadienoyl] pyrrolidine	355.47	3	0	4.51	0
52	1-[(2E,10E)-11-(3,4-methylenedioxyphenyl)-2,10-undecadienoyl] pyrrolidine	355.47	3	0	4.16	0
53	(2E,4E)-N-isobutyl-2,4-dodecadienamide	251.41	1	1	4.06	0
54	(2E,4E)-N-isobutyl-7-(3,4-methylenedioxyphenyl)-hepta-2,4-dienamide	301.38	3	1	3.67	0
55	(8E)-N-isobutyl-9-(3,4-methylenedioxyphenyl)nona-8-enamide	331.45	3	1	4.18	0
56	(2E,4E,8E)-N-isobutyl-11-(3,4-methylenedioxyphenyl)undeca-2,4,8-trienamide	355.47	3	1	4.42	0
57	N-trans-sinapoyltyramine	343.37	5	3	2.71	0
58	Dihydrocubebin, rel-	358.39	6	2	3.19	0
59	Justiflorinol	356.33	7	1	2.76	0
60	(−)-Sanguinolignan A	384.34	8	1	2.82	0
61	(−)-Sanguinolignan B	384.34	8	1	2.79	0
62	(−)-Sanguinolignan C	442.42	9	0	3.36	0

Continued

Table 4 The Lipinski rule calculation for the phytochemical compounds of the plant involved in the formulation of Nilavembu kudineer, Kabasura kudineer, and JACOM.—cont'd

S.·no	Compounds	MW	#H-bond acceptors	#H-bond donors	iLOGP	Lipinski #violations
63	(−)-Sanguinolignan D	426.37	9	0	2.74	0
64	(7′*S*)-parabenzlactone	370.35	7	1	3.12	0
65	Flavokawain B	284.31	4	1	2.63	0
66	Methyl 3,4-dihydroxy-5-(3′-methyl-2′-butenyl)benzoate	236.26	4	2	2.62	0
67	Pipercallosidine	303.4	3	1	3.76	0
68	Kadsurenin C	358.43	5	1	3.37	0
69	Kadsurenin K	342.39	5	1	3.02	0
70	Kadsurenin L	400.46	6	0	3.42	0
71	Pipercallosine	329.43	3	1	4.04	0
72	(*S*)-1′-methylhexyl caffeate	278.34	4	2	3.41	0
73	Futoenone	340.37	5	0	3.15	0
74	(−)-Cubebin	356.37	6	1	3.18	0
75	(−)-3,4-Dimethoxy-3,4-desmethylenedioxycubebin	372.41	6	1	3.06	0
76	(*S*)-1′-methylbutyl caffeate	250.29	4	2	2.81	0
77	(*S*)-1′-methyloctyl caffeate	306.4	4	2	3.66	0
78	Futokadsurin B	356.41	5	0	3.89	0
79	Futokadsurin C	356.41	5	0	3.77	0

80	Piperine	285.34	3	0	3.42	0
81	Futokadsurin A	358.43	5	1	3.46	0
82	Burchellin	340.37	5	0	3.26	0
83	Trans-cinnamic acid	148.16	2	1	1.55	0
84	Pinocembrin chalcone	256.25	4	3	1.09	0
85	Piperlactam S	295.29	4	1	2.54	0
86	Asebogenin	288.3	5	3	1.97	0
87	(+)-Sesamin	354.35	6	0	3.46	0
88	Benzoic acid	122.12	2	1	1.11	0
89	N-feruloyltyramine	313.35	4	3	2.58	0
90	Kadsurenin M	328.36	5	0	3.15	0
91	(−)-Epicubenol	222.37	1	1	3.11	0
92	Acacetin-8-C-neohesperidoside	592.55	14	8	1.49	3
93	Monocerin	308.33	6	1	2.99	0
94	(2S,3aR,9bR)-6,7-dihydroxy-8-methoxy-2-propyl-3,3a-dihydro-2H-furo[3,2-c]isochromen-5(9bH)-one	294.3	6	2	2.51	0
95	Fusarentin 6,7-dimethyl ether	310.34	6	2	2.85	0
96	Fusarentin 6-methyl ether	296.32	6	3	2.27	0
97	(3R,4R)-4,8-dihydroxy-3-((R)-2-hydroxypentyl)-6,7-dimethoxyisochroman-1-one	326.34	7	3	2.59	0

Continued

Table 4 The Lipinski rule calculation for the phytochemical compounds of the plant involved in the formulation of Nilavembu kudineer, Kabasura kudineer, and JACOM.—cont'd

S.-no	Compounds	MW	#H-bond acceptors	#H-bond donors	iLOGP	Lipinski #violations
98	Colletotrialide, (+)-	308.33	6	1	2.49	0
99	Galgravin	372.45	5	0	3.97	0
100	Piperonal	150.13	3	0	1.6	0
101	Pinocembrin	256.25	4	2	2.11	0
102	1,4-Cineole	154.25	1	0	2.68	0
103	Sodium benzoate	144.1	2	0	−11.15	0
104	(−)-Cubenol	222.37	1	1	3.24	0
105	(E,E)-piperic acid	218.21	4	1	2.2	0
106	Kavapyrone	244.24	4	0	2.6	0
107	Alpha-cubebene	204.35	0	0	3.4	1
108	Acuminatin	340.41	4	0	4.05	0
109	Betulin di(3-carboxybutanoate)	670.92	8	2	4.69	2
110	Dihydrobetulinic acid	458.72	3	2	3.91	1
111	Bevirimat	584.83	6	2	4.25	2
112	Platanic acid	458.67	4	2	3.52	1
113	Canophyllal	440.7	2	0	4.07	1
114	Cinnamic acid	148.16	2	1	1.55	0

115	Cinnamtannin D-1	864.76	18	14	2.04	3
116	Cassiatannin A	1153.01	24	19	2.75	3
117	Cinnamtannin B-1	864.76	18	14	2.25	3
118	Parameritannin A-1	1153.01	24	19	2.6	3
119	Methyl linolenate	292.46	2	0	4.94	1
120	Zerumbone	218.33	1	0	2.72	0
121	(Z)-3-phenyl-2-propenal	132.16	1	0	1.65	0
122	(E)-2-methoxycinnamic acid	178.18	3	1	1.76	0
123	Betulinic acid	456.7	3	2	3.81	1
124	Heptacosan-1-ol	396.73	1	1	6.93	1
125	Pyrethrin I	328.45	3	0	4.22	0
126	Pyrethrin II	372.45	5	0	4.2	0
127	14-Deoxy-11,12-didehydroandrographolide	332.43	4	2	3.07	0
128	Andrographolide	350.45	5	3	2.61	0
129	Dihydroferulic acid	196.2	4	2	1.62	0
130	Mesembryanthemoidigenic acid	472.7	4	3	3.56	1
131	Methyl *N*-methylanthranilate	165.19	2	1	2.02	0
132	Delta-elemene	204.35	0	0	3.43	1
133	Decussatin	302.28	6	1	2.96	0

Continued

Table 4 The Lipinski rule calculation for the phytochemical compounds of the plant involved in the formulation of Nilavembu kudineer, Kabasura kudineer, and JACOM.—cont'd

S.·no	Compounds	MW	#H-bond acceptors	#H-bond donors	iLOGP	Lipinski #violations
134	Capensinidin	345.32	7	3	−0.95	0
135	Termilignan B	294.34	3	1	2.74	0
136	(Z)-9-hydroxybenzo[*c*]oxepin-3(1H)-one	176.17	3	1	1.37	0
137	Cyclosordariolone, (rac)-	220.22	4	3	1.44	0
138	(R)-3-Hydroxy-1-[(R)-4-hydroxy-1,3-dihydroisobenzofuran-1-yl]butan-2-one	222.24	4	2	1.7	0
139	(R)-3-Hydroxy-1-[(S)-4-hydroxy-1,3-dihydroisobenzofuran-1-yl] butan-2-one	222.24	4	2	1.51	0
140	(E)-2-(Hydroxymethyl)-3-(4-hydroxypent-1-enyl)phenol	208.25	3	3	1.99	0
141	1-(3,9-Dihydroxy-1,3-dihydrobenzo[*c*]oxepin-3-yl)ethanone, (rac)-	220.22	4	2	1.53	0
142	Pestalospirane A	408.44	6	2	2.89	0
143	Pestalospirane B	408.44	6	2	3.02	0
144	Methyl 3,4,5-trihydroxybenzoate	184.15	5	3	0.96	0
145	Ursolic acid	456.7	3	2	3.95	1
146	Arecoline	155.19	3	0	2.26	0
147	Propyl benzoate	164.2	2	0	2.46	0
148	Plectranthol A	450.52	6	4	3.49	0

149	Plectranthol B	536.66	7	3	4.59	2
150	11,20-Dihydroxysugiol	332.43	4	3	2.87	0
151	11-Hydroxysugiol	316.43	3	2	3.15	0
152	Abietatriene	270.45	0	0	3.86	1
153	1beta-hydroxy arbusculin A	266.33	4	2	2.23	0
154	Alantolactone	232.32	2	0	2.71	0
155	Dehydrocostus lactone	230.3	2	0	2.67	0
156	Costunolide	232.32	2	0	2.72	0
157	Matairesinoside	520.53	11	5	2.37	2
158	3-Hydroxyhexane-2,5-dione	130.14	3	1	0.6	0
159	Heptacosan-1-ol	396.73	1	1	6.93	1
160	6-Hydroxyluteolin 7-O-laminaribioside	626.52	17	11	1.44	3
161	Cyclotetradecane	196.37	0	0	3.35	1
162	Cycloeucalenone	424.7	1	0	4.98	1
163	Rubimaillin	284.31	4	1	2.87	0
164	Canophyllal	440.7	2	0	4.07	1
165	Stepharanine	324.35	4	2	−0.53	0
166	(Z)-icos-13-enoic acid	310.51	2	1	4.26	1
167	Methyl 3,4,5-trihydroxybenzoate	184.15	5	3	0.96	0

Continued

Table 4 The Lipinski rule calculation for the phytochemical compounds of the plant involved in the formulation of Nilavembu kudineer, Kabasura kudineer, and JACOM.—cont'd

S.·no	Compounds	MW	#H-bond acceptors	#H-bond donors	iLOGP	Lipinski #violations
168	Serratol	290.48	1	1	3.82	1
169	Huperzine A	242.32	2	2	2.42	0
170	Maytansine	692.2	10	2	4.27	2
171	Mustakone	218.33	1	0	2.95	0
172	(+)-nootkatone	218.33	1	0	2.83	0
173	Pectenotoxin-11	875.05	15	4	4.09	2
174	Dinophysistoxin 2	805	13	5	6.22	2
175	1-Icosanoylglycerol	386.61	4	2	4.71	0
176	Swertisin	446.4	10	6	2.5	1
177	Serratol	290.48	1	1	3.82	1
178	Raffinose	504.44	16	11	1.12	3
179	PLX-4720	413.83	6	2	2.42	0
180	Huperzine A	242.32	2	2	2.42	0
181	Maytansine	692.2	10	2	4.27	2
182	N-(2-methoxyethyl)-4-{[6-(pyridin-4-yl)quinazolin-2-yl]amino}benzamide	399.45	5	2	2.8	0
183	Glucotropeolin	409.43	10	5	0.57	0

184	Glucotropeolin(1-)	408.42	10	4	1.04	0
185	Methyl 12-methyltetradecanoate	256.42	2	0	4.08	1
186	(3S,5R,6S)-beta-cryptoxanthin 5,6-epoxide	568.87	2	1	7.8	2
187	Carpaine	478.71	6	2	4.4	0
188	Beta-cryptoxanthin	552.87	1	1	7.6	2
189	24-Methylenecycloartanol	440.74	1	1	5.31	1
190	Helvolic acid methyl ester	582.72	8	0	4.5	1
191	Danielone	212.2	5	2	1.63	0
192	2-Isobutylthiazole	141.23	1	0	2.37	0
193	14-Deoxy-11,12-didehydroandrographolide	332.43	4	2	3.07	0
194	Andrographolide	350.45	5	3	2.61	0
195	Dihydroferulic acid	196.2	4	2	1.62	0
196	Mesembryanthemoidigenic acid	472.7	4	3	3.56	1
197	Methyl *N*-methylanthranilate	165.19	2	1	2.02	0
198	Delta-elemene	204.35	0	0	3.43	1
199	Decussatin	302.28	6	1	2.96	0
200	Linalool	154.25	1	1	2.7	0
201	Nevadensin	344.32	7	2	3	0
202	7-Epi-sesquithujene	204.35	0	0	3.37	1

Continued

Table 4 The Lipinski rule calculation for the phytochemical compounds of the plant involved in the formulation of Nilavembu kudineer, Kabasura kudineer, and JACOM.—cont'd

S.·no	Compounds	MW	#H-bond acceptors	#H-bond donors	iLOGP	Lipinski #violations
203	Selina-4(15),7(11)-diene	204.35	0	0	3.31	1
204	Codeine	299.36	4	1	2.67	0
205	Sulfanilamide	172.2	3	2	0.61	0
206	Avicularin	434.35	11	7	1.86	2
207	Pipercyclobutanamide A(rel)	570.68	6	0	5.26	1
208	(−)-Cubebin	356.37	6	1	3.18	0
209	(−)-3,4-Dimethoxy-3,4-desmethylenedioxycubebin	372.41	6	1	3.06	0
210	Piperine	285.34	3	0	3.42	0
211	Pellitorine	223.35	1	1	3.61	0
212	Gaudichaudianic acid, (−rac)	340.46	3	1	3.85	1
213	Sarmentosumin A	680.74	8	6	3.7	2
214	Sarmentosumin B	680.74	8	6	3.97	2
215	Sarmentosumin C	786.86	9	7	4.13	3
216	Sarmentosumin D	786.86	9	7	4.06	3
217	Isochamanetin	362.38	5	3	2.68	0
218	7-Methoxychamanetin	376.4	5	2	3.2	0
219	Dichamanetin	468.5	6	4	3.01	0

220	7-Methoxydichamanetin	482.52	6	3	3.58	0
221	5″-(2‴′-Hydroxybenzyl)uvarinol	694.77	8	5	4.16	1
222	2,4-Dodecadienamide	195.3	1	1	2.87	0
223	7-Methoxyisochamanetin	376.4	5	2	3.5	0
224	(2E,4E)-N-[2-(methylsulfinyl)ethyl]-2,4-decadienamide	257.39	2	1	3.05	0
225	(2E,4E)-N-[(4-hydroxy-3-methoxyphenyl)ethyl]-2,4-decadienamide	317.42	3	2	3.76	0
226	3-(4-Hydroxy-3,5-dimethoxyphenyl)propanoylpyrrole	275.3	4	1	2.63	0
227	3-(3,4,5-Timethoxyphenyl)propanoylpyrrole	289.33	4	0	3.21	0
228	1-[(2E,4E,6E)-2,4,6-dodecatrienoyl]pyrrolidine	247.38	1	0	3.86	0
229	1-[(2E,4Z,8E)-9-(3,4-methylenedioxyphenyl)-2,4,8-nonatrienoyl] pyrrolidine	325.4	3	0	3.88	0
230	1-[(2E,4E,10E)-10-(3,4-methylenedioxyphenyl)-2,4,10-undecatrienoyl]pyrrolidine	353.45	3	0	4.53	0
231	1-[(4E,10E)-11-(3,4-methylenedioxyphenyl)-4,10-undecadienoyl] pyrrolidine	355.47	3	0	4.6	0
232	1-[(9E)-10-(3,4-methylenedioxyphenyl)-9-decenoyl]pyrrolidine	343.46	3	0	4.43	0
233	1-[(2E,4E)-2,4-decadienoyl]pyrrolidine	221.34	1	0	3.45	0
234	1-[(2E,4E)-2,4-dodecadienoyl]pyrrolidine	249.39	1	0	3.95	0
235	1-[(2E)-7-(3,4-methylenedioxyphenyl)-2-heptenoyl]pyrrolidine	301.38	3	0	3.64	0

Continued

Table 4 The Lipinski rule calculation for the phytochemical compounds of the plant involved in the formulation of Nilavembu kudineer, Kabasura kudineer, and JACOM.—cont'd

S.·no	Compounds	MW	#H-bond acceptors	#H-bond donors	iLOGP	Lipinski #violations
236	1-[(2E,4E)-7-(3,4-methylenedioxyphenyl)-2,4-heptadienoyl] pyrrolidine	299.36	3	0	3.65	0
237	1-[(2E,8E)-9-(3,4-methylenedioxyphenyl)-2,8-nonadienoyl] pyrrolidine	327.42	3	0	4.04	0
238	1-[(8E)-9-(3,4-methylenedioxyphenyl)-8-nonenoyl]pyrrolidine	329.43	3	0	4.16	0
239	1-[(2E,4E,8E)-9-(3,4-methylenedioxyphenyl)-2,4,8-nonatrienoyl] pyrrolidine	325.4	3	0	3.88	0
240	1-[(2E,4E)-11-(3,4-methylenedioxyphenyl)-2,4-undecadienoyl] pyrrolidine	355.47	3	0	4.51	0
241	1-[(2E,10E)-11-(3,4-methylenedioxyphenyl)-2,10-undecadienoyl] pyrrolidine	355.47	3	0	4.16	0
242	(2E,4E)-N-isobutyl-2,4-dodecadienamide	251.41	1	1	4.06	0
243	(2E,4E)-N-isobutyl-7-(3,4-methylenedioxyphenyl)-hepta-2,4-dienamide	301.38	3	1	3.67	0
244	(8E)-N-isobutyl-9-(3,4-methylenedioxyphenyl)nona-8-enamide	331.45	3	1	4.18	0
245	(2E,4E,8E)-N-isobutyl-11-(3,4-methylenedioxyphenyl)undeca-2,4,8-trienamide	355.47	3	1	4.42	0
246	N-trans-sinapoyltyramine	343.37	5	3	2.71	0
247	Dihydrocubebin, rel-	358.39	6	2	3.19	0

248	Justiflorinol	356.33	7	1	2.76	0
249	(−)-Sanguinolignan A	384.34	8	1	2.82	0
250	(−)-Sanguinolignan B	384.34	8	1	2.79	0
251	(−)-Sanguinolignan C	442.42	9	0	3.36	0
252	(−)-Sanguinolignan D	426.37	9	0	2.74	0
253	(7′S)-parabenzlactone	370.35	7	1	3.12	0
254	Flavokawain B	284.31	4	1	2.63	0
255	Methyl 3,4-dihydroxy-5-(3′-methyl-2′-butenyl)benzoate	236.26	4	2	2.62	0
256	Pipercallosidine	303.4	3	1	3.76	0
257	Kadsurenin C	358.43	5	1	3.37	0
258	Kadsurenin K	342.39	5	1	3.02	0
259	Kadsurenin L	400.46	6	0	3.42	0
260	Pipercallosine	329.43	3	1	4.04	0
261	(S)-1′-methylhexyl caffeate	278.34	4	2	3.41	0
262	Futoenone	340.37	5	0	3.15	0
263	(S)-1′-methylbutyl caffeate	250.29	4	2	2.81	0
264	(S)-1′-methyloctyl caffeate	306.4	4	2	3.66	0
265	Futokadsurin B	356.41	5	0	3.89	0
266	Futokadsurin C	356.41	5	0	3.77	0

Continued

Table 4 The Lipinski rule calculation for the phytochemical compounds of the plant involved in the formulation of Nilavembu kudineer, Kabasura kudineer, and JACOM.—cont'd

S.·no	Compounds	MW	#H-bond acceptors	#H-bond donors	iLOGP	Lipinski #violations
267	Futokadsurin A	358.43	5	1	3.46	0
268	Burchellin	340.37	5	0	3.26	0
269	Piperonal	150.13	3	0	1.6	0
270	(−)-Cubenol	222.37	1	1	3.24	0
271	(E,E)-piperic acid	218.21	4	1	2.2	0
272	Trans-cinnamic acid	148.16	2	1	1.55	0
273	Pinocembrin chalcone	256.25	4	3	1.09	0
274	Piperlactam S	295.29	4	1	2.54	0
275	Asebogenin	288.3	5	3	1.97	0
276	Epicocconone	410.42	7	2	3.17	0
277	(+)-Sesamin	354.35	6	0	3.46	0
278	Pipataline	288.42	2	0	4.61	1
279	(−)-Antofine	363.45	4	0	3.8	0
280	Benzoic acid	122.12	2	1	1.11	0
281	Ethyl butyrate	116.16	2	0	1.9	0
282	N-feruloyltyramine	313.35	4	3	2.58	0
283	Kadsurenin M	328.36	5	0	3.15	0

284	(−)-Epicubenol	222.37	1	1	3.11	0
285	Acacetin-8-C-neohesperidoside	592.55	14	8	1.49	3
286	Monocerin	308.33	6	1	2.99	0
287	(2S,3aR,9bR)-6,7-dihydroxy-8-methoxy-2-propyl-3, 3a-dihydro-2H-furo[3,2-c]isochromen-5(9bH)-one	294.3	6	2	2.51	0
288	Fusarentin 6,7-dimethyl ether	310.34	6	2	2.85	0
289	Fusarentin 6-methyl ether	296.32	6	3	2.27	0
290	(3R,4R)-4,8-dihydroxy-3-((R)-2-hydroxypentyl)-6,7-dimethoxyisochroman-1-one	326.34	7	3	2.59	0
291	Colletotrialide, (+)-	308.33	6	1	2.49	0
292	Galgravin	372.45	5	0	3.97	0
293	9,10-Epoxy-18-hydroxyoctadecanoic acid	314.46	4	2	3.8	0
294	Pinocembrin	256.25	4	2	2.11	0
295	1,4-Cineole	154.25	1	0	2.68	0
296	Sodium benzoate	144.1	2	0	−11.15	0
297	9,10,18-Trihydroxyoctadecanoic acid	332.48	5	4	3.25	0
298	Kavapyrone	244.24	4	0	2.6	0
299	Alpha-cubebene	204.35	0	0	3.4	1
300	Acuminatin	340.41	4	0	4.05	0
301	Jerantinine E	384.47	5	2	3.52	0

Continued

Table 4 The Lipinski rule calculation for the phytochemical compounds of the plant involved in the formulation of Nilavembu kudineer, Kabasura kudineer, and JACOM.—cont'd

S.·no	Compounds	MW	#H-bond acceptors	#H-bond donors	iLOGP	Lipinski #violations
302	Jerantinine F	398.45	6	2	3.4	0
303	Jerantinine C	396.44	5	2	3.13	0
304	Jerantinine B	398.45	6	2	3.29	0
305	Jerantinine D	412.44	6	2	3.07	0
306	Jerantinine A	382.45	5	2	3.49	0
307	Tabernaemontanine	354.44	4	1	2.79	0
308	Mangiferin	422.34	11	8	0.89	2
309	Plectranthol A	450.52	6	4	3.49	0
310	Plectranthol B	536.66	7	3	4.59	2
311	11,20-Dihydroxysugiol	332.43	4	3	2.87	0
312	11-Hydroxysugiol	316.43	3	2	3.15	0
313	Abietatriene	270.45	0	0	3.86	1
314	7-oxo-10α-cucurbitadienol	440.7	2	1	4.65	1

Table 5 The list of phytochemical those satisfies the Lipinski rule of 5 and their structures in SMILES format.

S.·no	Compound	SMILES (chEBI)
1	Gingerenone B	C=1(C(=CC=C(C1)CCC(/C=C/CCC=2C=C(C(=C(C2)OC)O)OC)=O)O)OC
2	Beta-sesquiphellandrene	[H][C@@]1(CCC(=C)C=C1)[C@@H](C)CCC=C(C)C
3	3-(3,4-Dimethoxyphenyl)-4-[(Z)-3,4-dimethoxystyryl]cyclohex-1-ene	COc1ccc(\C=C/C2CCC=CC2c2ccc(OC)c(OC)c2)cc1OC
4	3-(3,4-Dimethoxyphenyl)-4-[(E)-3,4-dimethoxystyryl]cyclohex-1-ene	COc1ccc(\C=C\C2CCC=CC2c2ccc(OC)c(OC)c2)cc1OC
5	Pinocarveol	CC1(C)C2CC(O)C(=C)C1C2
6	Zerumboneoxide	[H][C@]12CC(C)(C)\C=C\C(=O)\C(C)=C\CC[C@@]1(C)O2
7	Ramonanin A, (rel)-	COc1cc(ccc1O)[C@H]1O[C@@H](c2ccc(O)c(OC)c2)[C@@]2(CCC3=C(C2)[C@H](O[C@H]3c2ccc(O)c(OC)c2)c2ccc(O)c(OC)c2)C1=C
8	Ramonanin B, (rel)-	COc1cc(ccc1O)[C@@H]1O[C@H](c2ccc(O)c(OC)c2)[C@@]2(CCC3=C(C2)[C@H](O[C@H]3c2ccc(O)c(OC)c2)c2ccc(O)c(OC)c2)C1=C
9	Ramonanin C, (rel)-	COc1cc(ccc1O)[C@H]1O[C@@H](c2ccc(O)c(OC)c2)[C@@]2(CCC3=C(C2)[C@@H](O[C@@H]3c2ccc(O)c(OC)c2)c2ccc(O)c(OC)c2)C1=C
10	Ramonanin D, (rel)-	COc1cc(ccc1O)[C@H]1O[C@@H](c2ccc(O)c(OC)c2)[C@@]2(CCC3=C(C2)[C@H](O[C@@H]3c2ccc(O)c(OC)c2)c2ccc(O)c(OC)c2)C1=C

Continued

Table 5 The list of phytochemical those satisfies the Lipinski rule of 5 and their structures in SMILES format.—cont'd

S.·no	Compound	SMILES (chEBI)
11	Zerumbone	C\C1=C/CC(C)(C)\C=C\C(=O)\C(C)=C\CC1
12	5-Hydroxyzerumbone	C\C1=C/CC(C)(C)\C=C\C(=O)\C(C)=C\C(O)C1
13	(2E,6E)-hedycaryol	C\C1=C/CC\C(C)=C\C[C@@H](CC1)C(C)(C)O
14	Dodecane	CCCCCCCCCCCC
15	7,4′-Dimethylkaempferol	C12=C(OC(C3=CC=C(OC)C=C3)=C(C1=O)O)C=C(OC) C=C2O
16	Zingiberene	[H][C@@]1(CC=C(C)C=C1)[C@@H](C)CCC=C(C)C
17	Glucoputranjivin(1-)	[C@H]1(O[C@@H]([C@@H](O)[C@@H]([C@H]1O)O)CO)S/C (=N\OS([O-])(=O)=O)/C(C)C
18	Glucoputranjivin	[C@H]1(O[C@@H]([C@@H](O)[C@@H]([C@H]1O)O)CO)S/C (=N\OS(O)(=O)=O)/C(C)C
19	All-cis-octadeca-6,9,12,15-tetraenoic acid	CC\C=C/C\C=C/C\C=C/C\C=C/CCCCC(O)=O
20	Pipataline	O1C=2C=C(\C=C\CCCCCCCCC)C=CC2OC1
21	Pipercyclobutanamide A(rel)	O=C(\C=C/[C@@H]1[C@@H](\C=C\c2ccc3OCOc3c2) [C@@H]([C@H]1c1ccc2OCOc2c1)C(=O)N1CCCCC1) N1CCCCC1
22	Pellitorine	CCCCC\C=C\C=C\C(=O)NCC(C)C
23	Gaudichaudianic acid, (−rac)	CC(C)=CCCC1(C)Oc2c(CC=C(C)C)cc(cc2C=C1)C(O)=O
24	Isochamanetin	Oc1ccccc1Cc1c(O)cc2O[C@@H](CC(=O)c2c1O)c1ccccc1

25	7-Methoxychamanetin	COc1cc(O)c2C(=O)C[C@H](Oc2c1Cc1ccccc1O)c1ccccc1
26	Dichamanetin	Oc1ccccc1Cc1c(O)c(Cc2ccccc2O)c2O[C@@H](CC(=O)c2c1O)c1ccccc1
27	7-Methoxydichamanetin	COc1c(Cc2ccccc2O)c(O)c2C(=O)C[C@H](Oc2c1Cc1ccccc1O)c1ccccc1
28	5″-(2‴′-Hydroxybenzyl)uvarinol	COc1c(Cc2cc(Cc3ccccc3O)ccc2O)c(O)c2C(=O)C[C@H](Oc2c1Cc1cc(Cc2ccccc2O)ccc1O)c1ccccc1
29	2,4-Dodecadienamide	CCCCCCC\C=C\C=C\C(N)=O
30	7-Methoxyisochamanetin	COc1cc2O[C@@H](CC(=O)c2c(O)c1Cc1ccccc1O)c1ccccc1
31	(2E,4E)-N-[2-(methylsulfinyl)ethyl]-2,4-decadienamide	CCCCC\C=C\C=C\C(=O)NCCS(C)=O
32	(2E,4E)-N-[(4-hydroxy-3-methoxyphenyl)ethyl]-2,4-decadienamide	CCCCC\C=C\C=C\C(=O)NCCc1ccc(O)c(OC)c1
33	3-(4-Hydroxy-3,5-dimethoxyphenyl)propanoylpyrrole	COc1cc(CCC(=O)n2cccc2)cc(OC)c1O
34	3-(3,4,5-Timethoxyphenyl)propanoylpyrrole	COc1cc(CCC(=O)n2cccc2)cc(OC)c1OC
35	1-[(2E,4E,6E)-2,4,6-dodecatrienoyl]pyrrolidine	CCCCC\C=C\C=C\C=C\C(=O)N1CCCC1
36	1-[(2E,4Z,8E)-9-(3,4-methylenedioxyphenyl)-2,4,8-nonatrienoyl]pyrrolidine	O=C(\C=C\C=C/CC\C=C\c1ccc2OCOc2c1)N1CCCC1
37	1-[(2E,4E,10E)-10-(3,4-methylenedioxyphenyl)-2,4,10-undecatrienoyl]pyrrolidine	O=C(\C=C\C=C\CCCC\C=C\c1ccc2OCOc2c1)N1CCCC1

Continued

Table 5 The list of phytochemical those satisfies the Lipinski rule of 5 and their structures in SMILES format.—cont'd

S.·no	Compound	SMILES (chEBI)
38	1-[(4E,10E)-11-(3,4-methylenedioxyphenyl)-4, 10-undecadienoyl]pyrrolidine	O=C(CC\C=C\CCCC\C=C\c1ccc2OCOc2c1)N1CCCC1
39	1-[(9E)-10-(3,4-methylenedioxyphenyl)-9-decenoyl]pyrrolidine	O=C(CCCCCCC\C=C\c1ccc2OCOc2c1)N1CCCC1
40	1-[(2E,4E)-2,4-decadienoyl]pyrrolidine	CCCCC\C=C\C=C\C(=O)N1CCCC1
41	1-[(2E,4E)-2,4-dodecadienoyl]pyrrolidine	CCCCCCC\C=C\C=C\C(=O)N1CCCC1
42	1-[(2E)-7-(3,4-methylenedioxyphenyl)-2-heptenoyl]pyrrolidine	O=C(\C=C\CCCCc1ccc2OCOc2c1)N1CCCC1
43	1-[(2E,4E)-7-(3,4-methylenedioxyphenyl)-2, 4-heptadienoyl]pyrrolidine	O=C(\C=C\C=C\CCc1ccc2OCOc2c1)N1CCCC1
44	1-[(2E,8E)-9-(3,4-methylenedioxyphenyl)-2, 8-nonadienoyl]pyrrolidine	O=C(\C=C\CCCC\C=C\c1ccc2OCOc2c1)N1CCCC1
45	1-[(8E)-9-(3,4-methylenedioxyphenyl)-8-nonenoyl]pyrrolidine	O=C(CCCCCC\C=C\c1ccc2OCOc2c1)N1CCCC1
46	1-[(2E,4E,8E)-9-(3,4-methylenedioxyphenyl)-2, 4,8-nonatrienoyl]pyrrolidine	O=C(\C=C\C=C\CC\C=C\c1ccc2OCOc2c1)N1CCCC1
47	1-[(2E,4E)-11-(3,4-methylenedioxyphenyl)-2, 4-undecadienoyl]pyrrolidine	O=C(\C=C\C=C\CCCCCCc1ccc2OCOc2c1)N1CCCC1
48	1-[(2E,10E)-11-(3,4-methylenedioxyphenyl)-2, 10-undecadienoyl]pyrrolidine	O=C(\C=C\CCCCCC\C=C\c1ccc2OCOc2c1)N1CCCC1

49	(2E,4E)-N-isobutyl-2,4-dodecadienamide	CCCCCCC\C=C\C=C\C(=O)NCC(C)C
50	(2E,4E)-N-isobutyl-7-(3,4-methylenedioxyphenyl)-hepta-2,4-dienamide	CC(C)CNC(=O)\C=C\C=C\CCc1ccc2OCOc2c1
51	(8E)-N-isobutyl-9-(3,4-methylenedioxyphenyl)nona-8-enamide	CC(C)CNC(=O)CCCCCC\C=C\c1ccc2OCOc2c1
52	(2E,4E,8E)-N-isobutyl-11-(3,4-methylenedioxyphenyl)undeca-2,4,8-trienamide	CC(C)CNC(=O)\C=C\C=C\CC\C=C\CCc1ccc2OCOc2c1
53	N-trans-sinapoyltyramine	COc1cc(cc(OC)c1O)\C=C\C(=O)NCCc1ccc(O)cc1
54	Dihydrocubebin, rel-	OC[C@@H](Cc1ccc2OCOc2c1)[C@@H](CO)Cc1ccc2OCOc2c1
55	Justiflorinol	OCC(CC(=O)c1ccc2OCOc2c1)C(=O)c1ccc2OCOc2c1
56	(−)-Sanguinolignan A	O[C@@H]([C@H]1COC(=O)[C@@H]1C(=O)c1ccc2OCOc2c1)c1ccc2OCOc2c1
57	(−)-Sanguinolignan B	O[C@@H]([C@@H]1COC(=O)[C@@H]1C(=O)c1ccc2OCOc2c1)c1ccc2OCOc2c1
58	(−)-Sanguinolignan C	COc1ccc(cc1OC)[C@@H](OC(C)=O)[C@H]1COC(=O)[C@@H]1C(=O)c1ccc2OCOc2c1
59	(−)-Sanguinolignan D	CC(=O)O[C@@H]([C@@H]1[C@H](COC1=O)C(=O)c1ccc2OCOc2c1)c1ccc2OCOc2c1
60	(7′S)-parabenzlactone	O[C@@H]([C@H]1COC(=O)[C@@H]1Cc1ccc2OCOc2c1)c1ccc2OCOc2c1
61	Flavokawain B	COc1cc(O)c(C(=O)\C=C\c2ccccc2)c(OC)c1

Continued

Table 5 The list of phytochemical those satisfies the Lipinski rule of 5 and their structures in SMILES format.—cont'd

S.·no	Compound	SMILES (chEBI)
62	Methyl 3,4-dihydroxy-5-(3′-methyl-2′-butenyl)benzoate	COC(=O)c1cc(O)c(O)c(CC=C(C)C)c1
63	Pipercallosidine	C(/C=C/CCCCC=1C=C2C(=CC1)OCO2)(NCC(C)C)=O
64	Kadsurenin C	O([C@@]12[C@@H]([C@H]([C@@]([C@H]1O)(C(=O)C(=C2)CC=C)[H])C3=CC(OC)=C(OC)C=C3)C)C
65	Kadsurenin K	O([C@@]12[C@@H]([C@H]([C@@](C1=O)(C(=O)C(=C2)CC=C)[H])C3=CC(OC)=C(O)C=C3)C)C
66	Kadsurenin L	O([C@@]12[C@@H]([C@H]([C@@]([C@@H]1OC(=O)C)(C(=O)C(=C2)CC=C)[H])C3=CC(OC)=C(OC)C=C3)C)C
67	Pipercallosine	C(/C=C/C=C/CCCCC1=CC=C2C(=C1)OCO2)(=O)NCC(C)C
68	(S)-1′-methylhexyl caffeate	CCCCC[C@H](C)OC(=O)\C=C\c1ccc(O)c(O)c1
69	Futoenone	O1[C@@H]2C[C@@]3([C@@H]([C@H](C2)C4=CC=5OCOC5C=C4)C)C1=CC(=O)C(OC)=C3
70	(−)-Cubebin	[H][C@@]1(CO[C@H](O)[C@]1([H])Cc1ccc2OCOc2c1)Cc1ccc2OCOc2c1
71	(−)-3,4-Dimethoxy-3,4-desmethylenedioxycubebin	[H][C@@]1(COC(O)[C@]1([H])Cc1ccc2OCOc2c1)Cc1ccc(OC)c(OC)c1
72	(S)-1′-methylbutyl caffeate	CCC[C@H](C)OC(=O)\C=C\c1ccc(O)c(O)c1
73	(S)-1′-methyloctyl caffeate	CCCCCCC[C@H](C)OC(=O)\C=C\c1ccc(O)c(O)c1

74	Futokadsurin B	COc1ccc(cc1OC)[C@@H]1O[C@@H]([C@H](C)[C@H]1C)c1ccc2OCOc2c1
75	Futokadsurin C	COc1ccc(cc1OC)[C@H]1O[C@H]([C@H](C)[C@@H]1C)c1ccc2OCOc2c1
76	Piperine	O=C(\C=C\C=C\c1ccc2OCOc2c1)N1CCCCC1
77	Futokadsurin A	COc1cc(ccc1O)[C@@H]1O[C@@H]([C@@H](C)[C@@H]1C)c1ccc(OC)c(OC)c1
78	Burchellin	O1[C@@H]([C@H]([C@]2(C1=CC(=O)C(OC)=C2)CC=C)C)C3=CC=4OCOC4C=C3
79	Trans-cinnamic acid	OC(=O)\C=C\c1ccccc1
80	Pinocembrin chalcone	Oc1cc(O)c(C(=O)\C=C\c2ccccc2)c(O)c1
81	Piperlactam S	O(C=1C2=C3C(=CC=4C2=CC=CC4)N(C(C3=CC1O)=O)OC)C
82	Asebogenin	COc1cc(O)c(C(=O)CCc2ccc(O)cc2)c(O)c1
83	(+)-Sesamin	[C@]12([C@@]([C@H](OC1)C3=CC4=C(C=C3)OCO4)(CO[C@@H]2C5=CC6=C(C=C5)OCO6)[H])[H]
84	Benzoic acid	OC(=O)c1ccccc1
85	N-feruloyltyramine	COc1cc(\C=C\C(=O)NCCc2ccc(O)cc2)ccc1O
86	Kadsurenin M	O1[C@@H]([C@H](C2=C1C(OC)=CC(=C2)C(=O)[H])C)C3=CC(OC)=C(OC)C=C3
87	(−)-Epicubenol	[C@]12([C@@H](CC[C@H]([C@]2(CCC(=C1)C)O)C)C(C)C)[H]

Continued

Table 5 The list of phytochemical those satisfies the Lipinski rule of 5 and their structures in SMILES format.—cont'd

S.·no	Compound	SMILES (chEBI)
88	Monocerin	[H][C@@]12C[C@H](CCC)O[C@]1([H])c1cc(OC)c(OC)c(O)c1C(=O)O2
89	(2S,3aR,9bR)-6,7-dihydroxy-8-methoxy-2-propyl-3,3a-dihydro-2H-furo[3,2-c]isochromen-5(9bH)-one	[H][C@@]12C[C@H](CCC)O[C@]1([H])c1cc(OC)c(O)c(O)c1C(=O)O2
90	Fusarentin 6,7-dimethyl ether	CCC[C@H](O)C[C@@H]1Cc2cc(OC)c(OC)c(O)c2C(=O)O1
91	Fusarentin 6-methyl ether	CCC[C@H](O)C[C@@H]1Cc2cc(OC)c(O)c(O)c2C(=O)O1
92	(3R,4R)-4,8-dihydroxy-3-((R)-2-hydroxypentyl)-6,7-dimethoxyisochroman-1-one	CCC[C@@H](O)C[C@H]1OC(=O)c2c(O)c(OC)c(OC)cc2[C@H]1O
93	Colletotrialide, (+)-	CCCC(=O)CC[C@H]1OC(=O)c2c(O)c(OC)c(OC)cc12
94	Galgravin	C1(=C(C=CC(=C1)[C@@H]2O[C@@H]([C@H]([C@H]2C)C)C3=CC=C(C(=C3)OC)OC)OC)OC
95	Piperonal	[H]C(=O)c1ccc2OCOc2c1
96	Pinocembrin	Oc1cc(O)c2C(=O)C[C@H](Oc2c1)c1ccccc1
97	1,4-Cineole	C1C[C@]2(CC[C@]1(C)O2)C(C)C
98	Sodium benzoate	C(C=1C=CC=CC1)([O-])=O.[Na+]
99	(−)-Cubenol	[C@]12([C@@H](CC[C@H]([C@@]2(CCC(=C1)C)O)C)C(C)C)[H]
100	(E,E)-piperic acid	OC(=O)\C=C\C=C\c1ccc2OCOc2c1

101	Kavapyrone	COc1cc(oc(=O)c1)[C@@H]1O[C@H]1c1ccccc1
102	Alpha-cubebene	C1C[C@H]([C@]2([C@]3([C@@H]1C)CC=C([C@]23[H])C)[H])C(C)C
103	Acuminatin	O1[C@H]([C@@H](C2=C1C(OC)=CC(=C2)/C=C/C)C)C3=CC(OC)=C(OC)C=C3
104	Dihydrobetulinic acid	[H][C@]12CC[C@]3([H])[C@@]4(C)CC[C@H](O)C(C)(C)[C@]4([H])CC[C@@]3(C)[C@]1(C)CC[C@]1(CC[C@@H](C(C)C)[C@]21[H])C(O)=O
105	Platanic acid	[H][C@]12CC[C@]3([H])[C@@]4(C)CC[C@H](O)C(C)(C)[C@]4([H])CC[C@@]3(C)[C@]1(C)CC[C@]1(CC[C@@H](C(C)=O)[C@]21[H])C(O)=O
106	Canophyllal	[C@]12([C@]([C@]3([C@@](CC1)(CCC(C3)(C)C)C(=O)[H])[H])(CC[C@@]4([C@@]2(CC[C@]5([C@]4(CCC([C@@H]5C)=O)[H])C)[H])C)C)C
107	Cinnamic acid	[H]C(=Cc1ccccc1)C(O)=O
108	Methyl linolenate	CC/C=C\C/C=C\C/C=C\CCCCCCCC(=O)OC
109	Zerumbone	C\C1=C/CC(C)(C)\C=C\C(=O)\C(C)=C\CC1
110	(Z)-3-phenyl-2-propenal	C1=CC=C(C=C1)/C=C\C=O
111	(E)-2-methoxycinnamic acid	COC=1C=CC=CC1/C=C/C(=O)O
112	Betulinic acid	[H][C@]12CC[C@]3([H])[C@@]4(C)CC[C@H](O)C(C)(C)[C@]4([H])CC[C@@]3(C)[C@]1(C)CC[C@]1(CC[C@@H](C(C)=C)[C@]21[H])C(O)=O

Continued

Table 5 The list of phytochemical those satisfies the Lipinski rule of 5 and their structures in SMILES format.—cont'd

S.·no	Compound	SMILES (chEBI)
113	Heptacosan-1-ol	C(CCCCCCCCCCCCCCCCO)CCCCCCCCCC
114	Pyrethrin I	CC(C)=C[C@@H]1[C@@H](C(=O)O[C@H]2CC(=O)C(C \C=C/C=C)=C2C)C1(C)C
115	Pyrethrin II	COC(=O)C(\C)=C\[C@@H]1[C@@H](C(=O)O[C@H]2CC (=O)C(C\C=C/C=C)=C2C)C1(C)C
116	14-Deoxy-11,12-didehydroandrographolide	[H][C@]12CCC(=C)[C@@H](\C=C\C3=CCOC3=O)[C@]1(C) CC[C@@H](O)[C@@]2(C)CO
117	Andrographolide	[H][C@]12CCC(=C)[C@@H](C\C=C3/[C@H](O)COC3=O) [C@]1(C)CC[C@@H](O)[C@@]2(C)CO
118	Dihydroferulic acid	C=1(C=C(C(=CC1)O)OC)CCC(=O)O
119	Mesembryanthemoidigenic acid	[C@@]12(C([C@@]3(C[C@@](CO)(C)CC[C@@]3(CC1)C(=O)O) [H])=CC[C@@]4([C@]5(CC[C@@H](C([C@@]5(CC[C@@] 24C)[H])(C)C)O)C)[H])C
120	Methyl *N*-methylanthranilate	CNC1=CC=CC=C1C(OC)=O
121	Delta-elemene	[C@@H]1(C=C(CC[C@@]1(C=C)C)C(C)C)C(C)=C
122	Decussatin	COc1cc(O)c2c(c1)oc1ccc(OC)c(OC)c1c2=O
123	Capensinidin	COc1cc(cc(OC)c1O)-c1[o+]c2cc(O)cc(OC)c2cc1O
124	Termilignan B	Oc1ccc(CC(=C)C(=C)Cc2ccc3OCOc3c2)cc1
125	(Z)-9-hydroxybenzo[*c*]oxepin-3(1H)-one	Oc1cccc2C=CC(=O)OCc12

126	Cyclosordariolone, (rac)-	CC1(O)C(=O)C=Cc2c(CO)c(O)ccc12
127	(R)-3-Hydroxy-1-[(R)-4-hydroxy-1, 3-dihydroisobenzofuran-1-yl]butan-2-one	C[C@@H](O)C(=O)C[C@H]1OCc2c(O)cccc12
128	(R)-3-Hydroxy-1-[(S)-4-hydroxy-1, 3-dihydroisobenzofuran-1-yl]butan-2-one	C[C@@H](O)C(=O)C[C@@H]1OCc2c(O)cccc12
129	(E)-2-(Hydroxymethyl)-3-(4-hydroxypent-1-enyl)phenol	C[C@@H](O)C\C=C\c1cccc(O)c1CO
130	1-(3,9-Dihydroxy-1, 3-dihydrobenzo[*c*]oxepin-3-yl)ethanone, (rac)-	CC(=O)C1(O)OCc2c(O)cccc2C=C1
131	Pestalospirane A	C[C@H]1O[C@]2(OCc3c(O)cccc3C=C2)[C@@H](C)O[C@@]11OCc2c(O)cccc2C=C1
132	Pestalospirane B	C[C@H]1O[C@]2(OCc3c(O)cccc3C=C2)[C@@H](C)O[C@]11OCc2c(O)cccc2C=C1
133	Methyl 3,4,5-trihydroxybenzoate	OC1=CC(=CC(O)=C1O)C(OC)=O
134	Ursolic acid	C[C@@H]1CC[C@@]2(CC[C@]3(C)C(=CC[C@@H]4[C@@]5(C)CC[C@H](O)C(C)(C)[C@@H]5CC[C@@]34C)[C@@H]2[C@H]1C)C(O)=O
135	Arecoline	COC(=O)C1=CCCN(C)C1
136	Propyl benzoate	C=1C=CC(=CC1)C(=O)OCCC
137	Plectranthol A	CC(C)c1cc2C=C[C@@]3(C)C(=CCC[C@]3(C)COC(=O)c3ccc(O)c(O)c3)c2c(O)c1O

Continued

Table 5 The list of phytochemical those satisfies the Lipinski rule of 5 and their structures in SMILES format.—cont'd

S.·no	Compound	SMILES (chEBI)
138	11,20-Dihydroxysugiol	C1=C(C(C)C)C(=C(C2=C1C(C[C@@]3([C@@]2(CCCC3(C)C)CO)[H])=O)O)O
139	11-Hydroxysugiol	C1=C(C(C)C)C(=C(C2=C1C(C[C@@]3([C@@]2(CCCC3(C)C)C)[H])=O)O)O
140	Abietatriene	CC(C)c1ccc2c(CC[C@H]3C(C)(C)CCC[C@]23C)c1
141	1beta-hydroxy arbusculin A	[H][C@@]12CC[C@@]3(C)[C@H](O)CC[C@@](C)(O)[C@]3([H])[C@@]1([H])OC(=O)C2=C
142	Alantolactone	[H][C@@]12C[C@@]3(C)CCC[C@H](C)C3=C[C@]1([H])C(=C)C(=O)O2
143	Dehydrocostus lactone	[H][C@@]12CCC(=C)[C@]1([H])[C@H]1OC(=O)C(=C)[C@@H]1CCC2=C
144	Costunolide	C\C1=C/CC\C(C)=C\[C@H]2OC(=O)C(=C)[C@@H]2CC1
145	3-Hydroxyhexane-2,5-dione	CC(CC(C(C)=O)O)=O
146	Heptacosan-1-ol	C(CCCCCCCCCCCCCCCCO)CCCCCCCCCC
147	Cyclotetradecane	C1CCCCCCCCCCCCC1
148	Cycloeucalenone	[C@]123CCC(=O)[C@@H](C)[C@]1([H])CC[C@]4([H])[C@@]2(C3)CC[C@@]5([C@](CC[C@@]45C)([C@@H](CCC(C(C)C)=C)C)[H])C
149	Rubimaillin	C12=CC=CC=C1C3=C(C(=C2O)C(OC)=O)C=CC(O3)(C)C

150	Canophyllal	[C@]12([C@]([C@]3([C@@](CC1)(CCC(C3)(C)C)C(=O)[H])[H])(CC[C@@]4([C@@]2(CC[C@]5([C@]4(CCC([C@@H]5C)=O)[H])C)[H])C)C)C
151	Stepharanine	O(C=1C2=C[N+]=3CCC=4C(C3C=C2C=CC1O)=CC(O)=C(OC)C4)C
152	(Z)-icos-13-enoic acid	CCCCCC/C=C\CCCCCCCCCCCC(O)=O
153	Methyl 3,4,5-trihydroxybenzoate	OC1=CC(=CC(O)=C1O)C(OC)=O
154	Serratol	C1C(=CCCC(=CCCC(=CC[C@@](C1)(C(C)C)O)C)C)C
155	Huperzine A	C\C=C1/[C@@H]2Cc3[nH]c(=O)ccc3[C@@]1(N)CC(C)=C2
156	Mustakone	CC(C)[C@@H]1CCC2(C)C3C1C2C(C)=CC3=O
157	(+)-Nootkatone	[C@@]12(C(CC[C@H](C1)C(C)=C)=CC(C[C@H]2C)=O)C
158	1-Icosanoylglycerol	C(CCCCCCCCCCCC(OCC(CO)O)=O)CCCCCCC
159	Swertisin	O1[C@@H](C2=C(O)C3=C(OC(=CC3=O)C4=CC=C(O)C=C4)C=C2OC)[C@H](O)[C@@H](O)[C@H](O)[C@H]1CO
160	Serratol	C1C(=CCCC(=CCCC(=CC[C@@](C1)(C(C)C)O)C)C)C
161	PLX-4720	C(CC)S(NC1=CC=C(C(=C1F)C(C=2C3=C(N=CC(=C3)Cl)NC2)=O)F)(=O)=O
162	Huperzine A	C\C=C1/[C@@H]2Cc3[nH]c(=O)ccc3[C@@]1(N)CC(C)=C2
163	N-(2-methoxyethyl)-4-{[6-(pyridin-4-yl)quinazolin-2-yl]amino}benzamide	O=C(C1=CC=C(NC2=NC3=CC=C(C4=CC=NC=C4)C=C3C=N2)C=C1)NCCOC

Continued

Table 5 The list of phytochemical those satisfies the Lipinski rule of 5 and their structures in SMILES format.—cont'd

S.·no	Compound	SMILES (chEBI)
164	Glucotropeolin	[C@H]1(O[C@@H]([C@@H](O)[C@@H]([C@H]1O)O)CO)S/C(=N\OS(O)(=O)=O)/CC=2C=CC=CC2
165	Glucotropeolin(1-)	[C@H]1(O[C@@H]([C@@H](O)[C@@H]([C@H]1O)O)CO)S/C(=N\OS([O-])(=O)=O)/CC=2C=CC=CC2
166	Methyl 12-methyltetradecanoate	O(C(=O)CCCCCCCCCCC(CC)C)C
167	Carpaine	C[C@@H]1N[C@H]2CC[C@@H]1OC(=O)CCCCCCC[C@@H]1CC[C@H](OC(=O)CCCCCCC2)[C@H](C)N1
168	24-Methylenecycloartanol	[C@]123[C@@]4([C@](C([C@@H](O)CC4)(C)C)(CC[C@]1([C@]5([C@]([C@@]([C@@H](CCC(C(C)C)=C)C)(CC5)[H])(C)CC2)C)[H])[H])C3
169	Helvolic acid methyl ester	C=1[C@@]2([C@@]3(CC[C@@]/4([C@@]([C@]3(C([C@H]([C@]2([C@@H](C(C1)=O)C)[H])OC(=O)C)=O)C)(C[C@@H](\C4=C(\CCC=C(C)C)/C(=O)OC)OC(=O)C)C)[H])[H])C
170	Danielone	COc1cc(cc(OC)c1O)C(=O)CO
171	2-Isobutylthiazole	N1=C(SC=C1)CC(C)C
172	14-Deoxy-11,12-didehydroandrographolide	[H][C@]12CCC(=C)[C@@H](\C=C\C3=CCOC3=O)[C@]1(C)CC[C@@H](O)[C@@]2(C)CO
173	Andrographolide	[H][C@]12CCC(=C)[C@@H](C\C=C3/[C@H](O)COC3=O)[C@]1(C)CC[C@@H](O)[C@@]2(C)CO
174	Dihydroferulic acid	C=1(C=C(C(=CC1)O)OC)CCC(=O)O

175	Mesembryanthemoidigenic acid	[C@@]12(C([C@@]3(C[C@@](CO)(C)CC[C@@]3(CC1)C(=O)O)[H])=CC[C@@]4([C@]5(CC[C@@H](C([C@@]5(CC[C@@]24C)[H])(C)C)O)C)[H])C
176	Methyl *N*-methylanthranilate	CNC1=CC=CC=C1C(OC)=O
177	Delta-elemene	[C@@H]1(C=C(CC[C@@]1(C=C)C)C(C)C)C(C)=C
178	Decussatin	COc1cc(O)c2c(c1)oc1ccc(OC)c(OC)c1c2=O
179	Linalool	CC(C)=CCCC(C)(O)C=C
180	Nevadensin	COc1ccc(cc1)-c1cc(=O)c2c(O)c(OC)c(O)c(OC)c2o1
181	7-Epi-sesquithujene	C[C@H](CCC=C(C)C)[C@@]12CC=C(C)[C@@H]1C2
182	Selina-4(15),7(11)-diene	C1CCC([C@]2([C@]1(CCC(C2)=C(C)C)C)[H])=C
183	Codeine	[H][C@]12C=C[C@H](O)[C@@H]3Oc4c(OC)ccc5C[C@H]1N(C)CC[C@@]23c45
184	Sulfanilamide	Nc1ccc(cc1)S(N)(=O)=O
185	Pipercyclobutanamide A(rel)	O=C(\C=C/[C@@H]1[C@@H](\C=C\c2ccc3OCOc3c2)[C@@H]([C@H]1c1ccc2OCOc2c1)C(=O)N1CCCCC1)N1CCCCC1
186	(−)-cubebin	[H][C@@]1(CO[C@H](O)[C@]1([H])Cc1ccc2OCOc2c1)Cc1ccc2OCOc2c1
187	(−)-3,4-dimethoxy-3,4-desmethylenedioxycubebin	[H][C@@]1(COC(O)[C@]1([H])Cc1ccc2OCOc2c1)Cc1ccc(OC)c(OC)c1
188	Piperine	O=C(\C=C\C=C\c1ccc2OCOc2c1)N1CCCCC1

Continued

Table 5 The list of phytochemical those satisfies the Lipinski rule of 5 and their structures in SMILES format.—cont'd

S.no	Compound	SMILES (chEBI)
189	Pellitorine	CCCCC\C=C\C=C\C(=O)NCC(C)C
190	Gaudichaudianic acid, (−rac)	CC(C)=CCCC1(C)Oc2c(CC=C(C)C)cc(cc2C=C1)C(O)=O
191	Isochamanetin	Oc1ccccc1Cc1c(O)cc2O[C@@H](CC(=O)c2c1O)c1ccccc1
192	7-Methoxychamanetin	COc1cc(O)c2C(=O)C[C@H](Oc2c1Cc1ccccc1O)c1ccccc1
193	Dichamanetin	Oc1ccccc1Cc1c(O)c(Cc2ccccc2O)c2O[C@@H](CC(=O)c2c1O)c1ccccc1
194	7-Methoxydichamanetin	COc1c(Cc2ccccc2O)c(O)c2C(=O)C[C@H](Oc2c1Cc1ccccc1O)c1ccccc1
195	5″-(2‴′-Hydroxybenzyl)uvarinol	COc1c(Cc2cc(Cc3ccccc3O)ccc2O)c(O)c2C(=O)C[C@H](Oc2c1Cc1cc(Cc2ccccc2O)ccc1O)c1ccccc1
196	2,4-Dodecadienamide	CCCCCCC\C=C\C=C\C(N)=O
197	7-Methoxyisochamanetin	COc1cc2O[C@@H](CC(=O)c2c(O)c1Cc1ccccc1O)c1ccccc1
198	(2E,4E)-N-[2-(methylsulfinyl)ethyl]-2,4-decadienamide	CCCCC\C=C\C=C\C(=O)NCCS(C)=O
199	(2E,4E)-N-[(4-hydroxy-3-methoxyphenyl)ethyl]-2,4-decadienamide	CCCCC\C=C\C=C\C(=O)NCCc1ccc(O)c(OC)c1
200	3-(4-Hydroxy-3,5-dimethoxyphenyl)propanoylpyrrole	COc1cc(CCC(=O)n2cccc2)cc(OC)c1O
201	3-(3,4,5-Timethoxyphenyl)propanoylpyrrole	COc1cc(CCC(=O)n2cccc2)cc(OC)c1OC

202	1-[(2E,4E,6E)-2,4,6-dodecatrienoyl]pyrrolidine	CCCCC\C=C\C=C\C=C\C(=O)N1CCCC1
203	1-[(2E,4Z,8E)-9-(3,4-methylenedioxyphenyl)-2,4,8-nonatrienoyl]pyrrolidine	O=C(\C=C\C=C/CC\C=C\c1ccc2OCOc2c1)N1CCCC1
204	1-[(2E,4E,10E)-10-(3,4-methylenedioxyphenyl)-2,4,10-undecatrienoyl]pyrrolidine	O=C(\C=C\C=C\CCCC\C=C\c1ccc2OCOc2c1)N1CCCC1
205	1-[(4E,10E)-11-(3,4-methylenedioxyphenyl)-4,10-undecadienoyl]pyrrolidine	O=C(CC\C=C\CCCC\C=C\c1ccc2OCOc2c1)N1CCCC1
206	1-[(9E)-10-(3,4-methylenedioxyphenyl)-9-decenoyl]pyrrolidine	O=C(CCCCCCC\C=C\c1ccc2OCOc2c1)N1CCCC1
207	1-[(2E,4E)-2,4-decadienoyl]pyrrolidine	CCCCC\C=C\C=C\C(=O)N1CCCC1
208	1-[(2E,4E)-2,4-dodecadienoyl]pyrrolidine	CCCCCCC\C=C\C=C\C(=O)N1CCCC1
209	1-[(2E)-7-(3,4-methylenedioxyphenyl)-2-heptenoyl]pyrrolidine	O=C(\C=C\CCCCc1ccc2OCOc2c1)N1CCCC1
210	1-[(2E,4E)-7-(3,4-methylenedioxyphenyl)-2,4-heptadienoyl]pyrrolidine	O=C(\C=C\C=C\CCc1ccc2OCOc2c1)N1CCCC1
211	1-[(2E,8E)-9-(3,4-methylenedioxyphenyl)-2,8-nonadienoyl]pyrrolidine	O=C(\C=C\CCCC\C=C\c1ccc2OCOc2c1)N1CCCC1
212	1-[(8E)-9-(3,4-methylenedioxyphenyl)-8-nonenoyl]pyrrolidine	O=C(CCCCCC\C=C\c1ccc2OCOc2c1)N1CCCC1
213	1-[(2E,4E,8E)-9-(3,4-methylenedioxyphenyl)-2,4,8-nonatrienoyl]pyrrolidine	O=C(\C=C\C=C\CC\C=C\c1ccc2OCOc2c1)N1CCCC1

Continued

Table 5 The list of phytochemical those satisfies the Lipinski rule of 5 and their structures in SMILES format.—cont'd

S.·no	Compound	SMILES (chEBI)
214	1-[(2E,4E)-11-(3,4-methylenedioxyphenyl)-2,4-undecadienoyl]pyrrolidine	O=C(\C=C\C=C\CCCCCCc1ccc2OCOc2c1)N1CCCC1
215	1-[(2E,10E)-11-(3,4-methylenedioxyphenyl)-2,10-undecadienoyl]pyrrolidine	O=C(\C=C\CCCCCC\C=C\c1ccc2OCOc2c1)N1CCCC1
216	(2E,4E)-N-isobutyl-2,4-dodecadienamide	CCCCCCC\C=C\C=C\C(=O)NCC(C)C
217	(2E,4E)-N-isobutyl-7-(3,4-methylenedioxyphenyl)-hepta-2,4-dienamide	CC(C)CNC(=O)\C=C\C=C\CCc1ccc2OCOc2c1
218	(8E)-N-isobutyl-9-(3,4-methylenedioxyphenyl)nona-8-enamide	CC(C)CNC(=O)CCCCCC\C=C\c1ccc2OCOc2c1
219	(2E,4E,8E)-N-isobutyl-11-(3,4-methylenedioxyphenyl)undeca-2,4,8-trienamide	CC(C)CNC(=O)\C=C\C=C\CC\C=C\CCc1ccc2OCOc2c1
220	N-trans-sinapoyltyramine	COc1cc(cc(OC)c1O)\C=C\C(=O)NCCc1ccc(O)cc1
221	Dihydrocubebin, rel-	OC[C@@H](Cc1ccc2OCOc2c1)[C@@H](CO)Cc1ccc2OCOc2c1
222	Justiflorinol	OCC(CC(=O)c1ccc2OCOc2c1)C(=O)c1ccc2OCOc2c1
223	(−)-Sanguinolignan A	O[C@@H]([C@H]1COC(=O)[C@@H]1C(=O)c1ccc2OCOc2c1)c1ccc2OCOc2c1
224	(−)-Sanguinolignan B	O[C@@H]([C@@H]1COC(=O)[C@@H]1C(=O)c1ccc2OCOc2c1)c1ccc2OCOc2c1
225	(−)-Sanguinolignan C	COc1ccc(cc1OC)[C@@H](OC(C)=O)[C@H]1COC(=O)[C@@H]1C(=O)c1ccc2OCOc2c1

226	(−)-Sanguinolignan D	CC(=O)O[C@@H]([C@@H]1[C@H](COC1=O)C(=O)c1ccc2OCOc2c1)c1ccc2OCOc2c1
227	(7'S)-Parabenzlactone	O[C@@H]([C@H]1COC(=O)[C@@H]1Cc1ccc2OCOc2c1)c1ccc2OCOc2c1
228	Flavokawain B	COc1cc(O)c(C(=O)\C=C\c2ccccc2)c(OC)c1
229	Methyl 3,4-dihydroxy-5-(3′-methyl-2′-butenyl)benzoate	COC(=O)c1cc(O)c(O)c(CC=C(C)C)c1
230	Pipercallosidine	C(/C=C/CCCCC=1C=C2C(=CC1)OCO2)(NCC(C)C)=O
231	Kadsurenin C	O([C@@]12[C@@H]([C@H]([C@@]([C@H]1O)(C(=O)C(=C2)CC=C)[H])C3=CC(OC)=C(OC)C=C3)C)C
232	Kadsurenin K	O([C@@]12[C@@H]([C@H]([C@@](C1=O)(C(=O)C(=C2)CC=C)[H])C3=CC(OC)=C(O)C=C3)C)C
233	Kadsurenin L	O([C@@]12[C@@H]([C@H]([C@@]([C@@H]1OC(=O)C)(C(=O)C(=C2)CC=C)[H])C3=CC(OC)=C(OC)C=C3)C)C
234	Pipercallosine	C(/C=C/C=C/CCCCC1=CC=C2C(=C1)OCO2)(=O)NCC(C)C
235	(S)-1′-methylhexyl caffeate	CCCCC[C@H](C)OC(=O)\C=C\c1ccc(O)c(O)c1
236	Futoenone	O1[C@@H]2C[C@@]3([C@@H]([C@H](C2)C4=CC=5OCOC5C=C4)C)C1=CC(=O)C(OC)=C3
237	(S)-1′-methylbutyl caffeate	CCC[C@H](C)OC(=O)\C=C\c1ccc(O)c(O)c1
238	(S)-1′-methyloctyl caffeate	CCCCCCC[C@H](C)OC(=O)\C=C\c1ccc(O)c(O)c1

Continued

Table 5 The list of phytochemical those satisfies the Lipinski rule of 5 and their structures in SMILES format.—cont'd

S.·no	Compound	SMILES (chEBI)
239	Futokadsurin B	COc1ccc(cc1OC)[C@@H]1O[C@@H]([C@H](C)[C@H]1C)c1ccc2OCOc2c1
240	Futokadsurin C	COc1ccc(cc1OC)[C@H]1O[C@H]([C@H](C)[C@@H]1C)c1ccc2OCOc2c1
241	Futokadsurin A	COc1cc(ccc1O)[C@@H]1O[C@@H]([C@@H](C)[C@@H]1C)c1ccc(OC)c(OC)c1
242	Burchellin	O1[C@@H]([C@H]([C@]2(C1=CC(=O)C(OC)=C2)CC=C)C)C3=CC=4OCOC4C=C3
243	Piperonal	[H]C(=O)c1ccc2OCOc2c1
244	(−)-Cubenol	[C@]12([C@@H](CC[C@H]([C@@]2(CCC(=C1)C)O)C)C(C)C)[H]
245	(E,E)-piperic acid	OC(=O)\C=C\C=C\c1ccc2OCOc2c1
246	Trans-cinnamic acid	OC(=O)\C=C\c1ccccc1
247	Pinocembrin chalcone	Oc1cc(O)c(C(=O)\C=C\c2ccccc2)c(O)c1
248	Piperlactam S	O(C=1C2=C3C(=CC=4C2=CC=CC4)N(C(C3=CC1O)=O)OC)C
249	Asebogenin	COc1cc(O)c(C(=O)CCc2ccc(O)cc2)c(O)c1
250	Epicocconone	C\C=C\C=C\C=C\C(=O)\C=C(O)\C1=C2C=C3C[C@@H](CO)OC=C3C(=O)[C@@]2(C)OC1=O
251	(+)-Sesamin	[C@]12([C@@]([C@H](OC1)C3=CC4=C(C=C3)OCO4)(CO[C@@H]2C5=CC6=C(C=C5)OCO6)[H])[H]

252	Pipataline	O1C=2C=C(\C=C\CCCCCCCCCC)C=CC2OC1
253	(−)-Antofine	C=1C2=C(C=C(C1OC)OC)C=3C=C(C=CC3C4=C2C[C@]5(CCCN5C4)[H])OC
254	Benzoic acid	OC(=O)c1ccccc1
255	Ethyl butyrate	CCCC(=O)OCC
256	N-feruloyltyramine	COc1cc(\C=C\C(=O)NCCc2ccc(O)cc2)ccc1O
257	Kadsurenin M	O1[C@@H]([C@H](C2=C1C(OC)=CC(=C2)C(=O)[H])C)C3=CC(OC)=C(OC)C=C3
258	(−)-Epicubenol	[C@]12([C@@H](CC[C@H]([C@]2(CCC(=C1)C)O)C)C(C)C)[H]
259	Monocerin	[H][C@@]12C[C@H](CCC)O[C@]1([H])c1cc(OC)c(OC)c(O)c1C(=O)O2
260	(2S,3aR,9bR)-6,7-dihydroxy-8-methoxy-2-propyl-3,3a-dihydro-2H-furo[3,2-c]isochromen-5(9bH)-one	[H][C@@]12C[C@H](CCC)O[C@]1([H])c1cc(OC)c(O)c(O)c1C(=O)O2
261	Fusarentin 6,7-dimethyl ether	CCC[C@H](O)C[C@@H]1Cc2cc(OC)c(OC)c(O)c2C(=O)O1
262	Fusarentin 6-methyl ether	CCC[C@H](O)C[C@@H]1Cc2cc(OC)c(O)c(O)c2C(=O)O1
263	(3R,4R)-4,8-dihydroxy-3-((R)-2-hydroxypentyl)-6,7-dimethoxyisochroman-1-one	CCC[C@@H](O)C[C@H]1OC(=O)c2c(O)c(OC)c(OC)cc2[C@H]1O
264	Colletotrialide, (+)-	CCCC(=O)CC[C@H]1OC(=O)c2c(O)c(OC)c(OC)cc12
265	Galgravin	C1(=C(C=CC(=C1)[C@@H]2O[C@@H]([C@H]([C@H]2C)C)C3=CC=C(C(=C3)OC)OC)OC)OC

Continued

Table 5 The list of phytochemical those satisfies the Lipinski rule of 5 and their structures in SMILES format.—cont'd

S.·no	Compound	SMILES (chEBI)
266	9,10-Epoxy-18-hydroxyoctadecanoic acid	C1(CCCCCCCC(=O)O)C(CCCCCCCCO)O1
267	Pinocembrin	Oc1cc(O)c2C(=O)C[C@H](Oc2c1)c1ccccc1
268	1,4-Cineole	C1C[C@]2(CC[C@]1(C)O2)C(C)C
269	Sodium benzoate	C(C=1C=CC=CC1)([O-])=O.[Na+]
270	9,10,18-Trihydroxyoctadecanoic acid	C(C(C(CCCCCCCCO)O)O)CCCCCCC(=O)O
271	Kavapyrone	COc1cc(oc(=O)c1)[C@@H]1O[C@H]1c1ccccc1
272	Alpha-cubebene	C1C[C@H]([C@]2([C@]3([C@@H]1C)CC=C([C@]23[H])C)[H])C(C)C
273	Acuminatin	O1[C@H]([C@@H](C2=C1C(OC)=CC(=C2)/C=C/C)C)C3=CC(OC)=C(OC)C=C3
274	Jerantinine E	[H][C@]12N3CCC[C@@]1(CC)CC(C(=O)OC)=C1Nc4cc(OC)c(O)cc4[C@]21CC3
275	Jerantinine F	[H][C@]12CCN3CC[C@@]45C(Nc6cc(OC)c(O)cc46)=C(C[C@@]1(CCO2)[C@]35[H])C(=O)OC
276	Jerantinine C	[H][C@@]12N3CC[C@]11C(Nc4cc(OC)c(O)cc14)=C(C[C@]2(CC)C=CC3=O)C(=O)OC
277	Jerantinine B	[H][C@]12CN3CC[C@@]45C(Nc6cc(OC)c(O)cc46)=C(C[C@](CC)([C@@]1([H])O2)[C@]35[H])C(=O)OC
278	Jerantinine D	[H][C@@]12O[C@]1([H])[C@@]1(CC)CC(C(=O)OC)=C3Nc4cc(OC)c(O)cc4[C@@]33CCN(C2=O)[C@@]13[H]

279	Jerantinine A	[H][C@@]12N3CC[C@]11C(Nc4cc(OC)c(O)cc14)=C(C[C@]2(CC)C=CC3)C(=O)OC
280	Tabernaemontanine	CC[C@@H]1CN(C)[C@]2(CC=3C4=CC=CC=C4NC3C(C[C@@]1([C@@]2(C(=O)OC)[H])[H])=O)[H]
281	Plectranthol A	CC(C)c1cc2C=C[C@@]3(C)C(=CCC[C@]3(C)COC(=O)c3ccc(O)c(O)c3)c2c(O)c1O
282	11,20-Dihydroxysugiol	C1=C(C(C)C)C(=C(C2=C1C(C[C@@]3([C@@]2(CCCC3(C)C)CO)[H])=O)O)O
283	11-Hydroxysugiol	C1=C(C(C)C)C(=C(C2=C1C(C[C@@]3([C@@]2(CCCC3(C)C)C)[H])=O)O)O
284	Abietatriene	CC(C)c1ccc2c(CC[C@H]3C(C)(C)CCC[C@]23C)c1
285	7-Oxo-10α-cucurbitadienol	[H][C@@]1(CC[C@@]2(C)[C@]3([H])C(=O)C=C4[C@@]([H])(CC[C@H](O)C4(C)C)[C@]3(C)CC[C@]12C)[C@H](C)CCC=C(C)C

3.5 Common target identification

The list of common targets between the COVID-19 infection (339 proteins) and phytochemical targets (24843) were analyzed using the Venn diagram (Fig. 1). The Venn diagram was plotted using a web server embedded in Bioinformatics & Evolutionary Genomics. From the analysis, 13 targets, namely, ACE, IMPDH2, EGFR, DPP4, RIPK1, HDAC2, CTSL, POLA1, CTSB, PABPC1, VEGFA, SIGMAR1, and IL6, were identified as potent protein targets.

3.6 Enrichment analysis

BP, CC, biological pathways, and MF were identified using the FunRich tool. Energy pathways, immune response, nucleoside, nucleic acid metabolism, nucleotide, and regulation of nucleobase were identified as the top biological process for the common receptor (Fig. 2A). Nucleus, plasma membrane, cytoplasm, exosome, lysosome, and extracellular were identified as the top cellular component for typical receptors (Fig. 2B). PAR1-mediated thrombin signaling events, IFN-gamma pathway, Nectin adhesion pathway, IL3-mediated pathway events, signaling events mediated by hepatocyte growth factor receptor (c-Met), and PDGF receptor signaling network were identified as the biological pathway for the typical receptors (Fig. 2C). Cytokine activity, transmembrane receptor protein tyrosine kinase activity, peptidase activity, transmembrane receptor activity, DNA-directed DNA polymerase activity, and cysteine-type peptidase activity (Fig. 2D). The complete results obtained from the FunRich analysis are tabulated in Supplementary Table 3 in the online version at https://doi.org/10.1016/bs.apcsb.2021.11.007.

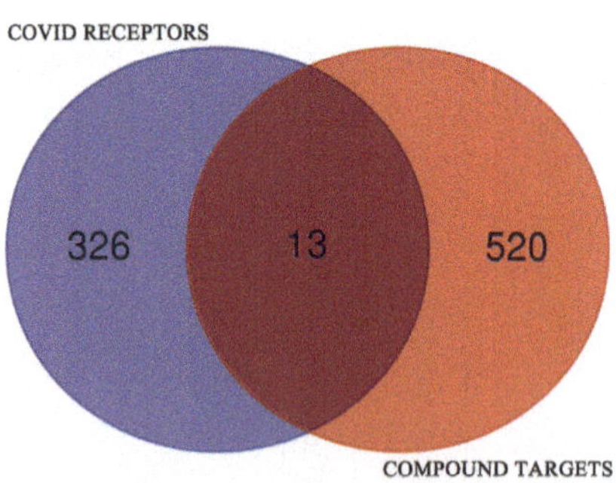

Fig. 1 Venn diagram showing the number of COVID-19 receptors, Compound targets and the common genes between the COVID-19 receptors and compound targets.

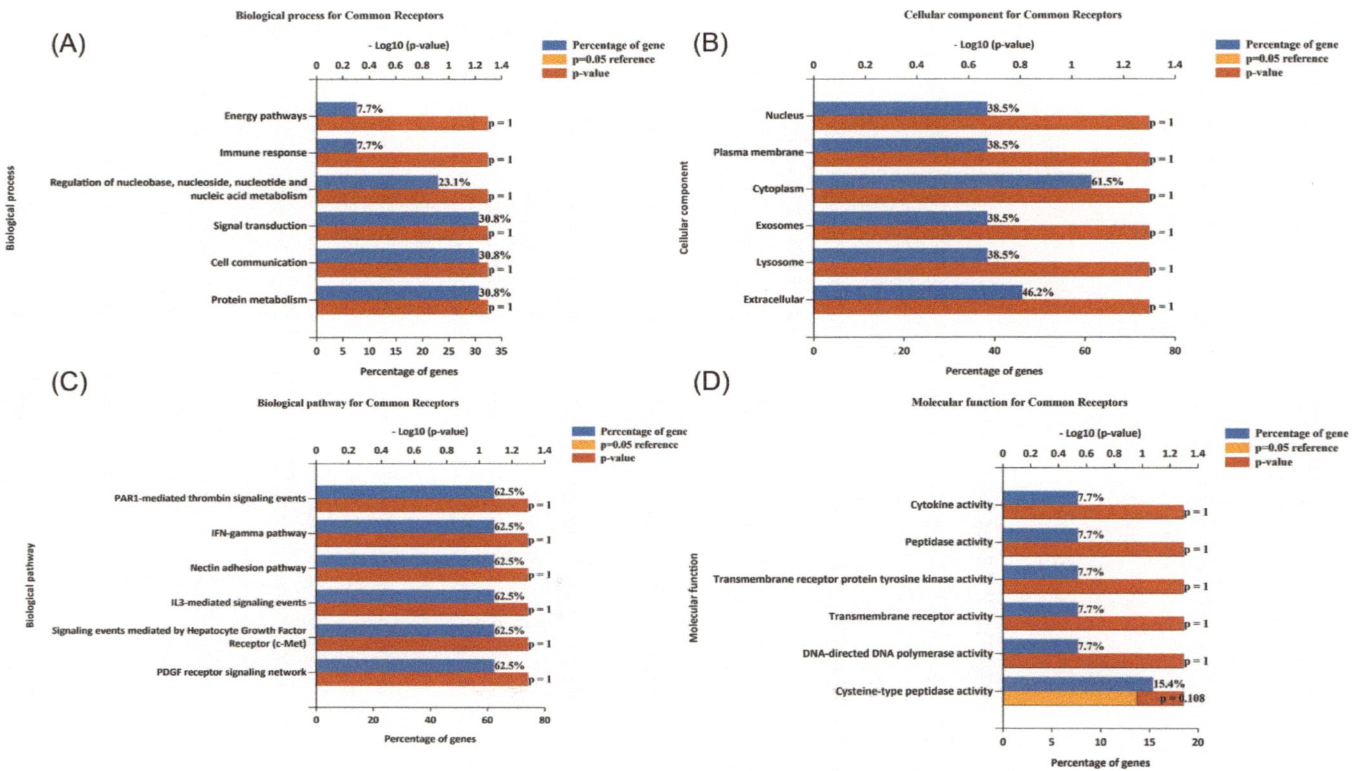

Fig. 2 Enrichment analysis using FunRich shows the most significant biological process, cellular component, biological pathway, and molecular function of the identified targets as the input. (A) Biological process for common receptors, (B) Cellular component for common receptors, (C) Biological pathway for common receptors, and (D) Molecular function for common receptors.

3.7 Pathway analysis

Pathway analysis was performed using the Reactome version 76 on 05/06/2021. All the 13 identifiers in the sample were found in Reactome, where 256 pathways were hit by at least one of them. The 25 most relevant significant pathways sorted by *P-value* are tabulated in Table 6. TFAP2 (AP-2) family regulates the transcription of growth factors and their receptors. Potential therapeutics for SARS and SARS-CoV (SARS-CoV-1 & SARS-CoV-2) Infections were identified as the top 3 significant pathways with 4/21, 5/84, and 6/203 entities found, respectively (Fig. 3A and B).

3.8 Gene–gene interaction analysis

The gene–gene interaction analysis helped to prioritize the study of genes in the pathway that helped to understand the underlying mechanism of COVID. The gene–gene interaction analysis was performed using the STRING online server. The number of nodes and edges were found to be 13 and 17, respectively. The average degree of nodes and clustering coefficient of local was found to be 2.62 and 0.566. The anticipated amount of edges was 9 with a PPI enrichment *P-value* of 0.0115. Except for IMPDH2, PABPC1, SIGMAR1, and POLA1, all the other genes interacted with at least one gene (Fig. 4).

3.9 Virtual screening and molecular interaction analysis

The identified targets such as VEGFA, CTSL, CTSB, EGFR, and IL6 have been identified as the target by more than one phytochemical compound (Table 7). Virtual screenings were performed to identify the most suitable phytochemical compound. The compound with the highest binding affinity was taken for the molecular interaction analysis using AutoDock (Supplementary Table 3 in the online version at https://doi.org/10.1016/bs.apcsb.2021.11.007). We observed that betulinic acid was found to interact best with the ACE protein with the binding energy −9.32 kcal/mol from the molecular docking results. The 5″-(2⁗-Hydroxybenzyl) uvarinol was found to interact best with Cathepsin L, Cathepsin B, Cathepsin K with the binding energy −6.71 kcal/mol and −6.06 kcal/mol, respectively. The (-)-antofine, S)-1′-methyloctyl caffeate, (Z)-3-phenyl-2-propenal, 7-oxo-10α-cucurbitadienol, PLX-4720, and 5″-(2⁗-Hydroxybenzyl) uvarinol were found to interact best to the DPP4, EGFR, HDAC2, IL6, RIPK1, and VEGFA with binding energies −8.25 kcal/mol, −4.76 kcal/mol, −4.39 kcal/mol, −6.9 kcal/mol, −5.91 kcal/mol, and −5.66 kcal/mol respectively. The interactions between the compounds and the proteins are shown in Fig. 5.

Table 6 The list of 25 most relevant pathways sorted by *P* value.

S.·no	Pathway name	Entities found	Ratio	*P* value	FDR[a]	Reactions found	Ratio
1	TFAP2 (AP-2) family regulates transcription of growth factors and their receptors	4/21	0.001	2.57E-08	6.88E-06	4/18	0.001
2	Potential therapeutics for SARS	5/84	0.006	0.000000122	0.0000163	6/32	0.002
3	SARS-CoV infections	6/203	0.014	0.000000338	0.0000301	9/254	0.019
4	Transcriptional regulation by the AP-2 (TFAP2) family of transcription factors	4/52	0.004	9.37e-07	6.28e-05	4/44	0.003
5	Trafficking and processing of endosomal TLR	3/16	0.001	1.75e-06	9.29e-05	2/7	5.21e-04
6	Infectious disease	9/1343	0.092	0.0000516	0.002	16/750	0.056
7	VEGF ligand-receptor interactions	2/8	0.000551	0.0000633	0.002	3/4	0.000298
8	VEGF binds to VEGFR leading to receptor dimerization	2/8	5.51e-04	6.33e-05	0.002	2/3	2.23e-04
9	RUNX1 regulates transcription of genes involved in differentiation of keratinocytes	2/11	7.57e-04	1.19e-04	0.003	1/8	5.96e-04
10	Assembly of collagen fibrils and other multimeric structures	3/67	0.005	1.23e-04	0.003	1/26	0.002
11	Toll-like receptor cascades	4/188	0.013	0.000141	0.003	14/185	0.014
12	Disease	11/2360	0.163	0.000148	0.003	99/1591	0.119

Continued

Table 6 The list of 25 most relevant pathways sorted by *P* value.—cont'd

S.·no	Pathway name	Entities found	Ratio	*P* value	FDR[a]	Reactions found	Ratio
13	Generic transcription pathway	9/1555	0.107	0.000163	0.003	17/824	0.061
14	Interleukin-4 and interleukin-13 signaling	4/211	0.015	2.19e-04	0.004	2/47	0.004
15	Regulation of gene expression by hypoxia-inducible factor	2/15	0.001	2.21e-04	0.004	1/7	5.21e-04
16	RNA polymerase II transcription	9/1694	0.117	0.000314	0.004	17/885	0.066
17	ERBB2 activates PTK6 signaling	2/18	0.001	0.000318	0.004	2/2	0.000149
18	ERBB2 regulates cell motility	2/19	0.001	0.000354	0.005	2/2	0.000149
19	Collagen formation	3/104	0.007	0.000444	0.006	1/77	0.006
20	PI3K events in ERBB2 signaling	2/22	0.002	0.000473	0.006	5/7	0.000521
21	Gene expression (transcription)	9/1855	0.128	0.000622	0.007	23/1000	0.074
22	Signaling by VEGF	3/137	0.009	0.000983	0.011	49/86	0.006
23	Signaling by ERBB2 KD Mutants	2/35	0.002	0.001	0.011	15/17	0.001
24	MHC class II antigen presentation	3/148	0.01	0.001	0.011	3/26	0.002
25	Degradation of the extracellular matrix	3/148	0.01	0.001	0.011	6/105	0.008

[a]False discovery rate.

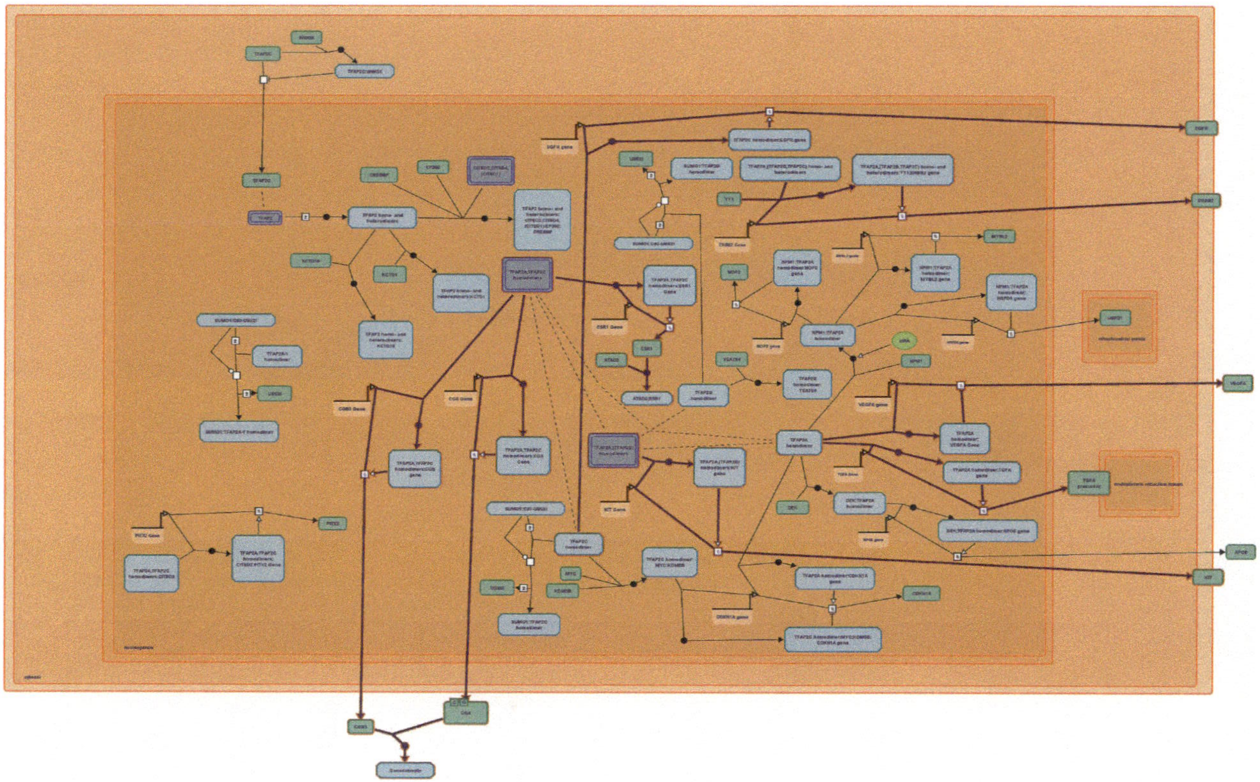

Fig. 3 (A) Pathway shows the transcriptional regulation by the AP-2 (TFAP2) transcription factor family. (B) Pathway showing the potential therapeutics for SARS.

(Continued)

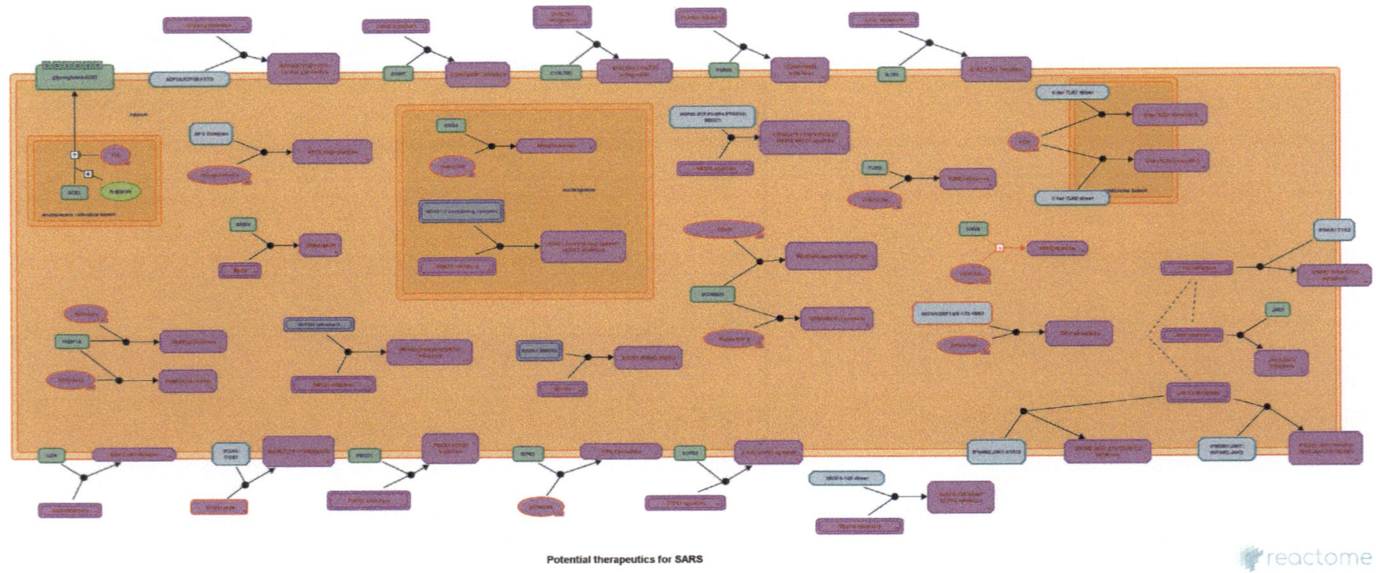

Fig. 3—Cont'd

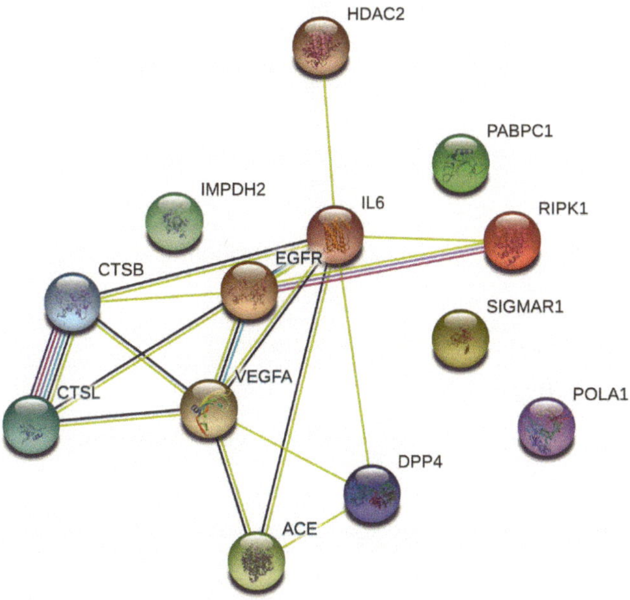

Fig. 4 The gene-gene interaction between the identified targets common to the COVID-19 receptors and compound targets.

Table 7 The list of selected receptors of the COVID-19 and the identified interacting phytochemical compounds.

S.·no	Name of the receptor	Name of the compound
1	Angiotensin-converting enzyme	Betulinic acid
2	Dipeptidyl peptidase IV	(−)-Antofine
3	Vascular endothelial growth factor A	Isochamanetin
		7-Methoxychamanetin
		Dichamanetin
		7-Methoxydichamanetin
		5″-(2⁗-Hydroxybenzyl)uvarinol
		7-Methoxyisochamanetin
		Pinocembrin
		1,4-Cineole
4	Cathepsin L	Isochamanetin
		7-Methoxychamanetin
		Dichamanetin
		7-Methoxydichamanetin

Continued

Table 7 The list of selected receptors of the COVID-19 and the identified interacting phytochemical compounds.—cont'd

S.·no	Name of the receptor	Name of the compound
		5″-(2⁗-Hydroxybenzyl)uvarinol
		7-Methoxyisochamanetin
		Pinocembrin chalcone
		Pinocembrin
		1,4-Cineole
		Carpaine
5	Cathepsin (B and K)	Isochamanetin
		7-Methoxychamanetin
		Dichamanetin
		7-Methoxydichamanetin
		5″-(2⁗-Hydroxybenzyl)uvarinol
		7-Methoxyisochamanetin
		Asebogenin
		Pinocembrin
		1,4-Cineole
		Carpaine
6	Epidermal growth factor receptor erbB1	7,4′-dimethylkaempferol
		Pipataline
		N-trans-sinapoyltyramine
		Flavokawain B
		(S)-1′-methylhexyl caffeate
		(S)-1′-methylbutyl caffeate
		(S)-1′-methyloctyl caffeate
		Trans-cinnamic acid
		Pinocembrin chalcone
		N-feruloyltyramine
		Cinnamic acid
		Dihydroferulic acid
		Capensinidin
		Nevadensin

Table 7 The list of selected receptors of the COVID-19 and the identified interacting phytochemical compounds.—cont'd

S.·no	Name of the receptor	Name of the compound
7	Interleukin-6	All-cis-octadeca-6,9,12,15-tetraenoic acid
		7-Oxo-10α-cucurbitadienol
8	Receptor-interacting serine/ threonine-protein kinase 1	PLX-4720
9	Histone deacetylase 2	(Z)-3-phenyl-2-propenal

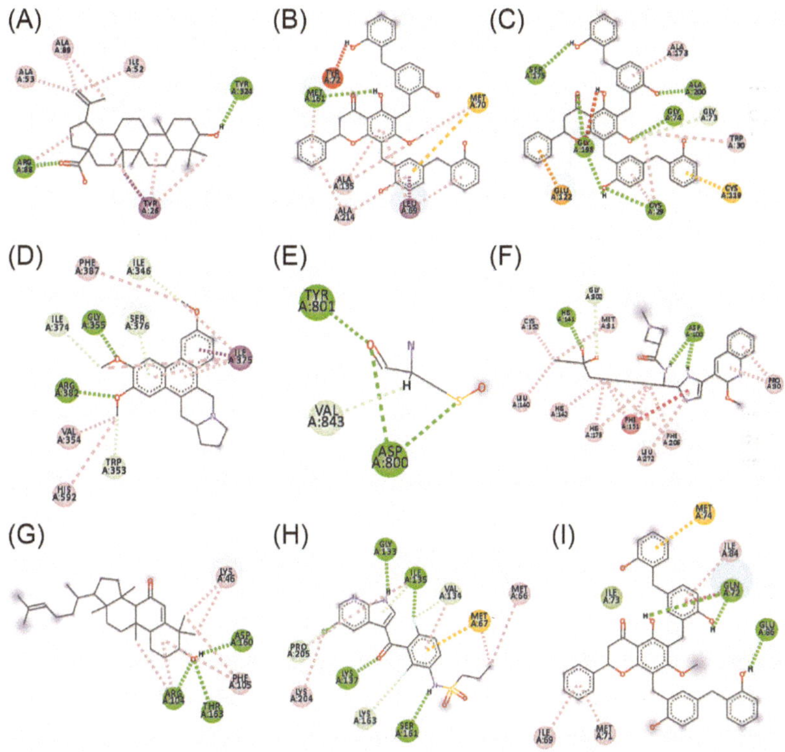

Fig. 5 Visualization of the best interacting phytochemical with the identified targets. (A) ACE receptor with betulinic acid, (B) Cathepsin L receptor with 5″-(2″″-Hydroxybenzyl)uvarinol, (C) Cathepsin B & K receptor with 5″-(2″″-Hydroxybenzyl) uvarinol, (D) DPPR receptor with (−)-antofine, (E) EGFR receptor with S)-1′-methyloctyl caffeate, (F) HDAC2 with (Z)-3-phenyl-2-propenal, (G) IL6 with 7-oxo-10α-cucurbitadienol, (H) RIPK1 with PLX-4720, and (I) VEGFA with 5″-(2″″-Hydroxybenzyl)uvarinol.

4. Discussion

SARS-CoV-2 belongs to a genetic group of viruses that cause respiratory sickness and was declared a worldwide pandemic on March 11th, 2020 (Cucinotta & Vanelli, 2020). As of September 6th, 2021, the reported cases are more than 220,563,227, and at least 4,565,483 people have died (WHO, 2021b). The number of cases across the globe is gradually increasing and decreasing in terms of the COVID-19 infection wave. The countries in America, Europe, and the South Asian region are highly infected with the second wave of COVID-19 infection. Few countries have started to face the third wave of COVID-19 (Taboada et al., 2021).

On the other hand, drug discovery is still far behind, and only drug repurposing serves as the day's choice (Lamontagne et al., 2020; Wu et al., 2020). Unfortunately, several repurposed drugs that showed promising results in the early studies were failed to treat COVID-19; this also includes hydroxychloroquine (Boulware et al., 2020; Geleris et al., 2020). With this less efficiency of repurposed drugs, the search for the vaccine was on a serious note, and several vaccines were developed across the globe, but none were 100% active. Further, there is always a chance of vaccine failure with the mutations and their evolution in the SARS-CoV-2 (Williams & Burgers, 2021). Developing an immune response against COVID-19 can only be a ray of hope in this scenario.

Traditional medicines have been immune boosters since the ancient days (Ravishankar & Shukla, 2007). Siddha herbal formulations with medicinal value effectively against various causative agents, including influenza, dengue fever, chikungunya, tuberculosis, and others (Jain et al., 2020; Jain, Narayanan, Chaturvedi, Pai, & Sunil, 2018; Jain, Pai, & Sunil, 2018). Currently, the ministry of AYUSH has also approved the use of Kabasura kudineer and Nilavembu kudineer against COVID-19 (Alagu Lakshmi, Shafreen, Priya, & Shunmugiah, 2020; Natarajan et al., 2020). This study is intended to understand the immune-boosting mechanism by the JACOM, Kabasura kudineer, and Nilavembu kudineer against COVID-19. A total of 339 human genes were found to be involved in COVID-19. This list was obtained from the GeneCards database (Table 1). Twenty-five plants were found in the Kabasura kudineer, Nilavembu kudineer, and JACOM formulations (Table 2). A list of 314 phytochemicals was obtained from these 25 plants from the ChEBI database (Table 3). The drug-likeness properties of the 314 phytochemical compounds were evaluated using the online

SwissADME server (Table 4). From the analysis, we observed 285 compounds to satisfy the Lipinski rule of 5, and these could be considered drug-like compounds (Table 5). The possible targets for these drug-like compounds were predicted using the online SwissTargetPrediction server. An overall of 24,839 targets was predicted for these 285 compounds (Supplementary Table 1 in the online version at https://doi.org/10.1016/bs.apcsb.2021.11.007). The targets were ranked based on the probability score. Out of 24,839 targets, 5129 genes (with repeats) were the promising target with a probability score of more than 0 (Supplementary Table 2 in the online version at https://doi.org/10.1016/bs.apcsb.2021.11.007). A Venn diagram was plotted to find the common targets between the genes involved in the COVID-19 and the compounds target (Tables 1 and 5). From the Venn diagram, 13 targets (ACE, IMPDH2, EGFR, DPP4, RIPK1, HDAC2, CTSL, POLA1, CTSB, PABPC1, VEGFA, SIGMAR1, and IL6) were found to be common between the causative and treatment for COVID-19. Three hundred and twenty-six and 520 genes were COVID-19 receptors and targets for compounds, respectively (Fig. 1). The enrichment analysis was performed to understand the role of identified targets in their BP, CC, biological pathways, and MF. Energy pathways, nucleoside, immune response, nucleotide, regulation of nucleobase, and nucleic acid metabolism were identified as the top biological process for the common receptor (Fig. 2A). Nucleus, plasma membrane, cytoplasm, exosome, lysosome, and extracellular were identified as the top cellular component for common receptors (Fig. 2B). PAR1-mediated thrombin signaling events, Nectin adhesion pathway, IL3-mediated signaling events, IFN-gamma pathway, cellular pathways engaged by hepatocyte growth factor receptor (c-Met), and PDGF receptor signaling network were identified as the biological pathway for the common receptors (Fig. 2C). Cytokine activity, peptidase activity, transmembrane receptor protein tyrosine kinase activity, transmembrane receptor activity, DNA-directed DNA polymerase activity, and cysteine-type peptidase activity (Fig. 2D). Further, the pathways involved in these gene targets were predicted using the Reactome database (Jassal et al., 2020). The pathways such as the TFAP2 (AP-2) family that controls the expression of growth factors and their receptors, Potential therapeutics for SARS, and SARS-CoV Infections were identified as the top 3 pathways out of significant 25 pathways (Table 6). The clear pathway of the TFAP2 (AP-2) family regulates the transcription of growth factors, and their receptors are shown in Fig. 3A. There are five transcription factors in the AP-2 (TFAP2) family in mammals: TFAP2A, TFAP2B, TFAP2C, TFAP2D, and TFAP2E (AP-2 epsilon). The AP-2 transcription factors have a helix-span-helix motif at the C-terminus, a core basic region, and a transactivation domain at the N-terminus, and are evolutionarily conserved in metazoans (Eckert, Buhl, Weber, Jäger, & Schorle, 2005). EGFR and VEGFA identified genes were found to be involved in this pathway. The identified second most significant pathway was potential therapeutics for SARS. The detailed pathway is shown in Fig. 3B. Based on their efficacy in treating infectious disease with other RNA viruses or in reducing cytokine storms and other illnesses caused by viruses are identical to SARS-CoV-1 and SARS-CoV-2. A significant number of intriguing therapeutic candidates have been found to be also similar. HDAC2, SIGMAR1, IMPDH2, VEGFA, and RIPK1 identified genes were found to be involved in this pathway. SARS-CoV-2 infection pathway is not well annotated. However, the viral infection pathways are curated based on the SARS-CoV-1 and SARS-CoV-2 infection processes and drug responses. Many of the steps in SARS-CoV-1 infections are well studied experimentally correlated with the steps involved in SARS-CoV-2 infection. In comparison with the other two significant pathways, a maximum of 6 targets (CTSL, RIPK1, HDAC2, SIGMAR1, IMPDH2, and VEGFA) are found to be in this pathway (Supplementary Data 1 and 2 in the online version at https://doi.org/10.1016/bs.apcsb.2021.11.007).

The common genes were subjected to the gene–gene interaction study using the online STRING database to identify the connected genes (Szklarczyk et al., 2019). Out of the identified targets, HDAC2, IL6, EGFR, DPP4, ACE, VEGFA, CTSL, CTSB, and RIPK1 were interlinked with another target (Fig. 4). These targets were already promising targets for COVID-19 treatment in several previous studies. Therefore, activating or suppressing one of these genes could impact the other and improve efficacy.

The Histone deacetylase 2 (HDAC2) plays a crucial role in monocytes' regular immune response function. These monocytes were found to maintain the immune system's regular role when combined with the macrophages. The HDAC2 drugs such as Theophylline, Macrolides, Nortriptyline, and many others have been shown to prevent the pathological response of the inflammatory monocytes and thereby regulate the normal lung function (HDAC2 Regulates Response of Inflammatory Monocytes: A COVID-19 Target, 2020). A study by Liu et al. in 2020 has also proposed that the inhibition of HDAC could be a promising target for the COVID-19 infection (Liu et al., 2020).

T-lymphocytes, adipose tissues, and macrophages all produce interleukin-6 (IL-6), a pro-inflammatory cytokine protein. It plays a significant role in atherogenesis and is linked to cardiovascular clinical outcomes (Gabay, 2006). The COVID-19 patients with more severe illnesses had higher inflammatory cytokines linked to pulmonary inflammation, lung destruction, and multiple organ failure. SARS-CoV-2 patients exhibit low concentrations of the modulator of cytokine signaling-3, which again is involved in the stimulation of the IL-6 negative feedback loop. Increased IL 6 has also been discovered to be an indicator of the change from a mild to a severe illness, limiting the severity if caught early (Vatansever & Becer, 2020).

The progression of SARS-CoV induces fibrosis, as well as a review of evidence indicating pulmonary fibrosis is generated by an overactive host response to lung damage driven through epidermal growth factor receptor (EGFR) signaling (Venkataraman & Frieman, 2017). Adeno-associated virus 2 (AAV2) second-strand DNA synthesis and transgene expression are inhibited by FK506-binding protein (FKBP52), which is phosphorylated at tyrosine residues by EGFR-PTK (Yano et al., 2003). Because EGFR is abundantly expressed in various solid tumors and its expression is linked to tumor progression, leading to chemotherapy resistance and poor prognosis. It is a promising strategy for the rational development of new anticancer drugs.

DPP4 is widely known to involve in type 2 diabetic conditions. DPP4 activity affects glucose homeostasis and inflammation in numerous ways. The change in expression levels of DPP4 in the early MERS infections was also studied several years ago (Chan et al., 2015). Currently, the SARS-CoV-2 being the same family of MERS infection, the role of DPP4 in SARS-CoV-2 conditions is also widely studied and found to be a promising target for COVID-19 treatment (Scheen, 2021; Solerte, Di Sabatino, Galli, & Fiorina, 2020).

ACE2 acts as a functional receptor on the cell's surface through which the SARS-CoV-2 can enter the human cell. This ACE2 is expressed highly in the heart, kidney, and lungs. The ACE/ACE2 balance disruption with RAAS (renin-angiotensin-aldosterone system) activation can lead to severe COVID-19, especially in diabetes, cardiovascular diseases, and hypertension (Beyerstedt, Casaro, & Rangel, 2021).

In ICU and non-ICU COVID-19 patients, VEGF concentrations were more significant than in healthy controls. Angiogenesis, neurogenesis, and neuroprotection are all aided by VEGFs, leading to vascular leakiness and permeability. ACE2 inhibits VEGF-A, which reduces vascular permeability in patients with acute lung damage. In the account of SARS-downregulation CoV-2's of ACE2, the VEGF-A antagonistic effect of ACE2 is expected to be canceled, resulting in overexpression of VEGF and increased vascular permeability and exacerbation of endothelial injury (Yazihan et al., 2021).

The *CTSL1* gene codes for a lysosomal cysteine protease involved in intracellular protein degradation. It impacts collagen and elastin, along with alpha-1 protease inhibitor, a key regulator of neutrophil elastase function. CatL is essential in degrading the extracellular matrix, a crucial mechanism for SARS-CoV-2 to enter host cells and is upregulated during chronic inflammation. CatL is also likely involved in the processing of SARS-CoV-2 spike protein. CatL could have been regarded as a valuable therapeutic target because its suppression damages SARS-CoV-2 infection and maybe egress from cells through late stages of infection (Gomes et al., 2020; Pišlar et al., 2020).

Because several patients had greater serum RIPK3 (a family of RIPK1) levels, it is possible that RIPK-3-mediated signaling may lead to necroptosis; it is implicated in the development of COVID-19 pneumonia-related acute lung injury. In ARDS patients, RIPK-3 levels were considerably more significant than in non-ARDS patients. In immunohistochemistry, all epithelial cell samples from COVID-19 individuals were confirmed for active phosphorylated RIPK1. Thus, ICD mediated by RIPK1 may have a part in the progression of SARS-CoV-2 infection and can be a new therapeutic target (Nakamura et al., 2020).

Several compounds targeting these genes can interact with more than one gene. Virtual screenings were performed to identify the best interaction of the compounds with these targets. From the virtual screening analysis, betulinic acid was found to interact with ACE, 5″-(2⁗-Hydroxybenzyl)uvarinol was found to interact with Cathepsin L, Cathepsin B, and Cathepsin K, -(-)-antofine was found to interact with DPP4, (S)-1′-methyloctyl caffeate was found to interact with EGFR, (Z)-3-phenyl-2-propenal was found to interact with HDAC2, 7-oxo-10α-cucurbitadienol was found to interact with IL6, PLX-4720 was found to interact with RIPK1, and 5″-(2⁗-Hydroxybenzyl)uvarinol was found to interact with VEGFA (Supplementary Table 3 in the online version at https://doi.org/10.1016/bs.apcsb.2021.11.007). The molecular dockings were performed using the AutoDock, and the interactions were studied using the Discovery Studio (Fig. 5). Thus collectively, these compounds targeting the identified targets could improve the immune system, fighting against the COVID-19 infection.

5. Conclusion

The COVID-19 infection is widely increasing across the globe. The drugs to treat COVID-19 are still in the clinical trials, which might take years for approval. Drugs such as Hydroxychloroquine, Remdesivir, Favipiravir, Lopinavir/ritonavir are widely repurposed to treat the COVID-19 during the earlier outbreak. On the other hand, various vaccine manufacturers such as COVID-19 Vaccine AstraZeneca, COVID-19 Vaccine Janssen, Sputnik, and Covaxin are in the race to develop vaccines, yet none seems to be 100% efficient. In addition, with the evolution of SARS-CoV-2 mutations, the efficiency of the vaccines is becoming a interrogation point every day.

Similarly, these mutations might lead to drug inefficacy and drug-resistant. The promising approach to be safe from COVID-19 is to develop the immune system. The traditional medicines are immune boosters, serving as a promising remedy for various diseases since ancient days. This study used the network pharmacology approach and analyzed the phytochemical compounds in Nilavembu kudineer, Kabasura kudineer, and JACOM and their interacting human protein targets, activating or suppressing the target. These identified compounds can be tested *in vivo* and *in vitro* to compare the toxicity and efficiency of the currently available formulations.

Conflict of interest

The authors have declared that no conflicts of interest exist.

Author contributions

D.T.K., S.U.K., S.P., S.D.M.S., S.R., and C.G.P.D. were involved in the study's design. D.T.K., A.S., A.M., L.M., R.G., M.R., and S.U.K. were involved in the data collection and experimentation. D.T.K., A.S., A.M., L.M., R.G., M.R., and S.U.K. were engaged in the acquisition, analysis, and interpretation results. D.T.K. and S.U.K. drafted the manuscript. C.G.P.D., S.R. and S.P. supervised the entire study and was involved in study design, the acquisition, analysis, understanding of the data, and critically reviewed the manuscript. All authors edited and approved the submitted version of the article.

Funding

No funding agency was involved in the present study.

Acknowledgments

The authors would like to take this opportunity to thank the management of Vellore Institute of Technology (VIT), Vellore, India, and Meenakshi Academy of Higher Education and Research, Chennai, for providing the necessary facilities and encouragement to carry out this work.

References

Alagu Lakshmi, S., Shafreen, R. M. B., Priya, A., & Shunmugiah, K. P. (2020). Ethnomedicines of Indian origin for combating COVID-19 infection by hampering the viral replication: Using structure-based drug discovery approach. *Journal of Biomolecular Structure and Dynamics*. https://doi.org/10.1080/07391102.2020.1778537.

Beyerstedt, S., Casaro, E. B., & Rangel, É. B. (2021). COVID-19: Angiotensin-converting enzyme 2 (ACE2) expression and tissue susceptibility to SARS-CoV-2 infection. *European Journal of Clinical Microbiology and Infectious Diseases*, *40*(5), 905–919. Springer Science and Business Media Deutschland GmbH https://doi.org/10.1007/s10096-020-04138-6.

Boulware, D. R., Pullen, M. F., Bangdiwala, A. S., Pastick, K. A., Lofgren, S. M., Okafor, E. C., et al. (2020). A randomized trial of hydroxychloroquine as postexposure prophylaxis for Covid-19. *New England Journal of Medicine, 383*(6), 517–525. https://doi.org/10.1056/nejmoa2016638.

Cain, D. W., & Cidlowski, J. A. (2020). After 62 years of regulating immunity, dexamethasone meets COVID-19. *Nature Reviews Immunology, 20*(10), 587–588. Nature Research https://doi.org/10.1038/s41577-020-00421-x.

Chan, J. F. W., Lau, S. K. P., To, K. K. W., Cheng, V. C. C., Woo, P. C. Y., & Yue, K. Y. (2015). Middle East Respiratory syndrome coronavirus: Another zoonotic betacoronavirus causing SARS-like disease. *Clinical Microbiology Reviews*. https://doi.org/10.1128/CMR.00102-14.

Corman, V. M., Landt, O., Kaiser, M., Molenkamp, R., Meijer, A., Chu, D. K. W., et al. (2020). Detection of 2019 novel coronavirus (2019-nCoV) by real-time RT-PCR. *Eurosurveillance, 25*(3), 2000045. https://doi.org/10.2807/1560-7917.ES.2020.25.3.2000045.

Cucinotta, D., & Vanelli, M. (2020). WHO declares COVID-19 a pandemic. *Acta Biomedica, 91*(1), 157–160. Mattioli 1885 https://doi.org/10.23750/abm.v91i1.9397.

Daina, A., Michielin, O., & Zoete, V. (2017). SwissADME: A free web tool to evaluate pharmacokinetics, drug-likeness and medicinal chemistry friendliness of small molecules. *Scientific Reports*, 7, 42717. https://doi.org/10.1038/srep42717.

Dallakyan, S., & Olson, A. J. (2015). Small-molecule library screening by docking with PyRx. *Methods in Molecular Biology, 1263*, 243–250. https://doi.org/10.1007/978-1-4939-2269-7_19.

Dhand, R., & Li, J. (2020). Coughs and sneezes: Their role in transmission of respiratory viral infections, including SARS-CoV-2. *American Journal of Respiratory and Critical Care Medicine, 202*(5), 651–659. American Thoracic Society https://doi.org/10.1164/rccm.202004-1263PP.

Eckert, D., Buhl, S., Weber, S., Jäger, R., & Schorle, H. (2005). The AP-2 family of transcription factors. *Genome Biology, 6*(13), 246. BioMed Central https://doi.org/10.1186/gb-2005-6-13-246.

Fabregat, A., Sidiropoulos, K., Garapati, P., Gillespie, M., Hausmann, K., Haw, R., et al. (2016). The reactome pathway knowledgebase. *Nucleic Acids Research, 44*(D1), D481–D487. https://doi.org/10.1093/nar/gkv1351.

Gabay, C. (2006). Interleukin-6 and chronic inflammation. *Arthritis Research and Therapy, 8*(Suppl. 2). https://doi.org/10.1186/ar1917.

Geleris, J., Sun, Y., Platt, J., Zucker, J., Baldwin, M., Hripcsak, G., et al. (2020). Observational study of hydroxychloroquine in hospitalized patients with Covid-19. *New England Journal of Medicine, 382*(25), 2411–2418. https://doi.org/10.1056/nejmoa2012410.

Gfeller, D., Grosdidier, A., Wirth, M., Daina, A., Michielin, O., & Zoete, V. (2014). SwissTargetPrediction: A web server for target prediction of bioactive small molecules. *Nucleic Acids Research, 42*(W1), W32–W38. https://doi.org/10.1093/nar/gku293.

Gomes, C. P., Fernandes, D. E., Casimiro, F., da Mata, G. F., Passos, M. T., Varela, P., et al. (2020). Cathepsin L in COVID-19: From pharmacological evidences to genetics. In *Vol. 10. Frontiers in cellular and infection microbiology* (p. 777). Frontiers Media S.A. https://doi.org/10.3389/fcimb.2020.589505.

Hastings, J., De Matos, P., Dekker, A., Ennis, M., Harsha, B., Kale, N., et al. (2013). The ChEBI reference database and ontology for biologically relevant chemistry: Enhancements for 2013. *Nucleic Acids Research, 41*(D1). https://doi.org/10.1093/nar/gks1146.

HDAC. (2020). *HDAC2 Regulates response of inflammatory monocytes: a COVID-19 target*. Retrieved June 12th, 2021, from https://www.drugtargetreview.com/article/60593/histone-deacetylase-2-hdac2-regulates-pathological-response-of-inflammatory-monocytes-a-potential-target-of-adjuvant-therapies-for-covid-19-infection/.

Jain, J., Kumar, A., Narayanan, V., Ramaswamy, R. S., Sathiyarajeswaran, P., Shree Devi, M. S., et al. (2020). Antiviral activity of ethanolic extract of Nilavembu Kudineer against dengue and chikungunya virus through in vitro evaluation. *Journal of Ayurveda and Integrative Medicine*, *11*(3), 329–335. https://doi.org/10.1016/j.jaim.2018.05.006.

Jain, J., Narayanan, V., Chaturvedi, S., Pai, S., & Sunil, S. (2018). In vivo evaluation of Withania somnifera–based Indian traditional formulation (amukkara choornam), against chikungunya virus–induced morbidity and arthralgia. *Journal of Evidence-Based Integrative Medicine*. https://doi.org/10.1177/2156587218757661.

Jain, J., Pai, S., & Sunil, S. (2018). Standardization of in vitro assays to evaluate the activity of polyherbal siddha formulations against Chikungunya virus infection. *VirusDisease*. https://doi.org/10.1007/s13337-018-0421-0.

Jassal, B., Matthews, L., Viteri, G., Gong, C., Lorente, P., Fabregat, A., et al. (2020). The reactome pathway knowledgebase. *Nucleic Acids Research*, *48*(D1), D498–D503. https://doi.org/10.1093/nar/gkz1031.

Kiran, G., Karthik, L., Shree Devi, M. S., Sathiyarajeswaran, P., Kanakavalli, K., Kumar, K. M., et al. (2020). In silico computational screening of Kabasura Kudineer—official siddha formulation and JACOM against SARS-CoV-2 spike protein. *Journal of Ayurveda and Integrative Medicine*. https://doi.org/10.1016/j.jaim.2020.05.009.

Kumar, D. T., Shaikh, N., Kumar, S. U., Doss, C. G. P., & Zayed, H. (2021). Structure-based virtual screening to identify novel potential compound as an alternative to remdesivir to overcome the RdRp protein mutations in SARS-CoV-2. *Frontiers in Molecular Biosciences*, *8*. https://doi.org/10.3389/fmolb.2021.645216.

Kumar, S. U., Kumar, D. T., Christopher, B. P., & Doss, C. G. P. (2020). The rise and impact of COVID-19 in India. *Frontiers in Medicine*, 7, 250. https://doi.org/10.3389/fmed.2020.00250.

Kumar, S. U., Kumar, D. T., Siva, R., & Doss, C. G. P. (2020). Kerala, India's front runner in novel coronavirus disease (COVID-19). *Frontiers in Medicine*, 7, 355. https://doi.org/10.3389/fmed.2020.00355.

Lamontagne, F., Agoritsas, T., MacDonald, H., Leo, Y. S., DIaz, J., Agarwal, A., et al. (2020). A living WHO guideline on drugs for covid-19. *The BMJ*, *370*, m3379. https://doi.org/10.1136/bmj.m3379.

Liu, K., Zou, R., Cui, W., Li, M., Wang, X., Dong, J., et al. (2020). Clinical HDAC inhibitors are effective drugs to prevent the entry of SARS-CoV2. *ACS Pharmacology and Translational Science*, *3*(6), 1361–1370. https://doi.org/10.1021/acsptsci.0c00163.

Morris, G. M., Goodsell, D. S., Halliday, R. S., Huey, R., Hart, W. E., Belew, R. K., et al. (1998). Automated docking using a Lamarckian genetic algorithm and an empirical binding free energy function. *Journal of Computational Chemistry*, *19*(14), 1639–1662. https://doi.org/10.1002/(SICI)1096-987X(19981115)19:14<1639::AID-JCC10>3.0.CO;2-B.

Morris, G. M., Huey, R., Lindstrom, W., Sanner, M. F., Belew, R. K., Goodsell, D. S., et al. (2009). AutoDock4 and AutoDockTools4: Automated docking with selective receptor flexibility. *Journal of Computational Chemistry*, *30*(16), 2785–2791. https://doi.org/10.1002/jcc.21256.

Nakamura, H., Kinjo, T., Arakaki, W., Miyagi, K., Tateyama, M., & Fujita, J. (2020). Serum levels of receptor-interacting protein kinase-3 in patients with COVID-19. *Critical Care*, *24*(1). BioMed Central https://doi.org/10.1186/s13054-020-03209-6.

Natarajan, S., Anbarasi, C., Sathiyarajeswaran, P., Manickam, P., Geetha, S., Kathiravan, R., et al. (2020). The efficacy of Siddha Medicine, Kabasura Kudineer (KSK) compared to Vitamin C & Zinc (CZ) supplementation in the management of asymptomatic COVID-19 cases: A structured summary of a study protocol for a randomised controlled trial. *Trials*, *21*(1). BioMed Central Ltd https://doi.org/10.1186/s13063-020-04823-z.

Pathan, M., Keerthikumar, S., Ang, C. S., Gangoda, L., Quek, C. Y. J., Williamson, N. A., et al. (2015). FunRich: An open access standalone functional enrichment and interaction network analysis tool. *Proteomics*, *15*(15), 2597–2601. https://doi.org/10.1002/pmic.201400515.

Patwardhan, B., Warude, D., Pushpangadan, P., & Bhatt, N. (2005). Ayurveda and traditional Chinese medicine: A comparative overview. *Evidence-Based Complementary and Alternative Medicine*, *2*(4), 465–473. https://doi.org/10.1093/ecam/neh140.

Pišlar, A., Mitrovic, A., Sabotič, J., Fonovic, U. P., Nanut, M. P., Jakoš, T., et al. (2020). The role of cysteine peptidases in coronavirus cell entry and replication: The therapeutic potential of cathepsin inhibitors. *PLoS Pathogens*, *16*(11), e1009013. Public Library of Science https://doi.org/10.1371/journal.ppat.1009013.

Plante, J. A., Mitchell, B. M., Plante, K. S., Debbink, K., Weaver, S. C., & Menachery, V. D. (2021). The variant gambit: COVID-19's next move. *Cell Host and Microbe*, *29*(4), 508–515. Cell Press https://doi.org/10.1016/j.chom.2021.02.020.

Rathinam, S., Muthiah, K., Parameswaran, S., Tamilarasan, K., Selvarajan, E., & Ayyasamy, U. (2020). Analogy of Kaba Suram with COVID-19 symptoms–A siddha literature review. *International Journal of Ayurvedic Medicine*, *11*(4), 616–621. https://doi.org/10.47552/ijam.v11i4.1674.

Ravishankar, B., & Shukla, V. J. (2007). Indian systems of medicine: A brief profile. *African Journal of Traditional, Complementary and Alternative Medicines*, *4*(3), 319–337. African Ethnomedicines Network https://doi.org/10.4314/ajtcam.v4i3.31226.

Rebhan, M., Chalifa-Caspi, V., Prilusky, J., & Lancet, D. (1997). GeneCards: Integrating information about genes, proteins and diseases. *Trends in Genetics*, *13*(4), 163. https://doi.org/10.1016/S0168-9525(97)01103-7.

Ross, C. L. (2009). Integral healthcare: The benefits and challenges of integrating complementary and alternative medicine with a conventional healthcare practice. *Integrative Medicine Insights*, *2009*(4), 13–20. SAGE Publications https://doi.org/10.4137/IMI.S2239.

Scheen, A. J. (2021). DPP-4 inhibition and COVID-19: From initial concerns to recent expectations. *Diabetes and Metabolism*, *47*(2), 101213. Elsevier Masson s.r.l https://doi.org/10.1016/j.diabet.2020.11.005.

Solerte, S. B., Di Sabatino, A., Galli, M., & Fiorina, P. (2020). Dipeptidyl peptidase-4 (DPP4) inhibition in COVID-19. *Acta Diabetologica*, *57*(7), 779–783. https://doi.org/10.1007/s00592-020-01539-z.

Srivastava, S., Banu, S., Singh, P., Sowpati, D. T., & Mishra, R. K. (2021). SARS-CoV-2 genomics: An Indian perspective on sequencing viral variants. *Journal of Biosciences*, *46*(1), 1–14. Springer https://doi.org/10.1007/s12038-021-00145-7.

Szklarczyk, D., Gable, A. L., Lyon, D., Junge, A., Wyder, S., Huerta-Cepas, J., et al. (2019). STRING v11: Protein-protein association networks with increased coverage, supporting functional discovery in genome-wide experimental datasets. *Nucleic Acids Research*, *47*(D1), D607–D613. https://doi.org/10.1093/nar/gky1131.

Taboada, M., González, M., Alvarez, A., Eiras, M., Costa, J., Álvarez, J., et al. (2021). First, second and third wave of COVID-19. What have we changed in the ICU management of these patients? *Journal of Infection*, *82*(6), e14–e15. W.B. Saunders Ltd https://doi.org/10.1016/j.jinf.2021.03.027.

Tai, W., He, L., Zhang, X., Pu, J., Voronin, D., Jiang, S., et al. (2020). Characterization of the receptor-binding domain (RBD) of 2019 novel coronavirus: Implication for development of RBD protein as a viral attachment inhibitor and vaccine. *Cellular and molecular immunology*, *17*(6), 613–620. https://doi.org/10.1038/s41423-020-0400-4.

Trott, O., & Olson, A. J. (2010). AutoDock Vina: Improving the speed and accuracy of docking with a new scoring function, efficient optimization, and multithreading. *Journal of Computational Chemistry*, *31*(2), 455–461. https://doi.org/10.1002/jcc.21334.

Vatansever, H. S., & Becer, E. (2020). Relationship between IL-6 and COVID-19: To be considered during treatment. *Future Virology*, *15*(12), 817–822. Future Medicine Ltd https://doi.org/10.2217/fvl-2020-0168.

Venkataraman, T., & Frieman, M. B. (2017). The role of epidermal growth factor receptor (EGFR) signaling in SARS coronavirus-induced pulmonary fibrosis. *Antiviral Research*, *143*, 142–150. Elsevier B.V https://doi.org/10.1016/j.antiviral.2017.03.022.

WHO. (2021a). *What is COVID-19 vaccine efficacy? | WHO | Regional Office for Africa.* Retrieved June 12th, 2021, from https://www.afro.who.int/news/what-covid-19-vaccine-efficacy.

WHO. (2021b). *WHO coronavirus (COVID-19) dashboard | WHO coronavirus (COVID-19) dashboard with vaccination data.* Retrieved June 8th, 2021, from https://covid19.who.int/.

Williams, T. C., & Burgers, W. A. (2021). SARS-CoV-2 evolution and vaccines: Cause for concern? *The Lancet Respiratory Medicine*, *9*(4), 333–335. Lancet Publishing Group https://doi.org/10.1016/S2213-2600(21)00075-8.

Wu, R., Wang, L., Kuo, H. C. D., Shannar, A., Peter, R., Chou, P. J., et al. (2020). An update on current therapeutic drugs treating COVID-19. *Current Pharmacology Reports*, *6*(3), 56–70. Springer https://doi.org/10.1007/s40495-020-00216-7.

Yano, S., Kondo, K., Yamaguchi, M., Richmond, G., Hutchison, M., Wakeling, A., et al. (2003). Distribution and function of EGFR in human tissue and the effect of EGFR tyrosine kinase inhibition. *Anticancer Research*, *Vol. 23*(5A), 3639–3650. https://europepmc.org/article/med/14666659.

Yazihan, N., Tanacan, A., Erol, S. A., Anuk, A. T., Sinaci, S., Biriken, D., et al. (2021). Comparison of VEGF-A values between pregnant women with COVID-19 and healthy pregnancies and its association with composite adverse outcomes. *Journal of Medical Virology*, *93*(4), 2204–2209. https://doi.org/10.1002/jmv.26631.

Youns, M., Hoheisel, J. D., & Efferth, T. (2010). Traditional Chinese Medicines (TCMs) for molecular targeted therapies of tumours. *Current Drug Discovery Technologies*, 7(1), 37–45. https://doi.org/10.2174/157016310791162730.

CHAPTER TEN

Immunotherapeutic interventions in Parkinson's disease: Focus on α-Synuclein

Upasana Ganguly[a], Sukhpal Singh[a], Sasanka Chakrabarti[a], Adesh K. Saini[b], and Reena V. Saini[b,*]

[a]Department of Biochemistry and Central Research Cell, Maharishi Markandeshwar Institute of Medical Sciences and Research, Maharishi Markandeshwar University (Deemed to be), Mullana, India
[b]Department of Biotechnology, Maharishi Markandeshwar Engineering College, Maharishi Markandeshwar (Deemed to be University), Mullana, India
*Corresponding author: e-mail address: reenavohra10@gmail.com

Contents

Abstract

Parkinson's disease (PD) is a neurodegenerative disorder characterized classically by motor manifestations. However, nonmotor symptoms appear early in the course of the disease progression, making both diagnosis and treatment difficult. The pathology of PD is complicated by the accumulation and aggregation of misfolded proteins in

Advances in Protein Chemistry and Structural Biology, Volume 129
ISSN 1876-1623
https://doi.org/10.1016/bs.apcsb.2021.11.010

intracellular cytoplasmic inclusions called Lewy bodies (LBs). The main toxic component of LBs is the protein α-Synuclein which plays a pivotal role in PD pathogenesis. α-Synuclein can propagate from cell-to-cell exhibiting prion-like properties and spread PD pathology throughout the central nervous system. Immunotherapeutic interventions in PD, both active and passive immunization, have targeted α-Synuclein in both experimental models and clinical trials. In addition, targeting the hyperactive inflammation in PD also holds promise in designing potential immunotherapeutics. The inflammatory and proteotoxic pathways are interlinked and contribute immensely to the disease pathology. In this chapter, we critically review the targets of immunotherapeutic interventions in PD, focusing on the pathogenetic mechanisms of PD, particularly neuroinflammation and α-Synuclein misfolding, aggregation, and propagation. We thoroughly summarized the various immunotherapeutic strategies designed to treat PD—*in vitro*, *in vivo*, and clinical trials. The development of these targeted immunotherapies could open a new avenue in the treatment of patients with PD.

1. Introduction

Parkinson's disease (PD) is a movement disorder with characteristic degeneration of dopaminergic neurons in the substantia nigra. PD is insidiously progressive, and when clinical symptoms manifest in full blow, the patients have already lost 50% of neurons in the substantia nigra and about 80% of the efferent termini (Surmeier, 2018). Dopamine depletion in PD is due to the loss of the presynaptic neurons and the upregulation of dopamine receptors in the postsynaptic neurons, which increases their sensitivities to dopaminergic ligands (Cox, Kerwin, & Lee, 1978). Therapeutic measures in PD, therefore, target to replenish the dopamine loss by synthetic dopamine replacement (levodopa/carbidopa) and the administration of dopamine agonists for the upregulated dopamine receptors. However, these treatment strategies do not offer a curative outcome, and treatment resistance is often a common feature in PD patients. Therefore, curative outcomes in PD patients need to address the loss of dopaminergic neurons by stem cell therapies, ameliorating neural injuries through treatment with growth factors, eliminating the misfolded protein accumulation, and modulating the diseased brain microenvironment (Jankovic & Aguilar, 2008).

Abnormal accumulation of proteins, particularly of α-Synuclein, is a central feature in PD pathogenesis. Unfortunately, this protein is an enigma for PD researchers—a protein whose exact physiological role and contribution to disease pathogenesis are challenging to decipher. However, the accumulated α-Synuclein is toxic, and it modulates the cells to a variety of degenerative proteotoxic pathways. In addition to α-Synuclein aggregation, oxidative stress, mitochondrial dysfunction, and neuroinflammation are also implicated in PD pathogenesis. These pathogenetic mechanisms are often interlinked and contribute widely to neurodegeneration in PD. Immunotherapeutic advances targeting α-Synuclein, therefore, will be an effective therapeutic advancement in PD. Another target of immunotherapy in PD is the hyperactive inflammation in the brain microenvironment. Abnormally aggregated α-Synuclein released into the extracellular space by dead or dying neurons often cause the activation of innate immune responses in the central nervous system (CNS), induction of reactive microglia, and release of proinflammatory mediators (Harms et al., 2018). In addition, adaptive immune responses (T-cell responses and antibody production) are activated to exacerbate the effects of microglial activation (Allen Reish & Standaert, 2015; Kannarkat, Boss, & Tansey, 2013).

Since conception, active and passive immunotherapies have been used to treat diseases; immunomodulatory therapies in PD either target α-Synuclein or the inflammatory cascade. This chapter reviews the immune regulation in PD, the pathogenic mechanisms of α-Synuclein accumulation, spread, and involvement in the neuroinflammatory cascade in PD neurodegeneration. Further, we will discuss the immunomodulatory strategies developed to treat PD and their mechanisms of action, especially active and passive immunization involving α-Synuclein.

2. Parkinson's disease: An overview

Parkinson's disease (PD) is a complex neurodegenerative disorder and affects primarily the "over 60" population. It was first described by British physician Sir James Parkinson in "An Essay on the Shaking Palsy" as "paralysis agitans" in 1817. Later Jean-Martin Charcot gave credit to Parkinson and characterized the disease as "Maladie de Parkinson" or Parkinson's disease. PD is currently the second most common neurodegenerative disorder, ranking next to Alzheimer's disease (Kalia & Lang, 2016). The disease prevalence is 0.5–1% in the age group of 65–69 years and 1–3% in 80 years and above (Kasten, Chade, & Tanner, 2007; Nussbaum & Ellis, 2003). The global burden of diseases (GBD) enterprise collaborators provided a systematic analysis of the global, regional and country-specific prevalence of PD from 1990 to 2016 (GBD 2016 Parkinson's Disease Collaborators, 2018). In 2016, the prevalence of PD was 6.1 million globally compared to 2.5 million in 1990. Moreover, the prevalence of PD is higher in men than women at the starting age of 50 years. Additionally, higher prevalence has been reported in Europe, North America, and South America than Asian, African, or Arabic countries.

PD is multifactorial with the involvement of both genetic and environmental factors. More than 90% of PD cases are sporadic, with onset later in life and uncertain etiology. Genetic involvement (familial PD) accounts for 5–10% of PD cases. To date, 19 disease-causing genes and 23 susceptibility loci have been identified (Deng, Wang, & Jankovic, 2018). The genetic studies have been linked to case-control and genome-wide association studies (GWAS). Mainly mutations in α-Synuclein (*PARK1/PARK4*), ubiquitin C-terminal hydrolase L1 (UCHL1/*PARK5*), and leucine-rich repeat kinase 2 (*LRRK2/PARK8*) result in autosomal dominant forms of PD. Mutations in Parkin (*PARK2*), PINK1 (*PARK6*), and DJ-1 (*PARK7*) result in autosomal recessive juvenile Parkinson's disease (AR-JP). Several candidate genes have been screened and not linked to PD like Apo-ε4, tyrosine hydroxylase, glutathione peroxidase, catalase, superoxide dismutase (SOD1 and SOD2), D2, D3, and D4 dopamine receptors (Gasser et al., 1994). Another important genetic risk factor is mutations in *GBA1* that encodes β-glucocerebrosidase, a lysosomal enzyme catalyzing the hydrolysis of glucocerebrosides (Nichols et al., 2009). In addition, there was an extensive search in the mitochondrial genome for the presence of any PD-linked mutation. However, no PD-specific mutation was identified, and most studies failed to detect a maternal inheritance pattern.

PD is pathologically characterized by the selective and profound loss of dopaminergic neurons in the substantia nigra, although neurodegeneration also affects cells located in other regions of the neural network. PD is a complex, heterogeneous disorder with a widespread pathology without a reliable diagnostic test. The current diagnosis relies on clinical symptoms with the criteria of the presence of two of the following clinical symptoms: resting tremor, bradykinesia, rigidity, and/or postural instability (Postuma et al., 2015). Clinical criteria can only diagnose a probable PD, while definitive diagnosis requires histopathological assessment of α-Synuclein containing Lewy bodies (LBs) or Lewy neurites (LNs). Nonmotor symptoms in early PD often precede motor manifestations. Early nonmotor symptoms manifest as impaired olfactory ability, autonomic dysfunction, pain, fatigue, sleep disorders, cognitive and psychiatric disturbances contributing to the patient's disability and life quality (Martinez-Martin, Rodriguez-Blasquez, Kurtis, & Ray Chaudhuri, 2011). Constipation and urinary incontinence are common, and dementia occurs in more than 80% of patients (Iranzo et al., 2013).

PD treatment focuses predominantly on symptomatic relief with drugs either restoring dopamine in the striatum or acting on the striatal postsynaptic dopamine receptors. However, dopamine is not the only neurotransmitter affected in PD neurodegeneration, and other drugs are being used to target symptoms such as dementia or depression. Still, the current treatment regime mainly includes levodopa (L-DOPA) as the gold standard of treatment in combination with other dopamine agonists, monoamine oxidase type B (MAO-B) inhibitors, aromatic amino acid decarboxylase (AADC) inhibitors, catechol-O-methyltransferase (COMT) inhibitors, and anticholinergics which has been listed in Table 1. However, levodopa therapy does not halt disease progression, and with prolonged use of the drug, patients develop resistance requiring higher doses of the medication and often exhibit drug-induced side effects like dyskinesia (Connolly & Lang, 2014). Moreover, levodopa does not alleviate the nonmotor symptoms in PD patients, immensely affecting their quality of life. As a result, the patients require hospital admission and further decline to death. Thus, we require more robust therapeutic measures that can slow down the disease progression and improve the quality of life in the affected patients.

Table 1 Pharmacological therapeutics in Parkinson's disease.

Treatment	Examples	Mechanism of action	Comments/ limitations
Levodopa (L-DOPA)	Levodopa	Dopamine precursor	▪ L-DOPA is still the gold standard in PD treatment ▪ Long-term use results in motor complications, motor oscillations and drug-induced dyskinesia
Aromatic amino acid decarboxylase (AADC) inhibitors	Benserazide, Carbidopa	Prevents peripheral metabolism of dopamine, enhances bioavailability	
Catechol-O-methyltransferase (COMT) inhibitors	Entacapone, Opicapone, Tolcapone, Nebicapone	Enhances bioavailability and half-life of L-DOPA	▪ Dependent on the concomitant use of L-DOPA ▪ Side effects include liver toxicity and diarrhea

Continued

Table 1 Pharmacological therapeutics in Parkinson's disease.—cont'd

Treatment	Examples	Mechanism of action	Comments/ limitations
Monoamine oxidase type B (MAO-B) inhibitors	Rasagiline, Selegiline, Safinamide	Increases and prolongs synaptic dopamine concentrations	▪ Side effects include nausea, dry mouth, light-headedness, constipation, confusion, hallucinations
Dopamine agonists	Pramipexole, Ropinirole, Piribedil	Dopamine receptor 2 (D2R) agonists	▪ Longer half-life than L-DOPA ▪ Induce drowsiness and impulse dyscontrol
	Apomorphine, Rotigotine	Dopamine receptor 1 (D1R) agonists	
Anticholinergics	Benztropine Trihexyphenidyl	Reduces acetylcholine activity	▪ Side effects include blurred vision, dry mouth, constipation, urinary retention, hallucinations

3. Parkinson's disease and the immune system

The blood-brain barrier (BBB) confers the status of immune privilege to the human brain. However, identification of the meningeal lymphatic vasculature has revealed that the CNS antigens can interact with the lymphocytes and antigen-presenting cells (APCs) (Brioschi & Colonna, 2019; Louveau et al., 2018). Moreover, increased leakage of the BBB has been reported in PD patients (Al-Bachari, Naish, Parker, Emsley, & Parkes, 2020; Gasca-Salas, Fernández-Rodríguez, Pineda-Pardo, et al., 2021). These findings suggest the involvement of the innate and adaptive immune responses in the neurodegenerative disorders of the CNS.

3.1 Neuroinflammation: A key player in Parkinson's disease neurodegeneration

The innate immune system consists of mechanisms encoded during embryogenesis in the germline, present by birth, and work nonspecific as a defense manager against various microbial infections. Increasing evidence from previous studies suggests innate immune responses in Parkinson's disease with variation in genetic, cellular, and biochemical markers. Innate immune system activation includes removal of foreign material by phagocytosis, recruitment of other cells to the site of infection, and production of cytokines and chemokines. In addition, the innate immune system acts in a nonspecific pathway, including generic recognition through pathogen-associated molecular patterns (PAMPs) by sharing common cell signaling pathways to endogenous and exogenous threats. Toll-like receptors (TLRs), expressed on microglial cells, astrocytes, oligodendrocytes, and neurons, recognize these PAMPs (Bsibsi et al., 2006; Bsibsi, Ravid, Gveric, & van Noort, 2002; Farina et al., 2005; Glezer, Simard, & Rivest, 2007; Lafon, Megret, Lafage, & Prehaud, 2006; Olson & Miller, 2004).

The brain and periphery of PD patients exhibit inflammatory changes (Kannarkat et al., 2013). Gagne and Power conducted a meta-analysis which revealed a 15% less incidence of PD in patients regularly using nonaspirin nonsteroidal anti-inflammatory drugs (NSAIDs)—highlighting a crucial role of inflammation in PD (Gagne & Power, 2010). Interestingly, the depletion of dopaminergic neurons in PD is more pronounced in the basal ganglia subdivision globus pallidus external (GPe) than globus pallidus internal (GPi) (Rajput et al., 2008). However, in addition to neurons, the GPe also contains three classes of glial cells regulating neuroinflammation—microglia, astrocytes, and oligodendrocytes. Thus, we must review the cells of the inflammatory pathway and their association to PD to get a clear understanding of the potential immunotherapeutic targets to halt PD neurodegeneration.

3.1.1 Role of microglial cells

Microglia are resident macrophages and primary cells of the immune system in the CNS. Microglia constitute 10% of the cell population in the human brain (Mittelbronn, Dietz, Schluesener, & Meyermann, 2001) and play a crucial role in maintaining synaptic activity (synaptic pruning), clearing apoptotic cells, and providing trophic support to the neurons (Paolicelli et al., 2011). Neuron-microglia maintains a bidirectional signaling pathway that is important for maintaining neural circuits and brain connectivity. Microglial cells form the first line of defense in the brain. In any pathological incident or injury, microglia activate a cascade of inflammatory processes that can act as a double-edged sword, either relieving or exacerbating the injury. Detection of pathogens [pathogen-associated molecular patterns (PAMPs)] or tissue damage [damage-associated molecular patterns (DAMPs)] by the microglial pattern recognition receptors (PRRs) mounts a complex immune response—increased expression of toll-like receptors (TLRs) and proinflammatory cytokines to restore tissue homeostasis (Block, Zecca, & Hong, 2007). McGeer et al. provided evidence of an increased expression of human leukocyte antigen D-related (HLA-DR) positive reactive microglia in the substantia nigra of postmortem PD brain (McGeer, Itagaki, Tago, & McGeer, 1988). Studies employing positron emission tomography (PET) also report a significant level of microglial activation in the PD brain (Bartels, De Schepper, & Hong, 2020; Gerhard et al., 2006). The substantia nigra and striatum also exhibit activated microglia in various animal models of PD (Subhramaniam & Federoff, 2017; Tansey & Goldberg, 2010).

Microglial cells are well-adapted for inflammatory responses, antibody-dependent cell-mediated cytotoxicity (ADCC), and T-cell response through antigen presentation (Aloisi, 2001). In postmortem studies as well as in animal models of PD receiving 1-methyl-4-phenyl-1,2,3,6-tetrahydropyridine (MPTP) injections, microglia demonstrate high IgG reactivity, increase in complement receptors, lymphocyte function associated antigen (LFA), cluster of differentiation (CD-54), (CD-106), (CD-1), and vascular adhesion cell molecule (VCAM) (Kreutzberg, 1996). Indeed, increased expression of intracellular adhesion molecule (ICAM-1) and its counter receptor LFA-1 were identified, and newly activated microglia aggregates with infiltrating LFA-1^+ leukocytes suggested sustained inflammation (Miklossy et al., 2019). Apart from killing pathogens, activated microglial cells are also known to produce some toxic substances like reactive oxygen species (ROS), reactive nitrogen species (RNS), prostaglandins, and proinflammatory cytokines. Several molecules increase microglial secretory responses: lipopolysaccharide (LPS), interferon-gamma (IFN-γ), amyloid-beta (Aβ), CD-40 ligand, chemokines, neurotransmitters, gangliosides, and proteases such as thrombin, tissue plasminogen activator (tPA), and matrix metalloprotein-3 (MMP-3) (Abd-el-Basset & Fedoroff, 1995; Aloisi, 1999; Kim et al., 2005; Moller, Hanisch, & Ransom, 2000; Suzumura, Sawada, & Takayanagi, 1998; Tan et al., 2000; Tsirka, 1997). The previous studies confirmed that many activated and human leukocyte antigen reactive microglia were found in substantia nigra of PD patients. Also, these reactive microglial cells are found in MPTP, 6-hydroxydopamine (6-OHDA) induced Parkinson's disease animal models (Brecknell, Dunnett, & Fawcett, 1995; Revuelta, Venero, Machado, & Cano, 1999; Sugama et al., 2003). The nigral region contains an increased number of microglial cells making this region highly vulnerable to altered microglial responses. Reactive microglial cells localize with cell bodies of degenerating dopaminergic neurons in substantia nigra, indicating retrograde mechanism for cell death (Kim et al., 2000; Lawson, Perry, Dri, & Gordon, 1990; Mirza, Hadberg, Thomsen, & Moos, 2000). During the initial phases of neurodegeneration, activation of microglia in PD was beneficial to the host. However, long-term hyperactivation of the microglia in the PD brain elevates the proinflammatory cytokines—TNF-α, IL-1β, IL-6, and IFN-γ, which accelerate dopaminergic neurodegeneration (Ferrari et al., 2006; McCoy et al., 2006). In addition to postmortem studies, various studies measured the levels of inflammatory markers in the blood and cerebrospinal fluid (CSF) of PD patients. Studies report elevated levels of proinflammatory cytokines—IL-1β, IL-2, IL-6, IFN-γ, TNF-α, and antiinflammatory cytokine IL-10 in the plasma and serum of PD patients (Brodacki et al., 2008; Karsarou et al., 2007). Elevated levels of IL-6 in the plasma could be associated with a high risk of developing PD (Pereira et al., 2016). Elevated macrophage migration inhibitory factor (MIF) levels were also observed (Nicoletti et al., 2011). Like blood, the CSF of PD patients also reflected a similar profile of cytokine elevation (Blum-Degen et al., 1995; Mogi, Harada, Kondo, et al., 1994; Mogi, Harada, Riederer, et al., 1994). Positron emission tomography (PET) studies using PK-1195 further confirmed microglial activation in PD. PK-1195 is a ligand of the peripheral benzodiazepine receptor expressed only by the reactive microglia (Gerhard et al., 2006). The degenerating dopaminergic neurons release α-Synuclein, ATP, and MMP-3, enhancing microglial activation (Saijo et al., 2009; Zhang et al., 2005). We will discuss the role of α-Synuclein mediated microglial activation in detail in the following sections.

3.1.2 Astrocyte activation in Parkinson's disease

Astrocytes are the most abundant glial cells of the CNS and sometimes adopt a "star-like" shape with multiple processes and ramifications. However, most astrocytes exhibit a heterogeneous morphology (Haim & Rowitch, 2017). The astrocytes are responsible for maintaining a variety of physiological functions in the CNS—secreting neurotrophic molecules, regulating synaptic transmission, maintaining ion balance, and regulating the blood-brain barrier (BBB) (Allaman, Belanger, & Magistretti, 2011; Sofroniew & Vinters, 2010; Wilton, Dissing-Olesen, & Stevens, 2019). In adult mammal brains, astrocytes also function as metabolic regulators synthesizing glycogen and supplying energy substrates (Belanger, Allaman, & Magistretti, 2011; Falkowska et al., 2015). In addition, astrocytes secrete a variety of neurotrophic agents—brain-derived neurotrophic factor (BDNF), glia-derived neurotrophic factor (GDNF), insulin-like growth factor (IGF), and other growth factors essential to promote the development and survival of dopaminergic neurons and glial cells (Cabezas et al., 2014; Zhang, Reichel, Han, Zuniga-Hertz, & Cai, 2017; Zhang, Wu, Peng, Zhao, & Gu, 2017). In healthy individuals, astrocytes are homogeneously distributed in the midbrain with low density in the substantia nigra. However, postmortem studies in PD brain report pathological astrocyte changes, increase in astrocyte number in the substantia nigra, change in astrocyte morphology and distribution pattern (Braak, Sastre, & Del Tredici, 2007; Damier, Hirsch, Zhang, Agid, & Javoy-Agid, 1993). Astrocytes maintain the blood-brain barrier (BBB) together with the endothelial cells and pericytes. In PD, increasing astrogliosis disrupts BBB, leading to increased infiltration of inflammatory mediators into the brain that causes activation of microglia and degeneration of the DA neurons (Gray & Woulfe, 2015). Astrocytes also maintain the aquaporin-4 (AQP4) water channels in the BBB, which modulates astrocyte-microglia communication. AQP4 knockout mice, when treated with MPTP, cause increased astrogliosis and microgliosis accompanied by an upregulated release of IL-1β and TNF-α in the midbrain (Sun et al., 2016).

Reactive astrocytes can be classified into two subtypes—A1 and A2. A1 astrocytes upregulate toxic genes and are neurotoxic; A2 astrocytes upregulate the trophic factors and are neuroprotective (Liddlelow et al., 2017). Reactive microglia promote the formation of A1 astrocytes via the secretion of various chemical mediators—IL-1α, TNF, and C1q. α-Synuclein also promotes the formation of A1 astrocytes and preventing α-Synuclein-mediated microglia, and A1 astrocyte activation is neuroprotective in a mouse model of sporadic PD (Yun et al., 2018). The A1 astrocytes upregulate proinflammatory cytokines—IL-1α, IL-1β, and TNF-α and lose their crucial functional properties of promoting neuronal survival, synaptogenesis, and phagocytosis. Glial fibrillary acidic protein (GFAP) is a crucial marker for reactive astrocytes following stress or injury in the CNS (Zhang, Reichel, et al., 2017; Zhang, Wu, et al., 2017). Animal models of PD using 6-hydroxydopamine or MPTP report typical reactive microglia with high GFAP immunoreactivity (Breidert et al., 2002; Carbone, Duty, & Rattray, 2012; Hasida et al., 2012; Rodrigues, Gomide, & Chadi, 2001; Smeyne et al., 2016). However, postmortem findings on GFAP expression are inconsistent. Some studies fail to report any astrocyte activation and increased GFAP expression in the substantia nigra, striatum, and frontal cortex of PD patients compared to healthy controls (Mirza et al., 2000; Song et al., 2009; Tong et al., 2015); other studies report the presence of reactive astrocytes and mildly increased levels of GFAP in the substantia nigra of PD patients (Damier et al., 1993; Knott, Wilkin, & Stern, 1999).

Astrocytes, like microglia, also contribute to neuroinflammation in PD. The activation of the nucleotide-binding oligomerization domain leucine-rich repeat and pyrin domain-containing protein 3 (NLRP3) inflammasome in both microglia and astrocytes lead to the release of proinflammatory cytokines IL-1β and IL-18 (Freeman et al., 2017; Song, Pei, Yao, Wu, & Shang, 2017). Elevated levels of IL-1β and IL-18 have been reported in the CSF of PD patients compared to healthy controls (Nagatsu, Mogi, Ichinose, & Togari, 2000). Subsequently, inhibiting the astrocyte-mediated NLRP3 inflammasome activation is neuroprotective (Wang et al., 2020; Zhu et al., 2018). Glutamate is the primary excitatory neurotransmitter in the CNS. Glutamate released in the synaptic cleft is internalized via the Na^{+}-dependent excitatory amino acid transporters—EAAT1 and EAAT2, highly expressed in astrocytes (Yi & Hazell, 2006). High levels of inflammatory mediators (TNF-α, IL-1β, and IFN-γ) can increase glutamate release or impair the glutamate uptake by astrocytes (Miller et al., 2017; Tilleux & Hermans, 2007; Zou & Crews, 2005). Other factors contributing to astrocyte-mediated neuroinflammation include the dopamine D3 receptor and NF-kB signaling (Hammond et al., 2020; Kirkley et al., 2019; Montoya et al., 2019). Genetic defects in dopamine D3 receptor signaling makes the astrocytes unresponsive to LPS-mediated neuroinflammation. In MPTP models of PD, NF-kB signaling contributes to astrocyte-mediated neuroinflammation and loss of dopaminergic neurons (Hammond et al., 2020; Kirkley et al., 2019).

3.2 Role of the complement system

Complement system activation, an immunochemical process that enhances pathogen clearance, can also affect neuronal functions. Classical complement, commonly activated by antibody deposition, is also considered an arm of the innate immune system (McGeer, Itagaki, Boyes, & McGeer, 1998). Apart from classical and alternative pathways, mannan-binding lecithin-mediated cascades have been described. The protein produced during the early phase act as chemotactic factors, anaphylatoxins, and opsonins (Cunnion, Zhang, & Frank, 2003; Erdei, Andrasfalvy, Peterfy, Toth, & Pecht, 2004; Grondahl, Hohannisson, Jensen-Waern, & Nilsson, 2001; Nolte, Moller, Walter, & Kettenmann, 1996; Yao, Harvath, Gilbert, & Colton, 1990). Due to the complete activation of any of these pathways, the C5b-9 membrane attack complex generated is neurotoxic (Shen, Halperin, & Lee, 1995). The complement proteins, e.g., C3d and C4d (early stage) and C7, C9 (late-stage), have been reported during immunocytochemical staining of Lewy bodies in the brainstem of patients with parkinsonism (Iseki, Marui, Akiyama, Ueda, & Kosaka, 2000). Immunoglobulin gamma (IgG) isolated from PD patients when introduced with human recombinant C5a in rat mesencephalic neuron-glia culture results in dopaminergic neurodegeneration by activating microglia (Wang, Yan, Lu, Stuart, & Chen, 2007). Activated complement cascades promote inflammation and opsonization, facilitating phagocytosis. Additionally, complement proteins lead to cell lysis by inserting themselves into a healthy cell membrane, making cytosolic leakage causing cell death. Leaky cell membrane causes outflow of inflammatory mediators and α-Synuclein to the extracellular environment to engage microglial cells for the propagation of inflammation could be a possible mechanism (Wang et al., 2007).

3.3 Role of monocytes

Monocytes are significant components of the peripheral innate immune system, having a variety of essential functions—migration to tissues, engulfment of pathogens and cellular debris, secretion of cytokines, complement proteins, antigen presentation to the cells of adaptive immune responses, and differentiation into macrophages and dendritic cells to take more specialized functions (Wong et al., 2012).

In PD, monocytes and macrophages penetrate the CNS via the meninges of the brain and choroid plexus (Harms et al., 2018; Herz, Filiano, Smith, Yogev, & Kipnis, 2017). These phagocytic cells may engulf misfolded protein aggregates from dying neurons (Weltzien & Pachter, 2000). Monocytes play an essential role in presenting misfolded pathogenic α-Synuclein with MHC complex PD-associated allele HLA-DR (Kannarkat et al., 2015; Sulzer et al., 2017). Experimental evidence on peripheral monocyte functions has reported decreased phagocytotic functions with higher α-Synuclein levels (Gardai et al., 2013; Grozdanov et al., 2014). In addition, monocytes uptake α-Synuclein fibrils and release proinflammatory cytokines such as IL-1β. On the other hand, oligomer uptake by monocytes was reported to decrease with age (Bliederhaeuser et al., 2016) which suggests that activated monocytes and microglial cells directly contribute to the pathobiology of Parkinson's disease.

3.4 Adaptive immune responses in PD

Adaptive immune responses play a crucial role in recognizing the pathogens or foreign materials and generate more robust responses due to immunological memory. Adaptive immune systems arise after the innate immune system ensures antigen presentation to lymphocytes. Antigen-presenting cells (APCs) with surface major histocompatibility molecules (MHC) recognize T-helper and T-cytotoxic cells having $CD4^+$ and $CD8^+$ markers (David, Ashley, Reynolds, Mosley, & Gendelmann, 2009). These T lymphocytes are activated by APCs and migrate toward the site of inflammation for proper immune responses and are also reported in the substantia nigra of PD patients. The activated microglial cells secrete particular factors that disrupt the blood-brain barrier and allow the entry of adaptive immune components (Benner et al., 2008; Fiszer, Mix, Fredrikson, Kostulas, & Link, 1994; Hisanaga, Asagi, Itoyama, & Iwasaki, 2001). Posttranslational modification of α-Synuclein, such as nitration, cause increased aggregation of the protein and has been found in postmortem PD brains (Giasson et al., 2000). This aggregated α-Synuclein further activates microglial cells, producing nitric oxide and superoxide in mice and inducible nitric oxide synthase (iNOS) in humans leading to the nitration of α-Synuclein. This nitrated α-Synuclein upregulates both cell-mediated and humoral responses in an MPTP model of PD (Benner et al., 2008; Gao et al., 2008; Hunot et al., 1996).

Autoimmune antibodies against dopaminergic neurons have also been found in serum and CSF of patients with Parkinson's disease showing the adaptive humoral role of the immune system (McRae-Degueurce et al., 1988). Autoantibodies against myelin basic protein, Aβ, neurofibrillary tangles, neuronal calcium channels, α-Synuclein, heat shock protein (HSP-65, HSP-70) are expressed in PD patients (Fiszer, Fredrikson, & Czlonkowska, 1996). In another study using immunohistochemistry, IgG but not IgM has been found in dopaminergic neurons of the substantia nigra in PD patients but not in age-matched control and non-nigral control (Orr, Rowe, Mizuno, Mori, & Halliday, 2005). Lewy body inclusions found in PD patients also shows IgG and α-Synuclein in it—a pattern consistent with dopaminergic neuron destruction. This report suggests that endogenous antibodies can cross the blood-brain barrier and bind to specific antigens expressed in dopaminergic neurons. Accumulating evidence suggests that drainage of α-Synuclein into lymphatic channels causes activation of the lymphocytes. α-Synuclein specific Th 1 and Th17 activated T cells infiltrate into the brains of PD patients and trigger inflammation and neurotoxic response by increasing cytokine production from microglia (Chen et al., 2017; Sommer et al., 2018).

3.4.1 Role of MHC proteins in the adaptive immune response

MHC molecules, also called human leukocyte antigen (HLA), are integral parts of adaptive immune responses. It plays a pivotal role in antigen presentation to the helper or cytotoxic T cells, subsequently leading to specific cellular immune responses. There are two major classes of MHC: MHC-I and MHC-II, and both are involved in antigen presentation to T cells. Class-I MHC is primarily present on nucleated cells and identified by $CD8^+$ cytotoxic T cells, while on the other hand, MHC-II is present on dendritic cells, macrophages, monocytes, and B cells and interact with $CD4^+$ helper T cells (David et al., 2009; McGeer et al., 1998). Neuronal cells, in general, are thought to be negative for MHC expression, but neurons from locus coeruleus and substantia nigra have been shown to express class-I MHC upon stimulation with IFN-γ (Cebrian et al., 2014; Neumann, Cavalié, Jenne, & Wekerle, 1995). In addition, MHC upregulation in dopaminergic neurons sensitive to inflammation makes them vulnerable to T cell infiltration than GABAergic neurons, which are more resistant (Garretti, Agalliu, Arlehamn, Sette, & Sulzer, 2019).

Polymorphism in the noncoding region of MHC-II gene: DRB5* 01, DRB1* 15:01 alleles increase the expression of MHC-II. Researchers have used the rodent model of Parkinson's disease to study the role of MHC in neurodegeneration (Arlehamn, Francesca, Sulzer, & Sette, 2019). Viral vector-mediated α-Synuclein overexpression in mice lacking MHC-II are reported to be protected against microglial cell activation, blood-brain barrier breakdown, and cell death (Subbarayan, Hudson, Moss, Nash, & Bickford, 2020). MPTP-treated mice lacking the MHC-II gene are protected against neuronal death due to inactivated microglial cells and the nonrelease of proinflammatory cytokines in contrast to wild-type animals. Accumulative evidence suggests that dopaminergic neurodegeneration may involve MHC-II expression, antigen presentation, and adaptive immune system (Harms et al., 2013).

3.4.2 Role of T cell in the adaptive immune response

T cells play a pivotal role in collaboration with B lymphocytes for adaptive immune activation (Kipnis, 2016). T lymphocytes recognize antigenic epitopes processed by MHC molecules, and B lymphocytes produce plasma cells, producing antibodies in surface-bound or secreted form (Filiano, Gadani, & Kipnis, 2017; Korn & Kallies, 2017). Previous studies have also explained that these $CD4^+$ T lymphocytes can be further differentiated into the following subtypes: Th1, Th2, Th17, and regulatory T cells (Tregs).

These subtypes of T cells are specialized in their function: Th1 cells produce interferon-gamma (IFN-γ) and IL-2 initiating type 1 proinflammatory responses; Th2 Cells produce IL-4, IL-5, and IL-13 and initiate the type-2 antiinflammatory responses (Korn & Kallies, 2017; Zhu & Paul, 2010). Thus, dysregulation between these subtypes leads to the development of neuroinflammation.

In 1994 Fiszer et al. showed the presence of $\gamma\delta$ T cells in peripheral blood and CSF in PD patients were high compared to patients with other kinds of neurological disease (Fiszer, Mix, Fredrikson, Kostulas, Olsson, et al., 1994). Also, altered TNF-α and IFN-γ receptor expression have been shown by activated T-lymphocytes, $CD4^{+}CD25^{+}$ lymphocytes, suggesting T-cell dysregulation in the peripheral circulation of PD patients (Bongioanni, Castagna, Maltinti, Boccardi, & Dadone, 1997; Bas, Calopa, Mestre, Mollevi, & Cutillas, 2001). A 2005 study done by Baba et al. shows increased ratios of IFN-γ to IL-4 producing T-cells and predominant expression of $CD8^{+}$ T cells (Baba, Kuroiwa, Uitti, Wszolek, & Yamada, 2005). A controversial study suggests decreased T cell subpopulations and cytokines levels in PD patients, but the authors also revealed that levodopa and pramipexole could reduce cytokine secretion from blood mononuclear cells; carbidopa could also suppress T cell response in PD patients (Rocha et al., 2018; Zhu et al., 2017). The most significant clinical observation of T cell dysregulation in Parkinson's is that the T cell populations match clinical manifestations. The elevated effector/memory T cells and decreased $CD31^{+}$ and $\alpha4\beta7^{+}CD4^{+}$ T cells were correlated with Unified Parkinson's Disease Rating Scale III (UPDRS-III) (Saunders et al., 2012).

Kustrimovic et al. (2016) found that in $CD4^{+}$ naïve T cells, the expression of D1-like and D2-like dopamine receptors was also correlated with the UPDRS-III scores (Kustrimovic et al., 2016). T cells have been shown to invade substantia nigra of PD subjects in addition to alteration in peripheral blood of PD patients suggesting that peripheral lymphocytes may be recruited into the brain of PD (Brochard et al., 2009). In MPTP and 6-OHDA toxin-based models, T cell activation and infiltration have been observed in substantia nigra (Chandra, Rangasamy, Roy, Kordower, & Pahan, 2016; Chandra, Roy, Rangasamy, & Pahan, 2017; Liu, Huang, Cao, Qiu, & Peng, 2017). As seen in 6-OHDA lesioned rats, dopaminergic neurodegeneration is mediated by $CD4^{+}$ T cells mediated through FasL expression (Brochard et al., 2009).

4. α-Synuclein in Parkinson's disease: Deciphering the cascade of proteotoxic pathways

A pathological hallmark of PD is the presence of cytoplasmic inclusions called Lewy bodies (LBs) and Lewy neurites (LNs) within the dopaminergic neurons of the substantia nigra, in both familial and sporadic forms of the disease. The major component of these LBs and LNs is the protein α-Synuclein, in its aggregated or insoluble form (Spillantini et al., 1997). LBs are intracellular, eosinophilic, round inclusions with the presence of a hyaline core and a peripheral halo, which are positive for both α-Synuclein and ubiquitin (Takahashi & Wakabayashi, 2001). LNs are abnormal neurites containing granular material and fibrillar α-Synuclein, similar to LBs. However, LNs are more abundant than LBs and are predominantly found in the striatum and amygdala of the PD brain (Braak, Del Tredici, et al., 2003; Duda, Lee, & Trojanowski, 2000). Due to its involvement in the pathogenic mechanisms of PD neurodegeneration, α-Synuclein has been the epicenter of PD research. The main aim is to understand how aggregation and accumulation of the toxic species of this protein can act as a driving force accounting for neuronal loss and disease development.

4.1 α-Synuclein mutations: Effect on PD neurodegeneration

Mutations in the *SNCA* gene (encoding α-Synuclein) are responsible for monogenic forms of early-onset PD (EOPD) with the age of onset <50 years. Various point mutations in *SNCA* gene have been identified in PD patients with familial etiology: A53T, A30P, E46K, H50Q, G51D, and A53E (Stefanis, 2012). Besides, duplication and triplication of the coding region of *SNCA* are also being reported. Duplications of *SNCA* locus are more common than triplications with milder clinical features and decreased penetrance (Chartier-Harlin et al., 2004; Singleton et al., 2003). Both familial and sporadic cases have been reported to harbor the duplication. The average age of onset is 50 years, and the clinical outcome is mainly heterogeneous, with typical late-onset motor dysfunction resembling sporadic PD (Kasten & Klein, 2013). Triplications of SNCA have been reported in three different families (Farrer et al., 2004; Ibanez et al., 2004; Singleton et al., 2003).

Interestingly, triplications of the gene are more detrimental (with an earlier onset, more severe phenotype, and increased disease progression) than duplications indicating an apparent dosage effect. *SNCA* triplications are typically associated with severe PD (average age of onset around 40) with concomitant cognitive decline. Nonmotor symptoms include RBD, psychiatric manifestations (mainly severe depression), and in some cases, rapid weight loss, dysautonomia, and fatigue precede the motor symptoms. These familial mutations of *SNCA* alter the fibrillation kinetics and promote the aggregation of α-Synuclein, suggesting a complex pathway of proteotoxicity in PD.

4.2 α-Synuclein structure and function

α-Synuclein is a small protein of 140 amino acids that can bind to phospholipid membranes (Davidson, Jonas, Clayton, & George, 1998). The native form of α-Synuclein is poorly defined; the protein can adopt intrinsically disordered conformation (Breydo, Wu, & Uversky, 2012; Theillet et al., 2016), helical forms (Bartels, Choi, & Selkoe, 2011; Wang et al., 2011) or a combination of the two (Burre et al., 2013). Structurally, α-Synuclein contains three domains: an N-terminal domain (1–65 amino acid) containing an apolipoprotein lipid-binding motif [seven repeats of an 11-residue consensus sequence (XKTKEGVXXXX)] with the propensity to form α-helical structures upon membrane binding, a central hydrophobic region called the nonamyloid β-component domain (NAC) domain (66–95 amino acid) which confers the β-sheet potential, and an unstructured C-terminal domain (96–140 amino acid) which is highly negatively charged and is prone to posttranslational modifications (Burre, Sharma, & Sudhof, 2018). Besides, there are three splice-variants of the protein (SNCA-126, -112, and -98), but their physiological and pathological functions are not well characterized (Maroteaux & Scheller, 1991). However, stoichiometric alterations in the isoforms of α-Synuclein have been linked to increased protein aggregation (Manda, Yedlapudi, Korukonda, Bojja, & Kalivendi, 2014). Interestingly, all pathogenic mutations are present in the N-terminal region, inferring this region's significance in the pathophysiological dysfunction of the protein (Dehay et al., 2015).

Although its role in PD neurodegeneration is widely established, the normal physiological function of α-Synuclein is still unclear. α-Synuclein is mainly a presynaptic protein involved in synaptic vesicle trafficking (Fortin et al., 2004; Kahle, Neumann, Ozmen, & Haass, 2000; Ross et al., 2008). It is widely abundant in the central nervous system (CNS), but its presence in non-neuronal tissues has also been reported (Aldeoca et al., 2015; Challis et al., 2020; Ma et al., 2019). Notably, a debate was initiated by the work of Braak, which demonstrated that the pathology of the protein might begin in the enteric nervous system (Braak, de Vos, Bohl, & Del Tredici, 2006). In addition, α-Synuclein interacts with a wide variety of intracellular proteins like phospholipase D (Gorbatyuk et al., 2010), rab3 (Chen et al., 2013), SNARE protein synaptobrevin-2 (Burre et al., 2010), tyrosine hydroxylase (TH) (Masliah et al., 2000; Perez et al., 2002), DJ-1 (Zondler et al., 2014), synphilin (Kawamata, Mclean, Sharma, & Hyman, 2001; Ribeiro, Carneiro, Ross, Menezes, & Engelender, 2002), tubulin (Lee, Khoshaghideh, Lee, & Lee, 2006), and tau (Haggerty et al., 2011; Qureshi & Paudel, 2011). However, the exact physiological significance of these interactions is still not clear.

4.3 Prion-like properties of α-Synuclein: Is it infectious?

The mechanism of α-Synuclein aggregation leading to neuronal dysfunction and death has been the subject of extensive research. Aggregated α-Synuclein in LBs in the individual neurons does not provide a clear perspective of PD pathology and symptomatic progression of the disease. Recent studies show that α-Synuclein is capable of self-propagating like prions—they can be released by one neuron and taken up by other neurons in the vicinity, pathologically spreading the disease. Furthermore, the neurons uptaking the protein from the extracellular space experience aggregation of their endogenous α-Synuclein, further contributing to the toxicity. Postmortem studies reveal that the neurodegeneration in the PD brain is not restricted to the substantia nigra; it extends to the brainstem and finally throughout the brain. An exciting study by Braak et al., suggested that PD pathology might initiate in the dorsal motor nucleus of the medulla or the olfactory bulb and then extend upward through the brainstem to the cortical regions (Braak, Del Tredici, et al., 2003). Therefore, the transmission of this toxic protein should be explored, and insights into its mechanism might be essential to develop therapeutic measures.

4.3.1 Evidence supporting the transmission of α-Synuclein

The first evidence of the pathological transmission of α-Synuclein reports the grafting of embryonic ventral mesencephalic neurons in the striatal region of the PD brain in an experimental attempt to replace the neuronal loss during disease. The treated patients reported improved motor symptoms; however, the grafted neurons developed LB pathology within 11–22 years (Kordower, Chu, Hauser, Freeman, & Olanow, 2008; Kordower, Chu, Hauser, Olanow, & Freeman, 2008; Li et al., 2008, 2010). This data suggest that the young neurons did not develop LB pathology themselves, instead, they were induced to develop the pathology by the PD brain microenvironment. More precisely, the extracellular space contained the misfolded α-Synuclein released by other LB-containing neurons in the brain; the transplanted neurons internalized the α-Synuclein and therefore developed LB pathology (Kordower, Chu, Hauser, Freeman, et al., 2008; Kurowska et al., 2011; Li et al., 2008, 2010). These pieces of evidence support the concept that the release of α-Synuclein from affected neurons into the extracellular space, propagation of α-Synuclein, and finally their uptake by vulnerable neurons forms the network of α-Synuclein spread, contributing to neuronal loss in PD (Kurowska et al., 2011).

Studies in cell culture and animal models of PD also tried to replicate the self-propagation of α-Synuclein. Exogenous α-Synuclein preformed fibrils (PFFs) were generated by sonicating monomeric α-Synuclein *in vitro*. Non-neuronal cells overexpressing α-Synuclein were treated with these fibrils augmented with a lipophilic agent to transfer the PFFs into the cytosol. These exogenous α-Synuclein PFFs induce the endogenous protein into misfolded aggregates exhibiting LB pathology (Luk et al., 2009; Nonaka, Watanabe, Iwatsubo, & Hasegawa, 2010). The α-Synuclein aggregates in the recipient neurons induced neuronal dysfunction and death in a time-dependent manner (Froula et al., 2018; Luna et al., 2018; Volpicelli-Daley et al., 2011). The extraction of α-Synuclein aggregates from the brain of PD and multiple system atrophy (MSA) patients also induced pathology in primary neurons (Peng et al., 2018). Similarly, brains of mice and rats injected with α-Synuclein PFFs induced rapid formation of LB pathology, motor deficits, and neurodegeneration (Koller, Brooks, Golde, Giasson, & Chakrabarty, 2017; Luk et al., 2012; Masuda-Suzukake et al., 2013; Paumier et al., 2015; Rey et al., 2018; Steiner, Quansah, & Brundin, 2018).

An essential question in this regard is which form of α-Synuclein is the most toxic? Recent evidence from several studies in animal and in vitro models have indicated that the toxic species of α-Synuclein contributing extensively to the process of neurodegeneration in PD is the prefibrillar or the soluble oligomeric form (Bousset et al., 2013; Choi et al., 2013; Colla et al., 2012; Emmanouilidou, Stefanis, & Vekrellis, 2010; Roberts & Brown, 2015; Winner et al., 2011). However, oligomeric α-Synuclein are inherently transient and rapidly recruit monomeric α-Synuclein to form fibrils. Kinetically trapped α-Synuclein oligomers have been biophysically characterized and were composed of 29 molecules (50 nm) with antiparallel β-sheet strands and are not seeding competent. Striatal injections of α-Synuclein oligomers can cause loss of DA neurons in the SN, confirming their toxicity. On the other hand, α-Synuclein fibrils are 70 nm in length with parallel β-sheet strands and are seeding competent. Striatal injections of α-Synuclein fibrils cause loss of DA neurons in the SN and lower the number of dopamine transporter-positive terminals, seeds α-Synuclein inclusions in SN, cortex, and amygdala, and induces motor dysfunction (Froula et al., 2019).

4.3.2 Mechanism of α-Synuclein transmission

The mechanism of α-Synuclein transmission can be categorized into three stages.

1. Uptake of α-Synuclein from the extracellular environment: The uptake of misfolded/aggregated α-Synuclein is not prone to any barrier. The most common methods include micropinocytosis and receptor-mediated endocytosis (Karpowicz et al., 2017; Kisos, Pukaβ, Ben-Hur, Richter-Landsberg, & Sharon, 2012). The heparan sulfate proteoglycan (HSP) is a critical cell surface receptor mediating the internalization of α-Synuclein fibrils (Holmes et al., 2013). Other specific cell surface receptors important in this regard are Na^+/K^+ ATPase (α3 subunit), LAG-3, amyloid-β precursor-like protein 1, neurexin 1β (Shrivastava et al., 2015). Interestingly, immune cells like T-cells and natural killer cells express LAG3, indicating the immune system's involvement in α-Synuclein propagation. Finally, the endo-lysosomal pathway transports the internalized α-Synuclein through the cell.
2. Processing of the internalized α-Synuclein: The internalized α-Synuclein in the lysosomes is subject to proteolytic processing by lysosome rupture, releasing into the cytosol (Guiney et al., 2020; Rodriguez, Marano, & Tandon, 2018; Sacino et al., 2017). Impaired lysosomal integrity with age might be a possible explanation of α-Synuclein accumulation in the aging brain. Misfolded exogenous α-Synuclein then recruits the endogenous α-Synuclein forming thread-like neuritic inclusions and transfers them to the neuronal soma where they undergo phosphorylation and ubiquitination to form the characteristic LB pathology (Volpicelli-Daley, Luk, & Lee, 2014).
3. Release of the pathogenic α-Synuclein into the extracellular space: The release of α-Synuclein is a regulated event. Impaired protein degrading machinery in the aging neurons releases lysosomal protein to unburden the lysosomal enzymes (Fussi et al., 2018). In addition, rapid cycles of exocytosis and endocytosis occur at the presynaptic terminus to release and retrieve neurotransmitters, making them more vulnerable to uptake α-Synuclein (released by the postsynaptic neuron).

4.4 Role of α-Synuclein in microglia and astrocyte-mediated neuroinflammation

Various PD risk factors like leucine-rich repeat kinase 2 (LRRK2), DJ-1, and α-Synuclein regulate the microglial inflammatory response. α-Synuclein and microglia-mediated neuroinflammation have a positive association. In the course of the disease, degenerating neurons secrete α-Synuclein into the extracellular space, detected in the body fluids like blood and CSF in PD patients (Majbour et al., 2016; Tokuda et al., 2010). The extracellular α-Synuclein is capable of directly activating microglia, which has significant consequences in PD pathology (Su et al., 2008). α-Synuclein causes the resting microglia to shift to a reactive M1 phenotype that secretes proinflammatory cytokines and are neurotoxic (Austin, Floden, Murphy, & Combs, 2006; Theodore, Cao, McLean, & Standaert, 2008). α-Synuclein can activate microglia via the toll-like receptor 2 (TLR2) (Kim et al., 2013). α-Synuclein-mediated activation of microglia further increases phagocytosis of α-Synuclein via the FcγR receptor, promotes nuclear translocation of the p65 subunit of NF-κβ, upregulates proinflammatory cytokine release, and causes loss of dopaminergic neurons (Cardona et al., 2006; Klegeris et al., 2008; Lee, Kim, & Lee, 2010; Lee, Woo, et al., 2010; Su et al., 2008). Recent studies show that microglial α-Synuclein uptake is regulated by Fyn kinase (linked to PD in GWAS) and CD36 (Nalls et al., 2019; Panicker et al., 2019). This signaling pathway causes priming and activation of the NLRP3 inflammasome, further causing an elevated release of IL-1β by microglia. However, the form of α-Synuclein that is most potent in causing microglial activation is still under debate. The typical consensus is that monomeric α-Synuclein causes phagocytosis, and oligomeric protein acts the opposite (Park, Paik, Jou, & Park, 2008). However, some studies report increased microglial α-Synuclein phagocytosis by fibrillar or truncated α-Synuclein (Sanchez-Guajardo, Tentillier, & Romero-Ramos, 2015). Toxic aggregates of α-Synuclein prevent microglial phagocytosis and cause increased release of IL-1β and TNF-α (Hoffmann et al., 2016). Point mutations in α-Synuclein (A53T, A30P, E46K) stimulate microglial cytokine release, activate MAP kinase pathways, and increase ROS production (Hoenen et al., 2016; Jiang et al., 2015; Klegeris et al., 2008; Zhang et al., 2007). Overexpression of human A53T in mice lacking microglial fractalkine receptor (CX3CR1) increases dopaminergic neurodegeneration and neuroinflammation in the substantia nigra (Castro-Sanchez et al., 2018). Injecting α-Synuclein preformed fibrils (PFFs) in the substantia nigra of rodents causes the advent of microgliosis before the loss of dopaminergic neurons, highlighting the importance of α-Synuclein-mediated microglial activation in PD pathology (Duffy et al., 2018).

The expression of α-Synuclein is low in astrocytes than neurons (Zhang et al., 2014; Zhang, Shao, et al., 2016; Zhang, Sloan, et al., 2016). However, depletion in α-Synuclein causes disruption in fatty acid uptake and trafficking in astrocytes implicating a physiological role of α-Synuclein in the process (Castagnet, Golovko, Barcelo-Coblijn, Nussbaum, & Murphy, 2005). The abundance of α-Synuclein in the neurons and its ability to self-propagate causes astrocytes to uptake large volumes of the protein from the extracellular space (Hansen et al., 2011). Co-culture of primary astrocytes and human neuroblastoma cells secreting α-Synuclein causes uptake of the protein and formation of inclusion bodies in the astrocytes (Lee, Kim, et al., 2010; Lee, Woo, et al., 2010). A study by Rostami et al. has shown that embryonic stem cell-derived human astrocytes can transfer aggregated α-Synuclein to the nearby astrocytes by direct contact via tunneling nanotubes (TNTs) (Rostami et al., 2017). The accumulation of α-Synuclein in astrocytes causes mitochondrial dysfunction (Lindström et al., 2017). Postmortem studies report that thc α-Synuclein pathology in the astrocytes resembles the intraneuronal LB pathology in PD patients (Braak et al., 2007; Fathy et al., 2019). However, ultrastructurally, the α-Synuclein in astrocytes and LBs in the brain differ—they are diffusely distributed throughout the astrocytic cell body and processes and exhibit a granular staining pattern similar to vesicles or lysosomes (Braak et al., 2007). Astrocytic α-Synuclein triggers the release of proinflammatory cytokines—IL-1, IL-6, TNF-α, and chemokines—C-X-C motif ligand 1 (CXCL1) (Lee, Kim, et al., 2010; Lee, Woo, et al., 2010). This proinflammatory response of astrocytic α-Synuclein is mediated by the toll-like 4 (TLR4) receptors (Fellner et al., 2013; Rannikko, Weber, & Kahle, 2015) although TLR4 is not required for α-Synuclein uptake. Mouse astrocytes that selectively express human α-Synuclein A53T cause noncell-autonomous killing of neurons and subsequent neurodegenerative movement disorder (Gu et al., 2010). Accumulated α-Synuclein in astrocytes causes toxicity via mitochondrial dysfunction, impaired autophagy, and mitophagy which leads to oxidative stress and apoptosis (Braidy et al., 2013; Cavaliere et al., 2017; Erustes et al., 2018; Lindström et al., 2017). Fig. 1 summarizes the interplay between α-Synuclein and neuroinflammation in the pathogenesis of PD, highlighting the involvement of microglia and astrocytes in the process.

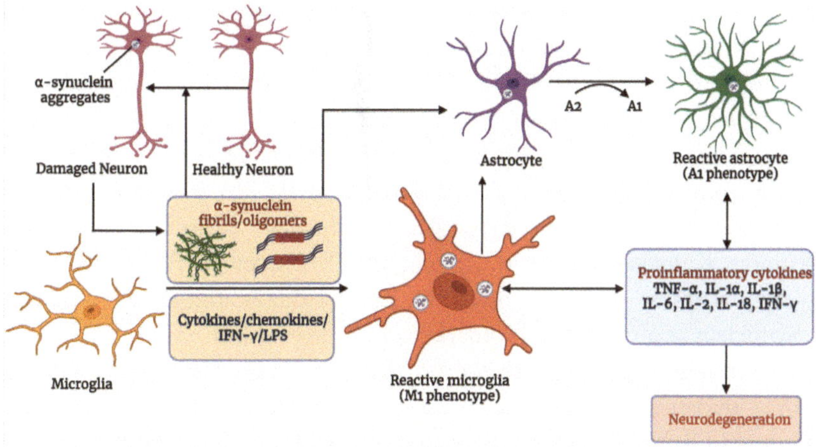

Fig. 1 α-Synuclein and neuroinflammation in Parkinson's disease (PD) neurodegeneration. Schematic diagram showing the mechanism of α-Synuclein, microglia, and astrocyte-mediated neuroinflammation in Parkinson's disease. Image was created using BioRender.com.

4.5 Gut inflammation and Parkinson's disease

Almost 80% of PD patients exhibit nonmotor symptoms (NMS), particularly gastrointestinal (GI) dysfunctions like constipation, dysbiosis, altered intestinal permeability, and inflammation (Ueki & Otsuka, 2004). Increased expressions of proinflammatory cytokines were observed in the colonic biopsies of PD patients compared to control patients (Bonaz & Bernstein, 2013), which points to a link between inflammation in the gut and neuroinflammation in PD pathology. The detection of LBs in the enteric nervous system (ENS) of PD patients was made very early in 1984 by Qualman et al., who identified these inclusions in the esophagus and colon of two PD patients, respectively (Qualman, Haupt, Yang, & Hamilton, 1984). Since then, the ENS has been the focus of research to detect the toxic aggregates of α-Synuclein in the gut of PD patients (Beach et al., 2010; Braak et al., 2006; Derkinderen et al., 2011). The aggregation of α-Synuclein in the human ENS was first reported in the vasoactive intestinal peptide (VIP)-positive neurons (Wakabayashi, Takahashi, Ohama, & Ikuta, 1990). A further interesting phenomenon is that the LB pathology in the ENS develops in the early course of disease (presymptomatic phase), even before the pathology could be detected in the neurons of the substantia nigra (Braak et al., 2006; Lebouvier et al., 2009). Several studies have identified the presence of α-Synuclein in different organs spanning the ENS (Gray, Munoz, Gray, Schlossmacher, & Woulfe, 2014; Hilton et al., 2014; Sanchez-Ferro et al., 2014). Enteric abnormality caused by aggregation of α-Synuclein is a clinical characteristic in PD patients. Intestinal α-Synuclein accumulation is more frequent and higher in PD patients than age-matched healthy controls (Braak et al., 2006; Forsyth et al., 2011; Gold, Turkalp, & Munoz, 2013; Shannon, Keshavarzian, Dodiya, Jakate, & Kordower, 2012; Shannon, Keshavarzian, Mutlu, et al., 2012). Therefore, routine gastric or colonic biopsies could be used in the histopathological assessment of α-Synuclein as a candidate biomarker in prodromal PD.

α-Synuclein pathology in the GI tract exists in a rostrocaudal gradient with a maximum number of pathological aggregates being observed in the upper GI tract, gradually decreasing downward. Beach et al. employed immunohistochemical techniques to detect phosphorylated α-Synuclein and confirmed the rostrocaudal gradient of protein aggregation with maximum aggregates in the lower esophagus followed by stomach and colon and rectum (Beach et al., 2010). This rostrocaudal gradient of α-Synuclein distribution could be attributed to the vagal innervations of the gut. An exciting hypothesis in this regard has been proposed by Braak et al., which suggests that PD pathology might begin by the entry of an exogenous pathogen in the GI tract causing the formation of α-Synuclein aggregates, which are retrogradely transported via axons of the dorsal motor nucleus of the vagus (DMV) from the ENS to the CNS (Braak, Del Tredici, et al., 2003; Braak, Rub, et al., 2003). The gut microbiota can affect brain activity via the gut-brain axis, inducing PD pathology through a prion-like mode of α-Synuclein transmission (Nair, Ramachandran, Joghee, Antony, & Ramalingam, 2018).

The association of increased risk of PD in patients with inflammatory bowel disease (IBD) also indicates a positive correlation between gut inflammation and PD (Brudek, 2019). Furthermore, colonic biopsies in PD patients report gut inflammation in the early stages of PD—elevation of proinflammatory cytokines like TNF-α, IF-γ, IL-6, and IL-1β, as well as glial markers GFAP and Sox-10 (Devos et al., 2013). Increased gut inflammation can alter the gut permeability, leaking microbiota and their metabolites from the gut lumen triggering α-Synuclein aggregation (Bhattacharya et al., 2019; Chorell et al., 2015). Dysbiosis in PD is widely reported, and gut microbiota can induce the aggregation of pathologic α-Synuclein (Cassani et al., 2015; Fitzgerald, Murphy, & Martinson, 2019; Scheperjans et al., 2015). The first mechanism points to *Escherichia coli* in the gut producing

Curli, a bacterial amyloid protein that can cause α-Synuclein aggregation. The second mechanism involves the LPS on the bacterial outer membrane that can trigger α-Synuclein aggregation (Bussell & Eliezer, 2003; Qin et al., 2007). Therefore, targeting gut inflammation could open a new therapeutic window in PD.

5. Immunotherapeutic strategies in Parkinson's disease

Immunotherapy is most commonly used in cancer treatment to modulate the immune system, thereby generating an enhanced immune response against the growing tumor (Banerjee et al., 2018). However, like other neuroinflammation-based neurodegenerative disorders, the immune system in PD patients is overactive and contributes to neurotoxicity and neurodegeneration. Therefore, mechanisms to attenuate neuroinflammation in PD would be beneficial to halt disease progression. The present immunotherapeutic advances in PD target mainly two processes—taming the hyperactive immune system or targeting α-Synuclein. Fig. 2 summarizes the different targets of immunotherapeutic modifications in PD.

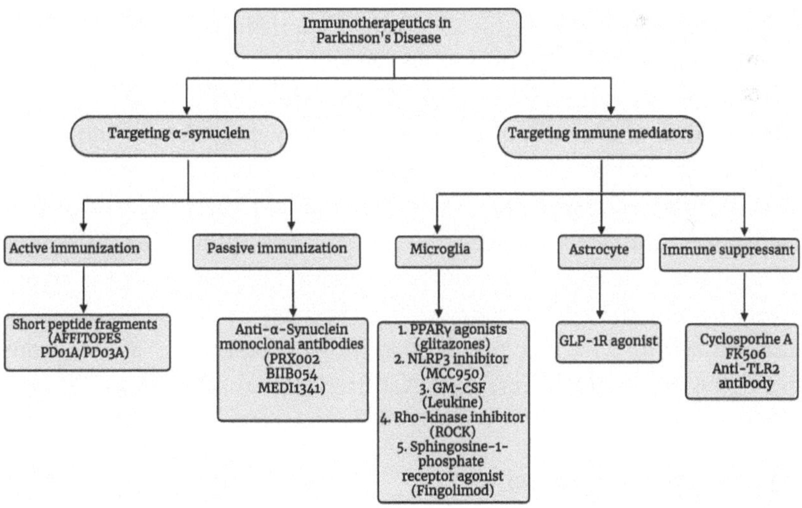

Fig. 2 Immunotherapeutic strategies in Parkinson's disease (PD). Flow-chart showing the different strategies developed in PD immunotherapy—with their targets and mode of action.

5.1 Targeting α-Synuclein

A significant amount of evidence indicates that α-Synuclein accumulation is a significant event in PD neurodegeneration. Therefore, immunotherapies targeting α-Synuclein hold promise in modifying PD progression in early-detected patients (Games et al., 2014). Furthermore, as discussed earlier, α-Synuclein exists in large amounts in the extracellular space. Studies have revealed mechanistic insights into how cells uptake the toxic species of α-Synuclein from the extracellular space. The lymphocyte activation gene 3 (LAG3) selectively interacts with α-Synuclein aggregates and not monomers *in vitro* and *in vivo* (Mao et al., 2016; Wood, 2016). Anti-LAG3 antibodies (C987W and 410C9) prevented the spread of α-Synuclein preformed fibrils (PFFs) in cell culture (Mao et al., 2016). Furthermore, α-Synuclein oligomers directly interact with gap junction protein connexin-32 (CX32), mediating an efficient cell-to-cell propagation (Reyes et al., 2019). The exhibition of prion-like properties highlighted the viability of immunization as a treatment avenue in PD. The current immunotherapies mainly aim to reduce the extracellular α-Synuclein load and to halt the disease progression.

Two types of α-Synuclein immunotherapy are present—active immunization or vaccination, which trains the immune system to generate antibodies against α-Synuclein, and passive immunization, which involves the administration of antibodies generated against different domains of the protein (Mandler et al., 2015; Sardi, Cedarbaum, & Brundin, 2018). Although most immunotherapeutic approaches target the extracellular α-Synuclein, targeting the intracellular protein is crucial as they initiate and perpetuate the disease pathology. Targeted deliveries of intracellular antibodies or antibody fragments might be difficult due to the selectively permeable plasma membrane. However, methods have been developed to achieve this intracellular delivery—trojan-horse intrabodies with signaling sequences that engage the endocytic membrane receptors for efficient delivery against the blood-brain barrier (Spencer et al., 2014); intrabody constructs within viral vectors for constitutive expression of the protein (Bhatt, Messer, & Kordower, 2013; Chatterjee et al., 2018).

5.1.1 Active immunization

Immunotherapy targeting α-Synuclein in animals was pioneered by Masliah et al. in 2005. Their study utilized transgenic mice overexpressing human α-Synuclein under the control of the platelet-derived growth factor-β

(PDGF-β) promoter (Masliah et al., 2005). These transgenic mice exhibited α-Synuclein aggregations in different parts of the brain—cortex, hippocampus, and olfactory system mimicking the typical LB neuropathology. The mice were vaccinated with human α-Synuclein leading to the production of high levels of affinity antibodies, and showed lesser accumulation of aggregated α-Synuclein in cell bodies and synapses of the neurons, and decreased neurodegeneration. Moreover, antibodies produced by the immunized mice detected the abnormal aggregates of α-Synuclein in the neuronal membrane and promoted their lysosomal degradation. Their results suggested the positive effects of vaccination in reducing α-Synuclein accumulation. Another kind of vaccination approach was performed by Sanchez-Guajardo et al. by first vaccinating rats with recombinant human α-Synuclein and then using recombinant adeno-associated viral vectors (AAV) to overexpress α-Synuclein in the substantia nigra of rats. They observed a high titer of anti-α-Synuclein antibodies, accumulation of reactive microglia in the substantia nigra, and infiltration of $CD4^+$, $Foxp3^+$ T cells throughout the nigrostriatal region. This vaccination strategy caused microglial activation and induced regulatory T-cell-mediated immune tolerance against α-Synuclein (Sanchez-Guajardo, Annibali, Jensen, & Romero-Ramos, 2013).

To avoid generating the risk of anti-human α-Synuclein T-helper cell responses, Ghochikyan et al. used novel approaches utilizing three-peptide--based epitope vaccines and used different B-cell epitopes of human α-Synuclein fused with Th epitope of tetanus toxin (P30). This method generated high titers of anti-α-Synuclein antibodies that recognized the LB and LN in the brain tissues of dementia with Lewy bodies (DLB) cases and mounted a robust Th cell response against P30 but not human α-Synuclein (Ghochikyan et al., 2014).

The next step involved the application of these strategies in clinical trials. Mandler et al. developed the next generation active vaccination technology—using human α-Synuclein short peptide fragments, AFFITOPEs® (AFF) that are too short to induce α-Synuclein specific T cell response. They screened an extensive peptide library mimicking the C-terminal region of α-Synuclein and developed AFF with the ability to identify α-Synuclein oligomers and immunized transgenic mice with these AFFs. This method produced specific, sustained, and noncross-reacting antibodies. These immunogens decreased α-Synuclein aggregates in axons and synapses, reduced degeneration of dopaminergic neurons, and improved motor and memory deficits. α-Synuclein clearance was further associated with elevated microglial activation and anti-inflammatory response (Mandler et al., 2014). Two of these immunogens

(PD01A and PD03A) have entered Phase 1 clinical trials by AFFiRiS, an Austrian company. PD01A is a short peptide of eight amino acids mimicking an epitope in the C-terminal region of α-Synuclein that acts as an active vaccine to α-Synuclein. The peptide-carrier conjugate is adsorbed into aluminum hydroxide adjuvant and is designed to stimulate antibody response by B-cells bypassing T-cell mobilization. Phase 1 clinical trial of PD01A (AFF008) began in February 2012. The clinical trial was randomized, single-center, parallel-group and recruited 32 individuals (age range 45–65 years) with early-stage sporadic PD but regular medication. Eight of these people served as the untreated, observational control group, and the rest 24 received 4 doses of PD01A (15–75 μg) 4 weeks apart and were observed for 1 year with an additional 1-year follow-up (AFF008E). A third trial extension (AFF008A) with 22 remaining participants received single booster doses of 15 or 75 μg with 24 weeks follow-up. A final extension of the trial (AFF008AA) having 21 participants received a second booster dose of 75 μg with a 1-year follow-up. The study goals included safety, tolerability, and immunogenicity with repeated subcutaneous injections. The primary endpoints were adverse effects and trial withdrawal. The secondary endpoints included measurements of antibody titers, motor and nonmotor symptoms, CSF levels of α-Synuclein, Aβ, and tau. The trials ended in 2017, and the results were published in 2020 by Volc et al., which reported PD01A to be safe after repeated administration and tolerated for a long duration. The most common adverse effects included transient local injection-site reaction, fatigue, headache, and nausea. They were about to enter Phase 2 trials with PD01A, but no registration has been done to date (Volc et al., 2020). Similar to PD01, AFFiRiS also entered into a clinical trial with its second candidate, PD03A. The trial included 36 people with PD and was administered 4 priming vaccinations (15, 75 μg, or placebo) 4 weeks apart with a booster dose at 9 months. Poewe et al. published the clinical trial results, which reported lower antibody titers with PD03A than PD01A (Poewe et al., 2021).

5.1.2 Passive immunization

The second strategy for α-Synuclein immunotherapy involves passive immunization with monoclonal antibodies directed against α-Synuclein. Masliah et al. treated α-Synuclein overexpressing transgenic mice with a novel α-Synuclein monoclonal antibody (9E4) that could cross the blood-brain barrier into the CNS and ameliorate the effects of aggregated α-Synuclein. 9E4 reduced behavioral deficits in the mice, caused calpain-mediated cleavage of α-Synuclein aggregates in axons and synapses of cortical and hippocampal neurons, and prevented the subsequent neurodegeneration highlighting the therapeutic benefits of the monoclonal antibodies in PD immunotherapy (Masliah et al., 2011). In a transgenic mouse model, Bae et al. used a different antibody (Ab274) (generated against the C-terminal of α-Synuclein). Antibodies against α-Synuclein specifically target the extracellular protein and aid in clearing them through the action of microglia via the Fcγ receptor. Direct administration of antibodies by stereotaxic method to the brain of transgenic mice prevented the neuron to astroglia transfer of the protein and caused increased localization of the protein and antibodies in the microglia. This passive immunization method reduced the accumulation of α-Synuclein in the neurons and glia and reduced behavioral deficits and neurodegeneration associated with α-Synuclein overexpression (Bae et al., 2012). Games et al. obtained similar effects using three C-terminal targeting anti-α-Synuclein antibodies—1H7, 5C1, and 5D12 in transgenic mice (Games et al., 2014).

Passive immunization approaches have also used anti-α-Synuclein antibodies targeting the N-terminal of the protein. For example, α-Synuclein monoclonal antibody (Syn303) targets misfolded α-Synuclein in a mouse model with intrastriatal injections of α-Synuclein PFFs that spread in the brain, causing motor deficits and neurodegeneration. The mAbs blocked the entry of PFFs, prevented the cell-to-cell transfer of α-Synuclein, reduced neurodegeneration in the substantia nigra, and improved motor impairments (Tran et al., 2014). In an AAV-α-Synuclein overexpressing rat model, administering antibodies (intraperitoneally) against the N-terminal (AB1) or central region (AB2) of α-Synuclein reduced loss of dopaminergic cells, decreased the number of reactive microglia, and lowered the levels of α-Synuclein in the substantia nigra (Shahaduzzaman et al., 2015).

Oligomeric and fibrillar forms of α-Synuclein are responsible for causing cellular toxicity. Therefore, antibodies against α-Synuclein have been designed that specifically target these toxic forms. A mouse PD model was developed by expressing human α-Synuclein (A30P mutant) under the Thy-1 promoter. These mice developed α-Synuclein fibrils, which caused the death of dopaminergic neurons accompanied by behavioral and motor deficits. Treatment with monoclonal antibody mAb47 targeting the oligomeric/fibrillar forms of α-Synuclein decreased the number of protofibrils in the spinal cord; microglial and astrocytic activation was unaffected (Lindström et al., 2014). A similar approach in Thy1 α-Synuclein mice models, weekly immunization with three antibodies (Syn-O1, Syn-O4, and Syn-F1) targeting the fibrils/oligomers of α-Synuclein for 3 months resulted in decreased α-Synuclein pathology, improved behavior deficits, reduced activation of microglia and astrocytes in the hippocampal region (El-Agnaf et al., 2017).

The *in vivo* studies provided enough evidence to translate passive immunization with α-Synuclein antibodies into clinical trials. The first α-Synuclein antibody to enter clinical trials was PRX002 (human IgG1 monoclonal version of the murine antibody 9E4 discussed above). This study was a randomized, double-blinded, placebo-controlled phase 1 trial sponsored by Prothena Biosciences Ltd. (California, USA) and F. Hoffmann-La Roche Ltd. (Basel, Switzerland). Forty healthy participants were randomized into five ascending dose cohorts (n = 8 in each cohort). Single intravenous injection of PRX002 (0.3, 1, 3, 10, 30 mg/kg; n = 6/cohort) or placebo (n = 2/cohort) was administered to the randomized participants. At the end of the trial, PRX002 was reported to be safe, tolerable with no immunogenicity or serious adverse effects (Schenk et al., 2016). Jankovic et al. performed the phase 1b trial of PRX002 to determine the safety and tolerability of multiple intravenous infusions (Jankovic, Goodman, & Safirstein, 2018). Eighty patients (aged 40–80 years) with idiopathic PD were treated with ascending doses (0.3, 1, 3, 10, 30, 60 mg/kg) or placebo in a multicentric, randomized, double-blinded, placebo-controlled trial across eight study centers in the USA from July 2014 to September 2016. The participants received placebo or three intravenous infusions per week and were monitored for 24 weeks. Single and multiple doses of PRX002 were safe and tolerable and reduced serum levels of free-to-total α-Synuclein. In collaboration with Prothena Biosciences Ltd., Roche initiated a Phase 2 clinical trial of PRX002 (PASADENA Study, ClinicalTrials.gov identifier NCT03100149). The results have not been published yet.

Anti-α-Synuclein antibody BIIB054 (IgG1 monoclonal antibody) targeted against N-terminal of the protein was developed by Neurimmune, a Swiss Biotech company, and evaluated in a clinical trial by Biogen (Brys et al., 2018a). This antibody was isolated from B-cell lines from neurologically healthy individuals and is specific for the aggregated forms of α-Synuclein. The trial recruited 18 PD patients (age 47–75 years) who were administered single, ascending doses of BIIB054 (up to 45 mg/kg) and were evaluated over 16 weeks. BIIB054 was safe and well-tolerated in PD patients. A Phase 2 trial was initiated in recently diagnosed PD patients (SPARK Study, ClinicalTrials.gov identifier NCT03318523). The study failed to meet the secondary outcome measures and was closed. In collaboration with Covance, MMS Holdings, and Catalent, AstraZeneca developed another monoclonal antibody, MEDI1341, that entered Phase 1 clinical trial in 2017 (NCT03272165), but no results have been posted to date. Table 2 contains an overview of the clinical trials involving active and passive immunization.

Table 2 Clinical trials with α-Synuclein: active and passive immunization strategies.

Type of immunization	Target site of α-Synuclein	Name of the molecule/antibody	Clinical trial no.	Sponsor/collaborator	Mode of application	Patient profile	Treatment duration	Trial status	Trial outcome	References
Active immunization	Short peptide of 8 amino acids targeting an epitope in the C-terminal of α-Synuclein	PD01A	EudraCT (2011-002650-31, 2013-001774-20, 2014-002489-54, and 2015-004854-16)	AFFiRiS and Michael J. Fox Foundation	Subcutaneous	32 PD patients (age 45–65 years) with early-stage idiopathic PD	52 weeks with a follow-up of 39 weeks Next single dose for 24 weeks with 52 weeks follow-up	Phase 1 completed	Safe and well tolerated for an extended period, generated substantial humoral response	Volc et al. (2020)
		PD03A	EudraCT (2014-000568-16)	AFFiRiS	Subcutaneous	36 patients (age 45–68 years) with early-stage idiopathic PD	52 weeks	Phase 1 completed	Safe and tolerable, generated humoral immune response	Poewe et al. (2021)
	Enhanced, synthetic, peptide-based vaccine	UB312	NCT04075318	United Neuroscience Ltd.	Subcutaneous	62 participants (age 42–80 years)	20 weeks treatment with 24 weeks follow-up	Phase 1	—	Study results not published yet

Continued

Table 2 Clinical trials with α-Synuclein: active and passive immunization strategies.—cont'd

Type of immunization	Target site of α-Synuclein	Name of the molecule/ antibody	Clinical trial no.	Sponsor/ collaborator	Mode of application	Patient profile	Treatment duration	Trial status	Trial outcome	References
	Humanized IgG1 monoclonal antibody	PRX002/ RG7935	NCT02157714	Prothena Biosciences and Hoffmann-La Roche	Intravenous (3 infusions every 4 weeks)	80 patients with mild to moderate idiopathic PD (age 40–80 years)	8 weeks with 16 weeks follow-up	Phase 1 completed	Safe and tolerable with no adverse effects	Jankovic et al. (2018)
	Humanized IgG1 monoclonal antibody	PRX002/ RO7046015	NCT03100149	Prothena Biosciences and Hoffmann-La Roche	Intravenous (Infusions every 4 weeks)	316 enrolled patients with mild idiopathic PD	52 weeks (Part 1) with further 52 weeks (Part 2)	Phase 2	—	Results not published
	Human IgG1 monoclonal antibody	BIIB054	NCT02459886	Biogen	Intravenous (single dose)	18 PD patients (age 47–75 years)	16 weeks	Phase 1	Safe and tolerable with no adverse effects	Brys et al. (2018a, 2018b)
	Human IgG1 monoclonal antibody	BIIB054	NCT03318523 (SPARK Study)	Biogen	Intravenous (every 4 weeks)	357 participants (age 40–80 years)	52 and 72 weeks with 12 week follow-up	Phase 2	Study was closed as it failed to meet the primary outcome measures	—
	Human IgG1 monoclonal antibody	MEDI1341	NCT04449484	AstraZeneca	Intravenous (3 infusicns every 4 weeks)	36 participants (age 40–85 years)	8 week treatment with 13 week follow-up	Phase 1	Ongoing	—

5.2 Targeting the immune mediators of inflammation

Besides targeting α-Synuclein, disease-modifying therapies can also modulate the immune mediators that protect against DA neurodegeneration. PPAR-γ agonists like rosiglitazone and pioglitazone hold promise which target reactive microglia and prevent them from releasing neurotoxic factors. These repurposed drugs entered a clinical trial for Parkinson's disease (NCT01280123) (Carta & Simuni, 2015). Phase 2 trials with pioglitazone in early PD patients did not modify disease progression at the prescribed doses (up to 45 mg) (NINDS Exploratory Trials in Parkinson Disease (NET-PD) FS-ZONE Investigators, 2015). Inhibiting the activation of NLRP3 inflammasome can also be another target. MCC950 is a small molecule inhibitor of the α-Synuclein-induced NLRP3 inflammasome activation in mice microglia and can be tried in more preclinical or clinical studies (Gordon et al., 2018). Granulocyte-macrophage colony-stimulating factor (GM-CSF) can increase the number or function of regulatory T-cells (Treg cells), a vital immunomodulator in PD. Recombinant human GM-CSF (sargramostim) is called Leukine, which entered into Phase 1 clinical trial in 2013 (NCT01882010). The study results were published by Gendelman et al., which showed modest improvement in cortical motor activities and Treg number and function in the sargramostim group after 6–8 weeks of treatment (Gendelman et al., 2017). Rho kinase (ROCK) activation can be linked to neuroinflammation and degeneration of DA neurons. ROCK inhibitors like fasudil have been reported to be neuroprotective in rodent PD models (Lopez-Lopez, Labandeira, Labandeira-Garcia, & Munoz, 2020; Roser, Tönges, & Lingor, 2017). Lipid signaling via the sphingosine-1-phosphate (S1P) pathway holds promise in the treatment of neurodegenerative disorders. Fingolimod (FTY720) is a sphingosine-1-phosphate receptor agonist and is an approved oral drug for multiple sclerosis. In rodent PD models (rotenone, MPTP, or 6-hydroxydopamine), fingolimod improved motor functions, loss of DA neurons, and striatal dopamine levels (Motyl, Przykaza, Boguszewski, Kosson, & Strosznajder, 2018; Zhao et al., 2017). Neuroprotective effects of fingolimod were also observed in SH-SY5Y cells treated with rotenone or 6-hydroxydopamine (Zhao et al., 2017).

Another class of repurposed drugs beneficial in PD is the glucagon-like peptide-1 receptor (GLP1R) agonists, an anti-diabetes medication with potential neuroprotective properties. GLP1R agonist NLY01 (pegylated exenatide) can cross the blood-brain barrier (BBB) and protect against α-Synuclein PFF-induced neurodegeneration in mice models of sporadic PD. In addition, NLY01 prevents the microglia-mediated conversion of astrocytes into the A1 toxic phenotype (Hinkle, Dawson, & Dawson, 2019; Yun et al., 2018). NLY01 entered into Phase 1 clinical trial (subcutaneous injections with single and multiple-dose, randomized, double-blinded, placebo-controlled trial) with ninety-six healthy participants (age 18–65 years), and the drug was found to be safe and tolerable (NCT03672604) after 4 weeks of treatment. NLY01 entered into Phase 2 trial (NCT04154072) in a multicentric, randomized, double-blinded, placebo-controlled study. The drug was tested on 240 patients (36 weeks treatment) with early, untreated PD (age 30–80 years); the results of this trial have not been published yet.

Immunosuppressants like cyclosporin also have potential benefits. For example, cyclosporin promotes the survival of fetal dopamine neurons transplanted in AAV-α-Synuclein rat models of PD (Tamburrino et al., 2015). In addition, Cyclosporine A can also prevent α-Synuclein aggregation in rats (Agrawal et al., 2014). Similarly, immunophilin ligand FK506 also reduced neurodegeneration in rAAV2/7-α-Synuclein-overexpressing rat PD models by preventing the infiltration of T-cells (helper and cytotoxic) and the number of reactive microglia (Van der Perren et al., 2015).

Another target of immunotherapy in PD is the Toll-like 2 receptor (TLR2). Elevated levels of TLR2 have been observed in the PD brain, and it is an essential mediator of the neurotoxic and proinflammatory actions of α-Synuclein (Dzamko et al., 2016). In addition, anti-TLR2 antibodies reduced α-Synuclein accumulation in neurons and astrocytes and improved motor deficits in mouse PD models (Kim et al., 2018). Anti-TLR2 antibodies, therefore, can be critical intracellular targets in PD therapeutics.

6. Conclusion

PD and other neurodegenerative disorders are complex diseases involving many causative factors, and treatment and early diagnosis are difficult. The advent of immunotherapeutic advances has greatly increased the hope associated with the treatment of PD. With the advancement of technology in antibody and peptide engineering, we can develop appropriate measures to halt the disease progression in PD. Immunotherapy holds promise in this area. Targeting either α-Synuclein or the inflammatory mediators or the cells of the neuroinflammatory cascade can lead us to a much-desired output in the therapeutic avenue of PD. Several molecules are already in different phases of clinical trials. However, the heterogeneity of the disease coupled with the complexity of the proteotoxic pathway and differential immune response makes the final outcome difficult. Nonetheless, immunotherapeutic targets in PD are an interesting research area, and we need to focus more on the *in vitro* and *in vivo* studies employing various PD models and tools to translate them into actual disease cases to provide a breakthrough in the current treatment regime.

References

Abd-el-Basset, E., & Fedoroff, S. (1995). Effect of bacterial wall lipopolysaccharide (LPS) on morphology, motility, and cytoskeletal organization of microglia in cultures. *Journal of Neuroscience Research*, *41*, 222–237.

Agrawal, S., Dixit, A., Singh, A., Tripathi, P., Singh, D., Patel, D. K., et al. (2014). Cyclosporine A and MnTMPyP alleviate α-synuclein expression and aggregation in cypermethrin-induced parkinsonism. *Molecular Neurobiology*, *52*, 1619–1628.

Al-Bachari, S., Naish, J. H., Parker, G. F. M., Emsley, H. C. A., & Parkes, L. M. (2020). Blood–brain barrier leakage is increased in Parkinson's disease. *Frontiers in Physiology*, *11*, 593026.

Aldeoca, I., Navarro-Otano, J., Stefanova, N., Sprenger, F. S., Seppi, K., Poewe, W., et al. (2015). Alpha-synuclein immunoreactivity patterns in the enteric nervous system. *Neuroscience Letters*, *602*, 145–149.

Allaman, I., Belanger, M., & Magistretti, P. J. (2011). Astrocyte-neuron metabolic relationships: For better and for worse. *Trends in Neurosciences*, *34*, 76–87.

Allen Reish, H. E., & Standaert, D. G. (2015). Role of α-synuclein in inducing innate and adaptive immunity in Parkinson's disease. *Journal of Parkinson's Disease*, *5*, 1–19.

Aloisi, F. (1999). The role of microglia and astrocytes in CNS immune surveillance and immunopathology. *Advances in Experimental Medical Biology*, *468*, 123–133.

Aloisi, F. (2001). Immune function of microglia. *Glia*, *36*, 165–179.

Arlehamn, C. S., Francesca, G., Sulzer, D., & Sette, A. (2019). Role for the adaptive immune system in Parkinson's and Alzheimer's disease. *Current Opinion in Immunology*, *59*, 115–120.

Austin, S. A., Floden, A. M., Murphy, E. J., & Combs, C. K. (2006). Alpha-synuclein expression modulates microglial activation phenotype. *Journal of Neuroscience*, *26*, 10558–10563.

Baba, Y., Kuroiwa, A., Uitti, R. J., Wszolek, Z. K., & Yamada, T. (2005). Alterations of T-lymphocyte populations in Parkinson disease. *Parkinsonism & Related Disorders*, *11*, 493–498.

Bae, E. J., Lee, H. J., Rockenstein, E., Ho, D. H., Bi, E., Yang, N. Y., et al. (2012). Antibody-aided clearance of extracellular α-synuclein prevents cell-to-cell aggregate transmission. *Journal of Neuroscience*, *32*, 13454–13469.

Banerjee, K., Kumar, S., Ross, K. A., Gautam, S., Poelaert, B., Nasser, M. W., et al. (2018). Emerging trends in the immunotherapy of pancreatic cancer. *Cancer Letters*, *417*, 35–46.

Bartels, T., Choi, J. G., & Selkoe, D. J. (2011). α-Synuclein occurs physiologically as a helically folded tetramer that resists aggregation. *Nature*, *477*, 107–110.

Bartels, T., De Schepper, S., & Hong, S. (2020). Microglia modulate neurodegeneration in Alzheimer's and Parkinson's disease. *Science*, *370*, 66–69.

Bas, J., Calopa, M., Mestre, M., Mollevi, D. G., & Cutillas, B. (2001). Lymphocyte populations in Parkinson's disease and in rat models of parkinsonism. *Journal of Neuroimmunology*, *113*, 146–152.

Beach, T. G., Adler, C. H., Sue, L. I., Vedders, L., Lue, L., White III, C. L., et al. (2010). Multi-organ distribution of phosphorylated alpha-synuclein histopathology in subjects with Lewy body disorders. *Acta Neuropathologica*, *119*, 689–702.

Belanger, M., Allaman, I., & Magistretti, P. (2011). Brain energy metabolism: Focus on astrocyte-neuron metabolic cooperation. *Cell Metabolism*, *14*, 724–738.

Benner, E. J., Banerjee, R., Reynolds, A. D., Sherman, S., Pisarev, V. M., Tsiperson, V., et al. (2008). Nitrated alpha-synuclein immunity accelerates degeneration of nigral dopaminergic neurons. *PLoS One*, *3*, e1376.

Bhatt, M. A., Messer, A., & Kordower, J. H. (2013). Can intrabodies serve as neuroprotective therapies for Parkinson's disease? Beginning thoughts. *Journal of Parkinson's Disease*, *3*, 581–591.

Bhattacharya, D., Mohite, G. M., Krishnamoorthy, J., Gayen, N., Mehra, S., Navalkar, A., et al. (2019). Lipopolysaccharide from gut microbiota modulates α-synuclein aggregation and alters its biological function. *ACS Chemical Neuroscience*, *10*, 2229–2236.

Bliederhaeuser, C., Grozdanov, V., Speidel, A., Zondler, L., Ruf, W. P., Bayer, H., et al. (2016). Age-dependent defects of alpha-synuclein oligomer uptake in microglia and monocytes. *Acta Neuropathologica*, *131*, 379–391.

Block, M. L., Zecca, L., & Hong, J. S. (2007). Microglia-mediated neurotoxicity: Uncovering the molecular mechanisms. *Nature Reviews. Neuroscience*, *8*, 57–69.

Blum-Degen, D., Muller, T., Kuhn, W., Gerlach, M., Przuntek, H., & Riederer, P. (1995). Interleukin-1 beta and interleukin-6 are elevated in the cerebrospinal fluid of Alzheimer's and de novo Parkinson's disease patients. *Neuroscience Letters*, *202*, 17–20.

Bonaz, B. L., & Bernstein, C. L. (2013). Brain-gut interactions in inflammatory bowel disease. *Gastroenterology*, *144*, 36–49.

Bongioanni, P., Castagna, M., Maltinti, S., Boccardi, B., & Dadone, F. (1997). T-lymphocyte tumor necrosis factor-alpha receptor binding in patients with Parkinson's disease. *Journal of the Neurological Sciences*, *149*, 41–45.

Bousset, L., Pieri, L., Ruiz-Arlandis, G., Gath, J., Jensen, P. H., Habenstein, B., et al. (2013). Structural and functional characterization of two alpha-synuclein strains. *Nature Communications*, *4*, 2575.

Braak, H., Del Tredici, K., Rub, U., de Vos, R. A. I., Jansen Steur, E. N. H., & Braak, E. (2003). Staging of brain pathology related to sporadic Parkinson's disease. *Neurobiology of Aging*, *24*, 197–211.

Braak, H., de Vos, R. A. I., Bohl, J., & Del Tredici, K. (2006). Gastric alpha-synuclein immunoreactive inclusions in Meissner's and Auerbach's plexuses in cases staged for Parkinson's disease-related brain pathology. *Neuroscience Letters*, *396*, 67–72.

Braak, H., Rub, U., & Del Tredici, K. (2003). Idiopathic Parkinson's disease: Possible routes by which vulnerable neuronal types may be subject to neuroinvasion by an unknown pathogen. *Journal of Neural Transmission (Vienna)*, *110*, 517–536.

Braak, H., Sastre, M., & Del Tredici, K. (2007). Development of alpha-synuclein immunoreactive astrocytes in the forebrain parallels stages of intraneuronal pathology in sporadic Parkinson's disease. *Acta Neuropathologica*, *114*, 231–241.

Braidy, N., Gai, W. P., Xu, Y. H., Sachdev, P., Guillemin, G. J., Jiang, X. M., et al. (2013). Uptake and mitochondrial dysfunction of alpha-synuclein in human astrocytes, cortical neurons and fibroblasts. *Translational Neurodegeneration*, *2*, 20.

Brecknell, J. E., Dunnett, S. B., & Fawcett, J. W. (1995). A quantitative study of cell death in the substantia nigra following a mechanical lesion of the medial forebrain bundle. *Neuroscience*, *64*, 219–227.

Breidert, T., Callebert, J., Heneka, M. T., Landreth, G., Launay, J. M., & Hirsch, E. C. (2002). Protective action of the peroxisome proliferator-activated receptor-gamma agonist pioglitazone in a mouse model of Parkinson's disease. *Journal of Neurochemistry*, *82*, 615–624.

Breydo, L., Wu, J. W., & Uversky, V. N. (2012). α-Synuclein misfolding and Parkinson's disease. *Biochimica et Biophysica Acta (BBA). Molecular Basis of Disease*, *1822*, 261–285.

Brioschi, S., & Colonna, M. (2019). The CNS immune-privilege goes down the drain. *Trends in Pharmacological Sciences*, *40*, 1–3.

Brochard, V., Combadiere, B., Prigent, A., Laouar, Y., Perrin, A., & Beray-Berthat, V. (2009). Infiltration of CD4+ lymphocytes into the brain contributes to neurodegeneration in a mouse model of Parkinson disease. *The Journal of Clinical Investigation*, *119*, 182–192.

Brodacki, B., Staszewski, J., Toczylowska, B., Kozlowska, E., Drela, N., Chalimoniuk, M., et al. (2008). Serum interleukin (IL-2, IL-10, IL-6, IL-4), TNF alpha, and IFN gamma concentrations are elevated in patients with atypical and idiopathic parkinsonism. *Neuroscience Letters*, *441*, 158–162.

Brudek, T. (2019). Inflammatory bowel disease and Parkinson's disease. *Journal of Parkinson's Disease*, *9*, 331–344.

Brys, M., Ellenbogen, A., Fanning, L., Penner, N., Yang, M., Welch, M., et al. (2018a). Randomized, double-blind, placebo-controlled, single ascending dose study of anti-alpha synuclein antibody BIIB054 in patients with Parkinson's disease (S26.001). *Neurology*, *90*.

Brys, I., Halje, P., Scheffer-Teixeira, R., Varney, M., Newman-Tancredi, A., & Petersson, P. (2018b). Neurophysiological effects in cortico-basal ganglia-thalamic circuits of antidyskinetic treatment with 5-HT_{1A} receptor biased agonists. *Experimental Neurology*, *302*, 155–168.

Bsibsi, M., Persoon-Deen, C., Verwer, R. W., Meeuwsen, S., Ravid, R., & Van Noort, J. M. (2006). Toll-like receptor 3 on adult human astrocytes triggers production of neuroprotective mediators. *Glia*, *53*, 688–695.

Bsibsi, M., Ravid, R., Gveric, D., & van Noort, J. M. (2002). Broad expression of Toll-like receptors in the human central nervous system. *Journal of Neuropathology and Experimental Neurology*, *61*, 1013–1021.

Burre, J., Sharma, M., & Sudhof, T. C. (2018). Cell biology and pathophysiology of α-synuclein. *Cold Spring Harbor Perspectives in Medicine*, *8*, a024091.

Burre, J., Sharma, M., Tsetsenis, T., Buchman, V., Etherton, M. R., & Sudhof, T. C. (2010). Alpha-synuclein promotes SNARE-complex assembly in vivo and in vitro. *Science*, *329*, 1663–1667.

Burre, J., Vivona, S., Diao, J., Sharma, M., Brunger, A. T., & Sudhof, T. C. (2013). Properties of native brain α-synuclein. *Nature*, *498*, 4–6.

Bussell, R., & Eliezer, D. (2003). A structural and functional role for 11-mer repeats in alpha-synuclein and other exchangeable lipid binding proteins. *Journal of Molecular Biology*, *329*, 763–778.

Cabezas, R., Avila, M., Gonzalez, J., El-Bacha, R. S., Baez, E., Garcia-Segura, L. M., et al. (2014). Astrocytic modulation of blood brain barrier: Perspectives on Parkinson's disease. *Frontiers in Cellular Neuroscience*, *8*, 211.

Carbone, M., Duty, S., & Rattray, M. (2012). Riluzole neuroprotection in a Parkinson's disease model involves suppression of reactive astrocytosis but not GLT-1 regulation. *BMC Neuroscience*, *13*, 38.

Cardona, A. E., Pioro, E. P., Sasse, M. E., Kostenko, V., Cardona, S. M., et al. (2006). Control of microglial neurotoxicity by the fractalkine receptor. *Nature Neuroscience*, *9*, 917–924.

Carta, A. R., & Simuni, T. (2015). Thiazolidinediones under preclinical and early clinical development for the treatment of Parkinson's disease. *Expert Opinion on Investigational Drugs*, *40*, 219–227.

Cassani, E., Barichella, M., Cancello, R., Cavanna, F., Lorio, K., Cereda, E., et al. (2015). Increased urinary indoxyl sulfate (indican): New insights into gut dysbiosis in Parkinson's disease. *Parkinsonism & Related Disorders*, *21*, 389–393.

Castagnet, P. I., Golovko, M. Y., Barcelo-Coblijn, G. C., Nussbaum, R. L., & Murphy, E. J. (2005). Fatty acid incorporation is decreased in astrocytes cultured from alpha-synuclein gene-ablated mice. *Journal of Neurochemistry*, *94*, 839–849.

Castro-Sanchez, S., Garcia-Yague, A. J., Lopez-Royo, T., Casarejos, M., Lanciego, J. L., & Lastres-Becker, I. (2018). Cx3cr1-deficiency exacerbates alpha-synuclein-A53T induced neuroinflammation and neurodegeneration in a mouse model of Parkinson's disease. *Glia*, *66*, 1752–1762.

Cavaliere, F., Cerf, L., Dehay, B., Ramos-Gonzalez, P., De Giorgi, F., Bourdenx, M., et al. (2017). In vitro α-synuclein neurotoxicity and spreading among neurons and astrocytes using Lewy body extracts from Parkinson disease brains. *Neurobiology of Disease*, *103*, 101–112.

Cebrian, C., Zucca, FA., Mauri, P., Steinbeck, JA., Studer, L., Scherzer, CR., et al. (2014). MHC-I expression renders catecholaminergic neurons susceptible to T-cell-mediated degeneration. *Nature Communication*, *5*, 3633.

Challis, C., Hori, A., Sampson, T. R., Yoo, B. B., Challis, R. C., Hamilton, A. M., et al. (2020). Gut-seeded α-synuclein fibrils promote gut dysfunction and brain pathology specifically in aged mice. *Nature Neuroscience*, *23*, 327–336.

Chandra, G., Rangasamy, S. B., Roy, A., Kordower, J. H., & Pahan, K. (2016). Neutralization of RANTES and eotaxin prevents the loss of dopaminergic neurons in a mouse model of Parkinson disease. *The Journal of Biological Chemistry*, *291*, 15267–15281.

Chandra, G., Roy, A., Rangasamy, S. B., & Pahan, K. (2017). Induction of adaptive immunity leads to nigrostriatal disease progression in MPTP mouse model of Parkinson's disease. *Journal of Immunology*, *198*, 4312–4326.

Chartier-Harlin, MC., Kachergus, J., Roumier, C., Mouroux, V., Douay, X., Lincoln, S., et al. (2004). Alpha-synuclein locus duplication as a cause of familial Parkinson's disease. *Lancet*, *364*, 1167–1169.

Chatterjee, D., Bhatt, M., Butler, D., Genst, E. D., Dobson, C. M., Messer, A., et al. (2018). Proteasome-targeted nanobodies alleviate pathology and functional decline in an α-synuclein-based Parkinson's disease model. *NPJ Parkinson's Disease*, *4*, 25.

Chen, S., Liu, Y., Niu, Y., Xu, Y., Zhou, Q., Xu, X., et al. (2017). Increased abundance of myeloid-derived suppressor cells and Th17 cells in peripheral blood of newly-diagnosed Parkinson's disease patients. *Neuroscience Letters*, *648*, 21–25.

Chen, R. H. C., Weslet-Gendebien, S., Samuel, F., Visanji, N. P., Zhang, G., Marsilio, D., et al. (2013). α-Synuclein membrane association is regulated by the Rab3a recycling machinery and presynaptic activity. *Journal of Biological Chemistry*, *288*, 7438–7449.

Choi, B. K., Choi, M. G., Kim, J. Y., Yang, Y., Lai, Y., Kweon, D. H., et al. (2013). Large α-synuclein oligomers inhibit neuronal SNARE-mediated vesicle docking. *Proceedings of the National Academy of Sciences of the United States of America*, *110*, 4087–4092.

Chorell, E., Andersson, E., Evans, M. L., Jain, N., Gotheson, A., Aden, J., et al. (2015). Bacterial chaperones CsgE and CsgC differentially modulate human α-synuclein amyloid formation via transient contacts. *PLoS One*, *10*, e0140194.

Colla, E., Jensen, P. H., Pletnikova, O., Troncosco, J. C., Glabe, C., & Lee, M. K. (2012). Accumulation of toxic α-synuclein oligomer within endoplasmic reticulum occurs in α-synucleinopathy in vivo. *Journal of Neuroscience*, *32*, 3301–3305.

Connolly, B. S., & Lang, A. E. (2014). Pharmacological treatment of Parkinson's disease: A review. *Journal of the American Medical Association*, *311*, 1670–1683.

Cox, B., Kerwin, R., & Lee, T. F. (1978). Dopamine receptors in the central thermoregulatory pathway of the rat. *The Journal of Physiology*, *282*, 471–483.

Cunnion, K. M., Zhang, H. M., & Frank, M. M. (2003). Availability of complement bound to Staphylococcus aureus to interact with membrane complement receptors influences efficiency of phagocytosis. *Infection and Immunity*, *71*, 656–662.

Damier, P., Hirsch, E. C., Zhang, P., Agid, Y., & Javoy-Agid, F. (1993). Glutathione peroxidase, glial cells and Parkinson's disease. *Neuroscience*, *52*, 1–6.

David, K. S., Ashley, D., Reynolds, R., Mosley, L., & Gendelmann, E. H. (2009). Innate and adaptive immunity for the pathobiology of Parkinson's disease. *Antioxidant & Redox Signalling*, *11*, 2151–2166.

Davidson, W. S., Jonas, A., Clayton, D. F., & George, J. M. (1998). Stabilization of α-synuclein secondary structure upon binding to synthetic membranes. *Journal of Biological Chemistry*, *273*, 9443–9449.

Dehay, B., Bourdenx, M., Gorry, P., Przedborski, S., Vila, M., Hunot, S., et al. (2015). Targeting α-synuclein for treatment of Parkinson's disease: Mechanistic and therapeutic considerations. *Lancet Neurology*, *14*, 855–866.

Deng, H., Wang, P., & Jankovic, J. (2018). The genetics of Parkinson's disease. *Ageing Research Reviews*, *42*, 72–85.

Derkinderen, P., Rouaud, T., Lebouvier, T., des Varannes, S. B., Neunlist, M., & De Giorgio, R. (2011). Parkinson disease: The enteric nervous system spills its guts. *Neurology*, *77*, 1761–1767.

Devos, D., Lebouvier, T., Lardeux, B., Biraud, M., Rouaud, T., Pouclet, H., et al. (2013). Colonic inflammation in Parkinson's disease. *Neurobiology of Disease*, *50*, 42–48.

Duda, J. E., Lee, L. M. Y., & Trojanowski, J. Q. (2000). Neuropathology of synuclein aggregates. *Journal of Neuroscience Research*, *61*, 121–127.

Duffy, M. F., Collier, T. J., Patterson, J. R., Kemp, C. J., Luk, K. C., Tansey, M. G., et al. (2018). Lewy body-like alpha-synuclein inclusions trigger reactive microgliosis prior to nigral degeneration. *Journal of Neuroinflammation*, *15*, 129.

Dzamko, N., Gysbers, A., Perera, G., Bahar, A., Shankar, A., Gao, J., et al. (2016). Toll-like receptor 2 is increased in neurons in Parkinson's disease brain and may contribute to alpha-synuclein pathology. *Acta Neuropathologica*, *133*, 303–319.

El-Agnaf, O., Overk, C., Rockenstein, E., Mante, M., Florio, J., Adame, A., et al. (2017). Differential effects of immunotherapy with antibodies targeting α-synuclein oligomers and fibrils in a transgenic model of synucleinopathy. *Neurobiology of Disease*, *104*, 85–96.

Emmanouilidou, E., Stefanis, L., & Vekrellis, E. (2010). Cell-produced alpha-synuclein oligomers are targeted to, and impair, the 26S proteasome. *Neurobiology of Aging*, *31*, 953–968.

Erdei, A., Andrasfalvy, M., Peterfy, H., Toth, G., & Pecht, I. (2004). Regulation of mast cell activation by complement-derived peptides. *Immunology Letters*, *92*, 39–42.

Erustes, A. G., Stefani, F. Y., Terashima, J. Y., Stilhano, R. S., Monteforte, P. T., da Silva Pereira, G. J., et al. (2018). Overexpression of α-synuclein in an astrocyte cell line promotes autophagy inhibition and apoptosis. *Journal of Neuroscience Research*, *96*, 160–171.

Falkowska, A., Gutowska, I., Goschorska, M., Nowacki, P., Chlubek, D., & Baranowska-Bosiacka, I. (2015). Energy metabolism of the brain, including the cooperation between astrocytes and neurons, especially in the context of glycogen metabolism. *International Journal of Molecular Sciences*, *16*, 25959–25981.

Farina, C., Krumbholz, M., Giese, T., Hartmann, G., Aloisi, F., & Meinl, E. (2005). Preferential expression and function of Toll-like receptor 3 in human astrocytes. *Journal of Neuroimmunology*, *159*, 12–19.

Farrer, M., Kachergus, J., Forno, L., Lincoln, S., Wang, D. S., Hulihan, M., et al. (2004). Comparison of kindreds with parkinsonism and alpha-synuclein genomic multiplications. *Annals of Neurology*, *55*, 174–179.

Fathy, Y. Y., Jonker, A. J., Oudejans, E., de Jong, F. J. J., van Dam, A. M. W., Rozemuller, A. J. M., et al. (2019). Differential insular cortex subregional vulnerability to α-synuclein pathology in Parkinson's disease and dementia with Lewy bodies. *Neuropathology and Applied Neurobiology*, *45*, 262–277.

Fellner, L., Irschick, R., Schanda, K., Reindl, M., Klimaschewski, L., Poewe, W., et al. (2013). Toll-like receptor 4 is required for α-synuclein dependent activation of microglia and astroglia. *Glia*, *61*, 349–360.

Ferrari, C. C., Godoy, M. C. P., Tarelli, R., Chertoff, M., Depino, A. M., et al. (2006). Progressive neurodegeneration and motor disabilities induced by chronic expression of IL-1beta in the substantia nigra. *Neurobiology of Disease*, *24*, 183–193.

Filiano, A. J., Gadani, S. P., & Kipnis, J. (2017). How and why do T cells and their derived cytokines affect the injured and healthy brain. *Nature Reviews. Neuroscience*, *18*, 375–384.

Fiszer, U., Fredrikson, S., & Czlonkowska, A. (1996). Humoral response to hsp 65 and hsp 70 in cerebrospinal fluid in Parkinson's disease. *Journal of the Neurological Sciences*, *139*, 66–70.

Fiszer, U., Mix, E., Fredrikson, S., Kostulas, V., & Link, H. (1994). Parkinson's disease and immunological abnormalities: Increase of HLA-DR expression on monocytes in cerebrospinal fluid and of CD45RO+ T cells in peripheral blood. *Acta Neurologica Scandinavica*, *90*, 160–166.

Fiszer, U., Mix, E., Fredrikson, S., Kostulas, V., Olsson, T., & Link, H. (1994). Gamma delta + T cells are increased in patients with Parkinson's disease. *Journal of the Neurological Sciences*, *121*, 39–45.

Fitzgerald, E., Murphy, S., & Martinson, H. A. (2019). Alpha-synuclein pathology and the role of the microbiota in Parkinson's disease. *Frontiers in Neuroscience*, *13*, 369.

Forsyth, C. B., Shannon, K. M., Kordower, J. H., Voigt, R. M., Shaikh, M., Jaglin, J. A., et al. (2011). Increased intestinal permeability correlates with sigmoid mucosa alpha-synuclein staining and endotoxin exposure markers in early Parkinson's disease. *PLoS One*, *6*, e28032.

Fortin, D. L., Troyer, M. D., Nakamura, K., Kubo, S. I., Anthony, M. D., & Edwards, R. H. (2004). Lipid rafts mediate the synaptic localization of alpha-synuclein. *Journal of Neuroscience*, *24*, 6715–6723.

Freeman, L., Guo, H., David, C. N., Brickey, W. J., Jha, S., & Ting, J. P. Y. (2017). NLR members NLRC4 and NLRP3 mediate sterile inflammasome activation in microglia and astrocytes. *Journal of Experimental Medicine*, *214*, 1351–1370.

Froula, J. M., Castellana-Cruz, M., Anabtawi, N. M., Camino, J. D., Chen, S. W., Thrasher, D. R., et al. (2019). Defining α-synuclein species responsible for Parkinson's disease phenotypes in mice. *Journal of Biological Chemistry*, *294*, 10392–10406.

Froula, J. M., Henderson, B. W., Gonzalez, J. C., Vaden, J. H., McLean, J. W., Wu, Y., et al. (2018). α-Synuclein fibril-induced paradoxical structural and functional defects in hippocampal neurons. *Acta Neuropathologica Communications*, *6*, 35.

Fussi, N., Höllerhage, M., Chakroun, T., Nykanen, N. P., Rösler, T. W., Koeglsperger, T., et al. (2018). Exosomal secretion of α-synuclein as protective mechanism after upstream blockage of macroautophagy. *Cell Death & Disease*, *9*, 757.

Gagne, J. J., & Power, M. C. (2010). Anti-inflammatory drugs and risk of Parkinson's disease. *Neurology*, *74*, 12.

Games, D., Valera, E., Spencer, B., Rockenstein, E., Mante, M., Adame, A., et al. (2014). Reducing C-terminal-truncated alpha-synuclein by immunotherapy attenuates neurodegeneration and propagation in Parkinson's disease-like models. *Journal of Neuroscience*, *34*, 9441–9454.

Gao, H. M., Kotzbauer, P. T., Uryu, K., Leight, S., Trojanowski, J. Q., & Lee, V. M. (2008). Neuroinflammation and oxidation/nitration of a-synuclein linked to dopaminergic neurodegeneration. *The Journal of Neuroscience*, *28*, 7687–7698.

Gardai, S. J., Mao, W., Schüle, B., Babcock, M., Schoebel, S., Lorenzana, C., et al. (2013). Elevated alpha-synuclein impairs innate immune cell function and provides a potential peripheral biomarker for Parkinson's disease. *PLoS One*, *8*, e71634.

Garretti, F., Agalliu, D., Arlehamn, C. S., Sette, A., & Sulzer, D. (2019). Autoimmunity in parkinson's disease: The role of alpha synuclein specific T cells. *Frontiers in Immunology*, *10*, 303.
Gasca-Salas, C., Fernández-Rodríguez, B., Pineda-Pardo, J. A., et al. (2021). Blood-brain barrier opening with focused ultrasound in Parkinson's disease dementia. *Nature Communications*, *12*, 779.
Gasser, T., Wszolek, Z. K., Trofatter, J., Ozelius, L., Uitti, R. J., Lee, C. S., et al. (1994). Genetic linkage studies in autosomal dominant parkinsonism: Evaluation of seven candidate genes. *Annals of Neurology*, *36*, 387–396.
GBD 2016 Parkinson's Disease Collaborators. (2018). Global, regional, and national burden of Parkinson's disease, 1990-2016: A systematic analysis for the Global Burden of Disease Study 2016. *The Lancet Neurology*, *17*, 939–953.
Gendelman, H. E., Zhang, Y., Santamaria, P., Olson, K. E., Schutt, C. R., Bhatti, D., et al. (2017). Evaluation of the safety and immunomodulatory effects of sargramostim in a randomized, double-blind phase 1 clinical Parkinson's disease trial. *NPJ Parkinsons Disease*, *3*, 10.
Gerhard, A., Pavese, N., Hotton, G., Turkheimer, F., Es, M., Hammers, A., et al. (2006). In vivo imaging of microglial activation with [^{11}C](*R*)-PK11195 PET in idiopathic Parkinson's disease. *Neurobiology of Disease*, *21*, 404–412.
Ghochikyan, A., Petrushina, I., Davtyan, H., Hovakimyan, A., Saing, T., Davtyan, A., et al. (2014). Immunogenicity of epitope vaccines targeting different B cell antigenic determinants of human α-synuclein: Feasibility study. *Neuroscience Letters*, *560*, 86–91.
Giasson, B. I., Duda, J. E., Murray, I. V., Chen, Q., Souza, J. M., Hurtig, H. I., et al. (2000). Oxidative damage linked to neurodegeneration by selective α-synuclein nitration in synucleinopathy lesions. *Science*, *290*, 985–989.
Glezer, I., Simard, A. R., & Rivest, S. (2007). Neuroprotective role of the innate immune system by microglia. *Neuroscience*, *147*, 867–883.
Gold, A., Turkalp, Z. T., & Munoz, D. G. (2013). Enteric alpha-synuclein expression is increased in Parkinson's disease but not Alzheimer's disease. *Movement Disorders*, *28*, 237–241.
Gorbatyuk, O. S., Li, S., Nguyen, F. N., Manfredsson, F. P., Kondrikova, G., Sullivan, L. F., et al. (2010). α-Synuclein expression in rat substantia nigra suppresses phospholipase D2 toxicity and nigral neurodegeneration. *Molecular Therapeutics*, *18*, 1758–1768.
Gordon, R., Albornoz, E. A., Christie, D. C., Langley, M. R., Kumar, V., Mantovani, S., et al. (2018). Inflammasome inhibition prevents α-synuclein pathology and dopaminergic neurodegeneration in mice. *Science Translational Medicine*, *10*, eaah4066.
Gray, M., Munoz, D. G., Gray, D. A., Schlossmacher, M. G., & Woulfe, J. M. (2014). Alpha-synuclein in the appendiceal mucosa of neurologically intact subjects. *Movement Disorders*, *29*, 991–998.
Gray, M. T., & Woulfe, J. M. (2015). Striatal blood-brain barrier permeability in Parkinson's disease. *Journal of Cerebral Blood Flow & Metabolism*, *35*, 747–750.
Grondahl, G., Hohannisson, A., Jensen-Waern, M., & Nilsson, E. K. (2001). Opsonization of yeast cells with equine iC3b, C3b, and IgG. *Veterinary Immunology and Immunopathology*, *80*, 209–223.
Grozdanov, V., Bliederhaeuser, C., Ruf, W. P., Roth, V., Fundel-Clemens, K., Zondler, L., et al. (2014). Inflammatory dysregulation of blood monocytes in Parkinson's disease patients. *Acta Neuropathologica*, *128*, 651–663.
Gu, X. L., Long, C. X., Sun, L., Xie, C., Lin, X., & Cai, H. (2010). Astrocytic expression of Parkinson's disease-related A53T alpha-synuclein causes neurodegeneration in mice. *Molecular Brain*, *3*, 12.
Guiney, S. J., Adlard, P. A., Lei, P., Mawal, C. H., Bush, A. I., Finkelstein, D. I., et al. (2020). Fibrillar α-synuclein toxicity depends on functional lysosomes. *Journal of Biological Chemistry*, *295*, 17497–17513.

Haggerty, T., Credle, J., Rodriguez, O., Wills, J., Oaks, A. W., Masliah, E., et al. (2011). Hyperphosphorylated Tau in an α-synuclein-overexpressing transgenic model of Parkinson's disease. *European Journal of Neuroscience*, *33*, 1598–1610.

Haim, L. B., & Rowitch, D. H. (2017). Functional diversity of astrocytes in neural circuit regulation. *Nature Reviews. Neuroscience*, *18*, 31–41.

Hammond, S. L., Bantle, C. M., Popichak, K. A., Wright, K. A., Thompson, D., Forero, C., et al. (2020). NF-κB signaling in astrocytes modulates brain inflammation and neuronal injury following sequential exposure to manganese and MPTP during development and aging. *Toxicological Sciences*, *177*, 506–520.

Hansen, C., Angot, E., Bergstrom, A. L., Steiner, J. A., Pieri, L., Paul, G., et al. (2011). α-Synuclein propagates from mouse brain to grafted dopaminergic neurons and seeds aggregation in cultured human cells. *Journal of Clinical Investigation*, *121*, 715–725.

Harms, A. S., Cao, S., Rowse, A. L., Thome, A. D., Li, X., Mangieri, L. R., et al. (2013). MHCII is required for alpha-synuclein-induced activation of microglia, CD4 T cell proliferation, and dopaminergic neurodegeneration. *The Journal of Neuroscience*, *33*, 9592–9600.

Harms, A. S., Thome, A. D., Yan, Z., Schonhoff, A. M., Williams, G. P., Li, X., et al. (2018). Peripheral monocyte entry is required for alpha-synuclein induced inflammation and neurodegeneration in a model of Parkinson disease. *Experimental Neurology*, *300*, 179–187.

Hasida, K., Kitao, Y., Sudo, H., Awa, Y., Maeda, S., Mori, K., et al. (2012). ATF6alpha promotes astroglial activation and neuronal survival in a chronic mouse model of Parkinson's disease. *PLoS One*, *7*, e47950.

Herz, J., Filiano, A. J., Smith, A., Yogev, N., & Kipnis, J. (2017). Myeloid cells in the central nervous system. *Immunity*, *46*, 943–956.

Hilton, D., Stephens, M., Kirk, L., Edwards, P., Potter, R., Zajicek, J., et al. (2014). Accumulation of α-synuclein in the bowel of patients in the pre-clinical phase of Parkinson's disease. *Acta Neuropathologica*, *127*, 235–241.

Hinkle, J. T., Dawson, V. L., & Dawson, T. D. (2019). The A1 astrocyte paradigm: New avenues for pharmacological intervention in neurodegeneration. *Movement Disorders*, *34*, 959–969.

Hisanaga, K., Asagi, M., Itoyama, Y., & Iwasaki, Y. (2001). Increase in peripheral CD4 bright+ CD8 dull+ T cells in Parkinson disease. *Archives in Neurology*, *58*, 1580–1583.

Hoenen, C., Gustin, A., Birck, C., Kirchmeyer, M., Beaume, N., Felten, P., et al. (2016). Alpha-synuclein proteins promote pro-inflammatory cascades in microglia: Stronger effects of the A53T mutant. *PLoS One*, *11*, e0162717.

Hoffmann, A., Ettle, B., Bruno, A., Kulinich, A., Hoffmann, A. C., Wittgenstein, J. V., et al. (2016). Alpha-synuclein activates BV2 microglia dependent on its aggregation state. *Biochemical and Biophysical Research Communications*, *479*, 881–886.

Holmes, B. B., DeVos, S. L., Kfoury, N., Li, M., Jacks, R., Yanamandra, K., et al. (2013). Heparan sulfate proteoglycans mediate internalization and propagation of specific proteopathic seeds. *Proceedings of the National Academy of Sciences of the United States of America*, *110*, 3138–3147.

Hunot, S., Boissiere, F., Faucheux, B., Brugg, B., Mouatt-Prigent, A., Agid, Y., et al. (1996). Nitric oxide synthase and neuronal vulnerability in Parkinson's disease. *Neuroscience*, *72*, 355–363.

Ibanez, P., Bonnet, A. M., Debarges, B., Lohmann, E., Tison, F., Pollak, P., et al. (2004). Causal relation between alpha-synuclein gene duplication and familial Parkinson's disease. *Lancet*, *364*, 1169–1171.

Iranzo, A., Tolosa, E., Gelpi, E., Molinuevo, J. L., Valldeoriola, F., Serradell, M., et al. (2013). Neurodegenerative disease status and post-mortem pathology in idiopathic rapid-eye-movement sleep behaviour disorder: An observational cohort study. *The Lancet Neurology*, *12*, 442–453.

Iseki, E., Marui, W., Akiyama, H., Ueda, K., & Kosaka, K. (2000). Degeneration process of Lewy bodies in the brains of patients with dementia with Lewy bodies using alpha-synuclein-immunohistochemistry. *Neuroscience Letters*, *286*, 69–73.

Jankovic, J., & Aguilar, L. G. (2008). Current approaches to the treatment of Parkinson's disease. *Neuropsychiatric Disease and Treatment*, *4*, 743–757.

Jankovic, J., Goodman, I., & Safirstein, B. (2018). Safety and tolerability of multiple ascending doses of PRX002/RG7935, an anti–α-synuclein monoclonal antibody, in patients with Parkinson disease: A randomized clinical trial. *JAMA Neurology*, *75*, 1206–1214.

Jiang, T., Hoekstra, J., Heng, X., Kang, W., Ding, J., Liu, J., et al. (2015). P2X7 receptor is critical in α-synuclein-mediated microglial NADPH oxidase activation. *Neurobiology of Aging*, *36*, 2304–2318.

Kahle, P. J., Neumann, M., Ozmen, L., & Haass, C. (2000). Physiology and pathophysiology of alpha-synuclein. Cell culture and transgenic animal models based on a Parkinson's disease-associated protein. *Annals of the New York Academy of Sciences*, *920*, 33–41.

Kalia, L. V., & Lang, A. E. (2016). Parkinson's disease in 2015: Evolving basic, pathological and clinical concepts in PD. *Nature Reviews. Neurology*, *12*, 65–66.

Kannarkat, G. T., Boss, J. M., & Tansey, M. G. (2013). The role of innate and adaptive immunity in Parkinson's disease. *Journal of Parkinson's Disease*, *3*, 493–514.

Kannarkat, G. T., Cook, D. A., Lee, J.-K., Chang, J., Chung, J., Sandy, E., et al. (2015). Common genetic variant association with altered HLA expression, synergy with pyrethroid exposure, and risk for Parkinson's disease: An observational and case–control study. *NPJ Parkinsons Disease*, *1*, 15002.

Karpowicz, R. J., Jr., Haney, C. M., Mihaila, T. S., Sandler, R. M., Petersson, E. J., & Lee, V. M. Y. (2017). Selective imaging of internalized proteopathic α-synuclein seeds in primary neurons reveals mechanistic insight into transmission of synucleinopathies. *Journal of Biological Chemistry*, *292*, 13482–13497.

Karsarou, Z., Bostantjoupoulou, S., Hatzizisi, O., Giza, E., Soler-Cardona, A., & Kyriazis, G. (2007). Immune factors or depression? Fatigue correlates in Parkinson's disease. *Revista de Neurologia*, *45*, 725–728.

Kasten, M., Chade, A., & Tanner, C. M. (2007). Epidemiology of Parkinson's disease. *Handbook of Clinical Neurology*, *83*, 129–151.

Kasten, M., & Klein, C. (2013). The many faces of alpha-synuclein mutations. *Movement Disorders*, *28*, 697–701.

Kawamata, H., Mclean, P. J., Sharma, N., & Hyman, B. T. (2001). Interaction of alpha-synuclein and synphilin-1: Effect of Parkinson's disease-associated mutations. *Journal of Neurochemistry*, 77, 929–934.

Kim, C., Ho, D. H., Suk, J. E., You, S., Michael, S., Kang, J., et al. (2013). Neuron-released oligomeric α-synuclein is an endogenous agonist of TLR2 for paracrine activation of microglia. *Nature Communications*, *4*, 1562.

Kim, Y. S., Kim, S. S., Cho, J. J., Choi, D. H., Hwang, O., Shin, D. H., et al. (2005). Matrix metalloproteinase-3: A novel signaling proteinase from apoptotic neuronal cells that activates microglia. *The Journal of Neuroscience*, *25*, 3701–3711.

Kim, W. G., Mohney, R. P., Wilson, B., Jeohn, G. H., Liu, B., & Hong, J. S. (2000). Regional difference in susceptibility to lipopolysaccharide-induced neurotoxicity in the rat brain: Role of microglia. *The Journal of Neuroscience*, *20*, 6309–6316.

Kim, C., Spencer, B., Rockenstein, E., Yamakado, H., Mante, M., Adame, A., et al. (2018). Immunotherapy targeting toll-like receptor 2 alleviates neurodegeneration in models of synucleinopathy by modulating α-synuclein transmission and neuroinflammation. *Molecular Neurodegeneration*, *13*, 43.

Kipnis, J. (2016). Multifaceted interactions between adaptive immunity and the central nervous system. *Science*, *353*, 766–771.

Kirkley, K. S., Popichak, K. A., Hammond, S. L., Davies, C., Hunt, L., & Tjalkens, R. B. (2019). Genetic suppression of IKK2/NF-κB in astrocytes inhibits neuroinflammation and reduces neuronal loss in the MPTP-Probenecid model of Parkinson's disease. *Neurobiology of Disease, 127*, 193–209.

Kisos, H., Pukaβ, K., Ben-Hur, T., Richter-Landsberg, C., & Sharon, R. (2012). Increased neuronal α-synuclein pathology associates with its accumulation in oligodendrocytes in mice modeling α-synucleinopathies. *PLoS One*, 7, e46817.

Klegeris, A., Pelech, S., Giasson, B. I., Maguire, J., Zhang, H., McGeer, E. G., et al. (2008). Alpha-synuclein activates stress signaling protein kinases in THP-1 cells and microglia. *Neurobiology of Aging, 29*, 739–752.

Knott, C., Wilkin, G. P., & Stern, G. (1999). Astrocytes and microglia in the substantia nigra and caudate-putamen in Parkinson's disease. *Parkinsonism & Related Disorders, 5*, 115–122.

Koller, E. J., Brooks, M. M. T., Golde, T. E., Giasson, B. I., & Chakrabarty, P. (2017). Inflammatory pre-conditioning restricts the seeded induction of α-synuclein pathology in wild type mice. *Molecular Neurodegeneration, 12*, 1.

Kordower, J. H., Chu, Y., Hauser, R. A., Freeman, T. B., & Olanow, C. W. (2008). Lewy body-like pathology in long-term embryonic nigral transplants in Parkinson's disease. *Nature Medicine, 14*, 504–506.

Kordower, J. H., Chu, Y., Hauser, R. A., Olanow, C. W., & Freeman, T. B. (2008). Transplanted dopaminergic neurons develop PD pathologic changes: A second case report. *Movement Disorders, 23*, 2303–2306.

Korn, T., & Kallies, A. (2017). T-cell responses in the central nervous system. *Nature Reviews. Immunology, 17*, 179–194.

Kreutzberg, G. W. (1996). Microglia: A sensor for pathological events in the CNS. *Trends in Neurosciences, 19*, 312–318.

Kurowska, Z., Englund, E., Widner, H., Lindvall, O., Li, J. Y., & Brundin, P. (2011). Signs of degeneration in 12-22-year old grafts of mesencephalic dopamine neurons in patients with Parkinson's disease. *Journal of Parkinson's Disease, 1*, 83–92.

Kustrimovic, N., Rasini, E., Legnaro, M., Bombelli, R., Aleksic, I., & Blandini, F. (2016). Dopaminergic receptors on CD4+ T naive and memory lymphocytes correlate with motor impairment in patients with Parkinson's disease. *Scientific Reports, 6*, 33738.

Lafon, M., Megret, F., Lafage, M., & Prehaud, C. (2006). The innate immune facet of brain: Human neurons express TLR-3 and sense viral dsRNA. *Journal of Molecular Neuroscience, 29*, 185–194.

Lawson, L. J., Perry, V. H., Dri, P., & Gordon, S. (1990). Heterogeneity in the distribution and morphology of microglia in the normal adult mouse brain. *Neuroscience, 39*, 151–170.

Lebouvier, T., Chaumette, T., Paillusson, S., Duyckaerts, C., des Varannes, S. B., Neunlist, M., et al. (2009). The second brain and Parkinson's disease. *European Journal of Neuroscience, 30*, 735–741.

Lee, H. J., Khoshaghideh, F., Lee, S., & Lee, S. J. (2006). Impairment of microtubule-dependent trafficking by overexpression of alpha-synuclein. *European Journal of Neuroscience, 24*, 3153–3162.

Lee, H. J., Kim, C., & Lee, S. J. (2010). Alpha-synuclein stimulation of astrocytes: Potential role for neuroinflammation and neuroprotection. *Oxidative Medicine and Cellular Longevity, 3*, 283–287.

Lee, E. J., Woo, M. S., Moon, P. G., Baek, M. C., Choi, I. Y., Kim, W. K., et al. (2010). Alpha-synuclein activates microglia by inducing the expressions of matrix metalloproteinases and the subsequent activation of protease-activated receptor-1. *Journal of Immunology, 185*, 615–623.

Li, J. Y., Englund, E., Holton, J. L., Soulet, D., Hagell, P., Lees, A. J., et al. (2008). Lewy bodies in grafted neurons in subjects with Parkinson's disease suggest host-to-graft disease propagation. *Nature Medicine, 14*, 501–503.

Li, J. Y., Englund, E., Widner, H., Rehncrona, S., Björklund, A., Lindvall, O., et al. (2010). Characterization of Lewy body pathology in 12- and 16-year-old intrastriatal mesencephalic grafts surviving in a patient with Parkinson's disease. *Movement Disorders*, *25*, 1091–1096.

Liddlelow, S. A., Guttenplan, K. A., Clarke, L. E., Bennett, F. C., Bohlen, C. J., Schirmer, L., et al. (2017). Neurotoxic reactive astrocytes are induced by activated microglia. *Nature*, *541*, 481–487.

Lindström, V., Fagerqvist, T., Nordström, E., Eriksson, F., Lord, A., Tucker, S., et al. (2014). Immunotherapy targeting α-synuclein protofibrils reduced pathology in (Thy-1)-h [A30P] α-synuclein mice. *Neurobiology of Disease*, *69*, 134–143.

Lindström, V., Gustafsson, G., Sanders, L. H., Howlett, E. H., Sigvardson, J., Kasrayan, A., et al. (2017). Extensive uptake of α-synuclein oligomers in astrocytes results in sustained intracellular deposits and mitochondrial damage. *Molecular and Cellular Neuroscience*, *82*, 143–156.

Liu, Z., Huang, Y., Cao, B. B., Qiu, Y. H., & Peng, Y. P. (2017). Th17 cells induce dopaminergic neuronal death via LFA-1/ICAM-1 interaction in a mouse model of Parkinson's disease. *Molecular Neurobiology*, *54*, 7762–7776.

Lopez-Lopez, A., Labandeira, C. M., Labandeira-Garcia, J. L., & Munoz, A. (2020). Rho kinase inhibitor fasudil reduces l-DOPA-induced dyskinesia in a rat model of Parkinson's disease. *British Journal of Pharmacology*, *177*, 5622–5641.

Louveau, A., Herz, J., Alme, M. N., Salvador, A. F., Dong, M. Q., Viar, K. E., et al. (2018). CNS lymphatic drainage and neuroinflammation are regulated by meningeal lymphatic vasculature. *Nature Neuroscience*, *21*, 1380–1391.

Luk, K. C., Kehm, V., Carroll, J., Zhang, B., O'Brien, P., Trojanowski, J. Q., et al. (2012). Pathological α-synuclein transmission initiates Parkinson-like neurodegeneration in nontransgenic mice. *Science*, *338*, 949–953.

Luk, K. C., Song, C., O'Brien, P., Stieber, A., Branch, J. R., Brunden, K. R., et al. (2009). Exogenous alpha-synuclein fibrils seed the formation of Lewy body-like intracellular inclusions in cultured cells. *Proceedings of the National Academy of Sciences of the United States of America*, *106*, 20051–20056.

Luna, E., Decker, S. C., Riddle, D. M., Caputo, A., Zhang, B., Cole, T., et al. (2018). Differential α-synuclein expression contributes to selective vulnerability of hippocampal neuron subpopulations to fibril-induced toxicity. *Acta Neuropathologica*, *135*, 855–875.

Ma, L. Y., Liu, G. L., Wang, D. X., Zhang, M. M., Kou, W. Y., & Feng, T. (2019). Alpha-synuclein in peripheral tissues in Parkinson's disease. *ACS Chemical Neuroscience*, *10*, 812–823.

Majbour, N. K., Vaikath, N. N., Eusebi, P., Chiasserini, D., Ardah, M., Varghese, S., et al. (2016). Longitudinal changes in CSF alpha-synuclein species reflect Parkinson's disease progression. *Movement Disorders*, *31*, 1535–1542.

Manda, KM., Yedlapudi, D., Korukonda, S., Bojja, S., & Kalivendi, SV. (2014). The chaperone-like activity of α-synuclein attenuates aggregation of its alternatively spliced isoform, 112-synuclein in vitro: Plausible cross-talk between isoforms in protein aggregation. *PLoS One*, *9*, e98657.

Mandler, M., Valera, E., Rockenstein, E., Mante, M., Weninger, H., Patrick, C., et al. (2015). Active immunization against alpha-synuclein ameliorates the degenerative pathology and prevents demyelination in a model of multiple system atrophy. *Molecular Neurodegeneration*, *10*, 10.

Mandler, M., Valera, E., Rockenstein, E., Weninger, H., Patrick, C., Adame, A., et al. (2014). Next-generation active immunization approach for synucleinopathies: Implications for Parkinson's disease clinical trials. *Acta Neuropathologica*, *127*, 861–879.

Mao, X., Ou, M. T., Karuppagounder, S. S., Kam, T. I., Yin, X., Xiong, Y., et al. (2016). Pathological a-synuclein transmission initiated by binding lymphocyte-activation gene 3. *Science*, *353*, aahh374.

Maroteaux, L., & Scheller, R. H. (1991). The rat brain synucleins; family of proteins transiently associated with neuronal membrane. *Molecular Brain Research, 11*, 335–343.

Martinez-Martin, P., Rodriguez-Blasquez, C., Kurtis, M. M., & Ray Chaudhuri, K. (2011). The impact of non-motor symptoms on health-related quality of life of patients with Parkinson's disease. *Movement Disorders, 26*, 399–406.

Masliah, E., Rockenstein, E., Adame, A., Alford, M., Crews, L., Hashimoto, M., et al. (2005). Effects of alpha-synuclein immunization in a mouse model of Parkinson's disease. *Neuron, 46*, 857–868.

Masliah, E., Rockenstein, E., Mante, M., Crews, L., Spencer, B., Adame, A., et al. (2011). Passive immunization reduces behavioral and neuropathological deficits in an alpha-synuclein transgenic model of Lewy body disease. *PLoS One, 6*, e19338.

Masliah, E., Rockenstein, E., Veinbergs, I., Mallory, M., Hashimoto, M., Takeda, A., et al. (2000). Dopaminergic loss and inclusion body formation in alpha-synuclein mice: Implications for neurodegenerative disorders. *Science, 287*, 1265–1269.

Masuda-Suzukake, M., Nonaka, T., Hosokawa, M., Oikawa, T., Arai, T., Akiyama, H., et al. (2013). Prion-like spreading of pathological α-synuclein in brain. *Brain, 136*, 1128–1138.

McCoy, M. K., Martinez, T. N., Ruhn, K. A., Szymkowski, D. E., Smith, C. G., Botterman, B. R., et al. (2006). Blocking soluble tumor necrosis factor signaling with dominant-negative tumor necrosis factor inhibitor attenuates loss of dopaminergic neurons in models of Parkinson's disease. *Journal of Neuroscience, 26*, 9365–9375.

McGeer, P. L., Itagaki, S., Boyes, B. E., & McGeer, E. G. (1998). Reactive microglia are positive for HLA-DR in the substantia nigra of Parkinson's and Alzheimer's disease brains. *Neurology, 38*, 1285–1291.

McGeer, P. L., Itagaki, S., Tago, H., & McGeer, E. G. (1988). Occurrence of HLA-DR reactive microglia in Alzheimer's disease. *Annals of the New York Academy of Sciences, 540*, 319–323.

McRae-Degueurce, A., Rosengren, L., Haglid, K., Booj, S., Gottfries, C. G., Granerus, A. C., et al. (1988). Immunocytochemical investigations on the presence of neuron-specific antibodies in the CSF of Parkinson's disease cases. *Neurochemical Research, 13*, 679–684.

Miklossy, J., Doudet, D. D., Schwab, C., Yu, S., McGeer, E. G., & McGeer, P. L. (2019). Role of ICAM-1 in persisting inflammation in Parkinson disease and MPTP monkeys. *Experimental Neurology, 197*, 275–283.

Miller, A. P., Shah, A. S., Aperi, B. V., Kurpad, S. N., Stemper, B. D., & Glavaski-Joksimovic, A. (2017). Acute death of astrocytes in blast-exposed rat organotypic hippocampal slice cultures. *PLoS One, 12*(3), e0173167.

Mirza, B., Hadberg, H., Thomsen, P., & Moos, T. (2000). The absence of reactive astrocytosis is indicative of a unique inflammatory process in Parkinson's disease. *Neuroscience, 95*, 425–432.

Mittelbronn, M., Dietz, K., Schluesener, H. J., & Meyermann, R. (2001). Local distribution of microglia in the normal adult central nervous system differs by up to one order of magnitude. *Acta Neuropathologica, 101*, 249–255.

Mogi, M., Harada, M., Kondo, T., Riederer, P., Inagaki, H., Minami, M., et al. (1994). Interleukin-1 beta, interleukin-6, epidermal growth factor and transforming growth factor-alpha are elevated in the brain from parkinsonian patients. *Neuroscience Letters, 180*, 147–150.

Mogi, M., Harada, M., Riederer, P., Narabayashi, H., Fujita, K., & Nagatsu, T. (1994). Tumor necrosis factor-alpha (TNF-alpha) increases both in the brain and in the cerebrospinal fluid from parkinsonian patients. *Neuroscience Letters, 165*, 208–210.

Moller, T., Hanisch, U. K., & Ransom, B. R. (2000). Thrombin-induced activation of cultured rodent microglia. *Journal of Neurochemistry, 75*, 1539–1547.

Montoya, A., Elgueta, D., Campos, J., Chovar, O., Falcon, P., Matus, S., et al. (2019). Dopamine receptor D3 signalling in astrocytes promotes neuroinflammation. *Journal of Neuroinflammation, 16*, 258.

Motyl, J., Przykaza, L., Boguszewski, P. M., Kosson, P., & Strosznajder, J. B. (2018). Pramipexole and fingolimod exert neuroprotection in a mouse model of Parkinson's disease by activation of sphingosine kinase 1 and Akt kinase. *Neuropharmacology, 135*, 139–150.

Nagatsu, T, Mogi, M, Ichinose, H, & Togari, A. (2000). Cytokines in Parkinson's disease. *Journal of Neural Transmission Supplementum, 58*, 143–151.

Nair, A. T., Ramachandran, V., Joghee, N. M., Antony, S., & Ramalingam, G. (2018). Gut microbiota dysfunction as reliable non-invasive early diagnostic biomarkers in the pathophysiology of Parkinson's disease: A critical review. *Journal of Neurogastroenterology and Motility, 24*, 30–42.

Nalls, M. A., Blauwendraat, C., Vallerga, C. L., Heilbron, K., Bandres-Ciga, S., Chang, D., et al. (2019). Identification of novel risk loci, causal insights, and heritable risk for Parkinson's disease: A meta-analysis of genome-wide association studies. *Lancet Neurology, 18*, 1091–1102.

Neumann, H., Cavalié, A., Jenne, D. E., & Wekerle, H. (1995). Induction of MHC class I genes in neurons. *Science, 28, 269*(5223), 549–552.

Nichols, W. C., Pankratz, N., Marek, D. K., Pauciulo, M. W., Elsaesser, V. E., Halter, C. A., et al. (2009). Mutations in GBA are associated with familial Parkinson disease susceptibility and age at onset. *Neurology, 72*, 310–316.

Nicoletti, A., Fagone, P., Donzusco, G., Magano, K., Dibilio, V., Caponnetto, S., et al. (2011). Parkinson's disease is associated with increased serum levels of macrophage migration inhibitory factor. *Cytokine, 55*, 165–167.

NINDS Exploratory Trials in Parkinson Disease (NET-PD) FS-ZONE Investigators. (2015). Pioglitazone in early Parkinson's disease: A phase 2, multicentre, double-blind, randomised trial. *Lancet Neurology, 14*, 795–803.

Nolte, C., Moller, T., Walter, T., & Kettenmann, H. (1996). Complement 5a controls motility of murine microglial cells in vitro via activation of an inhibitory G-protein and the rearrangement of the actin cytoskeleton. *Neuroscience, 73*, 1091–1107.

Nonaka, T., Watanabe, S. T., Iwatsubo, T., & Hasegawa, M. (2010). Seeded aggregation and toxicity of {alpha}-synuclein and tau: Cellular models of neurodegenerative diseases. *Journal of Biological Chemistry, 285*, 34885–34898.

Nussbaum, R. L., & Ellis, C. E. (2003). Alzheimer's disease and Parkinson's disease. *New England Journal of Medicine, 348*, 1356–1364.

Olson, J. K., & Miller, S. D. (2004). Microglia initiate central nervous system innate and adaptive immune responses through multiple TLRs. *Journal of Immunology, 173*, 3916–3924.

Orr, C. F., Rowe, D. B., Mizuno, Y., Mori, H., & Halliday, G. M. (2005). A possible role for humoral immunity in the pathogenesis of Parkinson's disease. *Brain, 128*, 2665–2674.

Panicker, N., Sarkar, S., Harischandra, D. S., Neal, M., Kam, T. I., Jin, H., et al. (2019). Fyn kinase regulates misfolded α-synuclein uptake and NLRP3 inflammasome activation in microglia. *Journal of Experimental Medicine, 216*, 1411–1430.

Paolicelli, R. C., Bolasco, G., Pagani, F., Maggi, L., Scianni, M., Panzanelli, P., et al. (2011). Synaptic pruning by microglia is necessary for normal brain development. *Science, 333*, 1456–1458.

Park, J. Y., Paik, S. R., Jou, I., & Park, S. M. (2008). Microglial phagocytosis is enhanced by monomeric alpha-synuclein, not aggregated alpha-synuclein: Implications for Parkinson's disease. *Glia, 56*, 1215–1223.

Paumier, K. L., Luk, K. C., Manfredsson, F. P., Kanaan, N. M., Lipton, J. W., Collier, T. J., et al. (2015). Intrastriatal injection of pre-formed mouse α-synuclein fibrils into rats triggers α-synuclein pathology and bilateral nigrostriatal degeneration. *Neurobiology of Disease, 82*, 185–199.

Peng, C., Gathagan, R. J., Covell, D. J., Medellin, C., Stieber, A., Robinson, J. L., et al. (2018). Cellular milieu imparts distinct pathological α-synuclein strains in α-synucleinopathies. *Nature*, *557*, 558–563.

Pereira, J. R., dos Santos, L. V., Silva Santos, R. M., Figueiredo Campos, A. L., Pimenta, A. L., de Oliveira, M. S., et al. (2016). IL-6 serum levels are elevated in Parkinson's disease patients with fatigue compared to patients without fatigue. *Journal of the Neurological Sciences*, *370*, 153–156.

Perez, R. G., Waymire, J. C., Lin, E., Liu, J. J., Guo, F., & Zigmond, M. J. (2002). A role for alpha-synuclein in the regulation of dopamine biosynthesis. *Journal of Neuroscience*, *22*, 3090–3099.

Poewe, W., Volc, D., Seppi, K., Medori, R., Luhrs, P., Kutzelnigg, A., et al. (2021). Safety and tolerability of active immunotherapy targeting α-synuclein with PD03A in patients with early Parkinson's disease: A randomized, placebo-controlled, phase 1 study. *Journal of Parkinson's Disease*, *11*, 1079–1089.

Postuma, R. B., Berg, D., Stern, M., Poewe, W., Olanow, C. W., Oertel, W., et al. (2015). MDS clinical diagnostic criteria for Parkinson's disease. *Movement Disorders*, *30*, 1591–1601.

Qin, L., Wu, X., Block, M. L., Liu, Y., Breese, G. R., Hong, J. S., et al. (2007). Systemic LPS causes chronic neuroinflammation and progressive neurodegeneration. *Glia*, *55*, 453–462.

Qualman, S. J., Haupt, H. M., Yang, P., & Hamilton, S. R. (1984). Esophageal Lewy bodies associated with ganglion cell loss in achalasia. Similarity to Parkinson's disease. *Gastroenterology*, *87*, 848–856.

Qureshi, H. Y., & Paudel, H. K. (2011). Parkinsonian neurotoxin 1-methyl-4-phenyl-1,2,3,6-tetrahydropyridine (MPTP) and alpha-synuclein mutations promote Tau protein phosphorylation at Ser262 and destabilize microtubule cytoskeleton in vitro. *Journal of Biological Chemistry*, *286*, 5055–5068.

Rajput, A. H., Sitte, H. H., Rajput, A., Fenton, M. E., Pifl, C., & Hornykiewicz, O. (2008). Globus pallidus dopamine and Parkinson motor subtypes: Clinical and brain biochemical correlation. *Neurology*, *70*, 6.

Rannikko, E. H., Weber, S. S., & Kahle, P. J. (2015). Exogenous α-synuclein induces toll-like receptor 4 dependent inflammatory responses in astrocytes. *BMC Neuroscience*, *16*, 57.

Revuelta, M., Venero, J. L., Machado, A., & Cano, J. (1999). Serotonin hyperinnervation in the adult rat ventral mesencephalon following unilateral transection of the medial forebrain bundle. Correlation with reactive microglial and astroglial populations. *Neuroscience*, *91*, 567–577.

Rey, N. R., George, S., Steiner, J. A., Madaj, Z., Luk, K. C., Trojanowski, J. Q., et al. (2018). Spread of aggregates after olfactory bulb injection of α-synuclein fibrils is associated with early neuronal loss and is reduced long term. *Acta Neuropathologica*, *135*, 65–83.

Reyes, J. F., Sackmann, C., Hoffmann, A., Svenningsson, P., Winkler, J., Ingelsson, M., et al. (2019). Binding of α-synuclein oligomers to Cx32 facilitates protein uptake and transfer in neurons and oligodendrocytes. *Acta Neuropathologica*, *138*, 23–47.

Ribeiro, C. S., Carneiro, K., Ross, C. A., Menezes, J. R. L., & Engelender, S. (2002). Synphilin-1 is developmentally localized to synaptic terminals, and its association with synaptic vesicles is modulated by alpha-synuclein. *Journal of Biological Chemistry*, *277*, 23927–23933.

Roberts, H. L., & Brown, D. R. (2015). Seeking a mechanism for the toxicity of oligomeric α-synuclein. *Biomolecules*, *5*, 282–305.

Rocha, N. P., Assis, F., Scalzo, P. L., Vieira, E. L. M., Barbosa, I. G., de Souza, M. S., et al. (2018). Reduced activated T lymphocytes (CD4+CD25+) and plasma levels of cytokines in Parkinson's disease. *Molecular Neurobiology*, *55*, 1488–1497.

Rodrigues, R. W., Gomide, V. C., & Chadi, G. (2001). Astroglial and microglial reaction after a partial nigrostriatal degeneration induced by the striatal injection of different doses of 6-hydroxydopamine. *International Journal of Neuroscience*, *109*, 91–126.

Rodriguez, L., Marano, M. M., & Tandon, A. (2018). Import and export of misfolded α-synuclein. *Frontiers in Neuroscience*, *12*, 344.

Roser, A. E., Tönges, L., & Lingor, P. (2017). Modulation of microglial activity by Rho-kinase (ROCK) inhibition as therapeutic strategy in Parkinson's disease and amyotrophic lateral sclerosis. *Frontiers in Aging Neuroscience*, *9*, 94.

Ross, O. A., Braithwaite, A. T., Skipper, L. M., Kachergus, J., Hulihan, M. M., Middleton, F. A., et al. (2008). Genomic investigation of alpha-synuclein multiplication and parkinsonism. *Annals of Neurology*, *63*, 743–750.

Rostami, J., Holmqvist, S., Lindström, V., Sigvardson, J., Westermark, G. T., Ingelsson, M., et al. (2017). Human astrocytes transfer aggregated alpha-synuclein via tunneling nanotubes. *Journal of Neuroscience*, *37*, 11835–11853.

Sacino, A. N., Brooks, M. M., Chakrabarty, P., Saha, K., Khoshbouei, H., Golde, T. E., et al. (2017). Proteolysis of α-synuclein fibrils in the lysosomal pathway limits induction of inclusion pathology. *Journal of Neurochemistry*, *140*, 662–678.

Saijo, K., Winner, B., Carson, C. T., Collier, J. G., Boyer, L., Rosenfeld, M. G., et al. (2009). A Nurr1/CoREST pathway in microglia and astrocytes protects dopaminergic neurons from inflammation-induced death. *Cell*, *137*, 47–59.

Sanchez-Ferro, A., Rabanno, A., Jose Catalan, M., Rodriguez-Valcarcel, F. C., Diez, S. F., Herreros-Rodriguez, J., et al. (2014). In vivo gastric detection of α-synuclein inclusions in Parkinson's disease. *Movement Disorders*, *30*, 517–524.

Sanchez-Guajardo, V., Annibali, A., Jensen, P. H., & Romero-Ramos, M. (2013). α-Synuclein vaccination prevents the accumulation of parkinson disease-like pathologic inclusions in striatum in association with regulatory T cell recruitment in a rat model. *Journal of Neuropathology & Experimental Neurology*, *72*, 624–645.

Sanchez-Guajardo, V., Tentillier, N., & Romero-Ramos, M. (2015). The relation between α-synuclein and microglia in Parkinson's disease: Recent developments. *Neuroscience*, *302*, 47–58.

Sardi, S. P., Cedarbaum, J. M., & Brundin, P. (2018). Targeted therapies for Parkinson's disease: From genetics to the clinic. *Movement Disorders*, *33*, 684–696.

Saunders, J. A., Estes, K. A., Kosloski, L. M., Allen, H. E., Dempsey, K. M., & Torres Russotto, D. R. (2012). CD4+ regulatory and effector/memory T cell subsets profile motor dysfunction in Parkinson's disease. *Journal of Neuroimmune Pharmacology: The Official Journal of the Society on NeuroImmune Pharmacology*, 7, 927–938.

Schenk, D. B., Koller, M., Ness, D. K., Griffith, S. G., Grundman, M., et al. (2016). First-in-human assessment of PRX002, an anti-α-synuclein monoclonal antibody, in healthy volunteers. *Movement Disorders*, *32*, 211–218.

Scheperjans, F., Aho, V., Pereira, P. A. B., Koshiken, K., Paulin, L., Pekkonen, E., et al. (2015). Gut microbiota are related to Parkinson's disease and clinical phenotype. *Movement Disorders*, *30*, 350–358.

Shahaduzzaman, M., Nash, K., Hudson, C., Sharif, M., Grimmig, B., Lin, X., et al. (2015). Anti-human α-synuclein N-terminal peptide antibody protects against dopaminergic cell death and ameliorates behavioral deficits in an AAV-α-synuclein rat model of Parkinson's disease. *PLoS One*, *10*, e0116841.

Shannon, K. M., Keshavarzian, A., Dodiya, H. B., Jakate, S., & Kordower, J. H. (2012). Is alpha-synuclein in the colon a biomarker for premotor Parkinson's disease? Evidence from 3 cases. *Movement Disorders*, *27*, 716–719.

Shannon, K. M., Keshavarzian, A., Mutlu, E., Dodiya, H. B., Daian, D., Jaglin, J. A., et al. (2012). Alpha-synuclein in colonic submucosa in early untreated Parkinson's disease. *Movement Disorders*, *27*, 709–715.

Shen, Y., Halperin, J. A., & Lee, C. M. (1995). Complement-mediated neurotoxicity is regulated by homologous restriction. *Brain Research, 671*, 282–292.

Shrivastava, A. N., Redeker, V., Fritz, N., Lieri, N., Almeida, L. G., Spolidoro, M., et al. (2015). α-Synuclein assemblies sequester neuronal α3-Na+/K+-ATPase and impair Na+ gradient. *EMBO Journal, 34*, 2408–2423.

Singleton, A. B., Farrer, M., Johnson, J., Singleton, A., Hague, S., Kachergus, J., et al. (2003). Alpha-synuclein locus triplication causes Parkinson's disease. *Science, 302*, 841.

Smeyne, R. J., Breckenridge, C. B., Beck, M., Jiao, Y., Butt, M. T., Wolf, J. C., et al. (2016). Assessment of the effects of MPTP and paraquat on dopaminergic neurons and microglia in the substantia nigra pars compacta of C57BL/6 mice. *PLoS One, 11*, e0164094.

Sofroniew, M. V., & Vinters, H. V. (2010). Astrocytes: Biology and pathology. *Acta Neuropathologica, 119*, 7–35.

Sommer, A., Maxreiter, F., Krach, F., Fadler, T., Grosch, J., Maroni, M., et al. (2018). Th17 lymphocytes induce neuronal cell death in a human iPSC-based model of Parkinson's disease. *Cell Stem Cell, 23*(123–31), e6.

Song, Y. J. C., Halliday, G. M., Holton, J. L., Lashley, T., O'Sullivan, S. S., McCann, H., et al. (2009). Degeneration in different parkinsonian syndromes relates to astrocyte type and astrocyte protein expression. *Journal of Neuropathology & Experimental Neurology, 68*, 1073–1083.

Song, L., Pei, L., Yao, S., Wu, Y., & Shang, Y. (2017). NLRP3 inflammasome in neurological diseases, from functions to therapies. *Frontiers in Cellular Neuroscience, 11*, 63.

Spencer, B., Emadi, S., Desplats, P., Eleuteri, S., Michael, S., Kosberg, K., et al. (2014). ESCRT-mediated uptake and degradation of brain-targeted α-synuclein single chain antibody attenuates neuronal degeneration in vivo. *Molecular Therapeutics, 22*, 1753–1767.

Spillantini, M. G., Schmidt, M. L., Lee, L. M. Y., Trojanowski, J. Q., Jakes, R., & Goedert, M. (1997). α-Synuclein in Lewy bodies. *Nature, 388*, 839–840.

Stefanis, L. (2012). α-Synuclein in Parkinson's disease. *Cold Spring Harbor Perspectives in Medicine, 2*, a009399.

Steiner, J. A., Quansah, E., & Brundin, P. (2018). The concept of alpha-synuclein as a prion-like protein: Ten years after. *Cell and Tissue Research, 373*, 161–173.

Su, X., Maguire-Zeiss, K. A., Giuliano, R., Prifti, L., Venkatesh, K., & Federoff, H. J. (2008). Synuclein activates microglia in a model of Parkinson's disease. *Neurobiology of Aging, 29*, 1690–1701.

Subbarayan, MS., Hudson, C., Moss, LD., Nash, KR., & Bickford, PC. (2020). T cell infiltration and upregulation of MHCII in microglia leads to accelerated neuronal loss in an α-synuclein rat model of Parkinson's disease. *Journal of Neuroinflammation, 17*, 242.

Subhramaniam, S. R., & Federoff, H. J. (2017). Targeting microglial activation states as a therapeutic avenue in Parkinson's disease. *Frontiers in Aging Neuroscience, 9*, 176.

Sugama, S., Yang, L., Cho, B. P., DeGiorgio, L. A., Lorenzl, S., Albers, D. S., et al. (2003). Age-related microglial activation in 1-methyl-4-phenyl-1,2,3,6- tetrahydropyridine (MPTP)-induced dopaminergic neurodegeneration in C57BL/6 mice. *Brain Research, 964*, 288–294.

Sulzer, D., Alcalay, R. N., Garretti, F., Cote, L., Kanter, E., Agin-Liebes, J., et al. (2017). T cells from patients with Parkinson's disease recognize α-synuclein peptides. *Nature, 546*, 656–661.

Sun, H., Liang, R., Yang, B., Zhou, Y., Liu, M., Fang, F., et al. (2016). Aquaporin-4 mediates communication between astrocyte and microglia: Implications of neuroinflammation in experimental Parkinson's disease. *Neuroscience, 317*, 65–75.

Surmeier, D. J. (2018). Determinants of dopaminergic neuron loss in Parkinson's disease. *The FEBS Journal, 285*, 3657–3668.

Suzumura, A., Sawada, M., & Takayanagi, T. (1998). Production of interleukin-12 and expression of its receptors by murine microglia. *Brain Research*, *787*, 139–142.
Takahashi, H., & Wakabayashi, K. (2001). The cellular pathology of Parkinson's disease. *Neuropathology*, *21*, 315–322.
Tamburrino, A., Churchill, M. J., Wan, O. W., Colino-Sanguino, Y., Ippolito, S., Bergstrand, S., et al. (2015). Cyclosporin promotes neurorestoration and cell replacement therapy in pre-clinical models of Parkinson's disease. *Acta Neuropathologica Communications*, *3*, 84.
Tan, J., Town, T., Mori, T., Wu, Y., Saxe, M., Crawford, F., et al. (2000). CD45 opposes beta-amyloid peptide-induced microglial activation via inhibition of p44/42 mitogen activated protein kinase. *Journal of Neuroscience*, *20*, 7587–7594.
Tansey, M. G., & Goldberg, M. S. (2010). Neuroinflammation in Parkinson's disease: Its role in neuronal death and implications for therapeutic intervention. *Neurobiology of Disease*, *37*, 510–518.
Theillet, F. X., Binolfi, A., Bekei, B., Martorana, A., Rose, H. M., Stuiver, M., et al. (2016). Structural disorder of monomeric α-synuclein persists in mammalian cells. *Nature*, *530*, 45–50.
Theodore, S., Cao, S., McLean, P. J., & Standaert, D. G. (2008). Targeted overexpression of human alpha-synuclein triggers microglial activation and an adaptive immune response in a mouse model of Parkinson disease. *Journal of Neuropathology & Experimental Neurology*, *67*, 1149–1158.
Tilleux, S., & Hermans, E. (2007). Neuroinflammation and regulation of glial glutamate uptake in neurological disorders. *Journal of Neuroscience Research*, *85*, 2059–2070.
Tokuda, T., Qureshi, M. M., Ardah, M. T., Varghese, S., Shehab, S. A. S., Kasai, T., et al. (2010). Detection of elevated levels of α-synuclein oligomers in CSF from patients with Parkinson disease. *Neurology*, *75*, 1766–1772.
Tong, J., Ang, L. C., Williams, B., Furukawa, Y., Fitzmaurice, P., Guttman, M., et al. (2015). Low levels of astroglial markers in Parkinson's disease: Relationship to α-synuclein accumulation. *Neurobiology of Disease*, *82*, 243–253.
Tran, H. T., Chung, C. H. Y., Iba, M., Zhang, B., Trojanowski, J. Q., Luk, K. C., et al. (2014). α-Synuclein immunotherapy blocks uptake and templated propagation of misfolded α-synuclein and neurodegeneration. *Cell Reports*, *7*, 2054–2065.
Tsirka, S. E. (1997). Clinical implications of the involvement of tPA in neuronal cell death. *Journal of Molecular Medicine*, *75*, 341–347.
Ueki, A., & Otsuka, M. (2004). Life style risks of Parkinson's disease: Association between decreased water intake and constipation. *Journal of Neurology*, *251*, 18–23.
Van der Perren, A., Macchi, F., Toelen, J., Carlon, M. S., Maris, M., de Loor, H., et al. (2015). FK506 reduces neuroinflammation and dopaminergic neurodegeneration in an α-synuclein-based rat model for Parkinson's disease. *Neurobiology of Disease*, *36*, 1559–1568.
Volc, D., Poewe, W., Kutzelnigg, A., Luhrs, P., Thun-Hohenstein, C., Schneeberger, A., et al. (2020). Safety and immunogenicity of the α-synuclein active immunotherapeutic PD01A in patients with Parkinson's disease: A randomised, single-blinded, phase 1 trial. *Lancet Neurology*, *19*, 591–600.
Volpicelli-Daley, L. A., Luk, K. C., & Lee, V. M. Y. (2014). Addition of exogenous α--synuclein preformed fibrils to primary neuronal cultures to seed recruitment of endogenous α-synuclein to Lewy body and Lewy neurite-like aggregates. *Nature Protocols*, *9*, 2135–2146.
Volpicelli-Daley, L. A., Luk, K. C., Patel, T. P., Tanik, S. A., Riddle, D. M., Stieber, A., et al. (2011). Exogenous α-synuclein fibrils induce Lewy body pathology leading to synaptic dysfunction and neuron death. *Neuron*, *72*, 57–71.

Wakabayashi, K., Takahashi, H., Ohama, E., & Ikuta, F. (1990). Parkinson's disease: An immunohistochemical study of Lewy body-containing neurons in the enteric nervous system. *Acta Neuropathologica*, *79*, 581–583.

Wang, W., Perovic, I., Chittuluru, J., Kaganovich, A., Nguyen, L. T. T., Liao, J., et al. (2011). A soluble α-synuclein construct forms a dynamic tetramer. *Proceedings of the National Academy of Sciences of the United States of America*, *108*, 17797–17802.

Wang, X. J., Yan, Z. Q., Lu, G. Q., Stuart, S., & Chen, S. D. (2007). Parkinson disease IgG and C5a-induced synergistic dopaminergic neurotoxicity: Role of microglia. *Neurochemistry International*, *50*, 39–50.

Wang, J., Zheng, B., Yang, S., Tang, X., Wang, J., & Wei, D. (2020). The protective effects of phoenixin-14 against lipopolysaccharide-induced inflammation and inflammasome activation in astrocytes. *Inflammation Research*, *69*, 779–787.

Weltzien, R. B., & Pachter, J. S. (2000). Visualization of beta-amyloid peptide (Abeta) phagocytosis by human mononuclear phagocytes: Dependency on Aβ aggregate size. *Journal of Neuroscience Research*, *59*, 522–527.

Wilton, D. K., Dissing-Olesen, L., & Stevens, B. (2019). Neuron-glia signaling in synapse elimination. *Annual Review of Neuroscience*, *42*, 102–107.

Winner, B., Jappelli, R., Maji, S. K., Desplats, P. A., Boyer, L., Aigner, S., et al. (2011). In vivo demonstration that alpha-synuclein oligomers are toxic. *Proceedings of the National Academy of Sciences of the United States of America*, *108*, 4194–4199.

Wong, K. L., Yeap, W. H., Tai, J. J. Y., Ong, S. M., Dang, T. M., & Wong, S. C. (2012). The three human monocyte subsets: Implications for health and disease. *Immunologic Research*, *53*, 41–57.

Wood, H. (2016). LAG3 facilitates cell-to-cell spread of α-synuclein pathology. *Nature Reviews. Neurology*, *12*, 678.

Yao, J., Harvath, L., Gilbert, D. L., & Colton, C. A. (1990). Chemotaxis by a CNS macrophage, the microglia. *Journal of Neuroscience Research*, *27*, 36–42.

Yi, J. H., & Hazell, A. (2006). Excitotoxic mechanisms and the role of astrocytic glutamate transporters in traumatic brain injury. *Neurochemistry International*, *48*, 394–403.

Yun, S. P., Kam, T. I., Panicker, N., Kim, S. M., Oh, Y., Park, J. S., et al. (2018). Block of A1 astrocyte conversion by microglia is neuroprotective in models of Parkinson's disease. *Nature Medicine*, *24*, 931–938.

Zhang, Y., Chen, K., Sloan, S. A., Bennett, M. L., Scholze, A. R., O'Keeffe, S., et al. (2014). An RNA-sequencing transcriptome and splicing database of glia, neurons, and vascular cells of the cerebral cortex. *Journal of Neuroscience*, *34*, 11929–11947.

Zhang, W., Dallas, S., Zhang, D., Guo, J. P., Pang, H., Wilson, B., et al. (2007). Microglial PHOX and Mac-1 are essential to the enhanced dopaminergic neurodegeneration elicited by A30P and A53T mutant alpha-synuclein. *Glia*, *55*, 1178–1188.

Zhang, Y., Reichel, J. M., Han, C., Zuniga-Hertz, J. P., & Cai, D. (2017). Astrocytic process plasticity and IKKb/NF-kB in central control of blood glucose, blood pressure, and body weight. *Cell Metabolism*, *25*, 1091–1102.

Zhang, P., Shao, X. Y., Qi, G. J., Chen, Q., Bu, L. L., Chen, L. J., et al. (2016). Cdk5-dependent activation of neuronal inflammasomes in Parkinson's disease. *Movement Disorders*, *31*, 366–376.

Zhang, Y., Sloan, S. A., Clarke, L. E., Caneda, C., Plaza, C. A., Blumenthal, P. D., et al. (2016). Purification and characterization of progenitor and mature human astrocytes reveals transcriptional and functional differences with mouse. *Neuron*, *89*, 37–53.

Zhang, W., Wang, T., Pei, Z., Miller, D. S., Wu, X., Block, M. L., et al. (2005). Aggregated α-synuclein activates microglia: A process leading to disease progression in Parkinson's disease. *The FASEB Journal*, *19*, 533–542.

Zhang, S., Wu, M., Peng, C., Zhao, G., & Gu, R. (2017). GFAP expression in injured astrocytes in rats. *Experimental and Therapeutic Medicine*, *14*, 1905–1908.

Zhao, P., Yang, X., Yang, L., Li, M., Wood, K., Liu, Q., et al. (2017). Neuroprotective effects of fingolimod in mouse models of Parkinson's disease. *The FASEB Journal*, *31*, 172–179.

Zhu, J., Hu, Z., Han, X., Wang, D., Jiang, Q., Ding, J., et al. (2018). Dopamine D2 receptor restricts astrocytic NLRP3 inflammasome activation via enhancing the interaction of β-arrestin2 and NLRP3. *Cell Death and Differentiation*, *25*, 2037–2049.

Zhu, H., Lemos, H., Bhatt, B., Islam, B. N., Singh, A., Gurav, A., et al. (2017). Carbidopa, a drug in use for management of Parkinson disease inhibits T cell activation and autoimmunity. *PLoS One*, *12*, e0183484.

Zhu, J., & Paul, W. E. (2010). Heterogeneity and plasticity of T helper cells. *Cell Research*, *20*, 4–12.

Zondler, L., Miller-Fleming, L., Repici, M., Goncalves, S., Tenreiro, S., Rosado-Ramos, R., et al. (2014). DJ-1 interactions with α-synuclein attenuate aggregation and cellular toxicity in models of Parkinson's disease. *Cell Death & Disease*, *5*, e1350.

Zou, J. Y., & Crews, F. T. (2005). TNFα potentiates glutamate neurotoxicity by inhibiting glutamate uptake in organotypic brain slice cultures: Neuroprotection by NFκB inhibition. *Brain Research*, *1034*, 11–24.

CHAPTER ELEVEN

Antivenom: An immunotherapy for the treatment of snakebite envenoming in sub-Saharan Africa

Mender M. Mender[a,b,*], Fiona Bolton[b], Colin Berry[a], and Mark Young[a]
[a]School of Bioscience, Cardiff University, Cardiff, United Kingdom
[b]Department of Research and Development, MicroPharm Ltd, Newcastle Emlyn, United Kingdom
*Corresponding author: e-mail address: mender.m.mender@micropharm.co.uk

Contents

Advances in Protein Chemistry and Structural Biology, Volume 129
ISSN 1876-1623
https://doi.org/10.1016/bs.apcsb.2021.11.004

Abstract

Snakebite envenoming (SBE) leads to significant morbidity and mortality, resulting in over 90,000 deaths and approximately 400,000 amputations annually. In sub-Saharan Africa (SSA) alone, SBE accounts for over 30,000 deaths *per annum*. Since 2017, SBE has been classified as a priority Neglected Tropical Disease (NTD) by the World Health Organisation (WHO).

The major species responsible for mortality from SBE within SSA are from the *Bitis, Dendroaspis, Echis* and *Naja* genera. Pharmacologically active toxins such as metalloproteinases, serine proteinases, 3-finger toxins, kunitz-type toxins, and phospholipase A_2s are the primary snake venom components. These toxins induce cytotoxicity, coagulopathy, hemorrhage, and neurotoxicity in envenomed victims. Antivenom is currently the only available venom-specific treatment for SBE and contains purified equine or ovine polyclonal antibodies, collected from donor animals repeatedly immunized with low doses of adjuvanted venom. The resulting plasma or serum contains a high titre of specific antibodies, which can then be collected and stored until required. The purified antibodies are either whole IgG, monovalent fragment antibody (Fab) or divalent fragment antibody $F(ab')_2$. Despite pharmacokinetic and pharmacodynamic differences, all three are effective in the treatment of SBE. No antivenom is without adverse reactions but, the level of their impact and severity varies from benign early adverse reactions to the rarely occurring fatal anaphylactic shock. However, the major side effects are largely reversible with immediate administration of adrenaline and corticosteroids. There are 16 different antivenoms marketed within SSA, but the efficacy and safety profiles are only published for less than 50% of these products.

1. Snakebite envenoming

SBE is an endemic and debilitating condition primarily affecting the remote subsistence farming communities of Africa, Asia, and Latin America (Kasturiratne et al., 2008). If untreated, there is significant morbidity and mortality and, depending on venom toxicity, death can occur within 1–25 h after envenoming (Baldé, Chippaux, Boiro, Stock, & Massougbodji, 2013). The WHO classified SBE as a Category A NTD in 2017 (Chippaux, 2017).

1.1 Geographical distribution and dietary acquisition of snakes

Venomous snakes occupy virtually all ecological niches (Vidal et al., 2007). Snakes such as *Bitis arietans, Bitis. gabonica, Echis leucogaster, Echis ocellatus, Naja haje, Naja nigricollis, Naja melanoleuca, Dendroaspis jamesoni, Dendroaspis polylepis* and *Dendroaspis viridis* are abundant in tropical and SSA (Tasoulis & Isbister, 2017). As shown in Fig. 1, these snakes occupy different regions of the African continent.

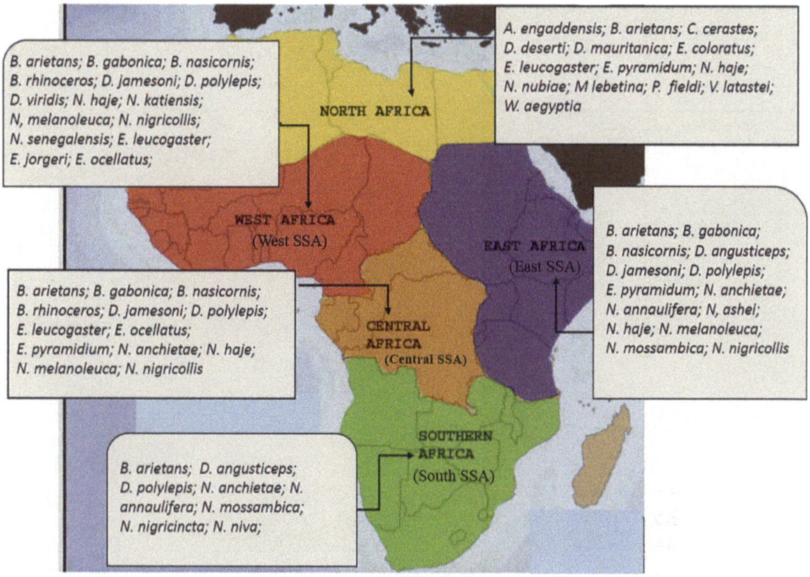

Fig. 1 Distribution of the most clinically relevant snake species within the African continent. Descriptions: *Atractaspis engaddensis; Bitis arietans; Bitis gabonica; Bitis nasicornis; Cerates cerastes; Dendroaspis angusticeps; Dendroaspis jamesoni; Dendroaspis polylepis; Dendroaspis viridis; Daboia deserti; Daboia Mauritania; Echis coloratus; Echis jorgeri; Echis leucogaster; Echis ocellatus; Echis pyramidin; Naja annaulifera; Naja ashei; Naja anchietae; Naja haje; Naja melanoleuca; Naja mossamica; Naja nigricincta; Naja nigricollis; Naja niva; Naja nubiae; Naja Senegalensis; Macrovipera lebetina; Pseudocerastes fieldi; Vipera latastei; Walterinnesia aegyptia. Reconstructed from WHO. 2016b. WHO guidelines for the production, Control and Regulation of Snake Antivenom Immunoglobulins.*

Venomous snakes feed on various prey, including small mammals, amphibians, reptiles, birds, snails, fish, chickens, lizards, rats and even other snakes (Kang et al., 2011). The diet of different snakes is largely influenced by environmental factors, predator and prey size, prey vulnerability and adaptive evolution (Barlow, Pook, Harrison, & Wuster, 2009).

1.2 The burden of snakebite envenoming

Venomous snakes are found throughout the world, with the exception of high altitudes, areas of permafrost and a few islands (Kasturiratne et al., 2008). However, as shown in Fig.2, over 95% of SBE occur in deprived, remote, rural areas of Southern Asia, Southeast Asia, SSA and Latin America (Benjamin, Chippaux, Sambo, & Massougbodji, 2018; Chippaux, 2011; Kasturiratne et al., 2008). Geographically, "sub-Saharan Africa" is an ambiguous term used to describe 46 of the 54 African countries that are fully or

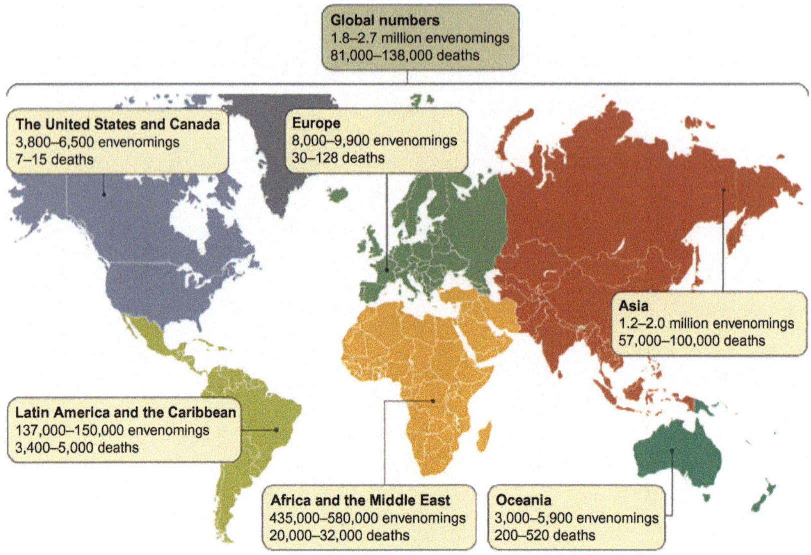

Fig. 2 Global distribution of estimated annual snakebite envenoming-related deaths. *Obtained from Gutierrez, J. M., Calvete, J. J., Habib, A. G., Harrison, R. A., Williams, D. J. and Warrell, D. A. 2017. Snakebite envenoming. Nature Reviews. Disease Primers 3, p. 17079. doi: 10.1038/nrdp.2017.79.*

partially located south of the Saharan desert [United Nations (UN) definition]. The exceptions being Algeria, Djibouti, Egypt, Morocco, Libya, Somalia, Sudan and Tunisia. Therefore, throughout this review, and in line with institutions such as the UN development program, the SSA term will be used to denominate these African countries.

1.3 Vulnerability to snakebites in the SSA region

Understanding snake species' behaviors and geographical distribution, availability of antivenoms and access to suitable clinical centres are key factors for assessing vulnerability of populations to SBE (Longbottom et al., 2018). It is estimated that 92.66 million people worldwide are vulnerable to snakebites, most of which occur in Asia and SSA (Longbottom et al., 2018). As shown in Fig. 3, populations living in tropical regions such as the Democratic Republic of Congo, are at highest risk. Factors such as non-mechanical and traditional farming, failure of farmers to use protective boots and gloves, heavy rainfall flooding snakes out of their burrows, and attraction of snakes to rodents and poultry found around grain stores in proximity to human habitation, increases potential human-snake interactions. Thus, agricultural

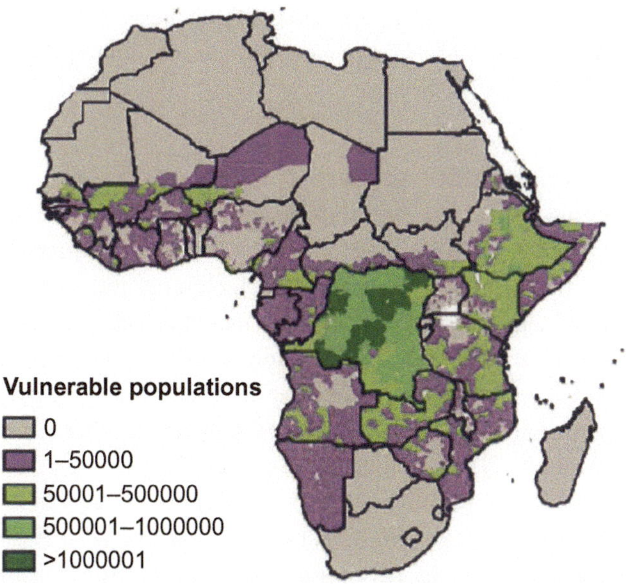

Fig. 3 Distribution of population vulnerability to snakebite in the African continent. *Adapted from Longbottom, J. et al. 2018. Vulnerability to snakebite envenoming: A global mapping of hotspots. Lancet 392(10148), pp. 673–684. doi: 10.1016/s0140-6736(18)31224-8.*

workers and nomadic farmers, particularly young males and children (Chippaux, 2011; Dalhat, 2015), pregnant women (Habib, 2015) and families living in poorly constructed houses are particularly vulnerable. Despite elegant studies which used the household survey to assess incidence of snakebite at a national level (Ediriweera et al., 2016; Mohapatra et al., 2011; Rahman et al., 2010), the global magnitude of snakebite and its clinical ramifications have not been fully elucidated (Longbottom et al., 2018).

1.4 Clinical presentation of SBE

Clinical manifestations of snakebite are species specific and can be broadly grouped as cytotoxic (Warrell, Greenwood, Davidson, Ormerod, & Prentice, 1976), neurotoxic (Baldé et al., 2013; Warrell, Barnes, & Piburn, 1976), and haemorrhagic (Chippaux, Massougbodji, Stock, & Alagon, 2007; Warrell et al., 1974, 1977; Warrell, Greenwood, et al., 1976). Clinical presentation is dependent on the arsenal of venom toxins delivered, each of which may act individually or synergistically. Individual venom composition is not only species specific, but influenced by factors such as geographical location, season, age and diet (Mackessy, 2009). Snakes from the *Bitis, Echis, Naja* and *Dendroaspis* genera are responsible for the majority of medically significant SBEs in SSA.

*1.4.1 African adders (*Bitis *species)*

Although some species in the *Bitis* genus are present in Saudi Arabia (Paixão-Cavalcante, Kuniyoshi, Portaro, da Silva, & Tambourgi, 2015), most of the medically important adders, including the puff adder (*Bitis arietans*), Gabon viper (*B. gabonica*), *B. gabonica rhinoceros* and *Bitis nasicornis,* are indigenous to Africa (Calvete, Escolano, & Sanz, 2007). Together with the saw-scaled viper and black-necked spitting cobra (see below), the puff adder is responsible for most SBEs in SSA (Currier, Harrison, Rowley, Laing, & Wagstaff, 2010; Habib, 2015). Their venom is rich in highly toxic components, such as snake venom serine proteinases (SVSPs); C-type lectins (CTLs) and snake venom metalloproteinases (SVMPs) (Calvete et al., 2007). Clinical symptoms of envenoming include: extensive tissue necrosis, cytotoxicity, hypotension, coagulopathy, hemorrhage and cardiotoxicity (Paixão-Cavalcante et al., 2015).

*1.4.2 Saw-scaled or carpet vipers (*Echis *species)*

The majority of publications describing venom composition of *Echis* spp. focus on *E. ocellatus* (Calvete et al., 2010; Casewell, Harrison, Wuster, & Wagstaff, 2009; Hamza et al., 2016; Rogalski et al., 2017; Wagstaff & Harrison, 2006; Wagstaff, Sanz, Juarez, Harrison, & Calvete, 2009) and *E. carinatus* (Patra, Kalita, Chanda, & Mukherjee, 2017), whereas those describing *E. leucogaster* (Escoriza, Metallinou, Donaire-Barroso, Amat, & Carranza, 2009) and *E. pyramidum* (Conlon et al., 2013) are limited. SVMPs are the major venom toxin component (Calvete et al., 2010; Wagstaff & Harrison, 2006). Headache, seizures, vomiting, systemic hemorrhage, incoagulable blood, swelling, blistering and necrosis are the hallmark signs and symptoms associated with *Echis* envenomings (Habib, 2015; Habib et al., 2008; Warrell et al., 1977). Extensive tissue necrosis may result in renal failure (Warrell et al., 1977). If treatment is delayed, death occurs most commonly due to intracranial hemorrhage (Habib & Abubakar, 2011). This genus of snakes is responsible for more deaths, globally, than any other (Casewell et al., 2009) and in Nigeria, *E. ocellatus* accounts for the highest mortality and more life-changing injuries than any other SBE (Habib et al., 2008; Warrell, 2008).

*1.4.3 Mambas (*Dendroaspis *species)*

Dendroaspis spp. are medically important snakes, which are present exclusively in SSA (Harvey & Robertson, 2004). The black mamba (*D. polylepis*) is highly venomous and very aggressive, envenoming resulting in potentially fatal neurological signs (Laustsen, Lomonte, Lohse, Fernandez, & Gutierrez, 2015). Mamba venom is highly toxic, dominated by postsynaptic and presynaptic neurotoxins, including Kunitz type toxins (KTT) and 3-finger toxins (3FTXs) (Ainsworth et al., 2018). Clinical signs include local tissue swelling, ptosis, fasciculation, dizziness, nausea, respiratory paralysis and cardiac arrhythmia that are potentially fatal (Ainsworth et al., 2018; Lauridsen, Laustsen, Lomonte, & Gutierrez, 2016).

*1.4.4 Cobras (*Naja *species)*

The Egyptian cobra (*N. haje*), the black-necked spitting cobra (*N. nigricollis*) and the forest cobra (*N. melanoleuca*) are the most medically important cobras in SSA. Proteomic studies (Lauridsen, Laustsen, Lomonte, & Gutierrez, 2017; Malih et al., 2014; Petras et al., 2011) demonstrate venoms dominated by 3FTx, neurotoxins (Warrell, Barnes, et al., 1976) and PLA_2 cytotoxins (Warrell, Greenwood, et al., 1976). Clinical signs of envenoming include local tissue swelling and necrosis, vomiting, drowsiness, impaired speech, ptosis and potentially fatal respiratory paralysis (Warrell, Barnes, et al., 1976; Warrell, Greenwood, et al., 1976). Interestingly, despite the fact that the venom of *N. nigricollis* is made up of over 70% 3FTXs (Petras et al., 2011), associated clinical signs are dominated by extensive tissue necrosis at the bite site (Habib, 2015; Warrell, Greenwood, et al., 1976).

1.5 Prevention and management of snakebite-associated morbidity and mortality

In SSA, integrated medical centres are scarce, often situated many miles from the vulnerable farming communities where snakebites are frequent (Kipanyula & Kimaro, 2015). The use of traditional and religious medicine is often the first port of call, despite being considered as time-wasting and potentially harmful treatments (Ghose & White, 2016; Hasan, Basher, Molla, Sultana, & Faiz, 2012; Yates, Lebas, Orpiay, & Bale, 2010). These treatments include: scarification, emetics, snakestones, suction at the bite site, tight arterial bands, and even digit amputation (Ghose & White, 2016; Habib, 2015; Yates et al., 2010).

Potentially, the most effective methods of preventing snakebite include: education, raising awareness within vulnerable communities and improving transportation to appropriate medical centres (Gutierrez, Calvete, et al., 2017). First Aid should include: seeking advice over the phone; hygiene at the bite site, pressure immobilisation, removal of constrictive items, placing the patient in lateral recumbency, administration of artificial respiration and rapid transfer to the nearest clinic (Ghose & White, 2016; Warrell, 2009). The use of nonsteroidal anti-inflammatory drugs (NSAIDs) should be avoided, unless otherwise advised by a medical specialist (Ghose & White, 2016). Admission to hospital should ideally occur within 24 h of the bite for assessment and appropriate treatment, including antivenom. It is estimated that a delay of 1 h increases mortality by 1% and 1 day by 23% (Habib & Warrell, 2013). Encouraging the financial support of local manufactures of effective antivenoms, from local and global institutions, is key to the control of SBE. This was demonstrated by the reduction in fatality from *E. ocellatus* bites from 11.1% to 1.3% after the introduction of geographically relevant antivenoms in West Africa (Habib & Warrell, 2013).

Identification of the snake responsible by bite configuration, evidence from witnesses and victim, geographical location and knowledge of clinical manifestations of species commonly found in the area, allows rapid administration of the correct antivenom (Dhananjaya, Menon, Joseph, Raveendran, & Oommen, 2015). Presentation of the killed snake responsible may allow identification, although misidentification is possible unless performed by an experienced herpetologist. Education of healthcare professionals and First Aiders; stocking up with geographically appropriate antivenom and associated medications must be the primary goal of hospitals providing care for victims of SBE. In this way, mortality may be significantly reduced (Ediriweera et al., 2016; Habib & Warrell, 2013; Visser et al., 2008). Medications such as antihistamine (chlorpheniramine), corticosteroids and adrenaline must be stocked to treat any adverse reactions to antivenom, including anaphylaxis (Habib, 2015).

Triage of SBE victims, where clinical presentations are spectra and time-dependent, with a thorough medical and laboratory investigation are key to establishing a prompt and appropriate treatment plan (Gutierrez, Calvete, et al., 2017). The 20-min whole blood clotting test (20WBCT) is a simple, inexpensive bedside test which may be performed to identify incoagulable blood of patients envenomed by snakes such as *Echis* spp., which induce a consumptive coagulopathy (Benjamin et al., 2018; Dhananjaya et al., 2015; Habib, 2015). Bites by *N. nigricollis* induce haematological abnormalities and Complement depletion (Warrell, Greenwood, et al., 1976). Immunological and molecular techniques, such as ELISA and PCR, may be used to identify venom more reliably in tertiary care centres (Habib, 2015; Williams et al., 2019). In Australia, there is a snake venom detection kit (SVDK) available at a cost of 150–200 USD (Valenza, Allavena, Haworth, Cochrane, & Henning, 2021), which is both unaffordable and unavailable in SSA.

The major effects of systemic envenoming are generally amenable to antivenom therapy (Gutierrez, Lomonte, Sanz, Calvete, & Pla, 2014; Habib, 2015), but exceptions include post-traumatic stress disorder (PTSD) and effects of local envenoming such as amputation (Halilu et al., 2019). Despite these discrepancies, antivenom remains the mainstay of SBE treatment (Chippaux et al., 2007; Habib & Warrell, 2013).

2. Snake venom: Overview

In venomous snakes the arsenal of toxins injected *via* a bite, or sprayed onto, is an effective bioweapon used primarily to incapacitate their prey or secondarily to deter predators (Mackessy, 2009). Snake venom consists of a mixture of enzymes, non-enzymatic proteins, peptides, and polypeptides, accounting for around 95% dry weight (Carregari et al., 2018), the remainder consists of non-proteinaceous compounds including inorganic salts, amines, carbohydrate, metal ions and lipids (Munawar, Ali, Akrem, & Betzel, 2018).

2.1 Constituents of snake venom

A single venom may contain more than 100 different proteins (Mackessy, 2009), one study identifying over 200 proteins (Petras, Heiss, Harrison, Sussmuth, & Calvete, 2016). Additionally, the complex mix of toxins of a single venom is reflected by the multitude of clinical manifestations observed in envenomed patients. Various venom proteins may present unique biological and pharmacological modes of action; but homologous structures between protein isoforms make classification into a small number of toxin families possible (Kang et al., 2011). Although the biochemical composition of venom is so variable (Chippaux, Williams, & White, 1991), it is dominated by the major protein families, which are consistent across the taxa (Mackessy, 2009; Mackessy & Saviola, 2016). External factors such as diet, age and geographical location may influence production of venom even within a species (Mackessy, 2009). Large numbers of protein families have been identified in venom from snakes found in SSA, using genomics and proteomics (Calvete et al., 2007; Casewell et al., 2009; Laustsen et al., 2015; Malih et al., 2014; Petras et al., 2016, 2011). Major toxin families identified are PLA_2s, SVMPs, SVSPs, cysteine-rich secretory proteins (CRISPs), 3FTXs, L-amino acid oxidases (LAAOs); disintegrins (DIS), CTLs, KTTs, dendrotoxins, Naturetic peptides (NPs), vascular endothelial growth factors (VEGFs) and nerve growth factors (NGFs). Further studies are required to elucidate the mechanisms governing the pharmacological actions of these proteins in order to achieve the milestone of discovering effective therapeutic strategies. Nevertheless, excellent reviews pertaining the current understanding structures and mechanisms of snake venom toxins are available, for example (Doley & Kini, 2009; Ferraz et al., 2019; Kang et al., 2011; Mackessy & Saviola, 2016; Munawar et al., 2018).

3. Antivenom

Antivenoms are purified polyclonal antibodies produced by immunization of a donor animal, which specifically bind to and neutralise venom toxins. Although various novel therapeutics have been proposed (Alvarenga et al., 2014; Pucca et al., 2019), heterologous antivenoms remain the only commercially available approved treatment for SBEs at present (Guidolin, Caricati, Marcelino, & da Silva, 2016). Antigen-specific antibody production is induced by repeatedly injecting small quantities of venom into animals, preferably larger farm animals such as horses and sheep, then collecting the resultant hyperimmunized plasma or serum. Immunoglobulin G (IgG) is collected by fractionation from smaller proteins with subsequent proteolytic digestion producing fragmented antibodies [Fab or $F(ab')_2$].

3.1 Brief history of the path to the discovery of antivenom

In 1894, Phisalix and Bertrand reported the first attempt to treat envenomed patients using whole blood from a sheep inoculated with detoxified *Viper aspis* (Dias da Silva & Tambourgi, 2011). In 1895, Calmette and colleagues produced the first serum antivenom against cobra snakes that was used to successfully treat a patient envenomed by *Naja kaouthia* (Pucca et al., 2019; Williams, Habib, & Warrell, 2018). At the time, the anti-cobra serum was presented as a universal antivenom capable of conferring protection against lethality induced by other snake species, even from different species and geographical regions. However, the claim was swiftly contested as the anti-sera failed to offer protection beyond the regional *Naja* spp. (Squaiella-Baptistao, Sant'Anna, Marcelino, & Tambourgi, 2018), emphasising the antivenom's specificity for the venom(s) used in its production.

Thereafter, various countries started producing regionally manufactured crude serum antivenoms effective at treating victims envenomed by local snakes, although adverse reactions were very common (Dias da Silva et al., 2011; Gutierrez, Guillermo, Bruno, & Yamileth, 2011).

Despite efficacy at treating envenomed patients, administrations of whole serum was associated with dangerous adverse effects attributed to the presence of heterologous proteins (Squaiella-Baptistao et al., 2018). This promoted the need to reduce contaminants, IgG and/or non-Ig serum proteins, where new protocols were proposed to manufacture antivenoms containing purified antibody instead of whole serum. Despite obtaining significant improvements in quality, adverse events were still prevalent leading to the conclusion that whole IgG was a contributory factor (Squaiella-Baptistao et al., 2018). This promoted the introduction of papain and pepsin digestions to cleave-off the Fragment Crystallisable (Fc) region from whole IgG to produce monomeric and dimeric antigen-binding fragments referred to as Fab and $F(ab')_2$, respectively. In the last two decades, several chromatographic techniques were introduced to capture specific antibodies and remove impurities, including the cleaved Fc fragment (Gutierrez et al., 2011).

3.2 Current antivenom formats

There are three different antivenom formats at present, based on their active substance (Fig.4): intact IgG, monovalent antigen-binding fragment (Fab),

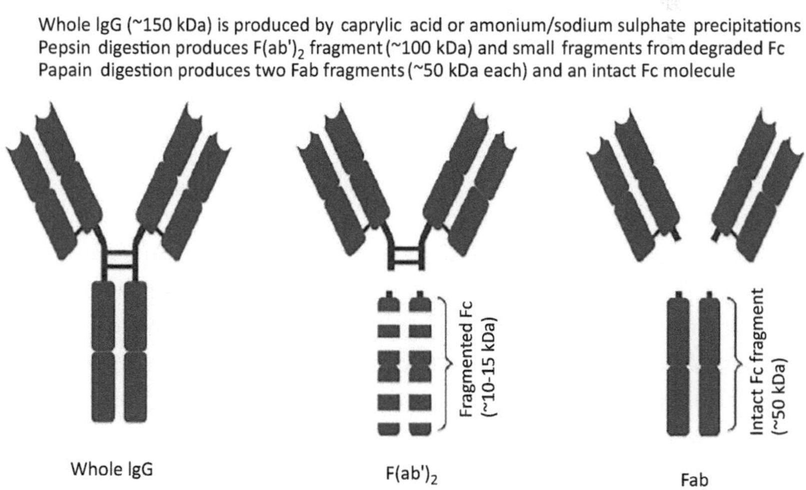

Fig. 4 Schematic representation of the three different antivenom formats currently available: whole IgG (left), $F(ab')_2$ (centre) and Fab (right). Created with BioRender.com.

and divalent antigen-binding fragment $F(ab')_2$. Most current antivenoms are horse derived (Laustsen et al., 2018), other larger farm animals such as sheep (Casewell, Al-Abdulla, Smith, Coxon, & Landon, 2014; Cresswell et al., 2005), and donkeys (Gutiérrez, Higashi, Wen, & Burnouf, 2007) are also used. Antivenom manufacturers use significantly different methods, but fundamental steps such as precipitation by ammonium (or sodium) sulfate/caprylic acid and pepsin/papain digestions are universally used.

3.2.1 Whole IgG antivenom

Some IgG-based antivenoms have excellent efficacy and safety profiles and are commercially globally available (Abubakar et al., 2010; Guidolin et al., 2016; Laustsen et al., 2018; Otero et al., 2006). Ammonium sulfate and caprylic acid precipitations are currently the processes of choice used to purify IgG (Otero et al., 1999; Otero-Patiño et al., 2012). The precipitating mechanisms differ significantly but the principles of fractionation to eliminate non-IgG proteins remains similar.

3.2.1.1 Ammonium sulfate

The mechanism by which ammonium sulfate ($(NH_4)_2SO_4$) separates proteins depend on altering ionic strength where, at increased salt concentration, globular solubility is enhanced, a process termed as "salting-out" (Duong-Ly & Gabelli, 2014). This will precipitate proteins, such as γ-globulins, that aggregate easily, leaving other plasma/serum proteins, predominantly albumin, in solution (Duong-Ly & Gabelli, 2014). Its main drawback is that contaminants can precipitate with proteins of interest, reflected by the higher occurrence of adverse reactions induced by antivenoms prepared in this way (Otero et al., 1999; Ryan, Kearney, Brown, & Isbister, 2016).

3.2.1.2 Caprylic acid (octanoic acid)

Caprylic acid precipitates non-IgG proteins including albumin, leaving IgG in solution. This IgG fractionation mechanism shows negligible detection of contaminants and, as such, is reflected by the lower incidence of adverse reactions (Abubakar et al., 2010; Guidolin et al., 2016; Otero-Patiño et al., 2012). Nonetheless, the precise mechanism involved in this separation remains unclear. However, it was suggested that caprylic acid can directly bind and destabilise the three-dimensional structure of monomeric and multimeric protein molecules, which is a pH and temperature-dependent phenomenon (Morais & Massaldi, 2012; Nudel, Perdomenico, Iacono, &

Cascone, 2012). The precipitation equilibrium is thermodynamically controlled, and protein partition relies on the hydrophobicity level. At lower pH the hydrophobicity of the octyl moiety dominates whereby the non-immunoglobulin molecules such as nucleic acids and host cell proteins (HCPs) precipitate whilst IgG, which has sufficient charge to overcome the hydrophobicity, remain in solution (Nudel et al., 2012). In addition to offering higher purity, caprylic acid is considered to be a viral inactivation step, by virtue of its ability to disrupt the integrity of enveloped viruses, despite no evidence of viral transmission associated with antivenom treatment (Burnouf et al., 2004).

3.2.2 Monovalent antigen-binding fragment; Fab-based antivenom

A small number of Fab-based antivenom formulations are available, including ViperaTAb®, and CroFab®. These commercially available antivenoms are made by papain digestion of caprylic acid/$(NH_4)_2SO_4$ fractionated sheep IgG (Casewell et al., 2014; Cresswell et al., 2005). Papain cleaves IgG at the amino-terminal side of the disulphide bond (Fig. 4), generating two separate Fab fragments and an intact Fc molecule (Janeway, Charles, Travers, Walport, & Shlomchik, 2001). Papain is a thiol protease that is activated by the presence of reducing agents such as L-cysteine, dithiothreitol (DDT) or mercaptoethanol (Goding, 1986), L-cysteine being used more frequently in biopharmaceuticals (Al-Abdulla, Casewell, & Landon, 2014; Cresswell et al., 2005). Digestion is terminated by addition of iodoacetamide (Al-Abdulla et al., 2014; Cresswell et al., 2005), which irreversibly alkylates the SH groups on the papain active site (Goding, 1986). Efficient cleavage of IgG is controlled by reaction time, temperature and concentration of papain and L-cysteine (Cresswell et al., 2005). The product is purified further by affinity chromatography, where antigen-specific antibodies are collected and formulated (Casewell et al., 2014; Cresswell et al., 2005).

3.2.3 Divalent antigen-binding fragment: $F(ab')_2$-based antivenom

The majority of current commercial antivenoms are $F(ab')_2$-based formulations (Bush et al., 2015; Kurtovic et al., 2016; Patra, Kalita, & Mukherjee, 2018; Pepin-Covatta, Lutsch, Grandgeorge, & Scherrmann, 1997). The pepsin digestion method was first introduced by Pope in the late 1930s (Pope, 1939a, 1939b). Pepsin cleaves IgG, at an acidic pH (Gutierrez et al., 2011), at the carboxylic-side of the disulfide bond to generate 2-covalently linked divalent molecules termed as $F(ab')_2$, with preserved antigenic binding activity (Boushaba, Kumpalume, & Slater, 2003;

Janeway et al., 2001). The digestion process is narrowly controlled by conditions including pepsin concentration, reaction time, pH, and temperature (Kinman & Pompano, 2019; Morais & Massaldi, 2005; Nudel et al., 2012). The pepsin-cleaved Fc region (Fig. 4) is degraded into smaller molecular weight molecules that are removed on subsequent steps, such as chromatography methods or $(NH_4)_2SO_4$ precipitation (Gutierrez et al., 2011). The $F(ab')_2$ containing product is then collected and formulated. It is suggested that the function of pepsin is to cleave selectively at amino acids L234-L235, within the residues responsible for FcγRs recognition (Brezski & Jordan, 2010).

3.3 Neutralization by antivenoms

Most of the current data on the pharmacokinetics of antivenoms were derived from animal studies (Gutiérrez, León, & Lomonte, 2003; Rojas et al., 2013), although limited data from clinical studies exist (Gutiérrez et al., 2003). These provide no clear evidence of which antivenom format has the best neutralising efficacy (Laustsen et al., 2018). Indeed, all forms of IgG and its fragments have been found to be effective at neutralising venom toxins, but with highly variable pharmacokinetic and pharmacodynamic properties (Gutiérrez et al., 2003). The success of antivenom is dependent on severity of envenoming, time lapse to administration, efficacy and dose (Silva & Isbister, 2020).

3.3.1 Pharmacokinetics (PK) of antivenoms

The PK of antivenoms is largely governed by their molecular size, which ranges from whole IgG at 150kDa to Fab at 50kDa (Gutiérrez et al., 2003). On the other hand, snake venoms have a larger range of molecular size (Kang et al., 2011; Sunagar et al., 2013; Zupunski, Kordis, & Gubensek, 2003), which together with charge, hydrophobicity and cellular permeability, governs their absorption and distribution to target tissues (Paniagua, Vergara, Boyer, & Alagón, 2015). Smaller MW venom toxins including 3FTXs and KTTs readily distribute to their target sites, despite a lack of evidence for the presence of neurotoxin in the vascular compartment (Baldé et al., 2013). In contrast, venom toxins with a larger MW, such as SVMPs and SVSPs, are characterized by biphasic pharmacokinetic profiles involved in initial rapid absorption followed by a slower absorption from injection site (Paniagua et al., 2015). Nevertheless, snake venoms are usually delivered subcutaneously and, to a lesser extent, intramuscularly whereas antivenoms are usually administered intravenously (I.V.) conferring an advantage over competition at target sites (Paniagua et al., 2015; Silva & Isbister, 2020). Effective neutralization of irreversible tissue damage is achieved if antivenoms are promptly administered into the circulating compartment at the same time as the venom toxins are absorbed into it. Unfortunately, there is almost always a delay from bite to antivenom administration, and additionally, small MW venom toxins may diffuse faster than antibody fragments to their targets (Silva & Isbister, 2020).

As shown in Table 1, the volume of distribution (Vd) is largely dependent on MW of antivenoms and, consequently, Fab and $F(ab')_2$ have a larger Vd than whole IgG antivenom (Gutiérrez et al., 2003). Vd is a measure of an agent's tendency to either remain in plasma or to access other compartments. In fact, IgG and $F(ab')_2$ have a similar distribution time and concentration ratio (Seifert & Boyer, 2001). Given its larger molecular weight, IgG has a poor absorption and distribution ratio (Silva & Isbister, 2020), which may limit neutralisation of venom toxins already bound to targets on target tissues. It is assumed that $F(ab')_2$ has intermediate pharmacokinetics (Gutiérrez et al., 2003), but intravenous injection of IgG and $F(ab')_2$ showed a biphasic decline in concentration, described as a rapid initial distribution phase (i.e., sum of distribution and elimination process) followed by a slower elimination phase (Gutiérrez et al., 2003; Silva & Isbister, 2020). During the early distribution phase, IgG and $F(ab')_2$, antivenoms are rapidly absorbed into target tissues and simultaneously eliminated, perhaps *via* phagocytosis

Table 1 Pharmacokinetic properties of the three antivenom formats.

Format	**Study subject**	**Vd (mL/kg)**	**$t_{1/2\text{-}\alpha}$ (h)**	**$t_{1/2\text{-}\beta}$ (h)**	**CL (mL/h/kg)**
Whole IgG	Human	58–118	0.22–5.62	34–72	1.1–1.6
	Rabbit	160–164	0.42–0.56[a]	40.62-44.7	0.30-0.42
F(ab')2	Human	177–387	0.2–2.31[a]	79-132	0.91-2.54
	Rabbit	207–225	0.17–0.27	32.91–34.77	3.22–3.9
Fab	Human	110	2.7	18	No data available
	Rabbit	228–32	0.68–0.72	7.7–8.3	52.5–53.5

[a]In the rabbit model, both IgG and $F(ab')_2$ antivenoms also have an intermediate phase ($t_{1/2\text{-}\alpha}$) of 5.98 ± 0.67 and 2.33 ± 0.45 h, respectively.
Abbreviations: Vd: volume of distribution; $t_{1/2\text{-}\alpha}$: distribution half-time; $t_{1/2\text{-}\beta}$: elimination half-life, CL: clearance.
Adapted from Gutiérrez, J. M., León, G. and Lomonte, B. 2003. Pharmacokinetic-pharmacodynamic relationships of immunoglobulin therapy for envenomation. Clinical Pharmacokinetics 42(8), pp. 721–741.

by the mononuclear phagocyte system, but once distribution equilibrium is reached, elimination becomes the only process causing concentration to decline (Gutiérrez et al., 2003).

Conversely, Fab fragments at 50 kDa, have a higher Vd and distribution ratio making it readily diffusible into extravascular compartments, conferring rapid restoration of damaged tissues (Gutiérrez et al., 2003; Seifert & Boyer, 2001). This may enable Fab antivenoms to protect against toxicity caused by smaller neurotoxins. Fab is rapidly eliminated *via* the renal glomerulus, unlike larger fragments, which remain within the intravascular compartment for longer (Gutiérrez et al., 2003). A clinical study demonstrated that Fab was eliminated 5–7 times faster than $F(ab')_2$ antivenom (Meyer et al., 1997). The pharmacokinetic implications of a short elimination half-life is that recurrence of envenoming is commonplace, and requires repeated antivenom administration (Seifert & Boyer, 2001). Furthermore, rapid glomerular elimination of Fab fragments may cause renal impairment, which would, in turn, affect elimination of Fab (Gutiérrez et al., 2003). See Table 2.

3.3.2 Pharmacodynamics (PD) of antivenoms

Pharmacodynamics (PD) is the triad interaction between the human/animal body, the toxin and the antivenom. The ability of antivenom to neutralise a given venom effectively is determined by its capacity to provide sufficient specific antibody molecules to neutralize venom toxins absorbed. The polyclonal nature of antivenoms offers the ability to recognise and neutralize multiple venom toxins simultaneously. Theoretically, IgG and $F(ab')_2$ based antivenoms may have higher avidity than Fab fragments, which only contain one antigen-binding fragment. Digestion by proteolytic enzymes, especially pepsin may affect avidity (Morais & Massaldi, 2005), but there is no evidence to suggest which antivenom type contains antibodies with the highest neutralising capacity (Laustsen et al., 2018). Several studies have demonstrated that IgG antivenom has a comparable efficacy profile to those containing Fab and $F(ab')_2$ despite the assumption otherwise (Leon, Monge, Rojas, Lomonte, & Gutierrez, 2001; Leon, Stiles, Alape, Rojas, & Gutierrez, 1999; Otero-Patiño et al., 2012). A clinical study comparing the efficacy of Fab and $F(ab')_2$ antivenoms found that 4 times the dose of Fab was required to induce comparable efficacy than that of $F(ab')_2$ (Boyer, Seifert, & Cain, 2001; Meyer et al., 1997).

The exact mechanism by which different antivenom formats neutralise various toxins is not fully understood. However, there is a consensus of

Table 2 Advantages and disadvantages of the different types of antivenom.

Format	Description	Advantage	Disadvantage
Fab	Monovalent antigen fragment produced by papain digestion at slightly acidic pH (5.0–5.5)	Has a rapid tissue penetration, larger volume of distribution and decreased risk of immunocomplex formation and complement-mediated anaphylactic reactions	Has lower bioavailability and excretion fate is unknown. It can cause renal toxicity. Requires a prolonged process involving in affinity purification adds to cost
F(ab')2	Most commonly manufactured antivenom, consisting of bivalent antigen fragment, produced by pepsin digestion at acidic pH (≤4.5)	It is relatively safe because over 30% of unwanted proteins have been removed during digestion. Highly desirable elimination and clearance ratio	Long process often involving chromatography for purification and removal of contaminants. Adds to cost
Whole IgG	Intact IgG produced by ammonium (or sodium) sulfate and/or caprylic acid precipitation	Easy to manufacture and cheap. Highly desirable elimination and clearance ratio	Poor tissue penetration with slower distribution rate Injection of greater quantities of exogenous proteins potentially causing adverse reactions

opinion that they occupy and stereotactically block toxin active sites, or cause conformational changes of enzymatic toxins, making toxin sites inaccessible (Gutiérrez et al., 2003; Laustsen et al., 2018; Nikapitiya & Maduwage, 2018). Additionally, antivenoms trap venoms in the central compartment, preventing access to pharmacological sites and preventing toxin dissociation (Laustsen et al., 2018). Being multivalent, IgG and $F(ab')_2$ are capable of forming antibody-venom complexes which may be eliminated by phagocytosis, diminishing the overall presence of toxins (Gutiérrez et al., 2003; Silva & Isbister, 2020).

3.4 Future direction of antivenoms

Antivenom, in its current format, is the only approved specific treatment for SBE (Silva & Isbister, 2020). Adverse effects, especially those related to excess foreign protein, acute anaphylaxis and lack of a universal antivenom led to efforts by the scientific community to derive more advanced therapies. Employing the latest biotechnology, monoclonal antibody (mAb) based antivenoms against several venom toxin types have been developed (Laustsen et al., 2018; Pucca et al., 2019). Preclinical studies demonstrated that these novel antivenoms were effective at neutralising toxins and preventing lethality in mice (Laustsen et al., 2018; Pucca et al., 2019). Phage display techniques were recruited for developing mAb fragments, with specificity for more than one toxin epitope curtailing the need of numerous mAbs or oligoclonal antivenom (Ledsgaard et al., 2018). It was envisaged that the development of a mAb, with multiple cross-reactivity, will pave the way to the production of mAb-based antivenoms that can be used against venoms from multiple species. An equine "experimental antivenom" was developed against biologically active recombinant consensus α-neurotoxin, the whole IgG-antivenom was preclinically effective at neutralisation of the neurotoxic effects of biogeographically diverse elapid snakes (De la Rosa et al., 2019). Experimental antivenom has been raised in rabbits against recombinant enzymatic venom toxins, including SVMPs, SVSPs and $SVPLA_2$s obtained from *Bothrops ammodytoides*. The anti-recombinant venom proteins successfully neutralised the lethal effect of *B. ammodytoides* in mice (Clement, Corrales-García, Bolaños, Corzo, & Villegas, 2019). Both approaches support the proof of concept for the use of recombinant enzymatic and/or non-enzymatic toxins to produce antibodies with superior titration, affinity and avidity, and thus herald the next generation antivenoms. However, the complexity and variability of snake toxins imposes a major hurdle for the translation of these strategies into a clinical setting. Therefore, at present, mAb based snake antivenoms have not even reached the clinical study stage, let alone as therapeutics.

3.5 Immunisation

In the 1890s, Calmette developed an immunization protocol for horses using *N. naja* venom which was quickly implemented globally by other institutions (Pucca et al., 2019). With this modification of immunization protocols, there was a simultaneous improvement of quality and purification methods. The importance of potentiating immune response of host animals to generate larger volumes of potent antisera was promptly recognised. In 1924, the South African Institute of Medical Research (SAIMR) added a tapioca substance to a pool of venom toxins as an "adjuvant" to enhance the immune response of the donor horse (Squaiella-Baptistao et al., 2018).

By 1926, aluminium was the first commercially used adjuvant to boost the immune response of diphtheria vaccine (Egli et al., 2014). Currently, Freund's complete and incomplete adjuvants are commonly used for initial immunization, and subsequently, with aluminium salts, bentonite or just emulsified in saline (Gutierrez et al., 2011). Different adjuvants enhance the immune response by different mechanisms, but slow release and sequestration of antigens remains an important feature (Awate, Babiuk, & Mutwiri, 2013; Egli et al., 2014). The ideal adjuvant must be safe, cheap and capable of slowly releasing antigens from sites of injection, to stimulate an effective immune response (León et al., 2011).

The diversity of snake venoms makes optimization of the immunisation protocol key to producing potent antibodies, without compromising the donor animals' welfare. The best protocol is to inject, slowly and repeatedly, small volumes of native, not detoxified, venoms, at several anatomical sites (Gutierrez et al., 2011). The subcutaneous route is most common, but intradermal, intramuscular routes as well as a liposomal delivery are also used to immunise animals for the production of antivenom (León et al., 2011). Venoms used for immunisation should be taken from multiple individuals of snake species relevant to the geographical region in which the antivenom is to be used (WHO, 2016b). The design of an immunization protocol requires careful evaluation of snake distribution and toxicity, epidemiological and clinical manifestations. A potent antivenom, designed based on this information, will be effective at neutralising the most clinically important venom toxins. Monospecific antivenoms are appropriate where there is only one species of venomous snake in a region, or if bites from one species are dominant in cases such as *E. ocellatus* bites in West Africa (Visser et al., 2008), particularly Nigeria (Habib et al., 2008). On the other hand, polyspecific antivenoms are preferable in cases where there are multiple clinically relevant species such as in SSA (Tasoulis & Isbister, 2017; WHO, 2016b).

3.6 Animals used for production of antivenom

Sheep were the first animals used to produce immunotherapeutic sera against diphtheria (Landon & Smith, 2003). The horse became the preferred source for production of antisera and was used for treating diphtheria, tuberculosis (TB) and pneumonia (Walther, Rusitzka, Diesterbeck, & Czerny, 2015). Although sheep (Al-Abdulla, Casewell, & Landon, 2013; Casewell et al., 2014) and donkey (Gutiérrez et al., 2007) are still used by some manufacturers, horses are most commonly used for manufacturing commercial antivenoms (Gutierrez et al., 2011). Nevertheless, chicken, rabbit, camels and goat are occasionally used (Sells, 2003).

Horses are expensive to maintain and, being regarded as companion animals, may not be permitted to be used for this purpose in some countries such as the UK. Avian and Camelid antibodies have been considered as alternatives for the conventional antivenom production due to their economic, ethical and stability advantages (Gutierrez et al., 2011). Promising efficacy results of IgY-based antivenom have been reported in a mouse model (Leiva et al., 2019). Experimental Camelid anti *E. ocellatus* IgG-antivenom demonstrated a comparable efficacy to EchiTabG® (Cook et al., 2010), a highly effective sheep-derived antivenom product currently marketed in Nigeria (Abubakar et al., 2010).

3.7 Future direction of immunizations

The current immunization protocol is based on injecting whole venom proteins with varying degrees of toxicity and immunogenicity (León et al., 2011). The immunogenicity of venom toxins will determine the immune response in the donor animal, and hence the antibody produced. Moreover, some pertinent snake venom toxins are weakly immunogenic because they are either expressed at lower concentration or contain few immunogenic epitopes (León et al., 2011). The α-neurotoxin, an abundant venom component found in cobras (Petras et al., 2011), and mambas (Ainsworth et al., 2018), is notorious for its profoundly lethal activity. Despite being an abundant and highly toxic venom component, these neurotoxins are characterised by significantly weak immunogenicity (León et al., 2011), thus producing antivenoms with low titres.

Sequence analysis has revealed that this tiny molecule (~6–7 kDa), encompasses largely nonimmunogenic residues (de la Rosa, Corrales-García, Rodriguez-Ruiz, López-Vera, & Corzo, 2018). A recombinant consensus short-chain α-neurotoxin (ScNtx) representing the most immunogenic residues within the molecule was identified and reconstructed. The horses produced a potent antibody, following a multisite immunization with the biologically active recombinant ScNtx adjuvanted with incomplete Freund's adjuvant (IFA) and alum initially, followed by no adjuvant (De la Rosa et al., 2019). Likewise, immunizing rabbits with reconstructed immunogenic recombinant enzymes (SVMPs, SVSPs and $SVPLA_2$) elicited an immune response resulting in the production of potent antisera (Clement et al., 2019). Evidently, this will encourage the search for better immunisation strategies, where potent antibodies are generated without inducing discomfort to host animals. Identifying the most medically important venom toxins for immunization protocols may pave the road for developing the next generation of antivenom.

3.8 Efficacy and effectiveness of antivenoms

Efficacy and effectiveness are two separate entities used to describe the safety of antivenom and its capability to treat the harmful effects of snake venoms successfully. Whilst efficacy is defined as the performance of antivenom under ideal and controlled circumstances, effectiveness refers to its response in the real-world (Isbister, 2010). To maintain a high standard of efficacy, effectiveness and safety, numerous preclinical assessments (Gutierrez et al., 2017), rigorous clinical trials (Chippaux et al., 2007), quality and regulatory control standards should be performed on newly developed antivenoms. Nevertheless, the efficacy and effectiveness of most antivenoms has been poorly studied. In clinical trials, antivenom efficacy is determined either by using placebo or by comparison with gold standard drug product. It is unethical to withhold life-saving treatment and choice of the gold standard is always difficult (Chippaux et al., 2007). This dilemma is not exclusive to antivenoms—other well-established medicinal drugs such as insulin and penicillin entered the market without placebo-controlled clinical trials (Williams et al., 2018).

The critical assessments may be circumvented by some manufacturers due to the cost implications of validation achieved from *in vitro* and animal studies (Warrell, 2008). Other manufacturers intentionally choose imprecise procedures so that they proceed with minimum requirements; for example, performance of low quality clinical trials such as non-randomized, one-armed and observational cohort study (Potet, Smith, & McIver, 2019). This is true especially for manufacturers targeting the SSA market (Brown, 2012). Despite the recommendation of the African Society of Venomology (ASV) for manufacturers to include meticulous preclinical tests and clinical trials (Chippaux, Stock, & Massougbodji, 2010), many antivenoms marketed in SSA are inherently ineffective at treating SBE by target snake venoms (Brown, 2012; Visser et al., 2008; Warrell, 2008). Influenced by corrupt policy makers and poor resources, manufacturers of lower quality and adulterated antivenoms continuously target SSA as a potential market. Nonetheless, despite its small proportion, ~3% of all antivenom stocks (Brown, 2012) have demonstrated a proven effectiveness for treating SBEs in SSA (Potet et al., 2019).

To prevent undesired clinical outcomes, replacement of existing antivenom with a new one, must follow critical assessments, such as comparison with the old product. A cheaper antivenom (Asna Antivenom-C; -C) was introduced as alternative to replace the well-renowned, but expensive Fav-Afrique™ (Alirol, Sharma, Bawaskar, Kuch, & Chappuis, 2010). Despite the manufacturer (Bharat Serum and Vaccine Ltd., India) having claimed it to be effective against *E. ocellatus* envenoming, clinical studies demonstrated that the -C antivenom offered limited benefits (Visser et al., 2008). In fact, the mortality rate was significantly higher and restoration of blood coagulation required twice the dose in patients who had received -C, compared to those treated with Fav-Afrique™ (Visser et al., 2008). Today, there are many antivenoms with a target specificity similar to that of Fav-Afrique; for example, Inoserp Pan-Africa and Antivipmyn-Africa, but there is little information on their safety and efficacy (Alirol et al., 2015).

3.9 Safety of antivenoms: Side effects associated with antivenoms

Antivenoms are heterologous antibodies produced in animals which have devolved from crude serum, which caused significant adverse events, to highly purified antibodies aided by the latest biotechnology. Despite this, side-effects still occur to all antivenoms, varying from 3% in high quality products to 88% in poorly prepared antivenoms (Leon et al., 2013; Williams et al., 2018). The severity of these events also varies from benign early allergic reactions to rare, life-threatening anaphylactic shock (Williams et al., 2018). Side effects are not exclusive to antivenoms, because no medicinal drug is entirely free from side effects (Alshammari, 2016). However, adverse reactions can be minimized with improved manufacturing processes and appropriate quality control measures (de Silva et al., 2011; Gutiérrez et al., 2003).

Despite the current advancement in biomedical research, the precise mechanism triggering adverse events following intravenous administrations of antivenoms is not fully understood. Poorly prepared impure and/or ineffective antivenom, the venom itself and patient's susceptibility to animal serum or plasma are strongly associated with antivenom-induced reactions (Isbister, Brown, MacDonald, White, & Currie, 2008; Stone et al., 2013). It was assumed that the IgG fragments $F(ab')_2$ and Fab were safer and more effective drugs, thus the therapeutic use of intact-IgG has fallen out of favour (Leon et al., 2013). Antivenom-related side effects are broadly classified as early adverse reactions (EARs), anaphylactic reactions, pyrogenic reactions, late adverse reactions (LARs) or serum sickness (Deshmukh et al., 2014).

3.9.1 *Early adverse reactions (EARs)*

EARs, usually occur within 2 h of administrations of antivenom (Chippaux, Stock, & Massougbodji, 2015; de Silva, Ryan, & de Silva, 2016), but may take slightly longer in some patients (Leon et al., 2013). Common symptoms of acute adverse reactions are generally benign such as urticaria, nausea, vomiting, headache and fever, but systemic anaphylaxis may also occur (Isbister et al., 2008; Malasit et al., 1986). Early reactions are very common and often clinicians need to deal with them as much as dealing with envenoming itself (Williams et al., 2018). However, the severity depends on the properties of triggering antigens, most importantly molecular size, but individual predisposition also plays a vital role in the reaction cascade (Chippaux et al., 2015). The mechanism of reaction to antivenom is poorly defined, but it is suggested that EARs are most likely to be due to a combination of type I hypersensitivity, complement activation, aggregates of immunoglobulin (Ig) and its fragmented molecules (Chippaux et al., 2015; de Silva et al., 2016; Lalloo & Theakston, 2003; Malasit et al., 1986; Morais, 2018). Depending on the underlying triggering mechanism, EARs are classified as anaphylactic (IgE- and non-IgE mediated) or pyrogenic reactions (Leon et al., 2013).

3.9.1.1 IgE-mediated anaphylactic reactions:

These occur rarely, but are the most severe and life-threatening reactions, which occur most commonly in known atopic individuals, previously exposed to antivenom products (Morais, 2018). However, the fact that it may also occur in individuals that have never been exposed to antivenoms before, requires further study (Lalloo & Theakston, 2003; Morais, 2018). IgE-mediated anaphylactic reactions are initiated when the allergic-specific IgE is crosslinked with mast cells, which, when activated by a subsequent episode, release potent proinflammatory mediators, which leads to anaphylactic shock (Stone et al., 2013). Anaphylaxis is characterised by oedema, cyanosis, hypotension, and vasodilation (Laustsen et al., 2015).

3.9.1.2 Non-IgE: Mediated anaphylactic reactions:

This type of EARs constitutes the vast majority of acute responses to intravenous administration in individuals with no previous exposure to antivenom (Leon et al., 2013). Clinical signs include nausea, vomiting, vasodilation, rash and pain (Morais, 2018). The mechanism triggering the non-IgE mediated reactions is not clearly defined.

It was assumed that the Fc fragment in whole IgG antivenoms activated the complement cascade in the classical manner (Leon et al., 2013). Activation is initiated by binding of Fc fragment to complement protein C1q (key subcomponent of complement classical pathway responsible for recognition and binding immune complex), leading to the subsequent generation of anaphylatoxins (Squaiella-Baptistao, Marcelino, da Cunha, Gutiérrez, & Tambourgi, 2014). Interestingly, this study also showed that antivenoms lacking the Fc domain such as $F(ab')_2$ also activate the complement system in the same manner as does whole IgG. Furthermore, administration of antivenom did not affect the concentration of C3, C4 and C5 (precursors for the potent anaphylatoxins known as C3a, C4a and C5a) in patients who experienced severe anaphylactic reactions (Stone et al., 2013). Despite experiencing significant acute anaphylaxis, absence of anaphylatoxins may suggest that aggregated Ig and macromolecular complexes interact with mast cells eliciting the degranulation process (Stone et al., 2013). Thus, the presence of aggregate proteins in antivenoms initiates the antivenom-induced anaphylactic reactions possibly by activating the complement system (Leon et al., 2013).

There is still a widespread belief that $F(ab')_2$ has a better safety profile compared to IgG antivenoms, mainly due to lack of the Fc portion, which is known to be involved in the complement activation *via* the classical pathways (Leon et al., 2013). However, preclinical (Squaiella-Baptistao et al., 2014) and substantial clinical studies, (Leon et al., 2013), demonstrated that $F(ab')_2$ also induces comparable early reactions. For instance, a clinical study has reported 23% incidence of acute anaphylactic reaction following administrations of Australian $F(ab')_2$ antivenoms (Ryan et al., 2016). Most importantly, in a randomized-clinical trial comparing the safety and efficacy of whole IgG *vs* $F(ab')_2$, the incidence of EARs was similar in both group of patients (Otero-Patino et al., 1998). Furthermore, a clinical trial reported 7.6% incidence of anaphylactic reactions with 3% death due to severe reactions in patients treated with -C antivenom, although administrations of premedication may possibly have skewed the outcome (Visser et al., 2008). Interestingly, patients treated with affinity purified CroFab® (sheep-derived product) that has had its Fc fragment removed chromatographically, induce acute hypersensitivity at a rate of 5.4% (Cannon, Ruha, & Kashani, 2008). The same antivenom (CroFab®) caused serum sickness in 14.3% of treated patients, although the outcome was skewed by a single lot of antivenom (Dart & McNally, 2001; Ruha et al., 2002). The implication of Fab and $F(ab')_2$ antivenoms in EARs, although occurring less commonly, may suggest that the non-IgE mediated anaphylactic reactions are beyond the Fc region, and thus further study is required.

3.9.1.3 Pyrogenic reactions

Pyrogenic substances such as microbial contaminants can be introduced in poorly prepared biologically derived products, mostly during the manufacturing process. Pyrogenic contaminants elicit pyrogenic reactions occurring during the first hour of infusion (de Silva et al., 2016). Pyrogenic reactions are characterized by shivering, sweating, pyrexia, fever, myalgia, tachycardia, hypotension, and vasodilation (Leon et al., 2013; Stone et al., 2013). Bacterial lipopolysaccharides (LPS), also known as endotoxins, are the most common contaminant present in biologically derived pharmaceutical products such as antivenoms (Leon et al., 2013).

While an adequate immune response is required to eradicate pathogenic invasion, if the LPS level exceeds the body's elimination capacity it may lead to aggregated inflammation that, in turn, may result in systemic inflammatory response syndrome (SIRS). SIRS is an over-reaction of the inflammatory response, characterised by hypothermia, tachycardia and tachypnoea, similar to symptoms of EARs (Stone et al., 2013). Preventative control measurements include strict asepsis at all stages of manufacture and handling (including administration), testing final product for endotoxins using either the rabbit pyrogenic test and/or the Limulus Amebocyte Lysate (LAL) test, recommended by WHO as a standard quality release test.

3.9.2 Serum sickness (delayed adverse reactions; DARs)

Serum sickness is classified as a type III hypersensitivity, where delayed IgG-mediated reactions appear 5–20 days after antivenom treatment (Leon et al., 2013; Ryan et al., 2016). Unlike anaphylaxis, it is not life-threatening but has a higher incidence rate (56%), of clinical signs, than anaphylaxis (Ryan, Downes, & Isbister, 2015). The incidence, characteristics and treatment of DARs are poorly characterised, mostly because affected patients do not return to medical centres after discharge (Ryan et al., 2016). These reactions are initiated when the immune system recognises the presence of heterologous proteins and responds by generating specific antibodies that react with the antivenom "foreign proteins," resulting in the formation of a soluble antigen-antibody complex (Morais, 2018). This immune complex mediates an immune response by activating the complement cascade, *via* the classical pathway. It causes ill-defined symptoms such as myalgia, fever, arthralgia, arthritis, urticaria and gastrointestinal disorders (Leon et al., 2013). Delayed reactions are expected to last for approximately 6 days (LoVecchio, Klemens, Roundy, & Klemens, 2003), although symptoms may persist for up to 6 weeks in some patients (Ryan et al., 2015). Most of the more serious serum sickness reactions are amenable to a week's course of corticosteroids, but in severe cases an incremental reducing dose for 2 or more weeks, with oral corticosteroids, is recommended (de Silva et al., 2016; Ryan et al., 2016).

A review of the literature suggested that the incidence of serum sickness varies greatly depending on the type of antivenoms, and perhaps quantity, used (Potet et al., 2019; Ryan et al., 2016). Theoretically, IgG causes more cases of serum-sickness than Fab and $F(ab')_2$ antivenoms because IgG contains highest quantity of non-specific proteins. This hypothesis was supported by experiments on mice (Leon et al., 2001). However, a randomized double-blinded clinical trial was conducted to assess the safety of EchiTAbG and EchiTAb-Plus (Abubakar et al., 2010). EchiTAbG is a monovalent intact IgG antivenom, targeting bites by *E. ocellatus* and EchiTAb-Plus a polyvalent, also an intact IgG antivenom, is directed against *E. ocellatus*, *N. nigricollis* and *B. arietans* venoms. Both products demonstrated efficacy against envenoming by *E. ocellatus*. The authors reported incidences of serum sickness of 5.2% on patients with EchiTAbG and 10.2% those received EchiTAb-Plus (Abubakar et al., 2010). In a clinical study, 56% of serum sickness were reported in patients treated with Crotalidae Polyvalent (ACP); an intact IgG antivenom used to treat envenoming by pit vipers (LoVecchio et al., 2003) where there was a greater occurrence of DARs at increased doses, suggesting that the incidence of serum sickness is proportional to the quantity of heterologous proteins (LoVecchio et al., 2003). In contrast, others found that the amount of antivenom is not associated with serum sickness (Ryan et al., 2016). Furthermore, the incidence of serum sickness from Fab and $F(ab)_2$-based antivenoms varies considerably depending on the type of antivenoms and geographical location. For example, the incidence of serum sickness from Australian antivenoms (all equine $F(ab')_2$), is 29% (Ryan et al., 2016), and 16% due to CroFab® (a sheep derived Fab antivenom), although the results were skewed by the inclusion of a partially purified lot (Ruha et al., 2002). Another clinical study to elucidate efficacy and safety of Antivipmyn® Africa, a polyvalent equine $F(ab')_2$ antivenom, reported 3% of treated patients developed serum sickness (Baldé, Chippaux, Boiro, Stock, & Massougbodji, 2012).

Unfortunately, there is a lack of published data reporting the incidences of DARs related to $F(ab')_2$ antivenoms in SSA, possibly influenced by the fact that follow-up by clinicians of SBE patients from rural regions is difficult because they are unable to return to medical centres after discharge.

In summary, the mechanism controlling early, and delayed adverse reactions is not well understood. Moreover, adverse reactions are generally related to heterologous proteins-even human IgG elicits complement activation (Squaiella-Baptistao et al., 2014). Furthermore, presence of antibodies transferred from horse treatment might be included in final formulations and this might enhance immune reactions (Leon et al., 1999). Thus, while removal of the Fc fragment lowers the MW of the administered protein, abiding by "good manufacturing practice" (GMP) and curtailing process-related impurities play a critical role in improving the safety of antivenoms. Despite the lack of rigorous supporting data, administration of antihistamine, corticosteroids and adrenaline are still widely used to treat and/or reduce adverse reactions (de Silva et al., 2011; Leon et al., 2013).

3.10 Current methods for assessing the safety and efficacy of antivenom

Extensive biological studies are essential to characterise each batch of antivenom manufactured, including safety and efficacy, throughout its production process. *In vivo* toxicity tests of a newly developed and/or existing antivenom are the "gold standard" potency test for antivenom safety before embarking on clinical studies. In cases where no robust and reliable clinical trial data exist, *in vivo* and/or *in vitro* studies can be used to assess the efficacy of antivenoms (Ainsworth, Menzies, Casewell, & Harrison, 2020; Theakston & Reid, 1983).

3.10.1 In vivo *tests for assessing LD_{50} and ED_{50}*

3.10.1.1 Murine LD_{50} and ED_{50}

Murine lethality assays are the "gold standard" pre-clinical assays of toxicity (LD_{50}) and treatment efficacy (ED_{50}), specifically in relation to neutralisation of venom toxicity with antivenoms (Parasuraman, 2011; Sells, 2003). Toxicity LD_{50} were first used in 1927 to determine the toxin concentration which kills 50% of a group of animals (Erhirhie, Ihekwereme, & Ilodigwe, 2018). These tests are recommended for every new batch of antivenom prior to their therapeutic use (WHO, 2016a). Despite physiological and anatomical differences, animal models (most commonly mouse) have been a satisfactory source for quantifying the effect of whole venom on a living system (WHO, 2016a).

The assays themselves are subject to wide variation, in both method and analysis (Gutiérrez et al., 2013), making comparison between laboratories challenging (Ainsworth et al., 2020). ED_{50} may be used to calculate a "human equivalent dose" as a starting point for clinical trials, in addition to measurement of antivenom efficacy (Saganuwan, 2016).

The standard test for LD_{50} of venoms is achieved by injecting intravenously or intramuscularly 5 to 6 mice with different venom concentrations (Theakston & Reid, 1979, 1983). The number of dead animals is recorded at 24 and 48 h, respectively (Gutiérrez et al., 2013). Likewise, ED_{50} of antivenom is determined by preparing different volumes of an antivenom pre-mixed with a challenge dose (two- to fivefold the determined LD_{50}) of relevant venom and intravenously or intraperitoneally injected into mice, and the number of surviving animals are recorded after 24 h (Sells, Laing, & Theakston, 2001). The venoms' LD_{50} and/or antivenoms' ED_{50} and 95% confidence interval (95% CI) values are calculated using Probit (Sells, Ioannou, & Theakston, 1998), Spearman-Kerber or other non-parametric test analyses (Gutierrez et al., 2017).

3.10.1.2 Minimum hemorrhagic dose (MHD)

This *in vivo* study is used to define the local hemorrhagic activities of venom toxins and the potency of antivenom to protect induced lesions, preferably in mice, but rabbits and rats may also be used (Gutierrez et al., 2017). The test is conducted by intradermally injecting a group of experimental animals with different test sample solutions (venom or venom-antivenom mixtures). The animals are sacrificed and dissected after 2 h incubation to assess haemorrhagic lesions. MHD is the quantity of venom which induces a 10 mm hemorrhagic lesion and the antivenom potency is defined as the amount (in mL) required to reduce 50% of the venom activity (Gutiérrez, Gené, Rojas, & Cerdas, 1985).

3.10.1.3 Minimum necrotising dose (MND)

Some snakes, for instance *N. nigricollis*, contain significant cytotoxic venom PLA_2 (Petras et al., 2011; Warrell, Greenwood, et al., 1976), where lethality is of less relevance. With that regard, studying the venom necrosis activity and its neutralisation by a biogeographically relevant antivenom is essential to establish its efficacy. The MND test is performed by injecting mice, or rats, subcutaneously, with venom to determine the lowest venom dose causing a necrotic lesion of 5 mm after 3 days (Gutierrez et al., 2005; Theakston & Reid, 1983). Similarly, the ED_{50} is expressed as the amount of antivenom; either as a volume or ratio of antivenom to venom, that results in a 50% reduction of venom toxicity (Gutierrez et al., 2005).

3.10.2 In vitro *assays for assessing toxicity and potency*

It is almost universally accepted that tests using animals should implement the 3Rs (refinement, reduction and replacement of animals in research) whenever possible with *in vitro* methods (Sells, 2003). At a WHO Workshop on standardisation and control of antivenoms in 2001, it was agreed that *in vivo* mouse assays inflict significant pain on the test animals, are expensive, and show little or no correlation with SBE therapy in humans (Theakston, Warrell, & Griffiths, 2003). The workshop highlighted the importance of devising alternative methods for replacing *in vivo* assays and assessing the antivenom efficacy at neutralizing its target venoms. Since then, several biological assays have been proposed as potential replacements for the murine assays (Lopes-de-Souza et al., 2019; Nalbantsoy et al., 2012; Ratanabanangkoon et al., 2018; Stransky et al., 2018). Furthermore, a panel of *in vivo* and *in vitro* techniques was compared to ascertain the toxic activities of *Bothrops* spp. venoms (de Souza et al., 2015). The authors reported a correlation between the two sets of results, proposing a possible replacement of animal tests by *in vitro* assays. More recently, proteomics has been introduced into antivenomics with the overall aim of revealing toxicological profile of venoms and the ability of antivenoms to recognise key venom components. The application of antivenomics is now considered a promising alternative to *in vivo* neutralisation tests (Calvete, Sanz, Pla, Lomonte, & Gutierrez, 2014). Despite such coherent efforts, *in vitro* assays have not yet been implemented universally as an accepted alternative to preclinical tests and the use of animal models is still recommended to determine the safety and efficacy of new and/or existing antivenoms (WHO, 2016a). To-date, there is no alternative *in vitro* assay that is independently used to assess the potency of antivenoms, and thus correlation to protective and/or therapeutic effect is required. Nevertheless, excellent reviews on the use of *in-vitro* assay are available (Gutierrez et al., 2017; Harrison et al., 2017). Herein, this review will cover some of the biological assays that are currently under evaluation as potential *in vivo* test replacement.

3.10.2.1 Small-scale affinity chromatography (SSAC)

Small-scale affinity chromatography (SSAC) is an immunoassay that allows venom-antivenom binding, using a gel matrix such as cyanogen bromide (CNBr)-activated sepharose (Smith, Reddi, Laing, Theakston, &

Landon, 1992). In this way, antigen-specific antibodies are captured and subsequently eluted by dropping the pH (Casewell et al., 2010). The method is cheap, rapid, and provides a direct measurement of antibody binding to target venoms. Furthermore, SSAC used by MicroPharm Ltd., UK for release and stability (i.e., quality control tests) of EchiTAbG provides consistent results with good correlation to *in vivo* ED_{50} results (Casewell et al., 2010). However, non-specific binding proves to be a significant limitation for quantitatively measuring specific antibody sing this method.

3.10.2.2 Enzyme-linked immunosorbent assay (ELISA)

Theakston, Lloyd-Jones, and Reid (1977) developed an Enzyme-linked Immunosorbent Assay (ELISA), initially to identify snakes responsible for envenoming SBE patients. A few years later, ELISA was introduced as a biological assay for assessing potency of antivenoms. Comparison of *in vivo* ED_{50} with that of *in vitro* ELISAs demonstrated an excellent correlation between the two methods. Since then, ELISA has been developed as a standard method to determine the potency of antivenoms and has been used as a substitute for many *in vivo* ED_{50} tests (Khaing et al., 2018; Maria, Pacheco, Barbosa, Velarde, & Chávez-Olórtegui, 2001; Ratanabanangkoon et al., 2017; Rial, Morais, Rossi, & Massaldi, 2006). The method uses samples of known *in vivo* ED_{50} to generate a standard curve, from which sample values are estimated from the linear range within the curve (Theakston & Reid, 1979). Despite promising results (Khaing et al., 2018; Rial et al., 2006), WHO still has reservations about the use of ELISA as an independent release test for potency assessment of antivenoms (WHO, 2016a). There are antivenom manufacturers who use ELISA standardised to ED_{50} as a product release assay. For example, Sanofi Pasteur, France has been using ELISA as a release and stability test for ViperFav®. As for most of the current *in vitro* assays, the main limitations of the ELISA assay are that, for use as a release test, it must be validated using *in vivo* studies; and that ELISA is very sensitive to experimental conditions, such as temperature, which may cause greater fluctuations in results compared to the SSAC method, however, such fluctuations would be still lower than that in *in vivo* testing.

3.10.2.3 Cytotoxicity assay (CTA) and immunocytotoxicity assay (ICTA)

Snake venom toxicity varies enormously from mild to extreme with different mechanisms of promoting cellular death. Such variations in venom components and their mechanism of action makes it challenging to develop a cell-based model that is universally acceptable to determine a suitable endpoint, such as 50% inhibitory concentration (IC_{50}). Various cell lines have been used and IC_{50} values reported for Latin American pit vipers (Lopes-de-Souza et al., 2019; Oliveira, De Oca, Duarte, Diniz, & Fortes-Dias, 2002), North Africa and Southeast Asia Viperidae; for example, (Ozverel et al., 2019) and Mediterranean vipers (Nalbantsoy et al., 2012). Recently, conducted studies (Chong, Tan, & Tan, 2020; Ozverel et al., 2019; Stransky et al., 2018), compared the response of normal (Keratinocytes), immortalized (HEK293 and VERO cells) and cancerous (HeLa, MCF-7) cell lines exposed to venom treatment. These studies found that, although each cell line exhibited a distinct level of cytotoxicity, snake venoms reduced the growth capacity of all tested cells disproportionally, albeit *via* various mechanisms.

The search for a cell-based assay as a promising *in vitro* method to screen antivenom potency is ongoing. Lopes-de-Souza et al. (2019) used VERO cells to show a concentration-dependent reduction in cell death, after treatment with $5 \times IC_{50}$. Similarly, Nalbantsoy et al. (2012) used L929 cells to assess antivenom efficacy compared to *in vivo* assays. They reported that antivenom-treatment reduced the venom toxic effects by 65% in cells incubated for up to 48h with a mixture of venom-antivenom. Cell-based assays are relatively inexpensive, rapid and, if optimized, a promizing *in vitro* candidate for replacement of the *in vivo* test with a venom neutralisation test. However, the inability of antivenoms to prevent death in some cells preincubated with venom toxins (Nalbantsoy et al., 2012), the nature of the assay being susceptible to higher variability, and variation of protocols between different laboratory settings are key limitations of these assays.

Antivenom potency assays remain a conundrum, weighing up the ethics of using *in vivo* studies with the development of *in vitro* assays, which can capture efficacy of antivenoms against all the complex arsenal of venom toxins (de Souza et al., 2015; Gutierrez et al., 2005; Gutiérrez et al., 2013; Sells, 2003; Theakston & Reid, 1983).

4. Conclusion

SBE is responsible for causing significant mortality and morbidity in the most disadvantaged regions of Africa, Asia, South and Central America. Snake venom contains a cocktail of toxins exhibiting highly variable biological and pharmacological activities. An in-depth knowledge and understanding of the pathophysiological mechanisms of snake venom is a prerequisite for developing appropriate antivenoms. Despite being successfully managed, the inherent prevalence of adverse effects associated with antivenom administration is increasingly becoming a regulatory issue.

There are few antivenoms with accessible and robust clinical data, with no reliable clinical data for over 43% of all antivenoms currently marketed in the SSA. Although promoted by its manufacturer as fit-for-purpose, at least one antivenom has been found ineffective against target venoms when independently assessed. This, therefore, highlights the necessity for independent scrutiny of antivenoms currently on the SSA market, which lack reliable and robust data. This will ensure only antivenoms with an acceptable efficacy and safety profile are available and is in line with WHO strategy for identification of effective and safe therapeutics for NTDs.

Acknowledgments

This study was funded by the European Social Fund (ESF) under the *Knowledge Economy Skills Scholarships* (KESS 2) Project.

Conflict of interest

There is no conflict of interest to declare.

Authors' contribution

This work was undertaken by Mender M. Mender and reviewed by Fiona Bolton and my PhD supervisors.

References

Abubakar, I. S., et al. (2010). Randomised controlled double-blind non-inferiority trial of two antivenoms for saw-scaled or carpet viper (Echis ocellatus) envenoming in Nigeria. *PLoS Neglected Tropical Diseases*, *4*(7), e767.

Ainsworth, S., Menzies, S. K., Casewell, N. R., & Harrison, R. A. (2020). An analysis of preclinical efficacy testing of antivenoms for sub-Saharan Africa: Inadequate independent scrutiny and poor-quality reporting are barriers to improving snakebite treatment and management. *PLoS Neglected Tropical Diseases*, *14*(8), e0008579.

Ainsworth, S., et al. (2018). The medical threat of mamba envenoming in sub-Saharan Africa revealed by genus-wide analysis of venom composition, toxicity and antivenomics profiling of available antivenoms. *Journal of Proteomics*, *172*, 173–189. https://doi.org/10.1016/j.jprot.2017.08.016.

Al-Abdulla, I., Casewell, N. R., & Landon, J. (2013). Long-term physicochemical and immunological stability of a liquid formulated intact ovine immunoglobulin-based antivenom. *Toxicon*, *64*, 38–42. https://doi.org/10.1016/j.toxicon.2012.12.022.

Al-Abdulla, I., Casewell, N. R., & Landon, J. (2014). Single-reagent one-step procedures for the purification of ovine IgG, F (ab') 2 and fab antivenoms by caprylic acid. *Journal of Immunological Methods*, *402*(1–2), 15–22.

Alirol, E., Lechevalier, P., Zamatto, F., Chappuis, F., Alcoba, G., & Potet, J. (2015). Antivenoms for snakebite envenoming: What is in the research pipeline? *PLoS Neglected Tropical Diseases*, *9*(9), e0003896.

Alirol, E., Sharma, S. K., Bawaskar, H. S., Kuch, U., & Chappuis, F. (2010). Snake bite in South Asia: A review. *PLoS Neglected Tropical Diseases, 4*(1).
Alshammari, T. M. (2016). Drug safety: The concept, inception and its importance in patients' health. *Saudi Pharmaceutical Journal, 24*(4), 405–412.
Alvarenga, L. M., Zahid, M., di Tommaso, A., Juste, M. O., Aubrey, N., Billiald, P., et al. (2014). Engineering venom's toxin-neutralizing antibody fragments and its therapeutic potential. *Toxins (Basel), 6*(8), 2541–2567. https://doi.org/10.3390/toxins6082541.
Awate, S., Babiuk, L. A. B., & Mutwiri, G. (2013). Mechanisms of action of adjuvants. *Frontiers in Immunology, 4*, 114.
Baldé, M., Chippaux, J.-P., Boiro, M., Stock, R., & Massougbodji, A. (2012). Étude clinique de la tolérance et de l'efficacité d'un sérum anti-ophidien polyvalent F (ab') 2 pour l'Afrique à Kindia, Guinée. *Bulletin de la Societe de Pathologie Exotique, 105*(3), 157–161.
Baldé, M. C., Chippaux, J.-P., Boiro, M. Y., Stock, R. P., & Massougbodji, A. (2013). Use of antivenoms for the treatment of envenomation by Elapidae snakes in Guinea, sub-Saharan Africa. *Journal of Venomous Animals and Toxins Including Tropical Diseases, 19*, 1–6.
Barlow, A., Pook, C. E., Harrison, R. A., & Wuster, W. (2009). Coevolution of diet and prey-specific venom activity supports the role of selection in snake venom evolution. *Proceedings of the Biological Sciences, 276*(1666), 2443–2449. https://doi.org/10.1098/rspb.2009.0048.
Benjamin, J. M., Chippaux, J.-P., Sambo, B. T., & Massougbodji, A. (2018). Delayed double reading of whole blood clotting test (WBCT) results at 20 and 30 minutes enhances diagnosis and treatment of viper evenomation. *Journal of Venomous Animals and Toxins Including Tropical Diseases, 24*.
Boushaba, R., Kumpalume, P., & Slater, N. K. (2003). Kinetics of whole serum and prepurified IgG digestion by pepsin for F (ab') 2 manufacture. *Biotechnology Progress, 19*(4), 1176–1182.
Boyer, L. V., Seifert, S. A., & Cain, J. S. (2001). Recurrence phenomena after immunoglobulin therapy for snake envenomations: Part 2. Guidelines for clinical management with crotaline fab antivenom. *Annals of Emergency Medicine, 37*(2), 196–201.
Brezski, R. J., & Jordan, R. E. (Eds.). (2010). *Cleavage of IgGs by proteases associated with invasive diseases: An evasion tactic against host immunity?*. MAbs: Taylor & Francis.
Brown, N. I. (2012). Consequences of neglect: Analysis of the sub-Saharan African snake antivenom market and the global context. *PLoS Neglected Tropical Diseases, 6*(6).
Burnouf, T., Griffiths, E., Padilla, A., Seddik, S., Stephano, M. A., & Gutierrez, J. M. (2004). Assessment of the viral safety of antivenoms fractionated from equine plasma. *Biologicals, 32*(3), 115–128. https://doi.org/10.1016/j.biologicals.2004.07.001.
Bush, S. P., et al. (2015). Comparison of F(ab')2 versus fab antivenom for pit viper envenomation: A prospective, blinded, multicenter, randomized clinical trial. *Clinical Toxicology (Philadelphia, Pa.), 53*(1), 37–45. https://doi.org/10.3109/15563650.2014.974263.
Calvete, J. J., Escolano, J., & Sanz, L. (2007). Snake venomics of Bitis species reveals large intragenus venom toxin composition variation: Application to taxonomy of congeneric taxa. *Journal of Proteome Research, 6*(7), 2732–2745. https://doi.org/10.1021/pr0701714.
Calvete, J. J., Sanz, L., Pla, D., Lomonte, B., & Gutierrez, J. M. (2014). Omics meets biology: Application to the design and preclinical assessment of antivenoms. *Toxins (Basel), 6*(12), 3388–3405. https://doi.org/10.3390/toxins6123388.
Calvete, J. J., et al. (2010). Antivenomic assessment of the immunological reactivity of EchiTAb-plus-ICP, an antivenom for the treatment of snakebite envenoming in sub-Saharan Africa. *The American Journal of Tropical Medicine and Hygiene, 82*(6), 1194–1201. https://doi.org/10.4269/ajtmh.2010.09-0733.

Cannon, R., Ruha, A.-M., & Kashani, J. (2008). Acute hypersensitivity reactions associated with administration of crotalidae polyvalent immune fab antivenom. *Annals of Emergency Medicine*, *51*(4), 407–411.

Carregari, V. C., Rosa-Fernandes, L., Baldasso, P., Bydlowski, S. P., Marangoni, S., Larsen, M. R., et al. (2018). Snake venom extracellular vesicles (SVEVs) reveal wide molecular and functional proteome diversity. *Scientific Reports*, *8*(1), 12067. https://doi.org/10.1038/s41598-018-30578-4.

Casewell, N. R., Al-Abdulla, I., Smith, D., Coxon, R., & Landon, J. (2014). Immunological cross-reactivity and neutralisation of European viper venoms with the monospecific Vipera berus antivenom ViperaTAb. *Toxins (Basel)*, *6*(8), 2471–2482.

Casewell, N. R., Cook, D. A., Wagstaff, S. C., Nasidi, A., Durfa, N., Wüster, W., et al. (2010). Pre-clinical assays predict pan-African Echis viper efficacy for a species-specific antivenom. *PLoS Neglected Tropical Diseases*, *4*(10), e851.

Casewell, N. R., Harrison, R. A., Wuster, W., & Wagstaff, S. C. (2009). Comparative venom gland transcriptome surveys of the saw-scaled vipers (Viperidae: Echis) reveal substantial intra-family gene diversity and novel venom transcripts. *BMC Genomics*, *10*, 564. https://doi.org/10.1186/1471-2164-10-564.

Chippaux, J. P. (2011). Estimate of the burden of snakebites in sub-Saharan Africa: A meta-analytic approach. *Toxicon*, *57*(4), 586–599. https://doi.org/10.1016/j.toxicon.2010.12.022.

Chippaux, J. P. (2017). Snakebite envenomation turns again into a neglected tropical disease! *Journal of Venomous Animals and Toxins Including Tropical Diseases*, *23*, 38. https://doi.org/10.1186/s40409-017-0127-6.

Chippaux, J.-P., Massougbodji, A., Stock, R., & Alagon, A. (2007). Clinical trial of an F (ab') 2 polyvalent equine antivenom for African snake bites in Benin. *The American Journal of Tropical Medicine and Hygiene*, 77(3), 538–546.

Chippaux, J.-P., Stock, R. P., & Massougbodji, A. (2010). Methodology of clinical studies dealing with the treatment of envenomation. *Toxicon*, *55*(7), 1195–1212.

Chippaux, J., Stock, R., & Massougbodji, A. (2015). Antivenom safety and tolerance for the strategy of snake envenomations management. In *Snake venoms–Toxinology* (pp. 1–16). Dordrecht: Springer Netherlands.

Chippaux, J. P., Williams, V., & White, J. (1991). Snake venom variability: Methods of study, results and interpretation. *Toxicon*, *29*(11), 1279–1303.

Chong, H. P., Tan, K. Y., & Tan, C. H. (2020). Cytotoxicity of Snake venoms and Cytotoxins from two southeast Asian cobras (Naja sumatrana, Naja kaouthia): Exploration of anticancer potential, selectivity, and cell death mechanism. *Frontiers in Molecular Biosciences*, 7.

Clement, H., Corrales-García, L. L., Bolaños, D., Corzo, G., & Villegas, E. (2019). Immunogenic properties of recombinant enzymes from Bothrops ammodytoides towards the generation of neutralizing antibodies against its own venom. *Toxins (Basel)*, *11*(12), 702.

Conlon, J. M., Attoub, S., Arafat, H., Mechkarska, M., Casewell, N. R., Harrison, R. A., et al. (2013). Cytotoxic activities of [Ser(4)(9)]phospholipase a(2) from the venom of the saw-scaled vipers Echis ocellatus, Echis pyramidum leakeyi, Echis carinatus sochureki, and Echis coloratus. *Toxicon*, *71*, 96–104. https://doi.org/10.1016/j.toxicon.2013.05.017.

Cook, D. A., Owen, T., Wagstaff, S. C., Kinne, J., Wernery, U., & Harrison, R. A. (2010). Analysis of camelid antibodies for antivenom development: Neutralisation of venom-induced pathology. *Toxicon*, *56*(3), 373–380.

Cresswell, C., Newcombe, A. R., Davies, S., Macpherson, I., Nelson, P., O'Donovan, K., et al. (2005). Optimal conditions for the papain digestion of polyclonal ovine IgG for the production of biotherapeutic fab fragments. *Biotechnology and Applied Biochemistry*, *42*(2), 163–168.

Currier, R. B., Harrison, R. A., Rowley, P. D., Laing, G. D., & Wagstaff, S. C. (2010). Intra-specific variation in venom of the African puff adder (Bitis arietans): Differential expression and activity of snake venom metalloproteinases (SVMPs). *Toxicon*, *55*(4), 864–873. https://doi.org/10.1016/j.toxicon.2009.12.009.

Dalhat, M. M. (2015). Socioeconomic aspects of snakebite in Africa and the tropics. In P. P. Gopalakrishnakone, A. Faiz, R. Fernando, C. A. Gnanathasan, A. G. Habib, & Chen-ChangYang (Eds.), *Clinical Toxinology in Asia Pacific and Africa* (pp. 300–308). Springer. https://doi.org/10.1007/978-94-007-6386-9.

Dart, R. C., & McNally, J. (2001). Efficacy, safety, and use of snake antivenoms in the United States. *Annals of Emergency Medicine*, *37*(2), 181–188.

de la Rosa, G., Corrales-García, L. L., Rodriguez-Ruiz, X., López-Vera, E., & Corzo, G. (2018). Short-chain consensus alpha-neurotoxin: A synthetic 60-mer peptide with generic traits and enhanced immunogenic properties. *Amino Acids*, *50*(7), 885–895.

De la Rosa, G., Olvera, F., Archundia, I. G., Lomonte, B., Alagón, A., & Corzo, G. (2019). Horse immunization with short-chain consensus α-neurotoxin generates antibodies against broad spectrum of elapid venomous species. *Nature Communications*, *10*(1), 1–8.

de Silva, H. A., Ryan, N. M., & de Silva, H. J. (2016). Adverse reactions to snake antivenom, and their prevention and treatment. *British Journal of Clinical Pharmacology*, *81*(3), 446–452. https://doi.org/10.1111/bcp.12739.

de Silva, H. A., et al. (2011). Low-dose adrenaline, promethazine, and hydrocortisone in the prevention of acute adverse reactions to antivenom following snakebite: A randomised, double-blind, placebo-controlled trial. *PLoS Medicine*, *8*(5), e1000435.

de Souza, L. L., Stransky, S., Guerra-Duarte, C., Flor-Sa, A., Schneider, F. S., Kalapothakis, E., et al. (2015). Determination of toxic activities in Bothrops spp. snake venoms using animal-free approaches: Correlation between in vitro versus in vivo assays. *Toxicological Sciences*, *147*(2), 458–465.

Deshmukh, V., Motghare, V., Gajbhiye, D., Sv, B., Deshpande, R., & Pise, H. (2014). Study on acute adverse drug reactions of antisnake venom in a rural tertiary care hospital. *Asian Journal of Pharmaceutical and Clinical Research*, 7(5).

Dhananjaya, B. L., Menon, J. C., Joseph, J. K., Raveendran, D. K., & Oommen, O. V. (2015). Snake venom detection kit (svdk): Update on current aspects and challenges. *Clinical Toxinology in Asia Pacific and Africa*, 379–400.

Dias da Silva, W., & Tambourgi, D. V. (2011). The humoral immune response induced by snake venom toxins. *Inflammation & Allergy-Drug Targets (Formerly Current Drug Targets-Inflammation & Allergy)*, *10*(5), 343–357.

Doley, R., & Kini, R. (2009). Protein complexes in snake venom. *Cellular and Molecular Life Sciences*, *66*(17), 2851–2871.

Duong-Ly, K. C., & Gabelli, S. B. (2014). Salting out of proteins using ammonium sulfate precipitation. *Methods in Enzymology*, *541*, 85–94. https://doi.org/10.1016/b978-0-12-420119-4.00007-0.

Ediriweera, D. S., et al. (2016). Mapping the risk of snakebite in Sri Lanka - a National Survey with geospatial analysis. *PLoS Neglected Tropical Diseases*, *10*(7), e0004813. https://doi.org/10.1371/journal.pntd.0004813.

Egli, A., et al. (2014). Vaccine adjuvants–understanding molecular mechanisms to improve vaccines. *Swiss Medical Weekly*, *144*(2122).

Erhirhie, E. O., Ihekwereme, C. P., & Ilodigwe, E. E. (2018). Advances in acute toxicity testing: Strengths, weaknesses and regulatory acceptance. *Interdisciplinary Toxicology*, *11*(1), 5–12.

Escoriza, D., Metallinou, M., Donaire-Barroso, D., Amat, F., & Carranza, S. (2009). Biogeography of the White-bellied carpet viper Echis leucogaster Roman, 1972 in Morocco, a study combining mitochondrial DNA data and ecological niche modeling. *Butll. Soc. Cat. Herp.*, *18*, 55–68.

Ferraz, C. R., Arrahman, A., Xie, C., Casewell, N. R., Lewis, R. J., Kool, J. and C. C. F. (2019). Multifunctional toxins in Snake venoms and therapeutic implications: From pain to hemorrhage and necrosis. *Frontiers in Ecology and Evolution*, 7:218. https://doi.org/10.3389/fevo.2019.00218.

Ghose, A., & White, J. (2016). Asian Snakes. In J. Brent, K. Burkhart, P. Dargan, B. Hatten, B. Megarbane, & R. Palmer (Eds.), *Critical care toxicology*. Cham: Springer. https://doi.org/10.1007/978-3-319-20790-2_95-1.

Goding, J. (1986). Purification, fragmentation and isotopic labelling of monoclonal antibodies. *Monoclonal antibodies: Principles and practice*, 108–109.

Guidolin, F. R., Caricati, C. P., Marcelino, J. R., & da Silva, W. D. (2016). Development of equine IgG Antivenoms against Major Snake groups in Mozambique. *PLoS Neglected Tropical Diseases*, *10*(1), e0004325. https://doi.org/10.1371/journal.pntd.0004325.

Gutierrez, J. M., Calvete, J. J., Habib, A. G., Harrison, R. A., Williams, D. J., & Warrell, D. A. (2017). Snakebite envenoming. *Nature Reviews. Disease Primers*, *3*, 17079. https://doi.org/10.1038/nrdp.2017.79.

Gutiérrez, J., Gené, J., Rojas, G., & Cerdas, L. (1985). Neutralization of proteolytic and hemorrhagic activities of costa Rican snake venoms by a polyvalent antivenom. *Toxicon*, *23*(6), 887–893.

Gutierrez, J. M., Guillermo, L., Bruno, L., & Yamileth, A. (2011). Antivenoms for snakebite envenomings. *Inflammation & Allergy-Drug Targets (Formerly Current Drug Targets-Inflammation & Allergy)*, *10*(5), 369–380.

Gutiérrez, J. M., Higashi, H. G., Wen, F. H., & Burnouf, T. (2007). Strengthening antivenom production in central and south American public laboratories: Report of a workshop. *Toxicon*, *49*(1), 30–35.

Gutiérrez, J. M., León, G., & Lomonte, B. (2003). Pharmacokinetic-pharmacodynamic relationships of immunoglobulin therapy for envenomation. *Clinical Pharmacokinetics*, *42*(8), 721–741.

Gutierrez, J. M., Lomonte, B., Sanz, L., Calvete, J. J., & Pla, D. (2014). Immunological profile of antivenoms: Preclinical analysis of the efficacy of a polyspecific antivenom through antivenomics and neutralization assays. *Journal of Proteomics*, *105*, 340–350. https://doi.org/10.1016/j.jprot.2014.02.021.

Gutierrez, J. M., et al. (2005). Pan-African polyspecific antivenom produced by caprylic acid purification of horse IgG: An alternative to the antivenom crisis in Africa. *Transactions of the Royal Society of Tropical Medicine and Hygiene*, *99*(6), 468–475. https://doi.org/10.1016/j.trstmh.2004.09.014.

Gutiérrez, J. M., et al. (2013). Assessing the preclinical efficacy of antivenoms: From the lethality neutralization assay to antivenomics. *Toxicon*, *69*, 168–179.

Gutierrez, J. M., et al. (2017). Preclinical evaluation of the efficacy of Antivenoms for snakebite envenoming: State-of-the-art and challenges ahead. *Toxins (Basel)*, *9*(5). https://doi.org/10.3390/toxins9050163.

Habib, A. G. (2015). Venomous snakes and Snake envenomation in Nigeria. In P. P. Gopalakrishnakone, A. Faiz, R. Fernando, C. A. Gnanathasan, A. G. Habib, & Chen-ChangYang (Eds.), *Clinical Toxinology in Asia Pacific and Africa* (pp. 275–295). *Springer*. https://doi.org/10.1007/978-94-007-6386-9.

Habib, A., & Abubakar, S. (2011). Factors affecting snakebite mortality in North-Eastern Nigeria. *International Health*, *3*(1), 50–55.

Habib, A. G., & Warrell, D. A. (2013). Antivenom therapy of carpet viper (Echis ocellatus) envenoming: Effectiveness and strategies for delivery in West Africa. *Toxicon*, *69*, 82–89.

Habib, A. G., et al. (2008). Envenoming after carpet viper (Echis ocellatus) bite during pregnancy: Timely use of effective antivenom improves maternal and foetal outcomes. *Tropical Medicine & International Health*, *13*(9), 1172–1175. https://doi.org/10.1111/j.1365-3156.2008.02122.x.

Halilu, S., Iliyasu, G., Hamza, M., Chippaux, J. P., Kuznik, A., & Habib, A. G. (2019). Snakebite burden in sub-Saharan Africa: Estimates from 41 countries. *Toxicon, 159*, 1–4. https://doi.org/10.1016/j.toxicon.2018.12.002.

Hamza, M., et al. (2016). Cost-effectiveness of Antivenoms for snakebite envenoming in 16 countries in West Africa. *PLoS Neglected Tropical Diseases, 10*(3), e0004568. https://doi.org/10.1371/journal.pntd.0004568.

Harrison, R. A., et al. (2017). Preclinical antivenom-efficacy testing reveals potentially disturbing deficiencies of snakebite treatment capability in East Africa. *PLoS Neglected Tropical Diseases, 11*(10), e0005969. https://doi.org/10.1371/journal.pntd.0005969.

Harvey, A. L., & Robertson, B. (2004). Dendrotoxins: Structure-activity relationships and effects on potassium ion channels. *Current Medicinal Chemistry, 11*(23), 3065–3072.

Hasan, S. M., Basher, A., Molla, A. A., Sultana, N. K., & Faiz, M. A. (2012). The impact of snake bite on household economy in Bangladesh. *Tropical Doctor, 42*(1), 41–43. https://doi.org/10.1258/td.2011.110137.

Isbister, G. K. (2010). Antivenom efficacy or effectiveness: The Australian experience. *Toxicology, 268*(3), 148–154.

Isbister, G. K., Brown, S. G., MacDonald, E., White, J., & Currie, B. J. (2008). Current use of Australian snake antivenoms and frequency of immediate-type hypersensitivity reactions and anaphylaxis. *Medical Journal of Australia, 188*(8), 473–476.

Janeway, J., Charles, A., Travers, P., Walport, M., & Shlomchik, M. J. (2001). The structure of a typical antibody molecule. In *Immunobiology: The immune system in health and disease (5th edition)*. Garland Science.

Kang, T. S., et al. (2011). Enzymatic toxins from snake venom: Structural characterization and mechanism of catalysis. *The FEBS Journal, 278*(23), 4544–4576. https://doi.org/10.1111/j.1742-4658.2011.08115.x.

Kasturiratne, A., et al. (2008). The global burden of snakebite: A literature analysis and modelling based on regional estimates of envenoming and deaths. *PLoS Medicine, 5*(11), e218. https://doi.org/10.1371/journal.pmed.0050218.

Khaing, E. M., et al. (2018). Development of an ELISA assay to determine neutralising capacity of horse serum following immunisation with Daboia siamensis venom in Myanmar. *Toxicon, 151*, 163–168.

Kinman, A. W., & Pompano, R. R. (2019). Optimization of enzymatic antibody fragmentation for yield, efficiency, and binding affinity. *Bioconjugate Chemistry, 30*(3), 800–807.

Kipanyula, M., & Kimaro, W. (2015). Snakes and snakebite envenoming in northern Tanzania: A neglected tropical health problem. *Journal of Venomous Animals and Toxins Including Tropical Diseases, 21*(1), 32.

Kurtovic, T., Brvar, M., Grenc, D., Lang Balija, M., Krizaj, I., & Halassy, B. (2016). A single dose of Viperfav(TM) may be inadequate for Vipera ammodytes Snake bite: A case report and pharmacokinetic evaluation. *Toxins (Basel), 8*(8). https://doi.org/10.3390/toxins8080244.

Lalloo, D. G., & Theakston, R. D. G. (2003). Snake antivenoms: antivenoms. *Journal of Toxicology. Clinical Toxicology, 41*(3), 277–290.

Landon, J., & Smith, D. S. (2003). Merits of sheep antisera for antivenom manufacture. *Journal of Toxicology - Toxin Reviews, 22*(1), 15–22.

Lauridsen, L. P., Laustsen, A. H., Lomonte, B., & Gutierrez, J. M. (2016). Toxicovenomics and antivenom profiling of the eastern green mamba snake (Dendroaspis angusticeps). *Journal of Proteomics, 136*, 248–261. https://doi.org/10.1016/j.jprot.2016.02.003.

Lauridsen, L. P., Laustsen, A. H., Lomonte, B., & Gutierrez, J. M. (2017). Exploring the venom of the forest cobra snake: Toxicovenomics and antivenom profiling of Naja melanoleuca. *Journal of Proteomics, 150*, 98–108. https://doi.org/10.1016/j.jprot.2016.08.024.

Laustsen, A. H., Lomonte, B., Lohse, B., Fernandez, J., & Gutierrez, J. M. (2015). Unveiling the nature of black mamba (Dendroaspis polylepis) venom through venomics and antivenom immunoprofiling: Identification of key toxin targets for antivenom development. *Journal of Proteomics, 119*, 126–142. https://doi.org/10.1016/j.jprot.2015.02.002.

Laustsen, A. H., et al. (2018). Pros and cons of different therapeutic antibody formats for recombinant antivenom development. *Toxicon, 146*, 151–175. https://doi.org/10.1016/j.toxicon.2018.03.004.

Ledsgaard, L., et al. (2018). Antibody cross-reactivity in antivenom research. *Toxins (Basel), 10*(10), 393.

Leon, G., Herrera, M., Segura, A., Villalta, M., Vargas, M., & Gutierrez, J. M. (2013). Pathogenic mechanisms underlying adverse reactions induced by intravenous administration of snake antivenoms. *Toxicon, 76*, 63–76. https://doi.org/10.1016/j.toxicon.2013.09.010.

Leon, G., Monge, M., Rojas, E., Lomonte, B., & Gutierrez, J. M. (2001). Comparison between IgG and F(ab')(2) polyvalent antivenoms: Neutralization of systemic effects induced by Bothrops asper venom in mice, extravasation to muscle tissue, and potential for induction of adverse reactions. *Toxicon, 39*(6), 793–801.

Leon, G., Stiles, B., Alape, A., Rojas, G., & Gutierrez, J. M. (1999). Comparative study on the ability of IgG and F(ab')2 antivenoms to neutralize lethal and myotoxic effects induced by Micrurus nigrocinctus (coral snake) venom. *The American Journal of Tropical Medicine and Hygiene, 61*(2), 266–271.

Leiva, C. L., et al. (2019). IgY-based antivenom against Bothrops alternatus: Production and neutralization efficacy. *Toxicon, 163*, 84–92.

León, G., et al. (2011). Immune response towards snake venoms. *Inflammation & Allergy-Drug Targets (Formerly Current Drug Targets-Inflammation & Allergy)(Discontinued), 10*(5), 381–398.

Longbottom, J., et al. (2018). Vulnerability to snakebite envenoming: A global mapping of hotspots. *Lancet, 392*(10148), 673–684. https://doi.org/10.1016/s0140-6736(18)31224-8.

Lopes-de-Souza, L., Costal-Oliveira, F., Stransky, S., de Freitas, C. F., Guerra-Duarte, C., Braga, V. M., et al. (2019). Development of a cell-based in vitro assay as a possible alternative for determining bothropic antivenom potency. *Toxicon, 170*, 68–76.

LoVecchio, F., Klemens, J., Roundy, E. B., & Klemens, A. (2003). Serum sickness following administration of antivenin (Crotalidae) polyvalent in 181 cases of presumed rattlesnake envenomation. *Wilderness & Environmental Medicine, 14*(4), 220–221.

Mackessy, S. P. (2009). The field of reptile toxinology: Snakes, lizards, and their venoms. In S. P. Mackessy (Ed.), *Handbook of venoms and toxins of reptiles* (pp. 3–23). Boca Raton: Taylor and Francis Group/CRC Press.

Mackessy, S. P., & Saviola, A. J. (2016). Understanding biological roles of venoms among the Caenophidia: The importance of rear-fanged snakes. *Integrative and Comparative Biology, 56*(5), 1004–1021. https://doi.org/10.1093/icb/icw110.

Malasit, P., Warrell, D. A., Chanthavanich, P., Viravan, C., Mongkolsapaya, J., Singhthong, B., et al. (1986). Prediction, prevention, and mechanism of early (anaphylactic) antivenom reactions in victims of snake bites. *British Medical Journal (Clinical Research Ed.), 292*(6512), 17–20.

Malih, I., Ahmad Rusmili, M. R., Tee, T. Y., Saile, R., Ghalim, N., & Othman, I. (2014). Proteomic analysis of Moroccan cobra Naja haje legionis venom using tandem mass spectrometry. *Journal of Proteomics, 96*, 240–252. https://doi.org/10.1016/j.jprot.2013.11.012.

Maria, W. S., Pacheco, B. G., Barbosa, C. F., Velarde, D. T., & Chávez-Olórtegui, C. (2001). Determination of the neutralizing potency of horse antibothropic and anticrotalic antivenoms in blood samples collected on filter paper. *Toxicon, 39*(10), 1607–1609.

Meyer, W., et al. (1997). First clinical experiences with a new ovine fab Echis ocellatus snake bite antivenom in Nigeria: Randomized comparative trial with institute Pasteur serum

(Ipser) Africa antivenom. *The American Journal of Tropical Medicine and Hygiene*, *56*(3), 291–300.

Mohapatra, B., et al. (2011). Snakebite mortality in India: A nationally representative mortality survey. *PLoS Neglected Tropical Diseases*, *5*(4), e1018. https://doi.org/10.1371/journal.pntd.0001018.

Morais, V. (2018). Antivenom therapy: Efficacy of premedication for the prevention of adverse reactions. *Journal of Venomous Animals and Toxins Including Tropical Diseases*, *24*.

Morais, V., & Massaldi, H. (2005). Effect of pepsin digestion on the antivenom activity of equine immunoglobulins. *Toxicon*, *46*(8), 876–882.

Morais, V., & Massaldi, H. (2012). A model mechanism for protein precipitation by caprylic acid: Application to plasma purification. *Biotechnology and Applied Biochemistry*, *59*(1), 50–54. https://doi.org/10.1002/bab.68.

Munawar, A., Ali, S. A., Akrem, A., & Betzel, C. (2018). Snake venom peptides: Tools of biodiscovery. *Toxins (Basel)*, *10*(11). https://doi.org/10.3390/toxins10110474.

Nalbantsoy, A., Karabay-Yavasoglu, N., Sayim, F., Deliloglu-Gurhan, I., Gocmen, B., Arikan, H., et al. (2012). Determination of in vivo toxicity and in vitro cytotoxicity of venom from the Cypriot blunt-nosed viper Macrovipera lebetina lebetina and antivenom production. *Journal of Venomous Animals and Toxins Including Tropical Diseases*, *18*(2), 208–216.

Nikapitiya, B., & Maduwage, K. (2018). Pharmacodynamics and pharmacokinetics of snake antivenom. *Sri Lanka Journal of Medicine*, *27*(1), 54–65.

Nudel, B. C., Perdomenico, C., Iacono, R., & Cascone, O. (2012). Optimization by factorial analysis of caprylic acid precipitation of non-immunoglobulins from hyperimmune equine plasma for antivenom preparation. *Toxicon*, *59*(1), 68–73. https://doi.org/10.1016/j.toxicon.2011.10.014.

Oliveira, J., De Oca, H. M., Duarte, M., Diniz, C., & Fortes-Dias, C. (2002). Toxicity of south American snake venoms measured by an in vitro cell culture assay. *Toxicon*, *40*(3), 321–325.

Otero, R., et al. (1999). A randomized blinded clinical trial of two antivenoms, prepared by caprylic acid or ammonium sulphate fractionation of IgG, in Bothrops and Porthidium snake bites in Colombia: Correlation between safety and biochemical characteristics of antivenoms. *Toxicon*, *37*(6), 895–908.

Otero, R., et al. (2006). Efficacy and safety of two whole IgG polyvalent antivenoms, refined by caprylic acid fractionation with or without β-propiolactone, in the treatment of Bothrops asper bites in Colombia. *Transactions of the Royal Society of Tropical Medicine and Hygiene*, *100*(12), 1173–1182.

Otero-Patino, R., et al. (1998). A randomized, blinded, comparative trial of one pepsin-digested and two whole IgG antivenoms for Bothrops snake bites in Uraba, Colombia. The regional group on Antivenom therapy research (REGATHER). *The American Journal of Tropical Medicine and Hygiene*, *58*(2), 183–189.

Otero-Patiño, R., et al. (2012). Comparative study of the efficacy and safety of two polyvalent, caprylic acid fractionated [IgG and F (ab') 2] antivenoms. *Bothrops asper bites in Colombia. Toxicon*, *59*(2), 344–355.

Ozverel, C. S., Damm, M., Hempel, B.-F., Göçmen, B., Sroka, R., Süssmuth, R. D., et al. (2019). Investigating the cytotoxic effects of the venom proteome of two species of the Viperidae family (Cerastes cerastes and Cryptelytrops purpureomaculatus) from various habitats. *Comparative Biochemistry and Physiology Part C: Toxicology & Pharmacology*, *220*, 20–30.

Paixão-Cavalcante, D., Kuniyoshi, A. K., Portaro, F. C., da Silva, W. D., & Tambourgi, D. V. (2015). African adders: Partial characterization of snake venoms from three Bitis species of medical importance and their neutralization by experimental equine antivenoms. *PLoS Neglected Tropical Diseases*, *9*(2).

Paniagua, D., Vergara, I., Boyer, L., & Alagón, A. (2015). Role of lymphatic system on snake venom absorption. In *Snake Venoms* (pp. 1–19). Netherlands: Springer.

Parasuraman, S. (2011). Toxicological screening. *The Journal of Pharmacy and Pharmacology, 2*(2), 74.

Patra, A., Kalita, B., Chanda, A., & Mukherjee, A. K. (2017). Proteomics and antivenomics of Echis carinatus carinatus venom: Correlation with pharmacological properties and pathophysiology of envenomation. *Scientific Reports,* 7(1), 17119. https://doi.org/10.1038/s41598-017-17227-y.

Patra, A., Kalita, B., & Mukherjee, A. K. (2018). Assessment of quality, safety, and pre-clinical toxicity of an equine polyvalent anti-snake venom (pan Africa): Determination of immunological cross-reactivity of antivenom against venom samples of Elapidae and Viperidae snakes of Africa. *Toxicon.* https://doi.org/10.1016/j.toxicon.2018.08.018.

Pepin-Covatta, S., Lutsch, C., Grandgeorge, M., & Scherrmann, J. M. (1997). Immunoreactivity of a new generation of horse F(ab')2 preparations against European viper venoms and the tetanus toxin. *Toxicon, 35*(3), 411–422.

Petras, D., Heiss, P., Harrison, R. A., Sussmuth, R. D., & Calvete, J. J. (2016). Top-down venomics of the east African green mamba, Dendroaspis angusticeps, and the black mamba, Dendroaspis polylepis, highlight the complexity of their toxin arsenals. *Journal of Proteomics, 146,* 148–164. https://doi.org/10.1016/j.jprot.2016.06.018.

Petras, D., et al. (2011). Snake venomics of African spitting cobras: Toxin composition and assessment of congeneric cross-reactivity of the pan-African EchiTAb-plus-ICP antivenom by antivenomics and neutralization approaches. *Journal of Proteome Research, 10*(3), 1266–1280. https://doi.org/10.1021/pr101040f.

Pope, C. (1939a). The action of proteolytic enzymes on the antitoxins and proteins in immune sera: I. true digestion of the proteins. *British Journal of Experimental Pathology, 20*(2), 132.

Pope, C. (1939b). The action of proteolytic enzymes on the antitoxins and proteins in immune sera: II. Heat denaturation after partial enzyme action. *British Journal of Experimental Pathology, 20*(3), 201.

Potet, J., Smith, J., & McIver, L. (2019). Reviewing evidence of the clinical effectiveness of commercially available antivenoms in sub-Saharan Africa identifies the need for a multi-Centre, multi-antivenom clinical trial. *PLoS Neglected Tropical Diseases, 13*(6), e0007551.

Pucca, M. B., Cerni, F. A., Janke, R., Bermúdez-Méndez, E., Ledsgaard, L., Barbosa, J. E., et al. (2019). History of envenoming therapy and current perspectives. *Frontiers in Immunology, 10,* 1598.

Rahman, R., et al. (2010). Annual incidence of snake bite in rural Bangladesh. *PLoS Neglected Tropical Diseases, 4*(10), e860. https://doi.org/10.1371/journal.pntd.0000860.

Ratanabanangkoon, K., Simsiriwong, P., Pruksaphon, K., Tan, K. Y., Chantrathonkul, B., Eursakun, S., et al. (2018). An in vitro potency assay using nicotinic acetylcholine receptor binding works well with antivenoms against Bungarus candidus and Naja naja. *Scientific Reports, 8*(1), 9716. https://doi.org/10.1038/s41598-018-27794-3.

Ratanabanangkoon, K., et al. (2017). A novel in vitro potency assay of antisera against Thai Naja kaouthia based on nicotinic acetylcholine receptor binding. *Scientific Reports,* 7(1), 8545. https://doi.org/10.1038/s41598-017-08962-3.

Rial, A., Morais, V., Rossi, S., & Massaldi, H. (2006). A new ELISA for determination of potency in snake antivenoms. *Toxicon, 48*(4), 462–466.

Rogalski, A., et al. (2017). Differential procoagulant effects of saw-scaled viper (Serpentes: Viperidae: Echis) snake venoms on human plasma and the narrow taxonomic ranges of antivenom efficacies. *Toxicology Letters*, *280*, 159–170. https://doi.org/10.1016/j.toxlet.2017.08.020.

Rojas, A., et al. (2013). Role of the animal model on the pharmacokinetics of equine-derived antivenoms. *Toxicon*, *70*, 9–14.

Ruha, A.-M., et al. (2002). Initial postmarketing experience with crotalidae polyvalent immune fab for treatment of rattlesnake envenomation. *Annals of Emergency Medicine*, *39*(6), 609–615.

Ryan, N. M., Downes, M. A., & Isbister, G. K. (2015). Clinical features of serum sickness after Australian snake antivenom. *Toxicon*, *108*, 181–183.

Ryan, N. M., Kearney, R. T., Brown, S. G., & Isbister, G. K. (2016). Incidence of serum sickness after the administration of Australian snake antivenom (ASP-22). *Clinical Toxicology*, *54*(1), 27–33.

Saganuwan, S. A. (2016). The new algorithm for calculation of median lethal dose (LD50) and effective dose fifty (ED50) of Micrarus fulvius venom and anti-venom in mice. *International Journal of Veterinary Science and Medicine*, *4*(1), 1–4.

Seifert, S. A., & Boyer, L. V. (2001). Recurrence phenomena after immunoglobulin therapy for snake envenomations: Part 1. Pharmacokinetics and pharmacodynamics of immunoglobulin antivenoms and related antibodies. *Annals of Emergency Medicine*, *37*(2), 189–195.

Sells, P. G. (2003). Animal experimentation in snake venom research and in vitro alternatives. *Toxicon*, *42*(2), 115–133.

Sells, P., Ioannou, P., & Theakston, R. (1998). A humane alternative to the measurement of the lethal effects (LD50) of non-neurotoxic venoms using hens' eggs. *Toxicon*, *36*(7), 985–991.

Sells, P., Laing, G., & Theakston, R. (2001). An in vivo but insensate model for the evaluation of antivenoms (ED50) using fertile hens' eggs. *Toxicon*, *39*(5), 665–668.

Silva, A., & Isbister, G. K. (2020). Current research into snake antivenoms, their mechanisms of action and applications. *Biochemical Society Transactions*, *48*(2), 537–546.

Smith, D. C., Reddi, K. R., Laing, G., Theakston, R. D., & Landon, J. (1992). An affinity purified ovine antivenom for the treatment of Vipera berus envenoming. *Toxicon*, *30*(8), 865–871.

Squaiella-Baptistao, C. C., Marcelino, J. R., da Cunha, L. E. R., Gutiérrez, J. M., & Tambourgi, D. V. (2014). Anticomplementary activity of horse IgG and F (ab') 2 antivenoms. *The American Journal of Tropical Medicine and Hygiene*, *90*(3), 574–584.

Squaiella-Baptistao, C. C., Sant'Anna, O. A., Marcelino, J. R., & Tambourgi, D. V. (2018). The history of antivenoms development: Beyond Calmette and vital Brazil. *Toxicon*, *150*, 86–95. https://doi.org/10.1016/j.toxicon.2018.05.008.

Stone, S. F., et al. (2013). Immune response to snake envenoming and treatment with antivenom; complement activation, cytokine production and mast cell degranulation. *PLoS Neglected Tropical Diseases*, 7(7), e2326.

Stransky, S., Costal-Oliveira, F., Lopes-de-Souza, L., Guerra-Duarte, C., Chávez-Olórtegui, C., & Braga, V. M. M. (2018). In vitro assessment of cytotoxic activities of Lachesis muta muta snake venom. *PLoS Neglected Tropical Diseases*, *12*(4), e0006427.

Sunagar, K., Jackson, T. N., Undheim, E. A., Ali, S. A., Antunes, A., & Fry, B. G. (2013). Three-fingered RAVERs: Rapid accumulation of variations in exposed residues of snake venom toxins. *Toxins (Basel)*, *5*(11), 2172–2208. https://doi.org/10.3390/toxins 5112172.

Tasoulis, T., & Isbister, G. K. (2017). A review and database of Snake venom proteomes. *Toxins (Basel)*, *9*(9). https://doi.org/10.3390/toxins9090290.

Theakston, R., Lloyd-Jones, M. J., & Reid, H. (1977). Micro-ELISA for detecting and assaying snake venom and venom-antibody. *The Lancet*, *310*(8039), 639–641.

Theakston, R., & Reid, H. (1979). Enzyme-linked immunosorbent assay (ELISA) in assessing antivenom potency. *Toxicon, 17*(5), 511–515.

Theakston, R., & Reid, H. (1983). Development of simple standard assay procedures for the characterization of snake venoms. *Bulletin of the World Health Organization, 61*(6), 949.

Theakston, R. D. G., Warrell, D., & Griffiths, E. (2003). Report of a WHO workshop on the standardization and control of antivenoms. *Toxicon, 41*(5), 541–557.

Valenza, L., Allavena, R., Haworth, M., Cochrane, J., & Henning, J. (2021). Diagnosis and treatment of Snake envenomation in dogs in Queensland. *Australia. Veterinary sciences, 8*(2), 14.

Vidal, N., Delmas, A. S., David, P., Cruaud, C., Couloux, A., & Hedges, S. B. (2007). The phylogeny and classification of caenophidian snakes inferred from seven nuclear protein-coding genes. *Comptes Rendus Biologies, 330*(2), 182–187.

Visser, L., Kyei-Faried, S., Belcher, D., Geelhoed, D., van Leeuwen, J. S., & Van Roosmalen, J. (2008). Failure of a new antivenom to treat Echis ocellatus snake bite in rural Ghana: The importance of quality surveillance. *Transactions of the Royal Society of Tropical Medicine and Hygiene, 102*(5), 445–450.

Wagstaff, S. C., & Harrison, R. A. (2006). Venom gland EST analysis of the saw-scaled viper, Echis ocellatus, reveals novel alpha9beta1 integrin-binding motifs in venom metalloproteinases and a new group of putative toxins, renin-like aspartic proteases. *Gene, 377*, 21–32. https://doi.org/10.1016/j.gene.2006.03.008.

Wagstaff, S. C., Sanz, L., Juarez, P., Harrison, R. A., & Calvete, J. J. (2009). Combined snake venomics and venom gland transcriptomic analysis of the ocellated carpet viper. *Echis ocellatus. J Proteomics, 71*(6), 609–623. https://doi.org/10.1016/j.jprot.2008.10.003.

Walther, S., Rusitzka, T. V., Diesterbeck, U. S., & Czerny, C. P. (2015). Equine immunoglobulins and organization of immunoglobulin genes. *Developmental and Comparative Immunology, 53*(2), 303–319. https://doi.org/10.1016/j.dci.2015.07.017.

Warrell, D. A. (2008). Unscrupulous marketing of snake bite antivenoms in Africa and Papua New Guinea: choosing the right product- -'what's in a name?'. *Transactions of the Royal Society of Tropical Medicine and Hygiene*, 102(5), 397–399. https://doi.org/10.1016/j.trstmh.2007.12.005.

Warrell, D. (2009). Commissioned article: Management of exotic snakebites. *QJM: An International Journal of Medicine, 102*(9), 593–601.

Warrell, D., Barnes, H., & Piburn, M. (1976). Neurotoxic effects of bites by the Egyptian cobra (Naja haje) in Nigeria. *Transactions of the Royal Society of Tropical Medicine and Hygiene, 70*(1), 78–79.

Warrell, D., Davidson, N. M., Greenwood, B., Ormerod, L., Pope, H. M., Watkins, B. J., et al. (1977). Poisoning by bites of the saw-scaled or carpet viper (Echis carinatus) in Nigeria. *QJM: An International Journal of Medicine, 46*(1), 33–62.

Warrell, D., Davidson, N. M., Omerod, L., Pope, H. M., Watkins, B. J., Greenwood, B., et al. (1974). Bites by the saw-scaled or carpet viper (Echis carinatus): Trial of two specific antivenoms. *British Medical Journal, 4*(5942), 437–440.

Warrell, D., Greenwood, B., Davidson, N. M., Ormerod, L., & Prentice, C. (1976). Necrosis, haemorrhage and complement depletion following bites by the spitting cobra (Naja nigricollis). *QJM: An International Journal of Medicine, 45*(1), 1–22.

WHO. (2016a). *Expert committee on biological standardization World Health Organization.* WHO Expert Committee on Biological Standardization: *Sixty-sixth Report.* World Health Organization.

WHO. (2016b). *WHO guidelines for the production.* Control and Regulation of Snake Antivenom: Immunoglobulins.

Williams, D. J., Habib, A. G., & Warrell, D. A. (2018). Clinical studies of the effectiveness and safety of antivenoms. *Toxicon, 150*, 1–10. https://doi.org/10.1016/j.toxicon.2018.05.001.

Williams, H. F., Layfield, H. J., Vallance, T., Patel, K., Bicknell, A. B., Trim, S. A., et al. (2019). The urgent need to develop novel strategies for the diagnosis and treatment of snakebites. *Toxins (Basel)*, *11*(6), 363.

Yates, V., Lebas, E., Orpiay, R., & Bale, B. (2010). Management of snakebites by the staff of a rural clinic: The impact of providing free antivenom in a nurse-led clinic in Meserani, Tanzania. *Annals of Tropical Medicine and Parasitology*, *104*(5), 439–448.

Zupunski, V., Kordis, D., & Gubensek, F. (2003). Adaptive evolution in the snake venom Kunitz/BPTI protein family. *FEBS Letters*, *547*(1–3), 131–136. https://doi.org/10.1016/s0014-5793(03)00693-8.